GW00600638

PRINCIPLES
OF
GENETICS

James W. Fristrom and Philip T. Spieth
University of California
Berkeley

Illustrations by Dianne K. Fristrom

CHIRON PRESS

INCORPORATED

New York & Concord

Distributed Outside North America
by
BLACKWELL SCIENTIFIC PUBLICATIONS
Oxford   London   Edinburgh   Melbourne

Sales: Chiron Press, Incorporated
            Publishers Storage & Shipping Corp.
            2352 Main Street
            Concord, Massachusetts 01742

Editorial: Chiron Press, Incorporated
            Post Office Box 1587
            Cathedral Station
            New York, New York 10025

Distributed throughout the World excluding North America by:
            Blackwell Scientific Publications
            Osney Mead, Oxford OX2 0EL
            8 John Street, London WC1N 2ES
            9 Forrest Road, Edinburgh EH1 2QH
            214 Berkeley Street, Carlton, Victoria 3053, Australia

Library of Congress Catalogue Card Number 80-65757

ISBN 0-913462-05-5   Blackwell ISBN 0-632-00647-1

First Printing

Additional drawings by Louise Garland

To our former colleagues
I. Michael Lerner and Curt Stern
In recognition of their excellence as teachers of genetics

Acknowledgements

First and foremost we wish to acknowledge the students and instructors in the genetics courses at Boston College, taught by William Petri, and at the University of San Francisco, taught by Carol Chihara, in 1979. They struggled heroically with photocopies of an earlier version of *Principles of Genetics,* and their many valuable comments have made this a better book than it would otherwise have been. We thank the members of the Department of Genetics at Berkeley, particularly Seymour Fogel, Roberta Palmour, Patricia St. Lawrence, and Glenys Thomson, for their help and criticism. Help at Berkeley also came from several members of the Department of Molecular Biology, particularly A. John Clark, Richard Calendar, Harrison Echols, and Robley Williams, Sr., and from David Wake of the Department of Zoology. We are indebted to Joseph Travis (Duke University), Melvin Green (University of California, Davis), John Lucchesi (University of North Carolina), John Gillespie (University of Pennsylvania), and Ernst Mayr (Harvard University) for their constructive comments. Bernadette Jaroch-Hagerman was not only a fine typist, but also a perceptive critic. We are grateful to the Literary Executor of the late Sir Ronald A. Fisher, F.R.S., to Dr. Frank Yates, F.R.S., and to Longman Group Ltd. London, for permission to reprint Table IV from their book *Statistical Tables for Biological, Agricultural and Medical Research* (6th edition, 1974). Finally, we are particularly grateful to Dianne Fristrom for her careful reading and rewriting of the manuscript and for her persistent, sometimes seemingly endless, but always useful criticism. Without her we would have had a book of fifteen hundred pages.

Preface

This book grew out of the General Genetics course that, together or separately, we have been teaching at Berkeley since 1965. The students are primarily biology majors in their junior year who have had one year of biology and two years of chemistry. We give approximately equal emphasis to molecular genetics, transmission genetics, and population genetics. Our approach is neither classical, starting with Mendel, nor "modern," starting with DNA → RNA → protein, but one born, we believe, of pedagogical logic. We start with the nature of the vehicles of genetic transmission (DNA and chromosomes), then discuss how these vehicles are transmitted from parents to progeny and, once transmitted, how they function, then, finally, consider how genetic variability is molded by evolutionary processes.

Throughout *Principles of Genetics* we take an evolutionary perspective. We emphasize evolution by devoting about thirty per cent of the book to population and evolutionary genetics (Chapters 19–25) and by using evolutionary theory to tie transmission genetics, functional genetics, and population genetics together. Although biological evolution is a theory rather than a fact of science such as the Copernican model of the solar system, evolutionary relatedness among living organisms is as much a fact as the Earth's yearly revolution about the sun. After all, in the words of Theodosius Dobzhansky, "Nothing in biology makes sense except in the light of evolution."

We wrote this book to create a text suitable for our own course and to create a text that describes experimental and theoretical foundations for modern genetics without hiding the principles in encyclopedic detail. Those who may want a text that is still more economical in length will realize that parts of chapters, or even whole chapters, may be omitted without affecting the flow of the book. For example, Chapter 9, dealing with somatic and non-nuclear genetics, might be bypassed. Many details of DNA replication or DNA renaturation could be skimmed. The sections on inbreeding and two-locus theory in Chapter 20 may be omitted without loss of continuity, as may either or both sections of Chapter 22. Other topics may undoubtedly be skimmed or omitted. We have attempted to organize the principles of genetics in a logical narrative that is reasonably adaptable to the individual taste of the instructor or, more important, to the needs of the student.

J. W. F. and P. T. S.
Berkeley, California

Contents

I

Nucleic Acids
and
Chromosomes

The Chemical Nature of Genes

Once a prehistoric blister
    In a prehistoric slime
Thought, "I'd like to have a sister,"
    And, to save himself some time
He wrote down the mystic secrets of his protoplasmic virus
On some stuff that he found handy—a sort of primitive papyrus.

Oh, alas; the fateful boding
    Oh, his early lack of vision—
The substratum for his coding
    He chose with rash decision,
For he thought it all quite simple his directions to dictate
To a convoluted molecule of deoxynucleate.

Oh, curse this lowbrow blister—he
    Incontrovertibly set
Our evolutionary history
    Though we haven't caught up yet:
While he added information to his nucleary log
It's perhaps misinformation that has all chemists now agog.

And—fantastic complications
    As each factor he encoiled,
Made inevitable implications
    As geneticists got embroiled
And traced conversions and inversions of these nucleotidic rolls
In various permutations of chromosomal pigeonholes.

Now, lest we be too placid—
    There might be other genes
That instead of nucleic acid
    Are made up of proteins.
How perverse our limitations—how incredibly tragic
That, for each and everyone of us, one or the other of these two
        Hypotheses should seem relatively acceptable and the
        Alternative one should seem like believing in magic.

"(R)Evolution(?)" by Rollin D. Hotchkiss was presented at the conclusion of a symposium on genetic recombination in 1954 at which the chemical nature of the gene was discussed.

INTRODUCTION

We are both victors and victims of evolution. We are victors because at this moment we exist in an evolutionary tributary that continues to flow. We are victims because the processes of evolution put us, as individuals, at risk from the moment of conception by making our mortality inevitable. For evolution has produced not only the collection of organisms that exist in the universe but also the mechanisms of heredity by which organisms pass genetic information from generation to generation. We expect that organisms on other planets will be physically and functionally different from those on Earth. We need only compare terrestrial with aquatic organisms on Earth to appreciate the proposition that physical environments influence the evolution of function and structure. Neither a fish out of water nor an elephant set down in the middle of the sea will survive. Even so, the structure of an elephant or a fish is the result of evolutionary processes whose rules apply on land and in the water. We expect the processes of evolution to be similar throughout the universe. We also expect mechanisms of heredity, because they are a result of evolutionary processes and rules, to have universal similarities. The general nature of evolutionary processes provides us with a basis for understanding genetic mechanisms, so that we may, in turn, understand why we are both victims and victors.

Three different kinds of evolution have occurred: the evolution of elements, the evolution of chemical compounds, and the evolution of organisms.

When we think of evolution, we often think of biological evolution. But biological evolution was preceded by and is dependent upon two other forms of evolution—the evolution of elements through changes in atomic structure and the evolution of chemical compounds through the bonding of different elements. Our own physical nature has resulted not only from biological evolution on Earth but also from the physical and chemical nature of the universe. Indeed, it is generally thought that life originated from subatomic particles and may, if the universe collapses back into the center from which it originated, return to subatomic particles.

The evolution of elements. Current cosmological findings reinforce the theory that evolution began about 10 billion (1×10^{10}) years ago (see Table 1-1). At that time, according to the theory, all of the matter in the universe existed in a gigantic sphere that had enormous density and a temperature in excess of 10^{11} degrees Kelvin (Kelvin (K) is the temperature on a centigrade scale measured from absolute zero, that is, from $-273°$ C). At such a temperature neither atoms nor atomic nuclei existed, only elementary particles such as electrons, positrons, and quarks (elementary particles from which protons and neutrons are constructed) were present. A cataclysmal explosion, called by George Gamow the "Big Bang," occurred. Matter was sent racing outwards. Now, billions of years later, galaxies still race away from each other as a result of the initial Big Bang.

Table 1-1   A Review of Exponents*

Number	Exponential Expression
0.000001	10^{-6}
0.001	10^{-3}
0.1	10^{-1}
1	10^{0}
10	10^{1}
100	10^{2}
1,000	10^{3}
1,000,000	10^{6}

* When dealing with very large numbers, as we will do throughout this book, it is convenient to use exponential expressions.

We believe that the formation of atomic nuclei began shortly after the Big Bang. That is, as the universe expanded, the temperature dropped rapidly, declining to approximately 10^{11} K in one-hundredth of a second and 3×10^{9} K in as little as 14 seconds. As the temperature dropped, protons (hydrogen nuclei) and neutrons formed from subatomic particles. About three minutes after the Big Bang, the temperature dropped sufficiently—to something less than 10^{9} K—so that atomic nuclei were able to form from protons and neutrons. Neutrons and protons bonded together to form deuterium (heavy hydrogen) nuclei (1 proton, 1 neutron). Pairs of deuterium nuclei came together to form helium nuclei (2 protons, 2 neutrons). The evolution of elements had begun.

Three to four minutes after the Big Bang, when the temperature became cool enough for deuterium nuclei to form, the ratio of protons to neutrons was about 87:13. Virtually all of the neutrons became incorporated into deuterium and then into helium nuclei so that, by weight, helium constituted about 26 per cent, and hydrogen 74 per cent, of the nuclei in the universe. According to one estimate 10^{80} nuclei were present, of which 93 per cent were hydrogen and 7 per cent were helium. The amount of any element present in the universe depends on the balance between its rate of formation and its rate of breakdown or conversion to another element. Thus, hydrogen and helium, formed in enormous amounts in the first few minutes after the Big Bang, are plentiful. As the average temperature of the universe continued to grow cooler, both temperature and concentration of hydrogen and helium nuclei remained high within stars and during stellar explosions, where the evolution of elements continued. Fusions of helium and other nuclei produced additional elements. Today, billions of years later, the universe is still composed of about 93 per cent hydrogen and 7 per cent helium nuclei. Carbon, nitrogen, and oxygen, the elements of which organisms on Earth are primarily constructed, constitute about 0.1 per cent of the universe's nuclei. All the other elements are present in the universe in negligible amounts. The evolution of elements continues today in stars, including our sun (which has an internal temperature of about $1.5 \times 10^{9\circ}$ K.) as heavier elements are formed through the helium connection.

The evolution of chemical compounds. According to the theory we have outlined, atoms composed of nuclei and surrounding electrons did not exist until approximately 700,000 years after the Big Bang, when the universe's average temperature had dropped to about 3000° K. and matter aggregated to form the first galaxies. The surfaces of planets cooled to temperatures far below those encountered on the surfaces of the stars. On Earth, the cool temperature opened up new possibilities for the continuation of evolution, and the composition of the elements on Earth is far different from the average composition of the universe. In weight (not the number of atoms present), the Earth's crust contains 48 per cent oxygen, 27 per cent silicon, and 8 per cent aluminum. The elements calcium, iron, magnesium, sodium, and potassium constitute 10 per cent of the total weight, with all the other elements contributing only 7 per cent of the total.

The cool surfaces of planets readily allow the formation of chemical compounds. On Earth, oxygen is present in silicates, water, nitrates, carbon dioxide, and carbon monoxide as well as in molecular oxygen. Elements no longer exist only in uncombined form but have evolved through the formation of chemical bonds into an enormous variety of compounds. The rules for the evolution of chemical compounds are analogous to those for the formation of the elements. Different compounds are formed from different precursors, for example, carbon dioxide from carbon and oxygen, and are, in turn, broken down or converted into other compounds that are stable under the physical conditions on Earth. Thus, as is true for the elements, the amount of any compound present is a result of its rate of formation less its rate of breakdown or conversion to other compounds. On Earth, where oxygen, silicon, and hydrogen are plentiful, substantial quantities of silicates and water are formed.

Chemical evolution was not limited to the formation of simple compounds. When the surface of the Earth eventually cooled, the oceans formed. We have good reason to believe that within these primeval oceans complicated compounds—such as amino acids, sugars, and purine and pyrimidine bases (Figure 1-1) that are used for the structure and function of modern organisms—already existed. These compounds were present as separate entities and some of them also joined together in short chains to form *polymers*. One polymer, a polysaccharide, is formed by linking individual sugar residues (Figure 1-1).

The formation of polymers from different kinds of individual building blocks, *monomers,* allows the creation of substances far more complex and varied than the individual monomers themselves. Imagine, for example, six different monomers (A, B, C, D, E, F) that can be linked together in a single linear polymer six residues long. Assume that any one of the six monomers can be present at any position. For example, a molecule could be A–A–A–A–A–A or A–B–C–D–E–F or any other combination. There are then 46,656 different molecules that could exist—$6 \times 6 \times 6 \times 6 \times 6 \times 6 = 6^6$, or 46,656. Depending on the chemical and physical properties of the monomers, the chemical and physical properties of the polymers could differ greatly. Thus, from only six building blocks an enormous number of different molecules with potentially different physical properties can be produced. The trick of

A

AMINO ACIDS   SUGARS   BASES

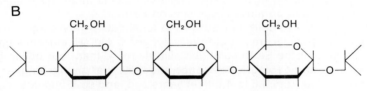

Figure 1-1   (A) Monomers commonly found in biological polymers. (B) A polysaccharide constructed from sugar monomers.

making giant polymeric molecules with highly diverse physical and chemical properties from monomers is used by modern organisms to provide the complex functional and synthetic machinery necessary for life.

   Biological evolution. We may imagine the existence of a special type of chemical reaction by which an existing molecule causes its own synthesis. One hypothetical scheme (Figure 1-2) by which such synthesis could occur involves a short molecule composed of two complementary linear polymers or strands, each of which is a template for the formation of the other (*Webster's* tells us that a template is a device for forming an accurate copy of an object or shape). The two polymeric strands become separated and each strand, using monomers found in the environment, causes the monomers to associate with it so as to create two new double-stranded polymers that are identical to the first one. The strands of such double-stranded molecules can separate again and again and direct the synthesis of additional copies of the original molecule. Such molecules possess one property of life: they are able to reproduce themselves. "Prehistoric blisters" containing such

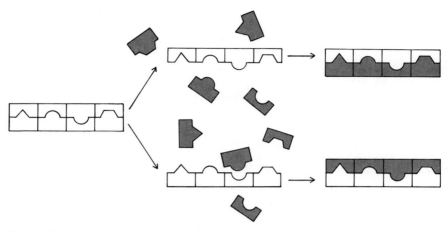

Figure 1-2   A template model for the replication of a polymer.

molecules and existing in the warm primeval oceans might have been the ancestors of the first primitive organisms on Earth.

The first primitive cells arrived on the local scene very early. Because the oldest known rocks are about four billion years old, we currently believe that the surface of the Earth solidified slightly more than four billion years ago. The fossil and geological evidence suggests that simple single-celled organisms existed from 3.8 to 4 billion years ago, and so it appears that life has existed on Earth during most of the time in which the planet has been reasonably cool. These simple cells presumably contained hereditary molecules that underwent duplication along with the cells. These hereditary molecules were probably able to direct the synthesis of other molecules necessary for the growth and reproduction of the cells and for the duplication of the hereditary molecules themselves, a process that geneticists call *replication*. The primitive cell evolved as a fancy machine to assure the replication of hereditary molecules and their transmission to new cells. We ourselves are far fancier machines that are still devoted to the process of making more copies of hereditary molecules. Indeed, organisms can be viewed as the hereditary molecules' means for making more hereditary molecules.

Biological evolution, in a sense, is a continuation of chemical evolution with the synthesis of specific chemical compounds greatly accelerated through the intervention of biological catalysts, the enzymes. The rules governing biological evolution, however, differ significantly from the rules of chemical evolution. Successful chemical compounds continue to exist because of their net formation, that is, more are formed than are broken down. Typically a compound cannot stimulate the production of more of itself, but successful species of organisms, human beings, for example, continue to exist because they are successful at making more copies of themselves, of *reproducing* themselves. Net synthesis is the key to the survival of chemical compounds; reproduction is the key to the survival of species of organisms.

Although others had glimpsed them, the basic rules of biological evolution were first proposed in detail by Charles Darwin and Alfred Russel Wallace—to wit, that species of organisms are not constant through time but are continually evolving into new forms. Such changes occur because they are advantageous; the new forms are able to reproduce more efficiently than the old forms or are able to occupy a niche in which an old form could not exist. If they reproduce more efficiently, the new forms will replace the old forms. Selection favoring reproductively improved organisms living in a particular environment is the driving force behind evolution. Darwin and Wallace believed that within any species of organisms inherited differences exist in different individuals. Because of these inherited differences some organisms are more adapted to a particular time and place and are more likely to survive, to reproduce, and to pass their *particular* hereditary endowment on to the next generation of organisms. But, living conditions themselves are continually changing, and some hereditary endowments are more suited to some conditions than to other conditions. There must be hereditary variety and heterogeneity that allows adaptation to the continually changing environment if a particular species is to survive. If a multiplicity of hereditary endowments did not exist, a species would be hard pressed to respond to environmental changes. Evolutionary history is littered with the remains of species of organisms that failed to adapt to environmental changes. The word "dinosaur" exists in our language not only as a term describing a group of reptiles which lived from 150 to 200 million years ago but also as a term denoting a person ill-suited for contemporary living. Dodo birds have come and gone because they were incapable of sensing a threat to their existence. Currently, with human population and pollution overspreading the Earth, species of organisms that do not possess sufficient numbers and hereditary variability to adapt to enormous environmental stress are threatened with extinction.

We now realize that the ultimate source of hereditary variability is *mutation,* a random change in hereditary information. Because mutations are random, they are usually deleterious and only rarely helpful in the same sense that randomly inserting a new part into a watch is usually not helpful to the function of the watch. If there were too much mutation, species could not continue to exist because of the enormous accumulation of deleterious effects. Yet the absence of mutation would eliminate the variability necessary for adaptation to a changing environment and thereby eliminate the possibility of evolution through selection favoring hereditary alternatives. Mutation is a necessary evil. Without it organisms would long ago have ceased to exist or would remain in very simple forms. Still, every species carries deleterious hereditary information within its genetic repertoire. Thus, the level of mutation must be controlled, neither too high nor too low. Mutation and the presence of mutant conditions have allowed the creation of the successful evolutionary tributary to which we belong. Mutation and the existence of mutant conditions also confront us as individuals but not as a species with that total risk from the moment of conception. Indeed, our inevitable mortality itself is a necessity of biological evolution. The displacement of old forms by new and better forms could not occur if organisms were immortal.

The major theme of biological existence is change not maintenance of the *status quo*. On an evolutionary time scale measured in hundreds of thousands, if not millions, of years, change predominates. Such change is not, however, rapid; it is ponderously slow. On a short-term time scale of mere hundreds of years, species are characterized by stability. Humans have human offspring instead of chimpanzee. Chimpanzees give birth to chimpanzees not to humans. Yet both human and chimpanzee hereditary endowments can be traced to common ancestors. Early observers of the nature of life could not have been aware of the presence of evolutionary flux but only of short-term reproductive stability. In the biblical story of the great flood, Noah was instructed by God to bring two of each kind of animal, one male and one female, aboard the ark. In the story of the ark, we see, in addition to an awareness of sexual reproduction, a vivid confirmation of the immutability of the rule that "like begets like." Not even God would instruct zebras to give birth to anything but more zebras. That like begets like is the basic observation on which genetics as a science is founded.

THE CHEMICAL NATURE OF GENES: HISTORICAL PERSPECTIVES

Early biologists did not appreciate the existence of hereditary molecules. It was not until near the end of the nineteenth century that the Swiss biologist Karl Nageli published an article in which he proposed the existence of hereditary molecules that carried hereditary information from one generation of organisms to the next. Nageli and his contemporaries believed the hereditary molecules were responsible for the physical appearances of the organisms containing them. Vigorous studies at the end of the nineteenth and beginning of the twentieth centuries revealed that these hereditary molecules were carried by chromosomes. We will review many of these studies in some depth later in this book. Several German biologists suggested that chromatin, a diffuse, fibrous network in cell nuclei that condensed into chromosomes during cell division, contained the hereditary molecules. In 1866 Gregor Mendel's work demonstrating that single physical traits in peas were inherited through indivisible hereditary factors was published. In 1900 Carl Correns and Hugo DeVries independently rediscovered Gregor Mendel's observations on the inheritance of single traits in peas. Subsequent light-microscopic studies of cells and the results from a research group headed by Thomas Hunt Morgan working with the small vinegar or fruit fly, *Drosophila melanogaster,* established unequivocally that chromosomes are the carriers of Mendel's traits, which is to say, of genetic information. In 1911

Wilhelm Johannsen, a Danish geneticist, first coined the word "gene" as a convenient term for the elementary unit of Mendelian genetics. Almost from the start there was interest in the molecular nature of the hereditary molecules, genes. As early as 1871, Fredrich Miescher had extracted a substance from cell nuclei that he called nuclein, composed of protein and nucleic acid. Chromosomes were soon found to be composed of nuclein. A well-known American cell biologist, Edmund B. Wilson, noting that chromosomes are composed of nuclein, that is, of nucleic acid and protein, suggested in 1895 that inheritance might result from "the physical transmission of a particular chemical compound." Wilson prophetically proposed in 1900 that nucleic acid, now called deoxyribonucleic acid or DNA, carries genetic information. Until the 1950's, however, virtually all geneticists believed that genes were constructed of proteins, not nucleic acid. We can understand why early geneticists believed genes were composed of protein by reviewing briefly the essential properties of genes and the state of protein and nucleic acid chemistry prior to 1940.

Genetic molecules must have the properties of replication, mutation, and functional complexity.

From our brief discussion of biological evolution we can already identify the major properties of genetic molecules. These include replication, a requirement for passing hereditary information from generation to generation; mutation, a requirement for the genetic variability necessary for evolution; and function and complexity, requirements for the production of the enormous range of inherited physical characteristics found in organisms. These requirements are discussed in the following paragraphs.

Replication. To pass genetic information from one generation to the next, there must be accurate, error-free replication of the entire body of genetic information so as to provide an extra copy of the genetic endowment that can be passed on to offspring. A means must exist of making exact copies of all genetic molecules. Failure to do so might result in dramatic loss of genetic content from one generation to the next. Because the genetic information is conserved from generation to generation, the genetic molecules themselves may also be physically conserved.

Mutation. A genetic molecule must also be able to undergo mutation, that is, change to a modified form. Furthermore, having undergone mutation, the modified form must be replicated with the same fidelity as the original form. We have already stressed the evolutionary necessity for mutation.

Function. The complete repertoire of genetic information in an organism is called the *genotype*; the physical characteristics of the organism constitute the *phenotype*. Mechanisms must exist by which the genetic constitutions, the genotypes, determine the physical characteristics, the phenotypes, of the organisms in which they exist.

Complexity. Early Mendelian genetics indicated that there are different genes for different traits, for example, genes for different eye colors, for different skin pigmentation, for hair color and texture, for height, weight, and also for a seemingly endless list of abnormalities: dwarfism, albinism (the absence of pigment), hemophilia (absence of blood clotting), and so on. Thus, there must be an array of genetic molecules equal in number to the array of different traits; genetic molecules must be able to exist in almost infinite variety.

The state of the knowledge of protein and nucleic acid chemistry in the 1930's greatly influenced early beliefs about the chemical nature of the genetic material. In brief, proteins were known to be highly complex polymeric molecules, but nucleic acids were thought to be comparatively simple. We will now consider some aspects of protein and nucleic acid chemistry.

Proteins contain linear polymers of amino acids.

Many of the salient facts about protein chemistry were known in the 1930's. Proteins are composed of linear polymers of amino acids, called polypeptides. Typically, amino acids have a basic amino ($-NH_2$) group and an acidic carboxyl ($-COOH$) group attached to a common carbon atom from which a variety of side chains extend (Figure 1-3). Twenty amino acids are commonly found in polypeptides. The amino acids are joined together in a linear polymer by peptide bonds between the amino group of one amino acid and the carboxyl group of an adjacent amino acid (Figure 1-4). As a result of these linkages, most polypeptides have a free, unbonded amino group at one end (the amino terminal or N-terminal amino acid) and a free carboxyl group at the other end (the carboxyl or C-terminal amino acid) of the linear polymer.

Physical properties of proteins depend on amino acid composition and sequence.

Proteins have different compositions and sequences of amino acids. The amino acid *composition* of a protein is the relative proportion of the different kinds of amino acids present in a protein. The amino acid *sequence* of a protein is the order in which the amino acids follow one another in the linear polypeptide. By convention the amino acid sequence of a polypeptide is listed starting with the N-terminal amino acid or residue and continues through to the C-terminal amino acid. The smallest proteins or polypeptides have fewer than one hundred amino acids; for example, the hormone insulin has only 51 amino acid residues. In contrast, some large proteins have polypeptide chains containing hundreds of amino acids. Myosin, a protein involved in muscle contraction, has a polypeptide chain with over 1,800 amino acid residues. In addition to being of many different sizes, proteins also have widely varying physical properties because of varying amino acid contents and the differing chemical and physical properties of the side chains of the amino acids. Most of the side chains are neutral, but three (lysine, histidine, and arginine) are basic, and two (aspartic acid and glutamic acid) are acidic. Furthermore, the various side chains have different affections for water. Some, in the basic and acidic amino acids, are water loving (hydrophilic) or water soluble (Figure 1-3). Others—in isoleucine, leucine, methionine, phenylalanine, proline, and valine—are water hating (hydrophobic) or water insoluble. But most, as in tyrosine, serine and glycine, are

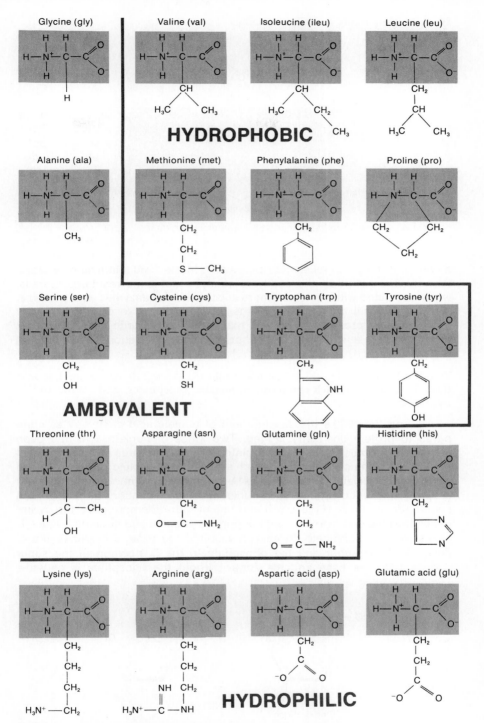

Figure 1-3   The twenty amino acids commonly found in proteins.

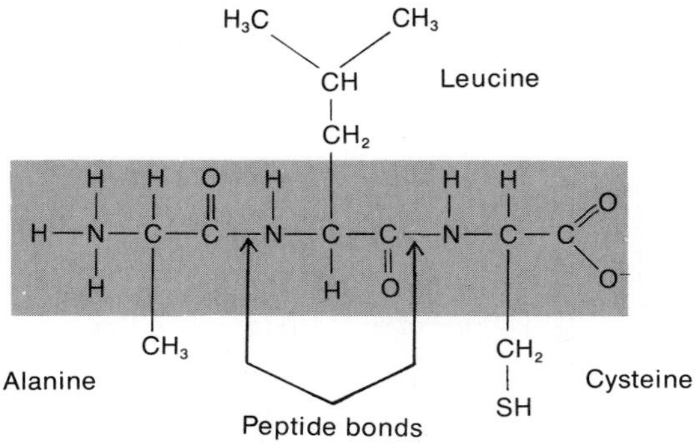

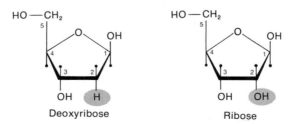

Figure 1-4   A polypeptide formed from three amino acids.

ambivalent; they can take water or leave it. In the 1930's biochemists knew that enzyme activities, the ability to catalyze metabolic reactions in cells, were associated with proteins—more succinctly, that enzymes were proteins. What emerged in the 1930's was the realization, still valid though greatly augmented by present knowledge, that proteins are large, highly complex molecules with a variety of physical properties and functional activities.

Two kinds of nucleic acid, ribonucleic acid (RNA) and deoxyribonucleic acid (DNA), are known, each composed of sugars, phosphates, and purine and pyrimidine bases.

During the 1930's the development of nucleic acid chemistry as compared to protein chemistry lagged. Two kinds of nucleic acid had been identified—thymus nucleic acid, isolated from calf thymus gland, and yeast nucleic acid. The calf thymus is a rich source of the nucleic acid now known as deoxyribonucleic acid or DNA. DNA is composed of nucleotides or, more properly, deoxynucleotides. Each nucleotide in DNA contains the five-carbon sugar 2-deoxyribose (Figure 1-5) and one of four different bases, the pyrimidines thymine and cytosine and the purines adenine and guanine (Figure 1-6), as well as phosphate. In each nucleotide the base is linked to the 1′ position of the sugar and the phosphate to the 5′ position of the sugar (Figure 1-7). The base and the sugar without the phosphate is called a

Figure 1-5   The two sugars in nucleic acids: 2-deoxyribose in DNA; ribose in RNA.

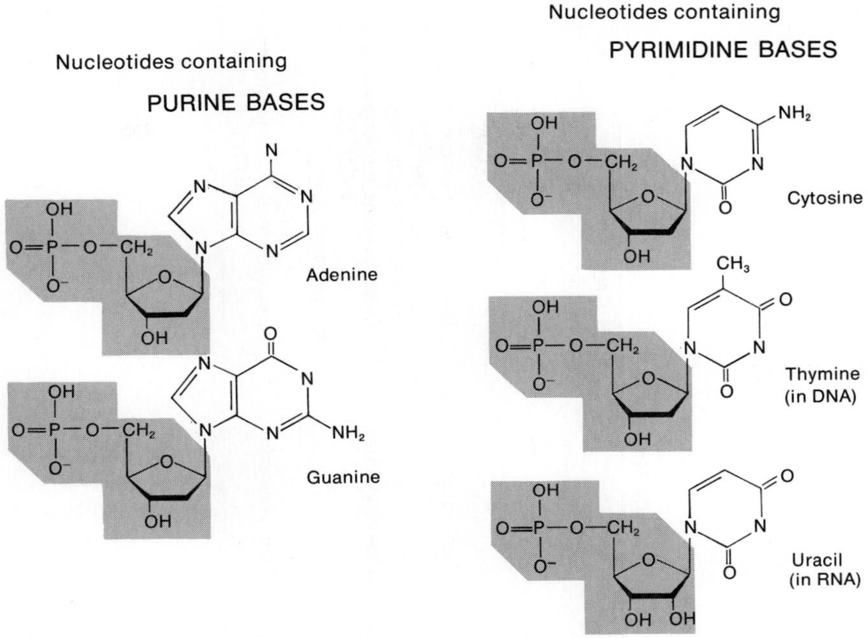

Nucleotides containing

PURINE BASES

Nucleotides containing

PYRIMIDINE BASES

Adenine

Guanine

Cytosine

Thymine
(in DNA)

Uracil
(in RNA)

Figure 1-6   The five nucleotides commonly found in nucleic acids. The names of the *bases* are given.

nucleoside. Yeast is a particularly rich source of another type of nucleic acid, ribonucleic acid or RNA, which is also composed of nucleotides. The nucleotides in RNA, however, contain ribose instead of 2-deoxyribose (Figure 1-5) and uracil instead of thymine (Figure 1-6) along with guanine, cytosine, adenine, and phosphate. Table 1-2 summarizes the different bases, nucleosides, and nucleotides.

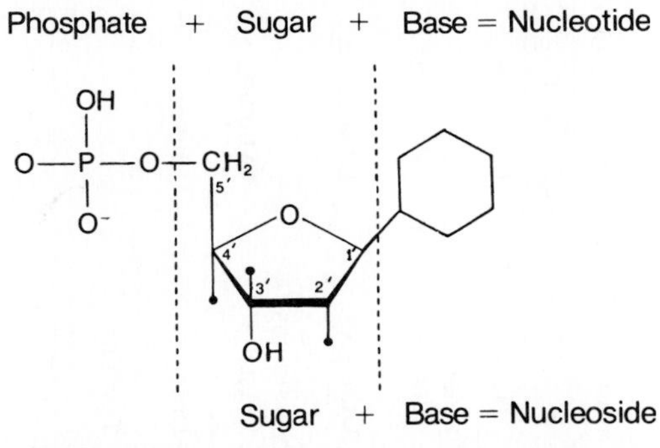

Phosphate   +   Sugar   +   Base = Nucleotide

Sugar   +   Base = Nucleoside

Figure 1-7   The general structure of a nucleoside and a nucleotide. The sugar carbon atoms are designated 1′ through 5′.

Table 1-2   Nomenclature of Purine and Pyrimidine Bases, Nucleosides and Nucleotides

Base	RNA		DNA	
	Nucleoside	Nucleotide	Nucleoside	Nucleotide
Purines:				
Adenine	Adenosine	Adenosine monophosphate (AMP)	Deoxyadenosine	Deoxyadenosine monophosphate (dAMP)
Guanine	Guanosine	Guanosine monophosphate (GMP)	Deoxyguanosine	Deoxyguanosine monophosphate (dGMP)
Pyrimidines:				
Cytosine	Cytidine	Cytidine monophosphate (CMP)	Deoxycytidine	Deoxycytidine monophosphate (dCMP)
Uracil	Uridine	Uridine monophosphate (UMP)		Not in DNA
Thymine		Not in RNA	Deoxythymidine	Deoxythymidine monophosphate (dTTP)

Nucleic acids are linear polymers of nucleotides linked together through 3′,5′ phosphodiester bonds.

Nucleic acids are linear polymers. The individual nucleotides are linked together by ester bonds between two sugar molecules and the phosphate molecule. One ester bond involves the 3′ carbon of one sugar. The second ester bond uses the 5′ carbon of another sugar to produce a diester. A linear series of nucleotides can be linked together through 3′,5′ phosphodiester bonds (Figure 1-8). The resulting polymer contains a sugar-phosphate "backbone" with the purines and pyrimidines sticking off to the side.

Early structural studies by Phoebus A. Levene indicated that DNA had approximately one molecule each of thymine, cytosine, guanine, and adenine, and RNA had one each of uracil, cytosine, guanine, and adenine. Levene assumed that nucleic acid is composed of four nucleotides linked together into a tetranucleotide (Figure 1-8) and that nucleic acid molecules are only four nucleotides long, resulting in at most only 256 different possible molecules ($4 \times 4 \times 4 \times 4 = 4^4$ or 256). It seemed unlikely that a molecule that existed in only 256 forms could account for the thousands of different genes suspected in higher organisms and, thus, unlikely that genes were

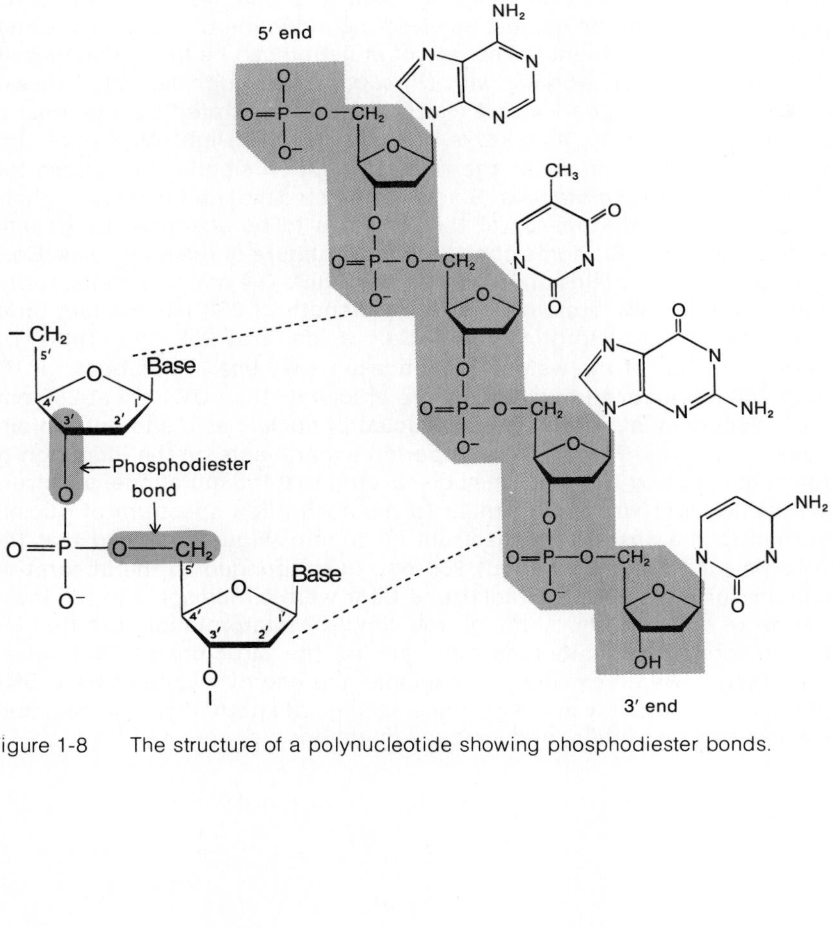

Figure 1-8   The structure of a polynucleotide showing phosphodiester bonds.

constructed of nucleic acid. The tetranucleotide hypothesis was, of course, wrong. Subsequent chemists with more accurate techniques demonstrated that in most organisms the ratio of the different nucleotides is not 1:1:1:1. Furthermore, we now know that DNA exists as enormous molecules composed of thousands of bases. But for the time the tetranucleotide hypothesis prevailed: if genes were not composed of nucleic acid, then they had to be composed of proteins. Nothing else was left. Proteins possessed large size, varied chemical and physical properties, and enzymatic actions and seemed eminently qualified to be the stuff from which genes are constructed.

EMPIRICAL TESTS OF THE CHEMICAL NATURE OF GENES

The "spectrum" for the induction of mutation by ultraviolet light is similar to the absorption spectrum of nucleic acids.

From the late 1930's through the 1950's experiments provided direct insight into the chemical nature of genes. These experiments correlated known properties of genes with physical or chemical properties of nucleic acids and proteins. One of the earliest involved the induction of mutations, which ordinarily occur infrequently. The rate of mutation can be greatly increased, as Hermann J. Muller, working with *Drosophila melanogaster,* and Lewis J. Stadler, working with barley and maize, first demonstrated by high-energy radiation such as X-rays, gamma rays, and ultraviolet (UV) light. All physical and chemical agents that increase the rate of mutation significantly above the normal rate are called *mutagens.* Stadler postulated that for ultraviolet light to act as a mutagen the energy of the light has to be absorbed by genetic molecules and produce some change in the structure of the molecules. Both nucleic acids and proteins absorb ultraviolet light. The maximum absorption of ultraviolet light by proteins is at a wavelength of 280 nanometers (nm), and the maximum absorption for nucleic acids is at 260 nm. (Table 1-3 provides a review of the weights and measures we use in this book.) If UV light at 260 nm induced mutations more efficiently than UV light at 280 nm, it might suggest that genes are constructed of nucleic acid and not protein. In 1942 Stadler and Fred M. Uber reported experiments on the induction of mutants in maize by UV light. They discovered that the mutational spectrum of ultraviolet light was more similar to the absorption spectrum of nucleic acid than to the absorption spectrum of protein (Figure 1-9) and that the most efficient wavelength, about 260 nm, corresponded to the absorption maximum for nucleic acid. Stadler and Uber were circumspect in the interpretation of their results. Although the simplest interpretation was that UV light was producing mutations by changing the structure of DNA, other interpretations were possible. For example, the energy trapped by the DNA might result secondarily in the chemical change of another type of molecule, perhaps a nearby protein, and produce mutations.

Table 1-3   Weights and Measures

Unit	Abbreviation	Equivalent
LENGTHS		
Meter	M	3.28 feet
Centimeter	cm	10^{-2} M
Millimeter	mm	10^{-3} M
Micrometer	μm	10^{-6} M
Nanometer	nm	10^{-9} M
Angstrom	A	10^{-10} M
WEIGHTS		
Kilogram	kg	2.2 pounds
Gram	g	0.0022 pounds
Milligram	mg	10^{-3} g
Microgram	microg	10^{-6} g
Nanogram	ng	10^{-9} g
Picogram	pg	10^{-12} g
Dalton*	D	3.32×10^{-24} g
VOLUMES		
Liter	l	1.0567 U.S. Quarts
Milliliter	ml	10^{-3} l

* The weight of one hydrogen nucleus. Molecular weights are expressed in Daltons.

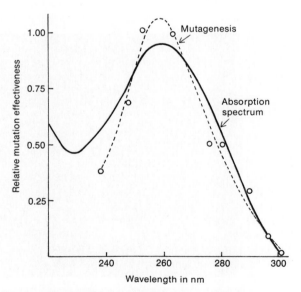

Figure 1-9   Comparison of the absorption spectrum of DNA with the mutagenic "spectrum" of ultraviolet light on corn pollen. After Stadler and Uber, *Genetics*, v. 27, p. 84, 1942.

Bacterial cells of one type can be gentically transformed into cells of a second type.

The property by which genes control the inherited physical attributes of cells was also utilized to investigate the nature of genetic molecules. In 1928 a British bacteriologist, Frederick Griffith, first reported experiments based on a phenomenon that came to be known as *transformation.* Griffith worked with *Diplococcus pneumoneae,* now renamed *Streptococcus pneumoniae,* Pneumococcus for short, the bacterium that causes one form of pneumonia. Griffith used two strains of Pneumococcus. The first, type IIIS, is a virulent form of the bacterium that possesses an extracellular polysaccharide coat and, when grown on a nutrient agar plate, produces large, smooth colonies (Figure 1-10). Each colony comprises the collected descendents of a single bacterial cell. The second, type IIR, is a non-virulent form of Pneumococcus that lacks the polysaccharide coat and produces small, rough colonies when grown on an agar plate (Figure 1-10). Because the cells of each strain maintain the same colony texture and size in generation after generation of growth, we can conclude that the difference between the two strains is inherited. Griffith, while investigating the presence of different strains of Pneumococcus in penumonia patients, injected mice with various combinations of live and heat-killed bacteria (Figure 1-11). When he injected live, virulent type IIIS pneumococci, many mice died. If he first killed the virulent bacteria by heat, the mice lived. When live, non-virulent type IIR bacteria were injected, the mice survived. But when live non-virulent type IIR bacteria were injected simultaneously with a large quantity of heat-killed, virulent

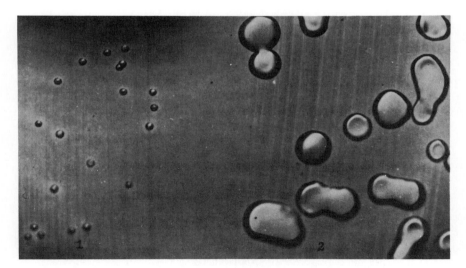

Figure 1-10   A photograph of rough (left) and smooth (right) colonies of *Pneumococcus.* Courtesy of Maclyn McCarty, Rockefeller University, New York City. From Avery, MacLeod, and McCarty, *Journal of Experimental Medicine,* v. 79, p. 137, 1944, with permission.

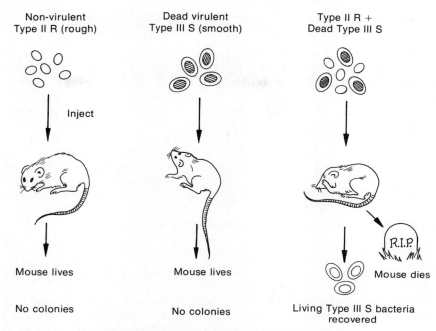

Non-virulent
Type II R (rough)

Dead virulent
Type III S (smooth)

Type II R +
Dead Type III S

Inject

Mouse lives

Mouse lives

R.I.P.

Mouse dies

No colonies

No colonies

Living Type III S bacteria
recovered

Figure 1-11   Griffith's transformation experiment using *Pneumococcus*.

type IIIS bacteria, many mice died! Bacteria recovered from the dead mice proved, when grown on nutrient agar plates, to be virulent type IIIS Pneumococci. This could be interpreted to mean that the live, non-virulent bacterial cells had been genetically transformed into virulent ones by something released from the dead cells. This "something" was called the *transforming factor.* The nature of the transforming factor was important because it had the ability to direct an inherited change in the bacteria, inasmuch as the descendants of the Pneumocci recovered from the dead mice were also type IIIS. In retrospect, the simplest interpretation of Griffith's results is that molecules carrying hereditary information, that is, genes, had been transferred from the dead virulent cells into the chromosomes of the live, non-virulent cells and had made the live cells virulent.

The transforming factor is DNA.

Following Griffith's discovery, several bacterial physiologists devoted much of their time to the identification of the chemical nature of the transforming factor. They demonstrated that transformation of Pneumococcus could be produced in the test tube using extracts of bacteria. Extracts from smooth Pneumococci caused the the transformation of a small percentage of living, non-virulent rough pneumococci into the virulent, smooth form (Figure 1-12). Transformation in a test tube with cell-free extracts facilitated

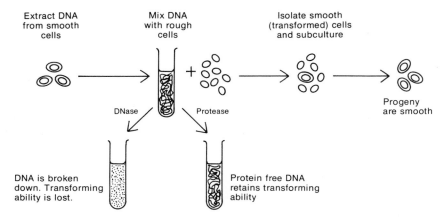

Figure 1-12   Genetic transformation of *Pneumococcus* in a test tube.

the purification and identification of the chemical nature of the transforming factor. In 1944, sixteen years after Griffith's publication, Oswald T. Avery, Colin M. MacLeod, and Maclyn McCarty published a remarkable paper proposing that the transforming factor is DNA. Their conclusion was based on several experiments. First, they demonstrated that the chemical composition and absorption spectrum of the purified transforming factor were indistinguishable from those of DNA. Their best preparations of transforming factor contained no detectable protein although subsequent, more sensitive tests showed that very pure preparations of transforming factor contained about 0.0002 mg of protein for every milligram of DNA, that is, the DNA was 99.98 per cent pure. Second, they studied the effects of different degradative enzymes. They found that trypsin and chymotrypsin, proteolytic enzymes that break proteins down into small polypeptides, had no effect on the transforming activity of the purified material. Nor did ribonuclease (RNase), an enzyme that breaks down RNA but not DNA to its nucleotide monomers. Because the molecules that caused transformation are not sensitive to these enzymes, they presumably are neither RNA nor protein. Deoxyribonuclease (DNase), an enzyme that breaks DNA down into deoxynucleotides, did destroy the transforming activity of the purified extract (Figure 1-12). The destruction of transforming activity by DNase indicates that the transforming factor is DNA. Third, Avery and his colleagues determined the molecular weight of the purified DNA in their preparations and found it to be about 500,000 Daltons (D). The average nucleotide has a molecular weight of about 310 D. A linear polymer of nucleotides with a molecular weight of 500,000 D would be much longer than a tetranucleotide; indeed, it could be 1,600 nucleotides long. Instead of only 256 possible tetranucleotides, there could be as many as 10^{960} ($4^{1,600}$) molecules. This would be many more possible forms of molecules than there are atoms in the universe. Because transformation involves an inherited change in bacterial cells, the demonstration that the transforming factor is DNA also proved that genes are DNA. Furthermore, the large size of the molecules showed that DNA is sufficiently complex to contain all of the genetic information of any organism.

The study of transformation indicated, more than any other experimental approach pursued during this period, that DNA carries genetic information. It is interesting, at least for those of us with an historical bent, that the significance of the transformation experiments was not immediately appreciated or accepted. Even in 1954, ten years after the original publication by Avery, MacLeod, and McCarty, some researchers, pointing to the small amount of protein present in the DNA preparations, still questioned whether the transforming factor was DNA, protein, or DNA and protein, prompting Rollin Hotchkiss to write the poem with which this book begins. Others accepted the idea that the transforming factor is DNA but suggested that transformation was a means of producing specific mutations and that DNA, rather than being a carrier of genetic information, actually was a unique form of a mutagen, perhaps more like ultraviolet light than like a gene.

Parental phage DNA is found in phage progeny following growth in bacteria.

In the experiments we have discussed, biologists have used two properties of the genetic material in their attempts to identify the chemical nature of genes. Gene mutation was used in the mutagenesis experiment with ultraviolet light, and control by genes of the inherited physical characteristics of cells or organisms in which they exist was used in the transformation experiments. We will now discuss an experiment in which replication of genes is used as an assay for determining the nature of genetic molecules. The conservation of genetic information from one generation to the next requires that genes be exactly replicated. Molecules carrying genetic information may themselves be physically conserved from generation to generation. For example, the original strands in the replicating molecule pictured in Figure 1-2 are still present after the molecule has become duplicated. If a means existed to identify molecules that are conserved from one generation to the next, we might discover something about the chemical nature of genetic molecules.

To this end, Alfred D. Hershey and Martha Chase conducted a series of experiments with bacteriophage T2, which they published in 1952. Bacteriophages, or, more simply, phages, are virus parasites that grow in bacteria and under some conditions destroy their hosts. Phage T2 is a bacterial virus with a chromosome contained within a complex protein structure that includes a head, a tail, and tail fibers (Figure 1-13). Phage T2 is a virulent phage, that is, it normally destroys its host, the common intestinal bacterium *Escherichia coli*. Phage T2 infects and ultimately destroys a bacterial cell with the production of more than one hundred new phages. This type of phage life cycle, called the *lytic* cycle, starts with the initial attachment of phages to specific sites on the surface of bacterial cells (Figure 1-13). One or several phages may attach to the same cell. Injection of the phage chromosome into the cell follows, leaving behind an empty phage head, or "ghost," on the outside of the cell. Once in the cell, the phage chromosome

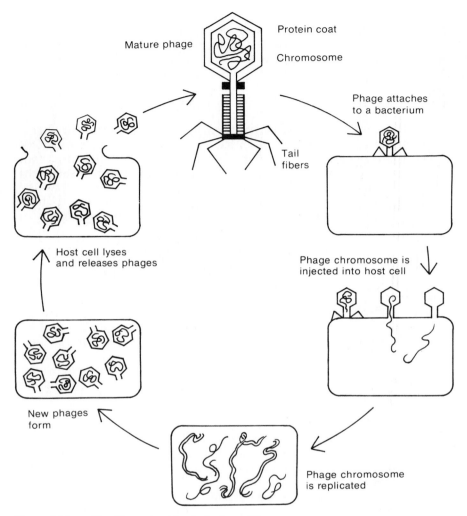

Figure 1-13   The life cycle of bacteriophage T2.

is replicated, and the components for making new phages are produced. The new phages, each containing a single chromosome, are assembled (Figure 1-14) and after lysis (breakdown) of the host cell membrane, are released from the broken cell. The whole process, from initial attachment to release of new phages, takes about 20 minutes at 37° C. The newly released phages are ready to infect other *E. coli* cells and so initiate another cycle of phage growth.

Hershey and Chase prepared phages that had either their DNA or protein labelled with a radioactive compound. The DNA was labelled with 32-phosphorous (^{32}P) in the phosphate of the nucleic acid. The protein was labelled

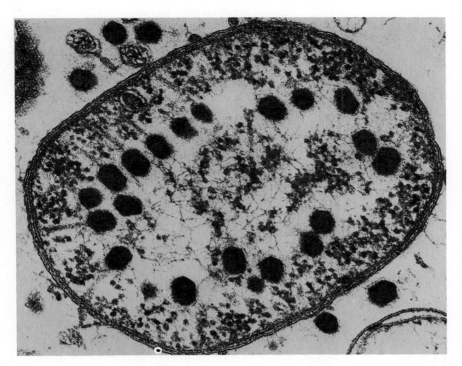

Figure 1-14   An electron micrograph of newly formed T2 phages in an *E. coli* cell shortly before lysis. Courtesy of Lee D. Simon, Department of Microbiology, Rutgers University, New Brunswick, New Jersey. From Simon, *Virology*, v. 38, p. 285, 1969, with permission from Academic Press, Inc.

with 35-sulfur (^{35}S). Sulfur is present in the amino acids methionine and cysteine (Figure 1-3). There is no phosphorous in proteins and no sulfur in nucleic acids. The radioactive phages (Figure 1-15) were allowed to infect *E. coli* cells, and the phage heads, empty of their chromosomes, were separated from the bacteria by first placing the phage-bacterial mixture in a blender to dislodge the phage ghosts from the surface of the bacteria. Next the mixture was gently centrifuged, forcing large bacteria to the bottom of a centrifuge tube while the small phage ghosts remained in suspension. The phage ghosts contained about 85 per cent of the original protein of the parent phages but only 15 per cent of the nucleic acid as revealed by the amount of ^{35}S and ^{32}P present. Thus most of the protein never entered the bacterial cells. Later, following lysis of the bacterial cells, Hershey and Chase recovered the next generation of newly produced phages and determined the amount of radioactive parental DNA and protein present. They found that less than 1 per cent of the original protein was present in the progeny phages, but about 30 per cent of the parental DNA was present. There was substantial conservation of DNA from generation to generation but almost

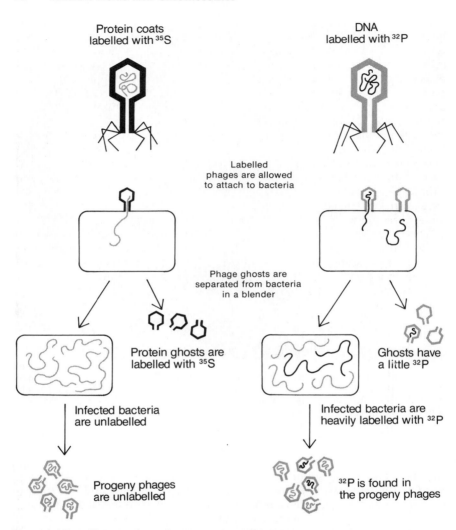

Figure 1-15   The experiment by Hershey and Chase demonstrating the preservation of parental DNA in phage progeny.

no conservation of protein. Hence it appeared that the genetic information was embodied in the DNA and not in the protein.

RNA can also carry genetic information. The nature of tobacco mosaic virus is determined by its RNA.

Viruses are by no means limited to bacterial hosts. Other kinds of viruses infect and parasitize animal and plant cells. One plant virus that contains RNA, not DNA, was an important tool for genetic experiments in the late

1940's and 1950's. Tobacco mosaic virus (TMV) (Figure 1-16), infects tobacco, causing the infected regions on leaves to become discolored and
blistered. Different strains of TMV produce recognizably different lesions on
infected leaves. The common virus (*TMV-common*) produces a green mosaic
disease, but a variant, Holmes Ribgrass (*TMV-HR*), produces ringspot lesions.
Moreover, the amino acid compositions of the proteins of the two strains
are different. Heinz Fraenkel-Conrat and Bea Singer found that they could
separate the RNA and the protein of TMV and put them back together to
make infectious viruses. They were able to reassemble viruses with the RNA
from *TMV-common* enclosed in *TMV-HR* protein and *TMV-HR* RNA with *TMV-
common* protein (Figure 1-17). These reassembled viruses produced infections in tobacco leaves and new progeny viruses. The reassembled viruses
with *TMV-common* RNA and *TMV-HR* protein produced a green mosaic disease characteristic of *TMV-common*. Recovered progeny viruses had proteins characteristic of *TMV-common*. The viruses containing *TMV-HR* RNA
with *TMV-common* protein produced ringspot lesions and the virus progeny
had proteins characteristic of *TMV-HR*. The source of the RNA determined
the nature of the mosaic infection and of the proteins present in the progeny
viruses, indicating again that nucleic acid, in this case RNA, carries hereditary information.

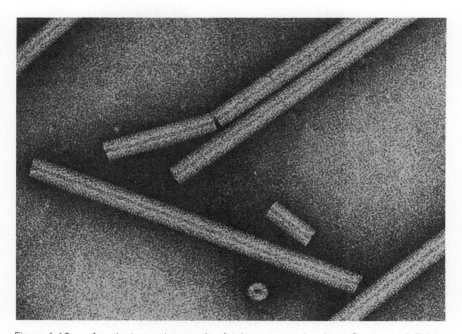

Figure 1-16   An electron micrograph of tobacco mosaic virus. Courtesy of Robley
Williams, Virus Laboratory, University of California, Berkeley. From Williams and Fisher,
An Electron Micrographic Atlas of Viruses, 1974, courtesy of Charles C Thomas, Publisher,
Springfield, Illinois.

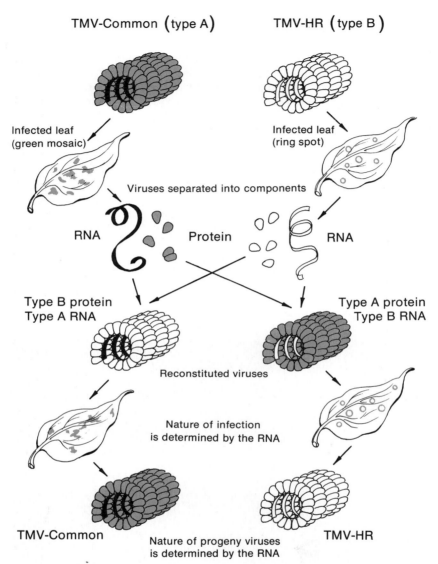

TMV-Common (type A)   TMV-HR (type B)

Infected leaf
(green mosaic)

Infected leaf
(ring spot)

Viruses separated into components

RNA   Protein   RNA

Type B protein
Type A RNA

Type A protein
Type B RNA

Reconstituted viruses

Nature of infection
is determined by the RNA

TMV-Common   TMV-HR

Nature of progeny viruses
is determined by the RNA

Figure 1-17   Hereditary effects of reconstituting tobacco mosaic viruses using RNA and protein from two strains.

THE STRUCTURE OF DNA

The results of the experiments we have described all point to nucleic acid, DNA or RNA, as the carrier of genetic information. Nucleic acid satisfies the four major characteristics expected of hypothetical genetic molecules: replication (the bacteriophage growth experiment); mutation (the ultraviolet light mutation experiment); function (the experiments on transformation and later on the reconstitution of tobacco mosaic virus); and complexity (the high molecular weight of the DNA in the transformation experiments). Nevertheless, as we have indicated, there was some resistance to the conclusion that genes are constructed of nucleic acids. Geneticists lacked sufficient knowledge of the structure of DNA to provide a theoretical framework for the replication, function, and mutation of DNA. Determination of the structure of DNA became a matter of great importance and in the early 1950's attracted many investigators, among them Rosalind Franklin and Maurice Wilkins, two X-ray crystallographers who worked at the University of London. It interested Francis H. C. Crick, who was working at Cambridge University on his doctoral dissertation. It had long occupied Erwin Chargaff, a biochemist at Columbia University in New York City. And it brought James Watson from Copenhagen, where he was in the midst of a fellowship working on phage growth in bacteria in the manner of Hershey and Chase, to Cambridge to work in association with Crick. Through the combined efforts of these five and others, the structure of DNA was elucidated. The foundations on which the structure of DNA were based included structural and theoretical chemistry, analyses of base compositions of DNA's from a variety of species and tissues, X-ray crystallographic data, and, perhaps most important, the ability of Watson and Crick to build structural models consistent with the chemical and physical data.

DNA is a double helix.

Levene had provided important information about DNA's chemical composition and demonstrated that it is a polymer of purine and pyrimidine nucleotides. Precise knowledge of the three-dimensional structure of DNA was lacking. Did the nucleotide polymers exist by themselves or as aggregates of two or more polymers? If there were aggregates, how were they held together? Did DNA molecules have specific three-dimensional structure? Watson and Crick attempted to answer these and other questions. An early X-ray crystallographic picture of DNA made by Rosalind Franklin helped Watson and Crick to propose a model for the structure of DNA. According to the model, DNA is composed of two long, unbranched polymers of deoxynucleotides lying side by side. The dexoynucleotides are linked together by phosphodiester bonds. The two strands of DNA are wound around each other to form a double helix.

The two strands of DNA are complementary. For every adenine on one strand there is a thymine on the other; for every cytosine on one, a guanine on the other.

Data from the laboratory of Erwin Chargaff helped Watson and Crick formulate their model of DNA. Chargaff conducted a series of exacting experiments to determine the base compositions, that is, the percentage of molecules or moles of adenine (A), thymine (T), guanine (G), and cytosine (C) in DNA's from several species. Chargaff noted that the amounts of A, T, C, and G are different in different species. In all of the DNA's he studied the amount of adenine equals the amount of thymine and the amount of guanine equals the amount of cytosine, G = C and A = T (Table 1-4). Watson and Crick's model showed one way to achieve Chargaff's "rules" that A = T and G = C. For every A on one strand of the double helix, there would be a T on the other strand (A = T), for every G on one strand, a C on the other (G = C), for every purine on one strand, a pyrimidine on the other. The two strands would have *complementary* base sequences with the base sequence on one strand being reflected by a complementary sequence on the other strand. The structure of DNA that Watson and Crick proposed showed the molecular bases for complementarity.

The two strands of the DNA are held together by hydrogen bonds between the specific complementary bases.

Watson and Crick suggested that the purine and pyrimidine bases occupy the inside of the double helix with the backbone of deoxyribose sugars held together by phosphodiester bonds occupying the outside (Figure 1-18). A purine on one strand could be bonded to a pyrimidine on the opposite strand by hydrogen bonds (Figure 1-19). Three hydrogen bonds hold G and C together, but there are only two bonds between A and T. Hydrogen bonds are weak bonds. The strength of the bonds would not be sufficient to hold

Table 1-4   Base Compositions of DNA from Some Organisms of Genetic Interest

Organism	Base Compositions (%)			
	A	T	G	C
Human	31.0	31.5	19.1	18.4
Cattle	28.7	27.2	22.2	22.0
Drosophila melanogaster (fruit fly)	27.3	27.6	22.5	22.5
Zea mays (corn or maize)	25.6	25.3	24.5	24.6
Neurospora crassa (pink bread mold)	23.0	23.3	27.1	26.6
Saccharomyces cerevisiae (baker's yeast)	31.3	32.9	18.7	17.1
Streptococcus pneumoniae	30.3	29.5	21.6	18.7
Salmonella typhimurium	22.9	23.0	27.1	27.0
Escherichia coli	24.6	24.3	25.5	25.6
Bacillus subtilis	28.4	29.0	21.0	21.6

Based on chemical methods of determining base compositions, similar to those of Chargaff.

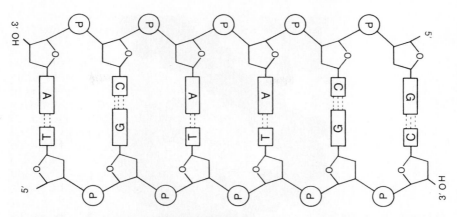

Figure 1-18   Schematic representation of base pairing and antiparallel orientation of polynucleotide strands in a double helix.

a single T together with a single A or a single G to a single C. The sum total of the hydrogen bonds between the two strands coupled with bonds between adjacent bases on the same polynucleotide (to be discussed shortly) assures that the two strands of the DNA double helix are firmly associated with each other under conditions commonly found in living cells.

The two strands of the double helix have opposite polarity.

Because of the configuration of the phosphodiester bonds between the 3′ and 5′ positions of adjacent deoxyribose molecules, every linear polynucleotide can have a free, unbonded 3′ hydroxyl group at one end of the polynucleotide (the 3′ end) and a free 5′ hydroxyl at the other end (the 5′ end) (Figure 1-18). There are then two possible ways for the two polynucleotides to be oriented in a double helix. They could have the same polarity, that is, be parallel, both strands having 3′ ends at one end and 5′ ends at the other end. Or, by rotating one strand 180 degrees with respect to the

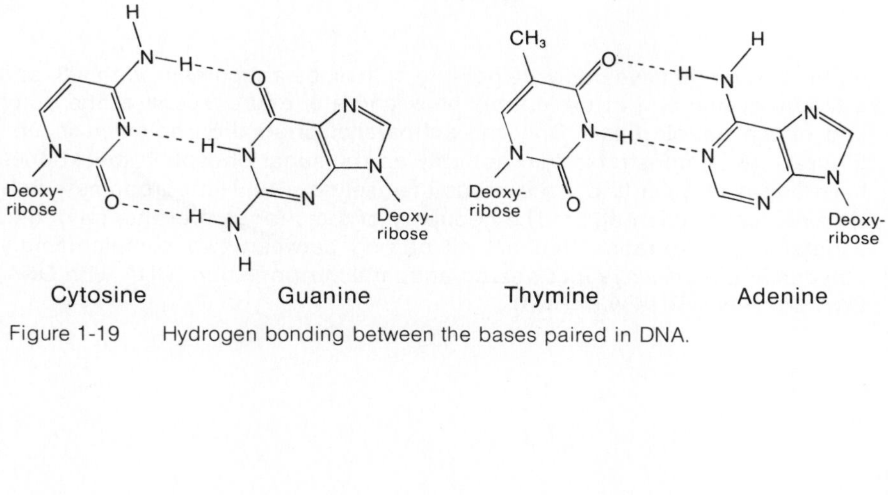

| Cytosine | Guanine | Thymine | Adenine |

Figure 1-19   Hydrogen bonding between the bases paired in DNA.

Figure 1-20   The space-filling model of DNA. Atoms in one sugar-phosphate backbone are white and those in the other are dark grey; the atoms in the paired bases are white. From "The Synthesis of DNA" by Arthur Kornberg, *Scientific American,* October 1968, with permission from W. H. Freeman and Co., San Francisco.

other, they could have opposite polarity, that is, be antiparallel, with a 3′ and a 5′ end at one end of the double helix and a 5′ and a 3′ end at the other end of the double helix. Only the antiparallel orientation actually occurs. Figure 1-18 demonstrates that not only do the sugar-phosphate backbones have opposite polarity but the bases themselves are inverted geometrically with respect to each other. The geometry of the strands prevents pairing in a parallel configuration. Indeed, all pairing between two complementary polynucleotides always involves an antiparallel orientation (DNA with DNA, DNA with RNA, RNA with RNA).

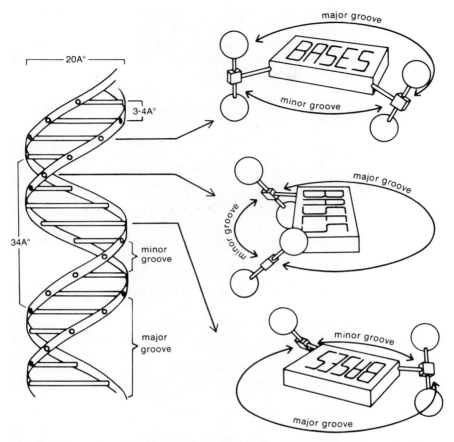

Figure 1-21   Left: A schematic representation of the DNA double helix. Horizontal rods represent base pairs connecting the two ribbon-like sugar-phosphate backbones. Right: Schematic representation of the rotation of the base pairs as a result of the helical structure of DNA. Cubes represent sugar molecules and spheres, phosphate molecules.

In double helical DNA the base pairs are stacked on each other like a pile of coins.

A model of DNA is presented in Figure 1-20. The two sugar-phosphate backbones, shown in two shades of grey, wind around each other. The atoms in the purine and pyrimidine bases, the white balls, lie between the sugar-phosphate backbones. The paired bases appear to be stacked on top of each other like a pile of coins. The model in Figure 1-20 is a "space-filling" model, showing the actual volume of the different atoms. The atoms in one pair of bases are in close association with those of the adjacent base pairs. Their closeness tells us that there are strong van der Waals forces—in this special case called "stacking" forces—between the stacked base pairs. These stacking forces are even more essential than the hydrogen bonds between the complementary bases for maintaining the double helical structure of DNA.

Each base pair is rotated 36° with respect to its neighbors, one complete turn in the double helix occurs every ten base pairs.

The dimensions of the double helix are indicated in Figure 1-21. The distance between the backbones is 20 A and each base pair is 3.4 A apart. The orientation of the bases rotates with the rotation of the double helix, each base pair being offset by 36° from the adjacent ones. Thus, a 360° rotation of the helix occurs every ten base pairs or every 34 A. We can also identify a major groove and a minor groove bounded by the sugar-phosphate backbones. The two grooves are of different size because the two attachment points of the base pairs to the backbones are not opposite each other (Figure 1-21).

The complementary structure of DNA suggests a mechanism for gene replication.

The structure of DNA proposed by Watson and Crick in 1954 is remarkable in that it remains essentially unchanged. Not surprisingly, it resembles the structure of the hypothetical replicating molecule presented in Figure 1-2. The specific pairing of the purine and pyrimidine bases suggests a means for the replication of DNA. In the double helix, when the identity of a base on one strand is known, the identity of its complement is also known. A particular base sequence, the order of nucleotides one after another on one strand, is necessarily associated with a complementary sequence on the other strand. For replication to occur (Figure 1-22), the two strands are separated, and each separated strand directs the synthesis of its complement. Appropriate nucleotides pair with their complements in the polymer in an antiparallel orientation. The individual nucleotides are then joined

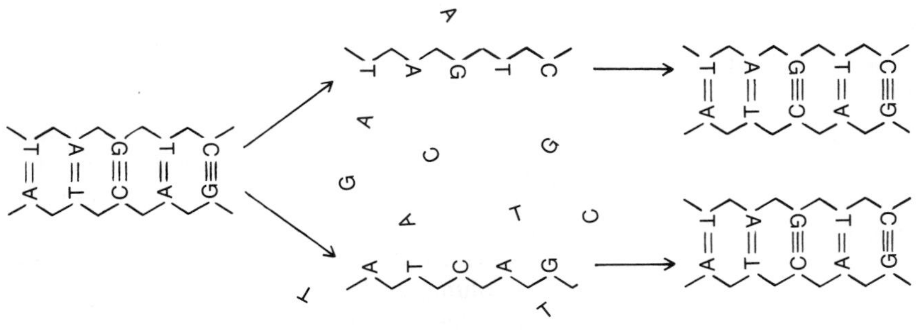

Figure 1-22   A scheme for the replication of DNA. Compare with Figure 1-2.

together by phosphodiester bonds to form two DNA helices with exactly the same sequences as the original double helix (Figure 1-22). A possible mechanism for the replication of genes is apparent.

SUMMARY

Evolution is responsible not only for the organisms that comprise the species of the universe but for the nature of hereditary mechanisms. We see that DNA, the terrestrial carrier of genetic information, is admirably suited to be a genetic molecule. The complementarity of the polynucleotide strands, resulting from the base pairing specificities (A with T, C with G) provides an evident mechanism for replication with separated strands acting as templates for the synthesis of their complements. DNA is economically constructed from only four building blocks held together by phosphodiester bonds. Yet, the long lengths of the polymeric molecules provide an almost infinite number of different possible sequences, thereby conferring on DNA the complexity necessary for containing information for the production of the wide range of physical characteristics found in terrestrial organisms. We can hypothesize that the different Mendelian factors, genes, that are responsible for different traits are composed of different nucleotide sequences.

2

The Nature of Chromosomes:
Chemical and Structural Properties

We can now assume that hereditary information is carried in DNA or RNA and that information for the vast range of structural characteristics found in organisms is encoded in different nucleotide sequences. The mechanisms by which hereditary information is packaged, replicated, and transmitted to new generations of organisms and by which the genetic endowment in the DNA (the genotype) produces the physical characteristics of organisms (the phenotype) are also highly efficient and intricate products of evolution. To understand genetic processes, we must understand the nature of chromosomes, the entities that carry genetic information. We will consider the structure of chromosomes and the variety and similarities of base compositions, base sequences, sizes, and shapes of the DNA molecules, which are integral parts of chromosomes. We will find both diversity and similarity in chromosome structures. Chromosomes of prokaryotes—bacteria and blue-green algae that lack nuclei in their cells—have many common properties. Chromosomes of eukaryotes—organisms with true nuclei—have different common properties. Indeed, all eukaryotes from simple fungi and green algae to large multicellular plants and animals have more in common with each other than the simplest of them has with bacteria. Before turning to chromosome structure, we will discuss the general properties of prokaryotic and eukaryotic cells.

Prokaryotic cells are smaller and less complex than eukaryotic cells.

Prokaryotes contain no nuclei within their cells; there is no large structure enclosed in a membrane that contains DNA. A bacterial cell is bounded by a membrane composed of lipids and proteins, which, in turn, is usually surrounded by a cell wall. Electron micrographs (Figure 2-1) show a clear nuclear area in which the chromosome of the cell is located. Generally no other internal subdivisions appear. Although different species of bacteria vary in size and shape, the rod-shaped bacterium *Escherichia coli,* shown in Figure 2-1, has overall dimensions characteristic of many bacteria.

Eukaryotes are either unicellular or multicellular. Multicellular organisms arise by a series of cell divisions from a single-celled zygote formed

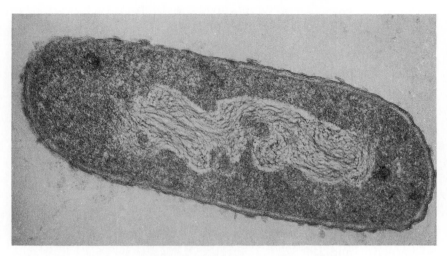

Figure 2-1   An electron micrograph of an *Escherichia coli* cell with a clear nuclear area in which chromosome fibers are seen. The cell is about 2 μm long. Courtesy of G. Cohen-Bazire, Institut Pasteur, Paris.

from the union of two germ cells. Eukaryotic cells, particularly those of multicellular organisms, are much more complicated than bacterial cells (Figure 2-2) and usually are much larger. Thousands of *E. coli* cells such as the one shown in Figure 2-1, could easily be contained within a single liver cell. Like bacterial cells, each eukaryotic cell is bounded by a cell membrane. All eukaryotic cells contain discrete, membrane-bound structures called organelles. Examples of organelles include mitochondria (Figure 2-3), which provide energy in both plant and animal cells and chloroplasts (Figure 2-4), the sites of photosynthesis in plant cells. Both mitochondria and chloroplasts have chromosomes composed of DNA and undergo division. Depending on the type of cell or species, the number of mitochondria or chloroplasts per cell varies from one to several hundred. These organelles double in number during cell growth and are distributed roughly equally between the two daughter cells formed by cell division. Another component, the ribosome, is present in both prokaryotic and eukaryotic cells. Ribosomes are small particles approximately 300 A in diameter that are the sites of protein synthesis in cells. All ribosomes in prokaryotic cells and some ribosomes in eukaryotic cells are unattached to other cell components. Some ribosomes in eukaryotes are attached to a network of membrane-bounded tubules, the endoplasmic reticulum. The membranes of the endoplasmic reticulum are continuous with the membrane bounding the most prominent organelle in the cell, the nucleus (Figure 2-2). Pores in the nuclear membrane connect the contents of the nucleus, the nucleoplasm, with contents of the cell outside of the nucleus, the cytoplasm. Most of the DNA in a eukaryotic cell is located in the nucleus. Between cell divisions, that is, during the so-called resting phase or interphase, the nucleus has a diffuse, granular appearance. Within

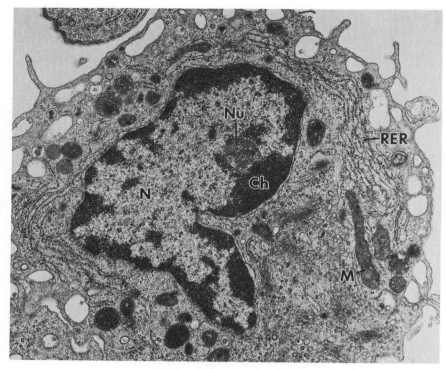

Figure 2-2   An electron micrograph of a mouse white blood cell with a diameter of about 20 μm. CH, condensed chromatin; M, mitochondrion; N, nucleus; Nu, nucleolus; RER, rough endoplasmic reticulum. Courtesy of Barbara Nichols, Proctor Foundation, University of California, San Francisco.

the nucleus, the nucleolus, a darkly staining body that is not bounded by a cell membrane, is seen. The nucleolus is the site of ribosome assembly. In most animal cells immediately outside the nuclear membrane is a centriole, which is involved in the separation of replicated chromosomes during cell division.

GENERAL CHARACTERISTICS OF CHROMOSOMES

Chromosomes are discrete entities that carry genetic information and contain DNA or RNA.

Cell biologists coined the term "chromosome," which literally means colored body, to describe the darkly staining bodies visible by light micros-copy in eukaryotic cells at the time of cell division. Geneticists at the begin-

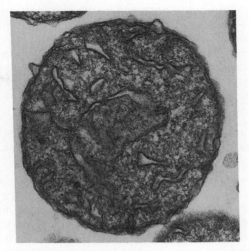

Figure 2-3   An electron micrograph of an isolated rat-liver mitochondrion. Courtesy of Lester Packer, Department of Anatomy and Physiology, University of California, Berkeley.

Figure 2-4   An electron micrograph of a chloroplast. Note the relatively small size of the mitochondrion below the chloroplast. Courtesy of Robert Buchanan, Department of Cell Physiology, University of California, Berkeley.

ning of the 20th century (see Chapters 6 and 8) demonstrated that eukaryotic chromosomes are carriers of genetic information. Today the term chromosome is also used for the molecules of DNA or RNA that carry genetic information in prokaryotes or in organelles such as chloroplasts and mitochondria even though these molecules are not visible by light microscopy. A general definition of a chromosome, then, is a structure containing DNA or RNA that carries genetic information from one generation to the next.

The nucleic acid in most chromosomes is double-helical DNA. Phages and viruses have chromosomes composed of single- or double-stranded DNA or RNA.

The DNA in every cellular organism and in most DNA-containing viruses is double helical. The DNA in a few viruses, such as the bacteriophage, ØX174, is single stranded. Others, such as tobacco mosaic virus, poliomyelitis virus, and some *oncogenic* (cancer-causing) viruses, have their genetic information encoded in RNA, and, as in most RNA viruses, the RNA is single-stranded. Still others, however, such as *reo* virus, which has been implicated as a cause of diabetes, have their genetic information carried in double-helical RNA molecules. The structure of double-helical RNA is in many ways similar to that of double-helical DNA. Like all RNA, the RNA in *reo* virus contains ribose instead of deoxyribose and uracil instead of thymine. The uracil in double-helical RNA, like thymine in DNA, pairs with adenine through two hydrogen bonds. From a genetic perspective, it is not a matter of great importance whether the nucleic acid in viruses is double or single stranded. The single-stranded molecules can serve as templates for the formation of their complements, creating double-stranded molecules. These molecules can then be replicated as part of a process, resulting in the production of more viruses. As we will discuss in Chapter 5, the double-stranded molecules used for replication become single-stranded ones again when new viruses are formed.

Bacterial cells contain a single large chromosome and small chromosomes called plasmids.

If on a short-term basis the transmission of hereditary information occurs essentially without error, two conditions should prevail. First, the amount of DNA should be the same in all or almost all cells of all members of a given species. Second, because hereditary information is directly or indirectly responsible for chromosome structure, the chromosome complement (with some exceptions), should also not vary essentially in cells of all members of a given species.

Bacterial cells contain one large chromosome composed essentially of a single, circular double-helical molecule of DNA.* In *E. coli*, the chromosome in each cell contains about 4×10^6 base pairs of DNA. Small circular

---

* We shall refer throughout this book to the large chromosome of a bacterial cell simply as the "chromosome."

chromosomes called *plasmids* may be present. A variety of different plasmids exist. Most plasmids contain about 5×10^3 base pairs of DNA and are much smaller than the chromosome, which encodes most or all of the information essential for the growth and division of the bacterium. Plasmids encode information essential for their own replication and sometimes for a few functions of the bacterium. Because plasmids are not always essential, some bacteria contain different plasmids or none at all.

Eukaryotic cells contain multiple chromosomes composed of DNA and structural proteins that undergo striking changes in appearance before and after cell division.

In contrast to bacteria, eukaryotes have multiple major chromosomes that contain, in addition to DNA, structural proteins called histones, which are rich in basic amino acids (lysine, arginine, and histidine—see Figure 1-3) and other, relatively acidic, non-histone proteins. The complex of DNA, histones, and non-histone proteins is called *chromatin*. In interphase nuclei the chromatin is uncondensed, and individual chromosomes are not visible. Prior to cell division, the chromosomes condense into several discrete elongated bodies that stain heavily with dyes that bind to DNA, for example, aceto-orcein or, using the Feulgen technique, acid-fuchsin following a brief hydrolysis of the DNA with hydrochloric acid. Thus, the chromosomes assume the characteristics of the heavily staining bodies from which their name derives. Prior to condensation, the chromosomes are replicated, and on condensation we can see that each chromosome is composed of two newly replicated identical structures called *chromatids*.

Condensed eukaryotic chromosomes have darkly staining (heterochromatic) regions adjacent to the unstaining centromere and lightly staining (euchromatic) "arms."

When viewed by light microscopy, condensed, stained eukaryotic chromosomes share a set of general characteristics. Almost all have a small unstaining constricted region, the *centromere,* which is located in different positions in different types of chromosomes (Figure 2-5). In *metacentric*

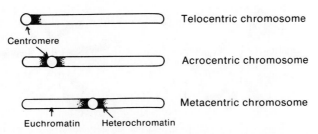

Figure 2-5   The classification of eukaryotic chromosomes according to the position of the centromere.

chromosomes the centromere is located near the center of the chromosome, so that the chromosome has two "arms" of equal length. In *acrocentric* chromosomes the centromere is located nearer one end of the chromosome, so that the arms are of unequal length. In *telocentric* chromosomes the centromere is located at or very near the end, the telomere, of a chromosome, so that the chromosome appears to have only one arm. The centromere contains the *kinetochore,* a specialized region to which fibers attach during nuclear divisions. These fibers separate the newly replicated, paired chromatids into daughter cells and so into two discrete chromosomes. More condensed parts of the chromosome, *heterochromatin,* stain more heavily than less condensed parts, *euchromatin.* Usually, regions adjacent to centromeres are highly condensed and constitute the centric heterochromatin of chromosomes.

The number of chromosomes varies in different species, but each species has a constant chromosome complement, the karyotype.

In terms of number and physical appearance, the chromosome complement, the *karyotype,* of most members of a given species is constant with one important exception. In animal species having two sexes, the chromosome complements of the sexes commonly differ. In mammals, females are the *homogametic* sex, having two identical sex chromosomes called X chromosomes, and males, the *heterogametic* sex, have one X chromosome and one Y chromosome. In birds and some insects, for example, the silk moth, the males are homogametic with two Z chromosomes, and females have two different sex chromosomes, Z and W. Chromosomes that are not sex chromosomes are called *autosomes.* The different chromosomes can be identified by their size, position of the centromere, and other physical characteristics and can be cataloged and numbered. In *Drosophila melanogaster,* the fruit fly that has been extensively utilized by geneticists, chromosome 1, the X chromosome, is a medium-sized telocentric, chromosomes 2 and 3 are large metacentrics, and chromosome 4 is a tiny acrocentric. The Y chromosome in males is a heavily heterochromatic acrocentric (Figure 2-6). The ability to distinguish different chromosomes allows us to demonstrate that

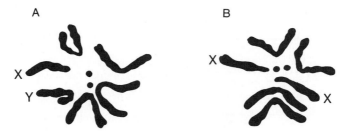

Figure 2-6   The karyotypes of (A) male and (B) female *Drosophila melanogaster.* The X and Y chromosomes are indicated. After Dobzhansky, in Morgan, *The Scientific Basis of Evolution,* W. W. Norton, 1932.

in the nuclei of most cells of most multicellular organisms two copies of each chromosome are present. The identical chromosomes are called homologous chromosomes, *homologs*. The symbol N is used to indicate the number of different chromosomes present. A complement comprised of one of each pair of homologs constitutes the haploid condition, N. A chromosomal complement with homologous pairs of each chromosome, typical of multicellular organisms, constitute the diploid condition, 2N. In some lower organisms, such as some fungi, only a single copy of each chromosome is present so that these organisms are haploid. Like higher animals, many higher plants are diploid. Corn (*Zea mays*) is an example. But many other higher plants are polyploid and have more than two copies of each chromosome. Commercially grown potatoes are tetraploid (four copies of each chromosome) and bread wheat is hexaploid (six copies of each chromosome). The haploid number of chromosomes varies from species to species (Table 2-1). Phages and bacteria have but one chromosome. *Drosophila* has a haploid number of 4, and the pink bread mold, *Neurospora crassa,* has 7. In plants, haploid chromosome numbers generally range from 10 to 20. Maize, for example, has 10. In mammals, haploid chromosome numbers are somewhat higher, mice having 20, and humans, 23.

Table 2-1   Haploid Chromosome Numbers in Some Eukaryotes of Genetic Interest

Species	Number of Chromosomes
Human *Homo sapiens*	23
Mouse *Mus musculus*	20
Cow *Bos taurus*	30
Fruit Fly *Drosophila melanogaster*	4
Pink Bread Mold *Neurospora crassa*	7
Unicellular alga *Chlamydamonas reinhardii*	16
Black Bread Mold *Aspergillus nidulans*	8
Garden Pea *Pisum sativum*	7
Corn (Maize) *Zea mays*	10
Wheat *Triticum aestivum*	21
Brewer's yeast *Saccharomyces cerevisiae*	18*
Nematode *Caenorhabditis elegans*	6

* Best Estimate.

Special techniques, banding techniques, for staining chromosomes are commonly used to distinguish different chromosomes in mammalian cells.

Because *Drosophila melanogaster* has a haploid number of 4 chromosomes in its karyotype, there is no great difficulty in identifying each chromosome. But in many organisms including mammals, although it is possible to demonstrate that all chromosomes are present in pairs, the large number and similar appearance of many chromosomes make it difficult to distinguish one pair from another. Techniques for preparing chromosomes for light microscopy and banding techniques for differentially staining chromosomes have greatly facilitated the cytological observation of chromosomes, particularly in mammalian cells.

For the routine light-microscopic observation of human chromosomes, cells are cultured under special conditions (pp. 96–97), so that newly replicated chromosomes composed of two chromatids remain condensed and are easy to see and photograph. The image of each chromosome may be cut out of the photograph and can then be arrayed side by side in a standard order with the largest chromosomes first and the smallest last. In the human karyotype with a haploid complement of 23 chromosomes, many of the chromosomes have similar sizes and shapes (for example, 4 and 5; 8 and 9) and cannot readily be distinguished from each other using procedures that stain the chromosomes uniformly. All of the chromosomes in the human karyotype can be distinguished by banding techniques, which result in limited regions of staining, bands, on the chromosome arms. One such procedure called G-banding utilizes a dye called Giemsa and is particularly effective in staining human chromosomes (Figure 2-7). The number of bands seen depends on the degree to which the chromosomes are condensed. In highly condensed human chromosomes, the cytogeneticist Jorge Yunis has found that the entire chromosome complement contains 300 to 500 bands. The G-banding patterns for fully condensed human chromosomes are depicted in Figure 2-8. Because of differences in banding patterns we can now distinguish chromosome 4 from chromosome 5 or chromosome 8 from 9. This ability to distinguish different human chromosomes is important because, as we shall see in Chapter 5, some inherited human abnormalities are the result of chromosomal abnormalities.

THE MOLECULAR ARCHITECTURE OF CHROMOSOMES

We have described some of the general attributes of chromosomes in prokaryotic and eukaryotic organisms and have seen that within any species that the chromosomal complement is constant. We now turn to the details of the molecular architecture of chromosomes, particularly eukaryotic chromosomes. We do this with some trepidation, because many aspects of chromosome structure are still in the process of being revealed. What we present is more a fair consensus than firmly established knowledge. We should

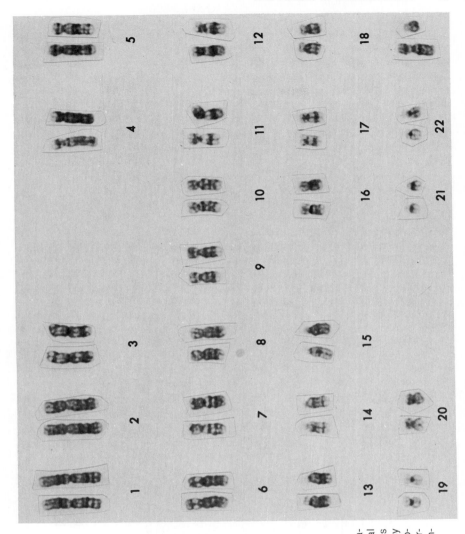

Figure 2-7 The Giemsa-banding pattern of a normal diploid set of chromosomes from a human male. Courtesy of William Loughman, Cytogenetics Laboratory, University of California, San Francisco.

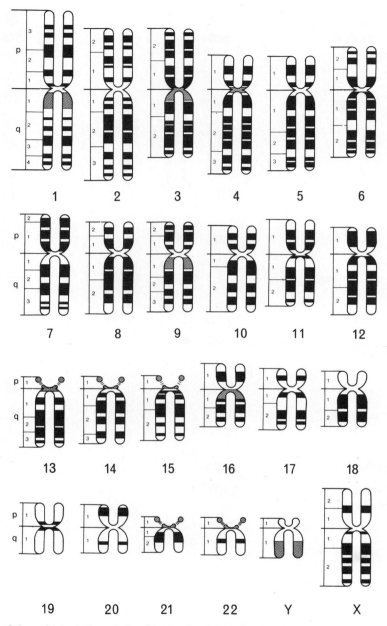

Figure 2-8   A depiction of the Giemsa-banding pattern of a haploid set of human chromosomes. *Paris Conference 1971, Standardization in Human Cytogenetics.* New York: National Foundation—March of Dimes, 1972.

emphasize, however, that biological structures are constructed in an orderly manner. The details of chromosome structure may be refined or changed in future, but we can be sure that orderly chromosome structure does exist. We will describe the existence of progressive hierarchies of organization within chromosomes, which result in the orderly packing of chromosomal DNA. The need for such packaging becomes clear when we consider the amount of DNA present in the chromosomal complement of different organisms.

Genome size, the amount of DNA present in the haploid set of chromosomes, varies widely from species to species.

The genome of an organism is defined as one each of the entire set of chromosomes, that is, the haploid complement. In prokaryotic cells and in their viruses, the genome size is equal to the total amount of DNA present, usually as a single chromosome. In eukaryotes the genome size is equal to the total amount of DNA in a haploid set of chromosomes. The amounts of DNA present in different species vary enormously over eight orders of magnitude from 10^3 to 10^{11} base pairs (Table 2-2) but are essentially constant within a given species. One of the smallest viruses known, the phage MS2, has its genome in a single RNA molecule composed of 3,000 nucleotides. Another small genome is found in the phage ∅X174, which contains 5,375 nucleotides in its circular chromosome of single-stranded DNA. Simian virus 40 (SV40), a cancer-causing virus, has a circular chromosome of double-helical DNA with 5,226 base pairs, measuring 1.7×10^{-6} meters in length. Among bacteria, *Escherichia coli* has a genome composed of 4×10^6 base pairs of DNA in its large circular chromosome. At the opposite extreme from bacteria and viruses is a plant in the lily family, *Lilium longiflorum*, that has a genome containing a total of 3×10^{11} base pairs of DNA in separate chromosomes. All of the DNA from a single lily cell laid end to end would be about 10^{12} A or 100 meters long, longer than an American football field. Among animals, *Amphiuma*, an amphibian commonly called the Congo eel, leads the way with about 50 meters, or 1.5×10^{11} base pairs of DNA in its genome.

The DNA within viruses, bacteria, and eukaryotic chromosomes must be highly condensed.

Despite the extreme lengths we have recounted, in the remarkable case of the lily 100 meters of DNA has a volume of only about 3.14×10^{-16} cubic meters (volume equals $\pi r^2 \times$ length or $3.14 \times 10^{-18} \times 100 = 3.14 \times 10^{-16}$ cubic meters) and would fit into a box measuring 0.01 mm on a side. To be contained within a limited volume, the DNA of both prokaryotes and eukaryotes must be tightly packaged. We can appreciate the variation in the amounts of DNA and the different degrees of close packaging required in different organisms by examining a series of electron micrographs that show SV40 viruses, a single SV40 chromosome (Figure 2-9A and B) and DNA spilling out of the confines of a T2 phage, a bacterial cell (Figure 2-9C and

Table 2-2   Haploid DNA Content

Organism	Amount of DNA 10^{-12} g.	Daltons	Base Pairs	Length (M)
Lily	343	2×10^{14}	3×10^{11}	100
Amphiuma	171	1×10^{14}	1.5×10^{11}	50
Maize	7.5	4.4×10^{12}	6.6×10^{9}	2.2
Mammals:				
Human	3.2	1.9×10^{12}	2.75×10^{9}	9.35×10^{-1}
Mouse	2.5	1.45×10^{12}	2.2×10^{9}	7.4×10^{-1}
Cow	2.8	1.6×10^{12}	2.45×10^{9}	8.3×10^{-1}
Lower Eukaryotes:				
Drosophila melanogaster	0.2	1.2×10^{11}	1.75×10^{8}	5.95×10^{-2}
Brewer's yeast	2×10^{-2}	1.2×10^{10}	1.75×10^{7}	6×10^{-3}
Bacteria:				
Salmonella typhimurium	1.3×10^{-2}	8×10^{9}	1.1×10^{7}	3.8×10^{-3}
Escherichia coli	4.7×10^{-3}	2.8×10^{9}	4.1×10^{6}	1.4×10^{-3}
Bacillus subtilis	3×10^{-3}	1.8×10^{9}	2.6×10^{6}	9×10^{-4}
Diplococcus pneumoniae	2×10^{-3}	1.2×10^{9}	1.75×10^{6}	6×10^{-4}
Viruses and Phages:				
T_2	2×10^{-4}	1.2×10^{8}	1.75×10^{5}	6×10^{-5}
Lambda	5.5×10^{-5}	3.3×10^{7}	4.65×10^{4}	1.6×10^{-5}
ØX-174*	6.2×10^{-6}	3.6×10^{6}	5375	1.8×10^{-6}
SV 40	5.9×10^{-6}	3.5×10^{6}	5226	1.7×10^{-6}

* Double-stranded form.

D), and a condensed mammalian chromosome (Figure 2-10). The enormous variation in the amount of DNA present in these complicated organisms is striking and the need for precise packaging of the DNA, particularly into the eukaryotic chromosome, is evident.

Chromosomes of eukaryotes probably contain a single uninterrupted molecule of DNA.

In the chromosomes of MS2, ØX174, and SV40 we can directly demonstrate by electron microscopy that the chromosome is a single uninterrupted molecule of RNA or DNA (for SV40, see Figure 2-9B). It is also possible to determine the genome size directly by measuring the length of the DNA because we know that there is about 3.4 A per base pair. (The length per base pair for DNA prepared for electron microscopy is slightly less than that of DNA determined from X-ray crystallography.) But, as we can see in the photographs of the DNA released from the bacterium *Hemophilus influenzae* and the eukaryotic chromosome pictured in Figures 2-9 and 2-10, we would be hard pressed to measure accurately the total length of the DNA present in the micrographs or to determine whether one or more DNA molecules are present. The DNA is too long and too convoluted to be traced from one end to the other. We must determine the size of DNA molecules in bacteria or within a single eukaryotic chromosome by some other method.

One property directly related to the enormous length of DNA molecules is the extreme viscosity of concentrated solutions owing to the frictional drag between the molecules. Measuring the viscosity of DNA solutions allows the determination of the size of DNA molecules. Large DNA molecules, however, are easily broken by physical shearing during extraction. Simply

Credits for Figures 2-9 and 2-10 (See pp. 50 and 51.) Figures 2-9 (A) and (B) courtesy of Robley Williams, Virus Laboratory, University of California, Berkeley. Figure (A) from Williams and Fisher, *An Electron Micrographic Atlas of Viruses,* courtesy of Charles C. Thomas, Publisher, Springfield, Illinois, 1974. Figure (C) courtesy of A. K. Kleinschmidt, Department of Microbiology, University of Ulm, Federal Republic of Germany. From Kleinschmidt, Lang, Jacherts, and Zahn, *Biochimica and Biophysica Acta,* v. 61, p. 857, 1962, with permission. Figure (D) courtesy of Lorne MacHattie, Department of Microbiology, University of Chicago and Charles Thomas, Department of Cell Biology, Scripps Institute, San Diego, California. From MacHattie, Ritchie, Thomas, and Richardson, *Journal of Molecular Biology,* v. 11, p. 648, 1965, with permission from Academic Press. Figure 2-10 courtesy of U.K. Laemmli, Department of Biological Sciences, Princeton University, New Jersey. From Paulson and Laemmli, *Cell,* v. 12, p. 817, 1977, copyright © M.I.T. Press.

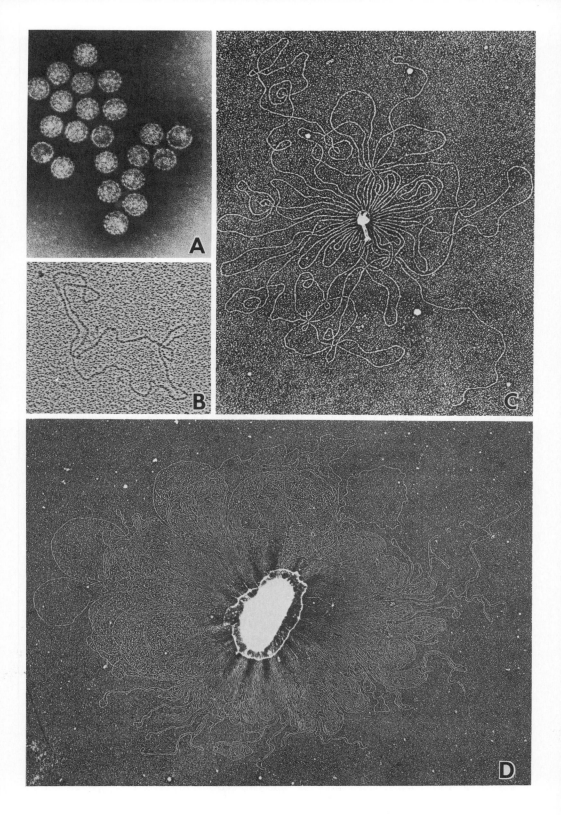

Figure 2-9 and 2-10   The enormous variation in DNA content in prokaryotes and eukaryotes is shown by electron microscopy. Figure 2-9 (A) Simian virus 40 (SV40); (B) The small, 5,226 base pair chromosome of SV40; (C) The linear, 175,000 base pair chromosome of DNA spilling from bacteriophage T2; (D) The 2.6 million base pair chromosome of *Hemophilus influenzae* released from a lysed bacterium. Figure 2-10 is a partial view of about 40 million base pairs of DNA released from a newly replicated and condensed mouse chromosome. Magnification is greatest in 2-9 (B) and smallest in 2-10.

drawing a solution of DNA into a pipet is sufficient to introduce breaks in DNA. Bruno Zimm and his colleagues have developed gentle procedures to release DNA from cells without shearing. They also developed techniques that relate the length of the DNA in solution to the viscosity of the solution, enabling them to determine the size of DNA molecules in several phages and bacteria. They have also applied their techniques to the DNA of *Drosophila.*

As a result of procedures that depend on the amount of dye bound to DNA or the absorption of DNA at 260 nm, we estimate that the total haploid genome of *Drosophila melanogaster* contains about 175×10^6 base pairs of DNA divided into four chromosomes, the largest of which is chromosome 3, a metacentric that contains 67×10^6 base pairs. It was not, however, known whether the DNA in a particular chromosome existed as a series of fragments or as one long molecule. Ruth Kavenoff and Zimm found DNA molecules about 67×10^6 nucleotides long in extracts of *Drosophila* cells. They then investigated an additional strain of *Drosophila* in which the X chromosome and the third chromosome were physically joined together to create a chromosome containing 92×10^6 base pairs of DNA. DNA molecules were extracted from cells of this stock and found to be about 91×10^6 base pairs long. According to the procedures Kavenoff and Zimm followed, both stocks contained DNA molecules equal in size to the largest expected. The easiest interpretation of the results is that the chromosomes of *Drosophila melanogaster* and presumably of other eukaryotes, particularly of animals, contain a single continuous molecule of DNA that passes from the tip of one chromosome arm through the centromere to the tip of the other arm. A chromosome containing a single DNA double helix is called a *unineme* chromosome. We shall see, in contrast, that some specialized chromosomes, *polytene* chromosomes, contain hundreds of similar double helices arranged side by side.

Eukaryotic chromosomes may be constructed in a hierarchy of helical structures.

The diversity in appearance of different chromosomes reflects to some extent the diversity of sequences within chromosomes, which, in turn, are responsible for the diversity of life itself. It is important to emphasize that for the DNA to determine the nature and developmental pathways of the cells and organisms in which it resides, it must be able to carry out its functional role. The mechanisms of function will be explored in detail in later chapters. For now, we should bear in mind that the packaging of DNA in a chromosome must permit the DNA to have an orderly arrangement but must also allow and indeed assist the DNA in its functional activities. Although we expect several features of the packaging of DNA to be similar from chromosome to chromosome and from species to species, we might also expect to find differences in packaging, that is, in chromosome structure, that result from differences in functional activity.

Molecular biologists have long recognized that the DNA in eukaryotes is associated with various types of proteins and that the DNA and the proteins must be packaged in some highly organized fashion in order to fit inside a nucleus and to condense during each cell division into structures that are invariant in appearance from cell to cell. It is only recently, however, that we have made substantial progress in understanding *how* the DNA is packaged in a chromosome. Chromosomal packaging of eukaryotic DNA appears to involve a succession of helical structures. In the smallest chromosomal helix the DNA is coiled around a series of flattened spheres called nucleosomes (Figure 2-11). The DNA is wound around the outside of the spheres that are connected in series by the continuous DNA molecule. Some molecular biologists believe that the string of nucleosomes may itself be coiled in a

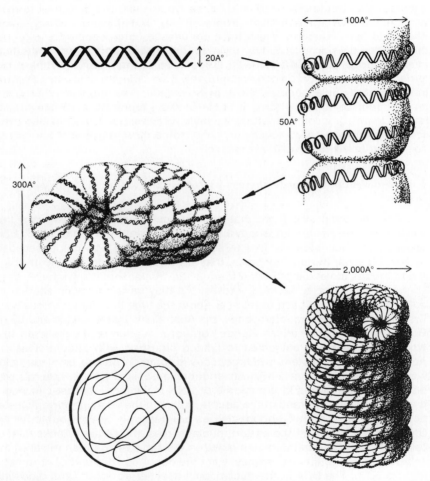

Figure 2-11   A model for the structure of a eukaryotic interphase chromosome. A hierarchy of helical structures is proposed. (A) DNA double helix. (B) DNA wound around outside of nucleosomes. (C) A string of nucleosomes coiled into a solenoid with a 500 Å diameter (D) The solenoid *may*, in turn, be coiled into a supersolenoid of 2000 Å diameter. (E) A representation of a chromosome coiled as a supersolenoid in an interphase nucleus.

helical manner to form a tube called a solenoid with a diameter of 300 to 500 A (Figure 2-11). As a result of the coiling of the DNA around nucleosomes and the arrangement of the nucleosomes into a solenoid, a DNA molecule 20 mm long, the length of the DNA molecule found in the standard chromosome 3 of *D. melanogaster*, is confined in a tube about 0.5 mm long. Our notions about the level of structure above the solenoid are very speculative. One view proposed by Arne Bak, Jesper Zeuthen, Francis H. C. Crick, and John Sedat and L. Manuelidis holds that the solenoid itself is coiled into yet another helix, a supersolenoid, with a diameter of 2000–4000 A that corresponds to the diameter of the uncondensed chromosome found in interphase nuclei (Figure 2-11). If this helical coil exists, the length of a chromosome containing 20 mm of DNA would be 0.01–0.05 mm, about a one-thousand-fold reduction in length. A condensed chromosome found at the time of cell division might involve yet another helical coil, bringing the total diameter to around 5000–10,000 A. Keep in mind that in an interphase cell different regions of chromosomes probably have different amounts of coiling ranging from nucleosomes up to supersolenoids, resulting from differences in function. Let us now look at some of the supporting evidence for the model for chromosome structure we have just described. We will see there is strong supporting evidence for nucleosome structure, but the evidence for higher levels of structure is sparse.

Nucleosomes are flattened spheres of DNA wound around a group of histones.

Chromosomal structures that are now called nucleosomes were first viewed with electron microscopy by Ada and Donald Olins and by C. L. F. Woodcock. They were the first to see that chromatin was composed of a series of beads, the nucleosomes, that are attached to each other through a continuous molecule of DNA (Figure 2-12). Nucleosomes have never been seen in bacteria or in bacterial viruses, but they have now been seen in all eukaryotes that have been examined. About the time that nucleosomes were first viewed by electron microscopy, the Australians Dean Hewish and Leigh Burgoyne and the American Roger Kornberg independently observed that when chromatin was subjected to limited digestion with endonuclease (a nuclease from the bacterium *Micrococcus* is commonly used), small particles were produced (Figure 2-13). An endonuclease is a DNase that cleaves phosphodiester bonds in the middle of DNA molecules. By first cleaving a diester bond in one strand of the double helix and then a neighboring diester bond in the other strand, endonucleases progressively cleave double-helical DNA into fragments. In the case of nucleosomes, the DNase attacks the so-called "linker" DNA in the region between the spheres and then releases the spheres, that is, the nucleosomes, from the continuous strand of chromatin (Figure 2-13). The DNA in the nucleosome itself is protected from digestion by the nuclease. When digestion by nuclease is limited, pieces of DNA about 200 base pairs long are produced, indicating that there are about 200 base pairs of DNA from linker to linker. When digestion by nuclease is extended, "core" nucleosomes are produced in which the linker DNA is digested,

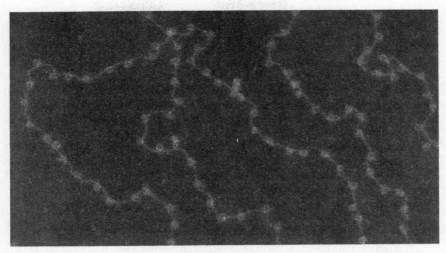

Figure 2-12   An electron micrograph of nucleosomes from a chicken red blood cell. The string of nucleosomes is stretched so that the "linker" DNA between the nucleosomes is seen (see Figure 2-13). Courtesy of Ada and Donald Olins, Biology Division, Oak Ridge National Laboratories, Tennessee.

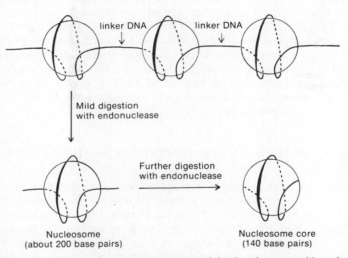

Figure 2-13   Production of nucleosome core particles by cleavage with endonuclease.

leaving a piece of DNA about 140 nucleotides long (Figure 2-13). Core nucleosomes are flattened spheres or discs with a diameter of 100 A and a thickness of about 50 A. The DNA is on the outside of the nucleosome and winds around it about $1\frac{3}{4}$ turns. The inside of the core is composed of the basic structural chromosomal proteins, the histones.

Nucleosome cores contain two molecules each of histones H2A, H2B, H3, and H4.

In most species of eukaryotes studied, five different histones—H1, H2A, H2B, H3, and H4—have been found. All of the histones are comparatively small as proteins go, containing 100 to 190 amino acids in their polypeptide chains. Each histone has a unique amino acid sequence and, thus, each histone is a unique protein. All histones, however, are rich in basic amino acids, particularly lysine and arginine, and have positive charges at the pH's found in living cells. The DNA, because of its phosphate groups, is negatively charged. Histones bind to DNA through ionic or salt bonds. Nucleosome cores contain two molecules each of histones H2A, H2B, H3, and H4. Histone H1 is absent from the core and may be associated with the linker region of the DNA between nucleosome spheres.

Nucleosomes self-assemble.

A particularly convincing demonstration of the role of histones in the formation of nucleosomes is provided by mixing protein-free DNA from any source, even bacterial DNA, with histones. When any three types of histones are mixed with DNA, no particles with the correct dimensions of nucleosomes are visible by electron microscopy. When the fourth histone is added, nucleosomes form as a function of the amount of the fourth histone added. Thus, in the presence of all four histones and DNA, nucleosomes will form spontaneously, that is, they will self-assemble. Because prokaryotic DNA can be used to form nucleosomes, we can conclude that specific DNA sequences unique to eukaryotes are not required. The four histones self-assemble into an octamer containing two molecules of each type of histone. The DNA, through salt linkages, winds around the aggregated histones to complete formation of the nucleosome.

The solenoid structures in chromosomes involve non-histone chromosomal proteins.

As we have seen the types of histones are limited in number and have specific structural roles. A large number of non-histone chromosomal proteins also have been identified. These proteins, as their name suggests, have properties different from those of histones. They are not rich in basic amino acids and have net negative charges at cellular pH's. Because of their negative charge, they are also called acidic chromosomal proteins. The roles of these proteins in chromosomal structure and function are largely unknown. Some probably have functional roles either in the replication of DNA preparatory for cell division or in gene function. Some may also have structural roles. For example, they may be involved in the formation of the 2,000–4,000 A supersolenoid (Figure 2-11). Chromosomes are not, however, composed of a uniform supersolenoid structure. Successive coils may be tightly packed in some regions and more open in others. Different coils could also vary in size, and the non-histone proteins may be involved in these structural differences. Evidence that there are hierarchies in coiling along the length of a chromosome has been provided by the particular electron micrograph from the laboratory of Ulrich Laemmli shown in Figure 2-9. All of the histones from a newly replicated, condensed eukaryotic chromosome have been removed by extraction in salt solution. The darkly staining region in the form of two newly replicated chromatids near the bottom of the micrograph is composed of non-histone proteins. The important observation, however, is that long loops of DNA, which now lack any superstructure, extend out from the central region. The loops have been measured and found to contain 30,000 to 90,000 nucleotide pairs, consistent with the amount of DNA estimated to be in a single coil of the supersolenoid. The loops exist in such preparations presumably because the non-histone proteins are still present as a type of scaffold holding the chromosome together. Although these preparations do not tell us how the non-histone proteins and DNA interact in the native chromosome, they do suggest a function for the non-histone proteins in maintaining the superstructure of DNA.

Lampbrush chromosomes are partially condensed chromosomes with a large number of chromatin loops extending from chromomeres.

Evidence exists to support a model of chromosome structure with differences in coiling along the length of the chromosome. This evidence stems from observations of specialized chromosomes, lampbrush chromosomes and polytene chromosomes, which are particularly suitable for observation by a light microscope. We will first discuss lampbrush chromosomes. During the maturation of the eggs of many organisms, particularly amphibians, and during the formation of sperm in others, such as some insects, we can see chromosomes (Figure 2-14) that looked to early cytologists like the brushes used to clean the chimneys of gas lamps. The structure of a loop in a lampbrush chromosome is depicted schematically (Figure 2-15) to stress that the chromosome has undergone replication and contains two double helices of DNA lying side by side. At points along the chromosome there are condensations of chromatin, called *chromomeres.* From many chromomeres loops extend, giving the chromosomes their characteristic lampbrush appearance. Fairly accurate chromomere counts can be made. In the haploid set of chromosomes of different species of *Triturus* (newts), 2,000 to 5,000 chromomeres are found. Loops extending from the main axis of lampbrush chromosomes contain from 50,000 to 200,000 base pairs of DNA, an amount similar to the amount of DNA in the histone-depleted loops photographed by Laemmli and his co-workers. We can speculate that in the case of the lampbrush chromosome, each loop represents one or two coils of the supersolenoid that have become unwound. Genes are actively functioning in the lampbrush loops. The unwinding of a coil in the supersolenoid may be a necessary prerequisite to gene function (a point we shall discuss again in Chapter 17).

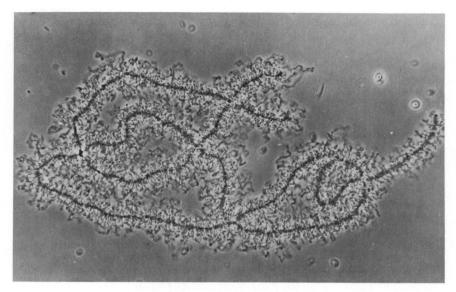

Figure 2-14   Light micrograph of a lampbrush chromosome from a maturing amphibian egg. Courtesy of Joseph Gall, Department of Biology, Yale University, New Haven, Connecticut.

Polytene chromosomes are interphase chromosomes in which multiple copies of the chromosomes remain aligned.

In the lampbrush chromosomes two replicates of a single chromosome are associated side by side prior to separation during nuclear division. A polytene chromosome is produced by endoreplication during which chromosomes undergo repeated replication without becoming separated. The replicated strands remain paired together, side by side in perfect register, so that a given region in one strand is paired with the same region in any

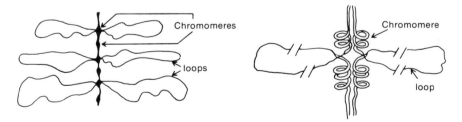

Figure 2-15   (A) A diagram of a short region of a lampbrush chromosome. (B) A possible interpretation of the structure of a lampbrush chromosome. Each chromomere is formed by coiling of the chromosomal solenoid into a supersolenoid. The lampbrush loop is a string of nucleosomes resulting from the uncoiling of the solenoid and supersolenoid.

other strand. All regions of a polytene chromosome do not become equally replicated. The centric heterochromatin undergoes little endoreplication, but the euchromatin becomes highly replicated. The number of chromosome strands in a chromosome becoming polytene increases geometrically. One pair of chromosomes with 2 double helices replicates to produce 4 double helices, each of which replicates to produce 8 double helices, then 16, 32, and so on, up to 1,024 (2^9) or more double helices. Polytene chromosomes containing 1,024 or more replicates are found in the larval salivary glands of *Drosophila* and some of its distant relatives such as the midge, *Chironomus tentans,* and the sciarid, *Rhynchosciara angelae.* Polytene chromosomes have also been found in some unicellular symbionts that live in the gut of Dipteran insects and in some plants. Beans, for example, have a form of polytene chromosome.

Chromomeres are easily identified in polytene chromosomes.

Each chromomere in a lampbrush chromosome is identified as a condensation of chromatin on the chromosomal axis from which, in some cases, a loop extends. In polytene chromosomes the chromatin loops are usually tightly coiled although there are exceptions. The result of endoreplication and the precise pairing of each replicated chromosome strand with all of the others is that the chromomeres in polytene chromosomes are all aligned with each other. Because the DNA in the chromomere is condensed, each chromomere is a region of high DNA concentration, which stains more heavily than the regions in between the chromomeres. The result in polytene chromosomes is that the aligned chromomeres form a series of identifiable, deeply-staining bands, interspersed with lightly-staining interband regions (Figure 2-16). The formation of the bands as a result of the juxtaposition of the chromomeres in separate chromosomal strands is depicted in Figure 2-17. In some of the bands of the polytene chromosomes, the DNA within the chromomeres is not tightly coiled. In these regions, called "puffs" or, in the case of particularly large examples, "Balbiani rings," the DNA is uncoiled and apparently has a structure similar to that in lampbrush chromosomes—except, of course, over a thousand double helices of DNA are present (Figure 2-18). Again we find evidence for the packaging of DNA into discrete condensed regions that can unfold to allow the extension of a loop or loops of DNA from the axis of the chromosome.

Polytene chromosomes can be used to identify specific chromosomal regions.

Because polytene chromosomes are so large, the individual bands or chromomeres can be easily counted. *Drosophila melanogaster* has about 5,000 bands. Moreover, different regions of the polytene chromosomes have distinctive banding patterns so that different chromosomes or regions of chromosomes can be recognized. Calvin Bridges produced a map of the bands of the polytene chromosomes of *Drosophila melanogaster* in which he gave every band an identifying number. The chromosomes were divided into different segments which were numbered from 1 to 100, each segment

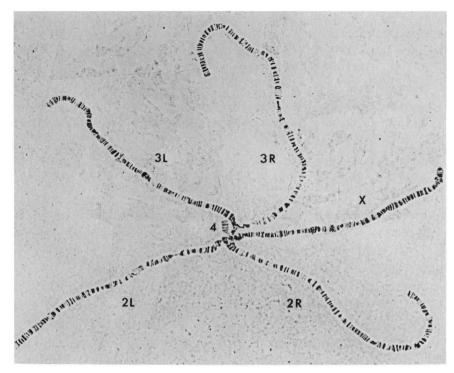

Figure 2-16   Polytene chromosomes from the larval salivary gland of *Drosophila melanogaster*. The left and right arms of chromosomes 2 and 3 are indicated. Courtesy of George Lefevre, Jr., Department of Biology, California State University, Northbridge. From, Ashburner and Novitski, Eds., *The Genetics and Biology of Drosophila*, v. 1a, p. 31, 1976, with permission from Academic Press.

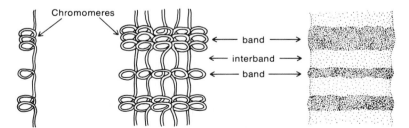

Figure 2-17   A model for the structure of a polytene chromosome. A chromosome containing chromomeres undergoes endoreplication. The replicated chromosomes remain together with the chromomeres arranged in register. The paired chromomeres form the bands of the polytene chromosome; the regions between the chromomeres, the interbands.

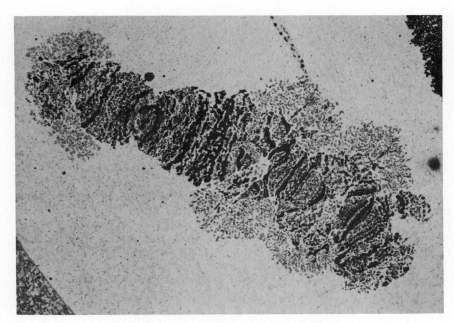

Figure 2-18   Light micrograph of a polytene chromosome from *Chironomus* with three distinct Balbiani rings. The Balbiani rings form from the uncoiling of the supersolenoid in chromomeres (bands). Courtesy of Bertil Daneholt, Department of Histology, Karolinska Institute, Stockholm. From Daneholt, *Cell*, v. 4, p. 1, 1975, copyright © MIT Press.

containing about fifty bands. The segments were divided into regions denoted by uppercase letters, and each band within a region was given an identifying number. Band 59B1 is the first band in the B region of the 59th segment and is located near the end of the right arm of chromosome number 2 (Figure 2-19). The precise identification of a particular region of a *Drosophila* chromosome, as we will see later, has important consequences for genetics. The bands themselves are heterogeneous. The average band in *Drosophila melanogaster* contains 20,000 to 30,000 base pairs per chromosomal strand, but some have as few as 5,000 to 10,000 base pairs and others contain over 60,000 base pairs. Polytene chromosomes in *Drosophila* have been extremely useful to geneticists. We shall refer to them frequently.

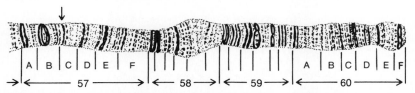

Figure 2-19   A drawing of the end of the right arm of polytene chromosome 2 of *Drosophila melanogaster*. Specific bands are indicated by numbers and a letter, for example, the arrow points to band 57C1. After Bridges, *Journal of Heredity*, v. 26, pp. 60–64, 1935.

SUMMARY

Both the amount of DNA per haploid genome and the chromosome complement vary enormously from species to species. Both quantities, however, are essentially constant in all members of a given species. Chromosomes of prokaryotic cells are composed effectively only of DNA; those of eukaryotes contain DNA and both histone and non-histone proteins. In both prokaryotes and eukaryotes, the DNA must be tightly packaged to fit into a limited space. In eukaryotes we believe that packaging of DNA into chromosomes involves a hierarchy of helices. The primary structure is the nucleosome in which the DNA is wound around the outside of a core of histones. The nucleosomes may themselves be coiled into a helical tube, a solenoid, and the solenoid itself may be coiled into a supersolenoid. Not all regions of chromosomes are coiled to the same degree. The DNA in the chromomeres of unineme chromosomes or bands of polytene chromosomes is in a highly coiled state and can become uncoiled to produce lampbrush loops in unineme chromosomes or puffs in polytene chromosomes.

The Structure of Chromosomes: DNA Sequence Structure

We learned in the previous chapter that the architectural organization of eukaryotic chromosomes is extremely complex compared to prokaryotic chromosomes. The complex organization of chromosomes in eukaryotes is necessary not only because of the enormous length of the DNA in the chromosomes but also for the function of these chromosomes in determining the phenotypes of the organisms in which they reside. In addition to differences in their chromosome architecture, there are differences in the number and distribution of nucleotide sequences in prokaryotes and eukaryotes. Furthermore, there are differences in sequences in the various regions of eukaryotic chromosomes, for example, between euchromatin and heterochromatin. We will consider in this chapter some of the differences in base compositions and base sequences that exist between prokaryotic and eukaryotic DNAs and among eukaryotic DNAs together with some of the methods used to identify these differences.

However, the reason for our interest in sequences transcends simple comparisons between prokaryotes and eukaryotes or between euchromatin and heterochromatin. We have already indicated that genetic information is embodied in discrete DNA sequences, genes. We have reached the stage at which genes and the nature of their products in simple organisms—viruses— can be identified by directly determining the base sequence in the DNA, and we are currently extending these techniques to all organisms. The future of genetical research is inseparably entwined with sequencing methodology. We can look forward, happily or unhappily, to the day when the information contained in parts of chromosomes of any human being will be accessible and, perhaps, to the Orwellian day when such information will be catalogued by an inefficient government agency—although not by 1984.

Before opening our discussion on sequencing techniques and their results, we should pause to emphasize the difference between base composition and base sequence. Base sequence is the order in which bases occur in a particular polynucleotide and is expressed, starting at a fixed point, by a listing of bases. For example, 5′-ACGTCTGCCATG-3′ represents a particular base sequence. Base composition is the relative amounts of bases found in DNA or RNA. For double-helical DNA, because the amount of A = T and G = C, base compositions are given as per cent G + C (written as GC), from which the percentage of each base can be calculated. For example, 30 per cent GC is also 70 per cent AT and contains 15 per cent G, 15 per cent C, 35 per cent A and 35 per cent T. DNA's with different base sequences can have the same base compositions, for example, 3' ATCG 5' and 3' CTAG 5'.

The density of DNA molecules depends on base composition.

One of our goals in this chapter is to understand how DNA sequences are distributed in chromosomes. We have already advertized that the sequences differ in the euchromatin and heterochromatin of eukaryotic chromosomes. Such differences, as we shall discuss presently, were first detected in the course of determining base compositions of sheared DNA fragments from various species. Having abandoned the methods for determining base compositions originally developed by Chargaff that measured the amount of each nucleotide present in a DNA hydroysate, molecular biologists now determine base composition from the density of the DNA. Density is defined as mass divided by volume, and a molecule of DNA, like any other object, has both mass and volume. The density of DNA or, more precisely, the buoyant density of DNA (that is, the density at which the molecules float in solution) increases as the GC composition increases. A double helix composed only of AT pairs has the lowest density and one composed entirely of GC pairs has the highest density.

Densities of DNA molecules are determined by equilibrium sedimentation in gradients of cesium salts.

DNA is extracted from cells and sheared so that fragments are produced. Buoyant densities of DNA molecules are determined with gradients of cesium salts. A solution of sheared DNA and a cesium salt (8 M CsCl is commonly used) is centrifuged at very high speeds to subject the DNA and the cesium salt to high sedimentation forces (100,000, or more, times the force of gravity, g). Cesium has a high atomic weight (140), and cesium ions have a high density. The cesium ions are pushed towards the bottom of the centrifuge tube. As the concentration of cesium ions increases near the bottom of the tube, there is increased diffusion of cesium ions back up the tube. Eventually an equilibrium is reached, the rate of sedimentation down the tube and the rate of diffusion up the tube becoming equal. A linear gradient of cesium ion concentrations is created with the concentration being highest at the bottom of the tube and lowest near the top (Figure 3-1). A linear gradient of densities, increasing from top to bottom, is also created. The DNA molecules move to a position in the gradient that is determined by their densities. Molecules that are below that position tend to float upwards, while those above that position are pushed downwards. At equilibrium, one or more "bands" of DNA are formed in the tube at positions at which the density of the solution equals the buoyant density of the DNA molecules. The position of the DNA in the tube is determined by its absorption of ultraviolet light at

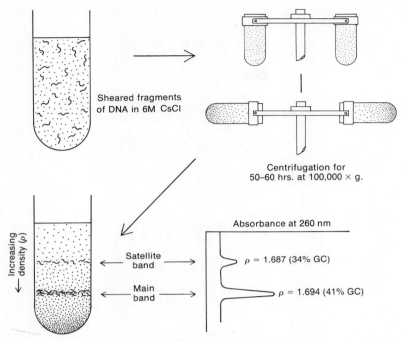

Figure 3-1   Determination of buoyant density (ρ) of DNA by equilibrium density gradient centrifugation.

260 nm. By determining the density of the solution at that point, we learn the buoyant density of the DNA, and, because buoyant density is a function of base composition, we also learn the base composition of the DNA.

Prokaryotic DNA's often form single bands in density gradients; eukaryotic nuclear DNA's often form a main band and one or more satellites.

When DNA's from phage or bacteria are subjected to density gradient centrifugation, a single band of DNA often results. DNA's from different species of bacteria have different base compositions (Table 3-1). However, when the DNA from certain bacteria is centrifuged, a minor, *satellite,* band of DNA separates from the major or main band. For example, the DNA extracted from the bacterium *Halobacterium salinarium,* a species that normally grows in soil, has main band DNA with a 67 per cent GC composition and a satellite band with a 58 per cent GC composition. The satellite constitutes 20 per cent of the DNA found in this organism. In eukaryotes, nuclear DNA, that is, DNA extracted from nuclei, often has one or more satellite bands (Table 3-1). The identification of such a satellite is presented in Figure 3-1 for the nuclear DNA of the mouse, which yields two bands of DNA, a main band with a 41 per cent GC composition and a satellite, composing 11

Table 3-1   Buoyant Densities and GC Compositions of Main Band and Satellite DNAs from Various Species

Species	Main Band DNA		Satellite DNA		
	Density (gm/ml)	Per cent GC	Density (gm/ml)	Per cent GC	Per cent of nuclear DNA in satellite
Mycoplasma mycoides	1.685	25			
Cancer antennarius	1.699	40	1.663	3	30
Cancer magister	1.700	41	1.663	3	12
Guinea pig	1.700	41	1. 1.698	39	5
			2. 1.704	45	4
Human	1.700	41	1. 1.692	33	0.5
			2. 1.694	35	2
			3. 1.698	39	15
Mouse	1.700	41	1.693	34	11
Drosophila melanogaster	1.701	42	1. 1.672	12	2.9
			2. 1.686	27	3.4
			3. 1.688	29	4.6
			4. 1.695	36	—
			5. 1.705	46	5.3
Bacillus subtilis	1.703	44			
Escherichia coli	1.709	50			
Halobacterium solinarium	1.726	67	1.717	58	20
Sarcina lutea	1.730	71			

Data compiled from several sources. Densities in CsCl and base compositions adjusted using the formula density = 1.660 + 0.098 × (mole fraction GC) after C. Schildkraut, J. Marmur and P. Doty, *Journal of Molecular Biology*, v. 4, 430–443, 1962.

per cent of the nuclear DNA, with a 34 per cent GC composition. The presence of multiple satellites is typical of the DNA from many animals. In humans, we have identified three satellites, each having a GC composition different from the main band DNA. In *Drosophila* five different satellites have been identified.

Satellite DNA's result from regional differences in base compositions in chromosomes.

To understand the presence of DNA satellites we must consider the steps taken to prepare DNA's for equilibrium gradient centrifugation and the possible differences in the distribution of DNA sequences within a genome. As we have noted the DNA is sheared prior to centrifugation. To produce DNA molecules of uniform size, solutions of DNA are passed at high speed through small openings or the DNA is broken in a high-speed blender. What is centrifuged is a collection of comparatively small DNA molecules (1,000–2,000 base pairs) created by the apparent random breakage of the enormous DNA molecules in chromosomes. If the base composition, as far as the individual molecules of DNA is concerned, is random, then the average molecule should have a base composition similar to that of the total DNA. No satellite DNA will be produced. A satellite could be produced, however, if there were non-random distributions of base compositions. If one region of a chromosome, larger than the DNA molecules produced by shearing, had a base composition different from the rest of the chromosome, a satellite would be produced (Figure 3-2). Such a regional difference in base composition could arise from a sequence with a base composition different from the average that is repeated over and over again. The sequence could be tandemly repeated; that is, one repeat would be followed by another, then another, and so on in a limited region of a chromosome. After shearing, a set of DNA molecules derived from parts of the tandemly repeated sequence would be produced and would have a base composition and density different

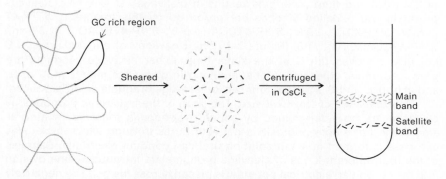

Figure 3-2   The origin of satellite DNA bands in density gradients.

from that of the majority of the DNA. We shall see (p. 85) that in eukaryotes the presence of satellite bands in the nuclear DNA is often the result of highly repeated tandem sequences that have base compositions different from the rest of the DNA.

Nucleotide sequences are determined through the selective cleavage of short DNA molecules.

The ultimate method for discovering the regional differences in DNA sequences in chromosomes is by the direct determination of the sequences. However, in contrast to the methods for determination of base compositions that have been available since the 1940's, effective methods for the rapid sequencing of DNA molecules are comparatively recent. It took over one year in the 1960's for Robert Holley and his co-workers to determine the sequence of an RNA molecule that was only 74 nucleotides long. Now methods are available that allow sequencing in one day of DNA molecules 100 to 200 nucleotides long. These methods, when coupled with procedures for producing a large series of small DNA molecules, have allowed the sequencing of DNA chromosomes with over 5,000 nucleotides. Methods are also available that allow sequencing of RNA molecules. One method for DNA sequencing developed by Allan Maxam and Walter Gilbert utilizes selective cleavage of a polydeoxynucleotide to produce different-sized fragments, from which the sequence is deduced according to fragment size.

Selective cleavage. The Maxam–Gilbert procedure utilizes purified, short (about 100–200 nucleotides long), single-stranded DNA molecules obtained by separating the complementary strands of double-helical molecules (described on p. 75). For our discussion let us consider a polynucleotide containing the sequence 5′TCCACCTGCTGA . . . 3′. Molecules with this sequence are incubated with ^{32}P-adenosine triphosphate in the presence of the enzyme polynucleotide kinase. The enzyme adds ^{32}P-bearing adenosine to the 5′ end of the molecule, producing 5′^{32}P-**A**TCCACCTGCTGA . . . 3′ (Figure 3-3). These DNA molecules with radiolabelled 5′ ends are then treated with reagents that produce occasional breaks at an A, a G, a C, or at both C and T. For example, treatment with solutions of the chemicals hydrazine in 2M NaCl and then piperidine results in cleavages at C's. Five cleavage products with labelled 5′ ends are produced by this treatment, 5′^{32}P-**A**T . . . 3′, 5′^{32}P-**A**TC . . . 3′, 5′^{32}P-**A**TCCA . . . 3′, 5′^{32}P-**A**TCCAC . . . 3′, and 5′^{32}P-**A**TCCACCTG . . . 3′ (Figure 3-3). The cleavage not only occurs at a C but also removes the C at the same time. Other cleavage products are produced under other conditions, resulting in cleavages mainly at A, at G, and at both C and T, a total of four sets of cleavage products being produced.

Determination of fragment size. The sizes of the fragments produced by cleavage are now determined by gel *electrophoresis* followed by *autoradiography.* For electrophoresis, a gel with pores through which molecules can pass is formed in a salt solution that can conduct electricity. Samples of the four different kinds of cleavage products are introduced side by side into the gel and an electrical potential is placed across the gel. The negatively charged DNA molecules migrate through the gel towards the positive elec-

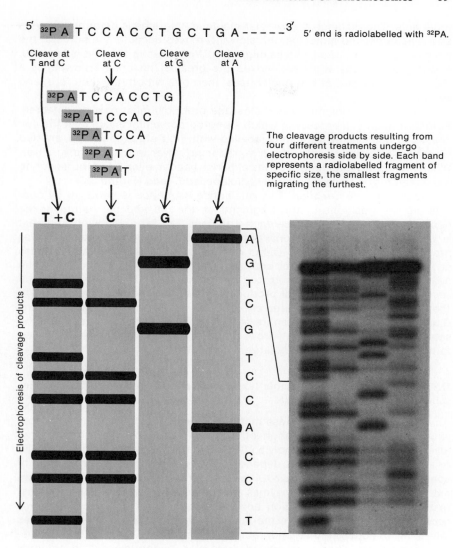

Figure 3-3 The Maxam-Gilbert method for sequencing DNA using gel electrophoresis to separate different-sized DNA fragments with radioactive 5′ ends. The sequence, 5′ to 3′, is then "read" from bottom to top. Photo of sequencing gel courtesy of Karen Sprague, Institute of Molecular Biology, University of Oregon, Eugene. From Sprague, Faulds and Smith, *Proceedings of the National Academy of Sciences, U.S.*, v. 75, p. 6182, 1978.

trical pole. The distance a particular molecule migrates is determined by its size. The larger the molecule, the slower it moves because the pores of the gel increasingly interfere with its migration. When electrophoresis is finished, a single nucleotide will have moved the greatest distance followed by a molecule composed of two nucleotides, then one with three, then four, and so on.

The different polynucleotide cleavage products must now be detected. Autoradiography, a process in which a photographic film is laid over the gel, is used to detect only fragments labelled with ^{32}P. Beta (β) particles, emitted when the ^{32}P decays, enter the film, reducing silver ions to metallic silver. When the film is developed, a series of bands appear, each band representing the position of a fragment with a radiolabelled 5' end (Figure 3-3).

Sequence determination. The nucleotide sequence is deduced by considering the positions of the fragments in the gel and the nature of the nucleotide cleaved to produce each fragment. The fastest-moving fragment is identified. It is composed only of the added ^{32}P-*A*, and we note that it was produced by cleavage at T + C, but not at C alone. Therefore, the first base at the 5' end of the DNA (not counting the one that was added) is T. The next fastest product was produced by cleavage at C, therefore C is the second base. The next fastest product was also cleaved at C, therefore C is the third base and the first three are 5'TCC . . . 3'. The next is A, then C, and C again, and so on until the entire sequence is determined. We see that the sequence can be read directly from the gel by identifying the base cleaved to produce each progressively larger polynucleotide fragment.

DNA fragments of a size suitable for sequencing are produced with enzymes called restriction endonucleases.

The successful use of these methods of sequencing of short polydeoxynucleotides for the elucidation of the DNA sequence of a chromosome (even a very small one with only 5,000 base pairs such as one finds in SV40) or a region of a chromosome depends on the production and purification of specific small fragments of DNA. Fortunately the production of these fragments is easy with a set of enzymes called *restriction endonucleases* or just restriction enzymes. Restriction enzymes are DNases that, like the *Micrococcal* nuclease discussed in the previous chapter, cleave DNA molecules internally. In addition, restriction endonucleases make the cuts only at specific sequences. The specificity of restriction endonucleases is not only critical for their normal biological activity but provides their name. Restriction endonucleases are found in bacteria where they cut up the DNA of some invading phages without cutting up the bacterial DNA. Furthermore, restriction endonucleases isolated from various sources have differing specificities and make cuts through different specific sequences. Most restriction endonucleases make cuts through DNA sequences that are *palindromes.* A palindrome is a word or a group of words that read the same both backwards and forwards. The word "madam" and the sentence "Able was I ere I saw Elba" are examples of palindromes. In the case of DNA, a palindrome is a base sequence, the complement of which has the same sequence. For example, the complement of the simple sequence 5′ AGCT 3′ is 5′ AGCT 3′, the two molecules being complementary when arranged in antiparallel fashion.

$$\frac{5'\ \text{AGCT}\ 3'}{3'\ \text{TCGA}\ 5'}$$

Restriction enzymes usually cut the DNA between two specific bases in a palindrome. For example, a restriction endonuclease, called *AluI* and isolated from the bacterium *Arthrobacter luteus,* makes cuts only between the G and C positions in the 5′AGCT 3′ palindrome, producing a cut at the same position in both strands of the DNA and cleaving the double helix. Table 3-2 provides a brief catalog of some restriction endonucleases. Treatment of DNA with any of these restriction endonucleases will produce fragments resulting from cuts at specific positions in the DNA. After gel electrophoresis to separate different-sized fragments, one can purify the fragments and determine the sequences of the different fragments following the procedures described on pp. 68–70.

The use of various restriction enzymes and analysis of the fragment produced can lead to the establishment of the base sequence of a small chromosome.

The foregoing procedure provides us with the base sequences of the restriction fragments. We must now determine the order in which the fragments were present in the intact chromosome. As an example, we will consider a short double helix, 500 base pairs in length, that contains sites for the restriction endonucleases *AluI* and *HaeIII* (Figure 3-4). When the original 500 base pair molecule is digested with *AluI* three fragments, 100, 150, and 250 base pairs long, are produced. The molecule is also cleaved by *HaeIII* (from *Haemophilus aegyptius,* it cuts between G and C in a GGCC palindrome) to produce three fragments 50, 150, and 300 base pairs long. The sequences of the various fragments are determined. The sequences in the two sets of fragments overlap (Figure 3-4). For example, the left *HaeIII* fragment contains sequences in common with the left *AluI* fragment (A–E) and part of the middle *AluI* fragment (FG). By comparing the sequences of the *AluI* fragments with those of the *HaeIII* fragments, we can determine that the 100 nucleotide *AluI* fragment is to the left of the 250 nucleotide *AluI* fragment. Similarly, we can ascertain the position of the other *AluI* fragment and the entire 500 nucleotide sequence will be known. It is usually not necessary to sequence all fragments produced by restriction endonucleases, as it is possible to determine positions of different restriction sites and fragments by studying the size of fragments produced and by exposing purified fragments produced by treatment with one restriction endonuclease to treatment with a second. For example, it could be demonstrated that the left *HaeIII* fragment has an *AluI* site in it. The result of such an analysis is a restriction map of a chromosome, that is, a map showing where different restriction sites are located along the chromosome. A restriction map for the chromosome of SV40 is depicted in Figure 3-5.

Table 3-2   Properties of Selected Restriction Endonucleases

Microorganism Source	Enzyme	Sequence Cut	Nature of Cut Ends	Number of Sites in SV40 DNA*
Arthrobacter luteus	*Alu*I	5'AGCT 3' 3'TCGA 5'	5' AG 3' TC	32
Bacillus amyloliquefaciens H	*Bam*HI	5'GGATCC 3' 3'CCTAGG 5'	5'G 3'CCTAG 5'	1
Bacillus globigii	*Bgl*I	5'AGATCT 3' 3'TCTAGA 5'	5'A 3'TCTAG 5'	0
Escherichia coli RY13	*Eco*RI	5'GAATTC 3' 3'CTTAAG 5'	5'G 3'CTTAA 5'	1
Haemophilus aegyptius	*Hae*III	5'GGCC 3' 3'CCGG 5'	5'GG 3'CC	18
Haemophilus influenzae R$_d$	*Hin*DIII	5'AAGCTT 3' 3'TTCGAA 5'	5'A 3'TTCGA 5'	6
Haemophilus parainfluenzae	*Hpa*I	5'GTTAAC 3' 3'CAATTG 5'	5'GTT 3'CAA	4
Providencia stuartii	*Pst*I	5'CTGCAG 3' 3'GACGTC 5'	5'CTGCA 3' 3'G	2
Streptomyces albus G	*Sal*I	5'GTCGAC 3' 3'CAGCTG 5'	5'G 3'CAGCT 5'	0

The arrows indicate the positions in each palindrome where a phosphodiester bond is cut by the restriction endonuclease. Only one of each pair of identical cut ends is shown for each endonuclease.

* See Figure 3-5.

Based on Appendix D, "Restriction Endonucleases," by R. J. Roberts, in *DNA Insertion Elements, Plasmids and Episomes,* Eds. A. I. Bukhari, J. A. Shapiro, and S. L. Adhya, Cold Spring Harbor Laboratories, 1977, and published nucleotide sequences for the SV40 chromosome.

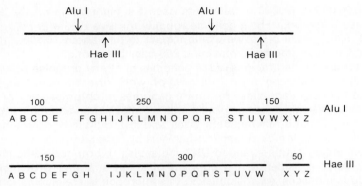

Figure 3-4   Determination of the linear order of DNA fragments produced by restriction endonucleases *Alu*I and *Hae*III by comparing regions of sequence overlap.

The entire base sequence of a large chromosome cannot be readily determined using fragments produced by restriction endonucleases.

Although base sequences of chromosomes containing 5,000 base pairs have been determined, determining the base sequence of the DNA in a large genome—for example, one with 100,000 or more base pairs—would be a formidable task. Even for a genome with *only* 100,000 base pairs, not to mention a mammalian genome with 10^9–10^{10} base pairs, restriction enzymes would produce an enormous number of small fragments for sequencing. The number of restriction sites is a function of the number of times a particular palindrome occurs by chance in the DNA. The frequency at which the *Alu*I site will be present in a genome is $(\frac{1}{4})^4$ ($1/256 = 0.0039$; the frequency at which a particular sequence occurs is the reciprocal of the total number of possible sequences of that length, which, for four bases, is 4^4 or 256). For a genome with only 100,000 base pairs, we expect to find about 390 *Alu*I

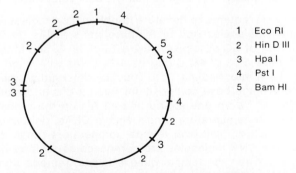

Figure 3-5   A partial restriction map of the SV40 chromosome. The sites at which various restriction enzymes cleave SV40 DNA are shown with the *Eco*RI site, by convention, at the top. The cleavage specificities of the endonucleases are given in Table 3-2.

sites (100,000 × 0.0039 = 390), with each fragment being on the average 256 base pairs long. For a genome roughly the size of the *E. coli* genome with 4 million base pairs, we would expect to find about 15,600 *Alu*I sites, with the average fragment remaining 256 base pairs long.

Although limited regions of the genomes of higher organisms have been and will be sequenced, we do not have or, with current procedures, expect to have the base sequences for entire genomes or chromosomes of higher organisms. We can, however, obtain indirect information about the sequence organization, that is, the numbers, sizes, and distributions of base sequences present in the genomes of higher organisms by measuring the rates at which complementary single-stranded DNA molecules reform double-helical molecules. After considering the creation of single-stranded DNA molecules from double helices, we will consider this approach to the elucidation of the sequence organization of eukaryotic genomes.

Heating converts double-helical DNA to single strands.

We have noted that the bonds conferring structural stability to the double helix are both hydrogen bonds between base pairs and, more important, the van der Waals forces between the stacked bases. Neither hydrogen bonds nor van der Waals forces are strong compared to covalent bonds, and heating DNA in solution can *denature* double-helical DNA by converting it to single-stranded molecules. As the double-helical molecules become separated into single-stranded ones, the absorption of ultraviolet light at 260 nm increases. Such hyperchromicity, the increased absorption of ultraviolet light, allows the convenient monitoring of the conversion of double-stranded to single-stranded molecules. As the double helices melt, that is, as the strands become separated at higher temperatures, the absorption of light at 260 nm increases until the double-helical DNA becomes completely converted into single-stranded DNA (Figure 3-6). The temperature at which half the DNA is single-stranded and half is double-stranded is called the melting temperature, T_M.

The melting temperature (T_M) depends on the base composition of the DNA.

More energy is required to denature double-helical molecules rich in GC base pairs than molecules rich in AT base pairs. DNAs (in 0.01 M NaCl) composed entirely of GC base pairs have a T_M of 82° C, and those composed entirely of AT have a T_M of 48° C (Figure 3-7). DNAs with intermediate base compositions have intermediate T_M's. Thus, by determining the T_M one can determine the average base composition. Such artificial double helices as molecules with all T's on one strand and all A's on the other strand melt over a very narrow temperature range. Native DNAs, double-helical DNAs recovered from nature, melt over a wider temperature range than artificial ones because the DNA molecules are composed of regions with differing base sequences. When the temperature is raised, regions rich in AT melt before regions rich in GC (Figure 3-6). During melting a native DNA molecule passes through a series of "blisters" as the temperature is increased. As each blister is created, the absorption of UV light increases.

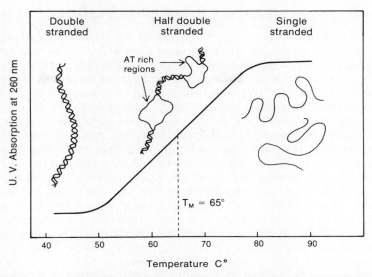

Figure 3-6   The melting (denaturation) of DNA. T_M, the melting temperature, is the temperature when half of the DNA is single-stranded.

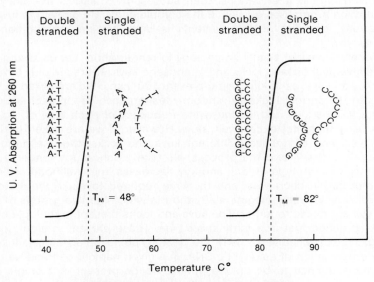

Figure 3-7   The relationship between T_M and base composition.

Denatured DNA can renature to form double-helical DNA again.

By discussing denaturation of DNA we set the stage for a discussion of the opposite process, the renaturation of DNA to reform double helices. Denaturation, the physical separation of two complementary polynucleotide strands, does not change the base sequence in either of the strands. Although separated, they remain complementary. If a solution of denatured DNA is returned to a temperature 10–20° C below the melting temperature, double-helical molecules will reform as complementary strands come together again. The process of renaturation depends, first, on complementary strands bumping into each other and, second, on the reformation of double helices between complementary molecules that have bumped together. Ideally, renaturation involves DNA molecules that are perfectly complementary. If the sequences are not complementary, two DNA molecules might bump into each other, but they would then separate. If only slight differences in complementarity exist between two single-stranded molecules that had bumped together and were considering renaturation, with say, 970 complementary bases and 30 mismatched ones, double-helical DNAs probably would form. Such DNA molecules would have lower T_M's than native molecules because of the mismatched regions, the degree to which T_M is lowered is an indication of the degree of mismatching.

Renaturation of DNA is a second-order reaction depending on the concentration of each type of complementary polynucleotide.

The collisions necessary for complementary denatured single-stranded DNA molecules to renature are accidental, success depending on the concentration of both complementary polynucleotides. If the complements are present in high concentrations, then successful collisions are likely, and renaturation progresses quickly. If the complements are present in low concentrations, collision of complements is unlikely, and renaturation progresses slowly.

We can consider briefly an analogy to renaturation. Let us assume that we are given a thousand locks and a thousand keys. There are only ten kinds of locks and keys, each kind being present 100 times. We are asked to fit the keys into the correct locks by randomly testing each key in each lock. The process will be slow but, because the frequency of each type of lock and key is fairly high, not excessively slow. However, if we are asked to do the same task with a thousand different keys and locks, the likelihood of a correct fit is small, and the process will take much, much longer. As the frequency of each type of lock and key decreases, the likelihood of finding a correct match decreases, and the time required to make the match increases. Now we may imagine a situation in which the two groups of locks and keys are mixed together. The keys and locks present in multiple copies will be matched first in a fairly short time. Those present in single copies will again be matched very slowly. The analogy points up an important aspect of the organization of eukaryotic DNA. In a given haploid genome we might expect a sequence to be unique, that is, to be present in a single copy.

Complements of single-copy sequences may have trouble "finding" each other during renaturation. However, we might also discover that other sequences are repeated and will, upon renaturation, "find" each other comparatively rapidly. Thus, differences in the rates of renaturation can reveal much about the number of copies of particular sequences present in genomes.

For the reader interested in or informed about kinetics, we are dealing with a second-order reaction where the rate of the reaction is a function of the concentration of two reactants (rate being equal to a rate constant (k) multiplied by the concentration of one molecule [X] and another molecule [Y], rate = k[X][Y]). For convenience in renaturation studies, the rate equation is expressed as follows:

$$\frac{C}{C_0} = \frac{1}{1 + kC_0t}$$

Initially all of the DNA present is single-stranded, the concentration of which is C_0. As a result of renaturation over time, t, the concentration, C, of double-helical DNA increases as a function of the rate constant, k. Thus, the formation of double-helical DNA is expressed as a fraction or percentage of the initial amount of total DNA present.

The renaturation of DNA is depicted using a "C_0t" plot.

To study the renaturation of DNA, double-helical DNA is sheared to constant size, about 400 base pairs (rates of renaturation depending on the size of the molecules), denatured, and then allowed to renature. The kinetics of renaturation are conveniently expressed in graphic form using a procedure introduced by Roy Britten and David Kohne. It is called the C_0t plot, the fraction of single-stranded molecules present being graphed as a logarithmic function of the initial concentration of DNA (C_0) multiplied by time (t) (= C_0t) (Figure 3-8). We have presented a theoretical curve, the dashed line, based on second-order kinetics, for DNA renaturation in which all of the sequences are unique (that is, they are single-copy sequences) and genome size is known. Data are also plotted from a renaturation experiment using DNA from the bacterium *Escherichia coli*. The experimental results agree with the theoretical curve. We can conclude that DNA renaturation kinetics follow second-order kinetics as advertised and that the genome of *E. coli* is composed, as far as we can tell, of single-copy sequences.

The rate of renaturation depends on genome size, becoming slower as genome size increases.

The size of the genome determines how frequently a single-copy sequence will be present in a solution of DNA. For example, at a given concentration of DNA, the concentration of a particular single-copy sequence will be greater if the DNA is from an organism with a small genome than if the DNA is from an organism with a large genome. Recalling our lock and

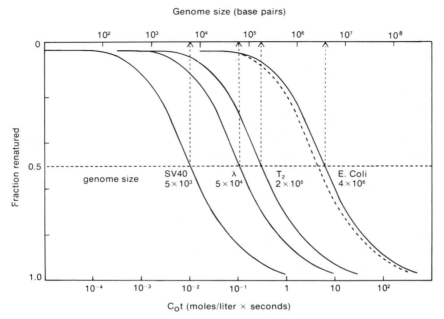

Figure 3-8 C_0t (initial DNA concentration × time) plots depicting the renaturation of DNA's from organisms with small genomes. The dashed line represents a theoretical plot for the renaturation of single-copy DNA from an organism with a genome size equal to that of *Escherichia coli*.

key analogy, we expect the DNA from the organism with the small genome will renature faster than the DNA from the organism with the large genome because each complementary sequence is present in a higher concentration. Different rates of renaturation of DNA from viruses and bacteria with varying genome sizes are depicted in Figure 3-8. All of the depicted renaturation plots have the shape expected for the renaturation of single-copy DNA. Within the limits of the experiments, we can conclude that the genomes of these organisms are constructed of single-copy sequences. We also see the effect of increased genome size on the rate of renaturation of the single-copy sequences. As the genome size becomes larger, the C_0t value at which half of the denatured DNA has renatured, the $C_0t_{1/2}$ value, becomes larger. Furthermore, the $C_0t_{1/2}$ values increase in a completely predictable manner as a function of genome size, which is plotted across the top of the graph in Figure 3-8. Because the results from renaturation fit expectations so perfectly, the results can be used in reverse. DNA from an organism with an unknown genome size can be isolated, sheared, denatured, and renatured and the $C_0t_{1/2}$ determined. The genome size of the organism can then be determined from the $C_0t_{1/2}$ value.

DNA's from eukaryotes contain repeated sequences.

Let us take the DNA's of two of the organisms depicted in Figure 3-8, SV40 and *E. coli*. Equal amounts of the two DNA's are mixed together and allowed to renature. The C_0t plot no longer has the shape characteristic of a population of single-copy sequences from a single organism. The virus DNA, coming from a small genome, renatures first, so that half of the total DNA has become renatured. At a later time the bacterial DNA renatures. The C_0t plot (Figure 3-9) is broken into two separate phases—one when the SV40 DNA renatures, a second when the bacterial DNA renatures. The SV40 DNA renatures first because the concentration of SV40 single-copy sequences is greater than the concentration of bacterial single-copy sequences.

Now let us look at the C_0t plot for the renaturation of cow DNA which is superimposed on a renaturation curve for *E. coli* DNA (Figure 3-10). The renaturation plot for the cow DNA does not have the simple shape of the renaturation plot of prokaryote DNA's but has a shape similar to the plot derived from mixing the DNA from SV40 and *E. coli* together. Part of the cow DNA renatures extremely rapidly, much faster than the *E. coli* DNA, even though the genome size of the cow is about 500 times larger than that of *E. coli*. A second fraction of cow DNA renatures faster than expected but slower than the extremely rapidly renaturing fraction. Finally, a third fraction of cow DNA, about 55 per cent of the total, renatures at the rate expected for single-copy sequences. Because the rate of renaturation is dependent on the concentration of complementary sequences, the sequences in the extremely rapidly renaturing fraction must be more concentrated than those that renature at a rate expected for single-copy sequences. Because the $C_0t_{1/2}$ value

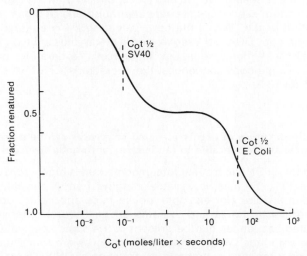

Figure 3-9   The C_0t plot expected for the renaturation of an equal mixture of SV40 and *E. coli* DNA.

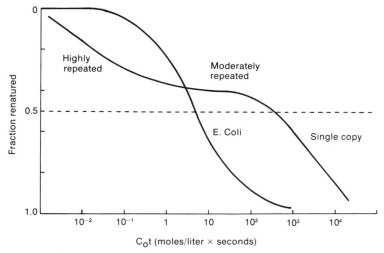

Figure 3-10   Comparison of C_0t plots for the renaturation of DNA from *E. coli* and from the cow. After Britten and Kohne, *Science*, v. 161, p. 529, 1968.

for those sequences that are renaturing extremely rapidly is over 10,000 times smaller than $C_0t_{1/2}$ value of the single-copy sequences, the rapid fraction must be present in a more than 10,000-fold greater concentration than the single-copy sequences. Because all of the DNA comes from the same organism, from a single genome, we conclude that some sequences in the cow genome are repeated over 10,000 times. Similarly, in the second rapidly renaturing fraction, sequences are repeated hundreds of times. Still, a substantial fraction of the DNA in the cow, indeed in every eukaryote, is composed of single-copy DNA. DNA sequences in eukaryotes are typically divided into three classes—highly repeated sequences, moderately repeated sequences, and single-copy sequences, the characteristics of which are summarized in Table 3-3.

Up to ten million copies of highly repeated sequences can be present in a haploid genome and are located in tandem in continuous stretches of DNA.

The fraction of DNA in a eukaryote genome that is highly repeated, being present from 10^5–10^7 times, is variable, ranging from 5 per cent to 10 per cent in several species (for example, house mice and cattle) up to 30 per cent in the domestic horse and in some crabs. Several different kinds of highly repeated sequences can be present in the same species (*Drosophila* has five different kinds). Highly repeated sequences are found together in a series of tandem repeats in continuous stretches of DNA. The length of the repeated sequence varies somewhat but is often very short. One of the

Table 3-3   Characteristics of Sequences in Eukaryotic Genomes

Type of Sequence	Copies per Genome	Arrangement	Average Length (base pairs)	Chromosomal Location
Highly Repeated	10^5–10^7	Tandem	5–10	Heterochromatin
Moderately Repeated	30–10,000	Short-period Interspersion	200–400	Euchromatin
		Long-period Interspersion	3,000–5,000	Euchromatin
		Tandem	100–10,000	Euchromatin
Single-copy	1	Short-period Interspersion	1,000–3,000	Euchromatin
		Long-period Interspersion	10,000–30,000	Euchromatin

simplest repeated sequences is found in the guinea pig and is composed of only six base pairs that are repeated over and over in tandem:

$$5' \ldots CCCTAA \ldots 3'$$
$$3' \ldots GGGATT \ldots 5'$$

Highly repeated DNA found in some horseshoe crabs is also very simple, containing only 1–3 GC base pairs for every 100 base pairs and, thus, is almost a polymer of AT. Because the copies of the highly repeated sequences are identical to each other, they usually renature to produce double-helical DNA with a T_M only slightly below that of native DNA, that is, with little or no mismatching.

Most moderately repeated sequences, those present 30 to 10,000 times per haploid genome, are interspersed with single-copy sequences.

Moderately repeated sequences, in addition to the lower frequency of their presence (30 to 10,000 copies per genome), differ from highly repeated sequences in other important ways. Not just one kind or a few kinds of repeated sequences but a large number of kinds or *families* are present. One family with one type of sequence might be present 500 times, a second, unrelated to the first, could be present 3,000 times, a third, unrelated to the first two, could be present 50 times, and so on. In general there are two classes of moderately repeated sequences, one class with 30 to 200 repeats per family and a second class with 1,000 to 10,000 repeats per family. Also in contrast to highly repeated sequences, moderately repeated sequences are not present as short, simple sequences 5 to 10 base pairs long. The size of the average moderately repeated sequence in many organisms is from 200 to 400 base pairs in length. Furthermore, again in contrast to the highly repeated sequences, related sequences within a family have often evolved from each other so that they are not identical. Upon renaturation, the DNAs from such families renature with substantial amounts of mismatching. The repeated sequences are no longer identical, the family being composed of distant cousins rather than identical twins.

By using DNA's sheared to different lengths, Eric Davidson and Roy Britten have developed procedures that provide information about the distribution of moderately repeated sequences in a genome. They, their colleagues, and others have found two basic types of distribution of moderately repeated sequences, one (short-sequence periodicity) typified by *Xenopus,* an amphibian, and the other (long-sequence periodicity) typified by *Drosophila.* In short-sequence periodicity, on average, there are repeated sequences, 200 to 400 base pairs long, interspersed between single-copy sequences, 1,000 to 3,000 base pairs long (Figure 3-11). In *Drosophila* the average repeated sequence is 3,000–5,000 base pairs long instead of 200 to 400 base pairs long. These repeated sequences are interspersed between unique sequences, on average, 10,000–30,000 base pairs long. Short-period interspersion is far more prevalent than long-period interspersion and is found in most plants and animals. Long period-interspersion, found in *Drosophila,* its relative *Chironomus,* the honey bee, and the water mold *Aclya* is

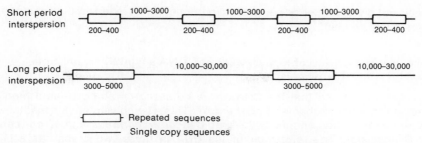

Figure 3-11   Short- and long-period interspersion of single-copy and moderately repeated sequences.

not found in all other insects—the moth *Antheraea pernyi* has short-period interspersion—nor even in all members of the insect order *Diptera* (two winged insects) to which *Drosophila* and *Chironomus* belong. The house fly, *Musca domestica,* has short-period interspersion. Chickens are so far unique, having approximately equal amounts of long- and short-period interspersion. The significance of the presence of alternating repeated and single-copy sequences is not understood nor is the significance of short-period versus long-period interspersion. We assume, because of the universality of interspersion in eukaryotes, that it has some basic biological significance. All moderately repeated sequences are not, however, interspersed with unique sequences. We will discuss some in Chapter 17 that occur in tandem repeats and are limited to specific regions of the genome.

The single-copy sequences of eukaryotes renature at the expected rate relative to genome size, providing evidence for the presence of a single double helix of DNA in a chromosome.

As we have noted, the genome of a cow contains about 55 per cent single-copy DNA. Keeping in mind that the haploid genome of a cow contains about 2.8×10^9 base pairs of DNA, we find over 250 times more single-copy DNA in the cow genome than in the genome of *E. coli*. Thus, the cow genome has the potential of encoding an enormous amount of genetic information. Even *D. melanogaster,* which has a small genome for an animal, has about 10^8 base pairs of single-copy DNA in its haploid genome. Even those organisms with immense amounts of DNA, such as the Congo eel, *Amphiuma,* contain single-copy DNA. Furthermore, Charles Laird has demonstrated that the rates at which single-copy DNA's from higher eukaryotes renature occur as expected from estimated haploid genome sizes of these organisms. Laird's observations demonstrate that for much or most of the DNA of eukaryotes, sequences are present only once per haploid genome. The existence of single-copy DNA in eukaryotes is consistent with the observation that eukaryotic chromosomes contain a single double helix of DNA. The possibility is eliminated that chromosomes typically contain two identical double-helical molecules. If such were the case, there would be at least two copies of every sequence per haploid genome, whereas in fact only one copy of many is present.

The density satellites of higher eukaryotic DNA's typically result from highly repeated sequences.

We have noted that the DNA of the horseshoe crab contains a highly repeated sequence that is 97 per cent AT. This AT-rich sequence constitutes 30 per cent of the genome of the crab *Cancer antennarius* and 12 per cent of *C. magister*. The remainder of the DNA in these two crabs has a GC composition of about 40 per cent. We can imagine the result of extracting the DNA from *C. antennarius* and subjecting it to analysis by equilibrium centrifugation in a CsCl gradient. We would find main-band DNA, constituting 70 per cent of the nuclear DNA, with a density of 1.699 gm/ml, corresponding to 40 per cent GC. We would also find a satellite band, constituting 30 per cent of the nuclear DNA, with a low density of 1.660 gm/ml, corresponding to 3 per cent GC. These two DNA bands would be easily resolved in CsCl gradients. The satellite band results from the presence of highly repeated tandem DNA sequences with a base composition distinct from the moderately repeated and single-copy DNA present in the crab genome. Indeed, the presence of satellites in eukaryotic nuclear DNA is typically the result of the presence of repeated sequences. Repeated sequences may, however, have the same base composition as the main band DNA and thus would not be detected as a separate satellite by conventional density gradient centrifugation. Although satellites that are readily detected often constitute 10 per cent to 30 per cent of the nuclear DNA, it is possible to detect others that constitute only a small percentage of the nuclear DNA. For example, using specialized techniques Douglas Brutlag and James Peacock have identified five different satellites in the DNA of *Drosophila melanogaster,* each of which constitutes only 3 to 5 per cent of the total nuclear DNA.

The location of specific DNA sequences in a chromosome can be determined by *in situ* hybridization.

Highly repeated sequences occur together as tandem repeats. Therefore, highly repeated sequences must be localized in specific regions of chromosomes. If we knew the location and distribution of highly repeated sequences, we might have some idea as to their function in eukaryotic chromosomes. The location of a specific type of sequence in a chromosome can be determined using a procedure developed by Mary Lou Pardue and Joseph Gall called *in situ* hybridization. In *in situ* hybridization, double-helical molecules, *hybrids,* are formed between the DNA in a set of condensed or polytene chromosomes mounted on a slide with radioactive DNA (or, as we shall see later, radioactive RNA) added in solution on top of the chromosomes. The chromosomes are mounted on a slide so that they can be viewed by light microscopy. The DNA on the mounted chromosomes is then denatured without destroying the general physical appearance of the chromosomes. Radioactive DNA, usually labelled with tritium (an unstable isotope of hydrogen containing three neutrons), is denatured and added to the slide in a drop of saline. The slide is then incubated so that renaturation can occur. Double helices form between the DNA in the chromosome and complementary radioactive strands present in the fluid drop. Such hybrid double helices will be limited to those regions of the chromosome where the chromosomal DNA is complementary to the added radioactive DNA. The resulting double helices will be radioactive. Excess radioactive DNA that has not renatured with the chromosomal DNA is washed away, and an autoradiograph is made of the slide. A photographic emulsion is layered over the slide, which is then placed in the dark for periods of a week or two up to several months. During this time β particles are given off from the DNA as the tritium decays. The β particles enter the film immediately above the position of the hybrid DNA in the chromosomes. Silver grains are formed in the film by the β particles. Eventually a large number of silver grains are produced in the film immediately above the region in the chromosomes where the radioactive hybrid DNA is located. The film is developed and the positions of the silver grains are determined using a microscope. The chromosomes can also be seen, and the region of the chromosome above which the silver grains are located can be identified. Because the radioactive sequences form hybrids only with complementary DNA, the location of a hybrid, as revealed by the presence of silver grains, must be the location in the chromosome of a particular complementary sequence. If a specific, purified DNA is used for *in situ* hybridization, it is possible to find the chromosomal location of particular DNA.

Satellite DNA's are typically located in centric heterochromatin but are also found in limited amounts near the tips of chromosomes.

Many satellite DNA's found in eukaryotes are composed of highly repeated simple sequences. Because these satellites are separated from the main-band DNA upon equilibrium gradient centrifugation, they can be purified. A purified, radiolabelled satellite can then be used in *in situ* hybridization to determine the chromosomal location of the satellite. One of the first satellite DNA's used by Pardue and Gall for *in situ* hybridization was the mouse satellite (Figure 3-1), which was shown to hybridize to the centric heterochromatin of mouse chromosomes (Figure 3-12). We can conclude that the repeated DNA sequences of the mouse satellite are located in the centric heterochromatin of the mouse chromosomes.

James Peacock and his colleagues in Canberra, Australia, have studied the chromosomal distribution of five different satellite DNA's of *Drosophila melanogaster*. These five satellites constitute 80 per cent of the highly repeated sequences, or 16 per cent of the total nuclear DNA of this organism. The satellites, each with a different repeated sequence, have been found to hybridize mainly to the centric heterochromatin of the four different *Drosophila* chromosomes, although one satellite, that with a density of 1.672 gm/ml, also hybridizes near the tip of one arm of chromosome number two and near the telomeric end of the X chromosome (Figure 3-13). Each of the satellite DNA's is located in the centric heterochromatin of at least three

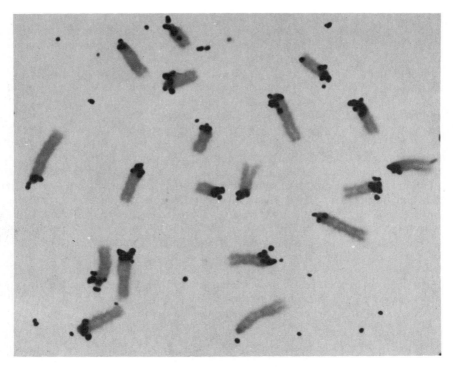

Figure 3-12   The *in situ* hybridization of mouse satellite DNA to the centric heterochromatin of condensed mouse chromosomes. Courtesy of Mary Lou Pardue, Department of
Biology, M.I.T., Cambridge, Mass. From Pardue and Gall, *Chromosomes Today*, v. 3,
p. 47, 1972, Darlington and Lewis, Eds., New York: Hafner Publishing Co.

different chromosomes. The 1.682 gm/ml and the 1.688 gm/ml satellites are
found in both the X and Y chromosomes as well as in the three autosomes.
Note that the Y is entirely heterochromatic and composed mainly of highly
repeated sequences.

We see that centric heterochromatin, also called constitutive heterochromatin because it is present in all cells, is predominantly composed of
highly repeated sequences although some single-copy and moderately repeated sequences also exist. We also know that the centric heterochromatin
is more deeply staining than the euchromatin, indicating that the DNA in the
centric heterochromatin is more tightly coiled than the DNA in the euchromatin. We can hypothesize that the presence of the highly repeated sequences is related to the high condensation of the DNA. We also see that,
in the case of *Drosophila*, different highly repeated sequences are not limited
to single chromosomes but are found in tandem arrays in various chromosomes. The functional significance of sharing single highly repeated sequences between different chromosomes remains, for now, a mystery.

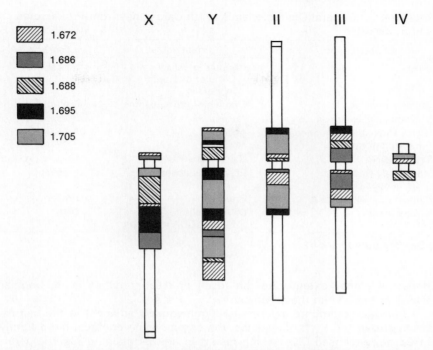

Figure 3-13   The distribution of highly repeated sequences in the heterochromatin of chromosomes of *Drosophila melanogaster*. From James Peacock.

SUMMARY

There are differences, summarized in Table 3-4, in prokaryotic and eukaryotic chromosomes not only at the level of structure but also at the level of DNA sequences. Prokaryotic chromosomes are small, often circular, and lack structural chromosomal proteins. They are predominantly composed of single-copy sequences. Eukaryotic chromosomes contain comparatively large, linear DNA molecules and are also constructed of chromosomal proteins. The DNA in eukaryotic nuclear chromosomes is present in different amounts—highly repeated (10^5–10^7 copies per genome), moderately repeated (30 to 10,000 copies per genome), and single-copy sequences. We can now take a general view of the structure of a eukaryotic chromosome. There are four points to be made.

1. Eukaryotic chromosomes are composed of a single linear double helix of DNA. A hierarchy of helical structures are present. The elementary helical structure involves the nucleosomes, containing histones around which the DNA is wrapped. Structures of higher order are also present, possibly larger helical structures. Chromomeres, regional condensations of DNA from which

Table 3-4   General Characteristics of Chromosomes from Prokaryotes and Eukaryotes

Characteristic	Prokaryotes	Eukaryotes
Shape	Circular in cells; Circular or linear in viruses	Linear
Number per Cell	One major chromosome	Multiple
Nucleosomes/Histones	Absent	Present
Highly Repeated Sequences, Heterochromatin	Absent	Present
Satellite DNA	Usually absent	Usually present
Moderately Repeated Sequences (10)	Essentially absent*	Present
Single-copy Sequences	Present	Present
Nucleic Acid	DNA in cells, RNA or DNA in viruses	DNA

* See Chapters 10 and 12.

loops sometimes extend, are the result of discontinuities in the regular coiling of the DNA in the chromosome.

2. In the heterochromatic regions immediately adjacent to the centromeres and in the Y chromosomes, the chromosome contains more tightly coiled or condensed DNA, which results in heavier staining. Centric heterochromatin is rich in highly repeated sequences (Figure 3-14). A particular highly repeated sequence is not limited to a particular chromosome but is shared by several different chromosomes.

3. In other regions, the DNA is less tightly coiled. These regions are less heavily staining and constitute the euchromatin. Euchromatin is composed

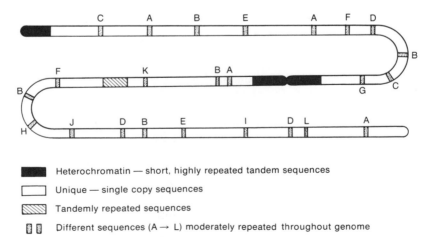

Figure 3-14   A summary of the distribution of sequences in a eukaryotic chromosome.

of single-copy and moderately repeated sequences, with the repeated sequences interspersed between the single-copy sequences. Two kinds of interspersion have been found, short- and long-period interspersion.

4. The interspersion of single-copy and moderately repeated sequences requires that every chromosome that contains euchromatic regions also contains single-copy sequences, that is, no chromosome or set of chromosomes can be composed entirely of repeated sequences. Because each chromosome in a particular karyotype carries single-copy sequences or, to use another term, unique sequences, it is necessary that every chromosome type have its own unique set of sequences. As an example, in *Drosophila* each of the four different kinds of chromosomes has its own set of unique sequences. If, as we already have suggested, genetic information is encoded in the DNA and, thus, in the DNA sequences, it follows that because each chromosome has its own set of unique sequences, each chromosome will carry a unique inventory of genetic information.

II

Gene Transmission

The Maintenance of Genetic Continuity I: The Replication of DNA

Over a few generations, genetic information is transmitted with fidelity. Offspring resemble their parents or sometimes are identical copies of them. Two primary conditions must be met if genetic information is to be transmitted from parent to offspring without error. First, the hereditary information encoded in DNA or RNA sequences must be accurately replicated so that exact copies of the entire set of base sequences are made. Second, an entire set of sequences must then be transmitted to the offspring. In this chapter we shall review the general mechanisms by which DNA replication occurs. In Chapter 5 we will review the mechanisms by which replicated chromosomes are then transmitted from parents to offspring.

The "semi-conservative" replication proposed by Watson and Crick predicts that newly replicated DNA will contain one old parental strand and one newly synthesized strand.

As we noted in Chapter 1, the double-helical structure of DNA proposed by Watson and Crick provides a means by which the genetic endowment can be accurately replicated. The two complementary strands become separated, and each serves as a template for the synthesis of its complement, resulting in the production of two double helices with identical base sequences. Each new complementary strand is made by linking nucleotides together with an old, that is, parental, strand as a template. The two newly formed double helices will each contain one parental strand and one newly synthesized strand. Because one strand in the newly formed double helix is conserved from the parental DNA and one strand is new, this method of replication is called "semi-conservative." If replication occurs again, four double helices will be produced. Two of these double helices will each contain an original parental DNA strand, but two will contain only newly synthesized strands of DNA. If these four double helices are again replicated, eight double helices will be produced, two of these will contain an original parental strand, but now six double helices will contain newly synthesized DNA. As long as replication continues in this manner, two double helices will always contain single parental polynucleotide strands. The fraction of the double helices containing the original parental strand would be reduced after each replication.

Density labeling of DNA with a heavy, stable isotope of nitrogen (^{15}N) allows identification of parental strands and newly synthesized ones.

To demonstrate that DNA replication is, in fact, semi-conservative, it is necessary to have a technique that distinguishes between DNA double helices containing one or two parental strands and those containing only newly formed strands. In 1958 Mathew Meselson and Frank Stahl, graduate students at the California Institute of Technology, reported a procedure that allowed them to distinguish parental DNA strands from newly synthesized ones and to demonstrate that DNA replication in *Escherichia coli* is semi-conservative. Meselson and Stahl grew bacterial cells in a medium that contained the stable, non-radioactive, and heavier form of nitrogen, ^{15}N, in place of the common, lighter form of nitrogen, ^{14}N. The nucleus of ^{15}N contains one more neutron than ^{14}N. ^{15}N was present in the growth medium as ammonium chloride. Normal *E. coli* cells can synthesize their purines and pyrimidines from ammonium and a carbon source, for example, a sugar. Bacteria grown for many generations in the ^{15}N medium have DNA that contains purines and pyrimidines containing ^{15}N and not ^{14}N. Because ^{15}N is heavier than ^{14}N, DNA containing ^{15}N (^{15}N-DNA) is denser than DNA containing ^{14}N (^{14}N-DNA). ^{15}N-DNA, therefore, goes to a lower position in a cesium chloride gradient during equilibrium centrifugation and can be distinguished from ^{14}N-DNA.

DNA replication in bacteria is semi-conservative.

Bacterial cells were grown for many generations in the ^{15}N medium so that both strands of virtually every DNA double helix contained ^{15}N. Meselson and Stahl then grew bacterial cells containing two parental strands of ^{15}N-DNA in ^{14}N medium for three generations, during which time the DNA molecules were replicated three times. Only ^{14}N DNA was synthesized in the ^{14}N medium. Samples of bacteria were removed after one, two, and three generations of growth in ^{14}N medium, and the DNA's were extracted and centrifuged to determine their buoyant densities. If semi-conservative replication had been occurring, after one generation of growth all of the DNA molecules should have replicated once, and each double helix should have contained one strand of ^{15}N-DNA, the parental strand, paired with one strand of ^{14}N-DNA, the newly synthesized strand. Such ^{15}N^{14}N-DNA should have occupied a position in the density gradient halfway between ^{15}N^{15}N-DNA and ^{14}N^{14}N-DNA. That is exactly what Meselson and Stahl observed (Figure 4-1). After the second generation of cell division and the second replication of DNA in the ^{14}N medium, two kinds of DNA molecules were detected. One, constituting half of the DNA, was composed only of ^{14}N, but the second contained one strand of ^{15}N-DNA and one strand of ^{14}N-DNA. This also was the result expected from semi-conservative replication of DNA. After three generations of DNA replication, DNA composed mainly of two strands of ^{14}N was found,

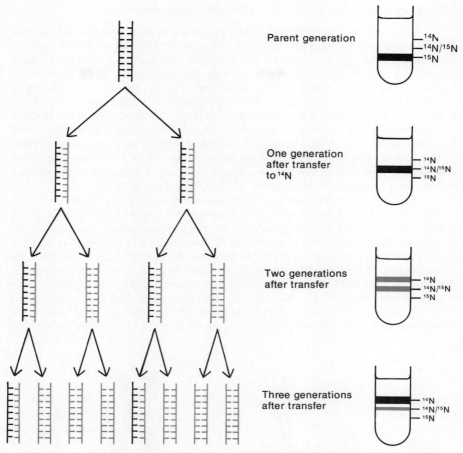

Figure 4-1   The experiment of Meselson and Stahl demonstrating semiconservative replication of bacterial DNA.

but molecules composed of one strand of ^{15}N-DNA and one strand of ^{14}N-DNA were still present in the amount expected, one fourth of the total, for semi-conservative replication (Figure 4-1). Indeed, if the methods used by Meselson and Stahl had been sensitive enough, they could have detected ^{15}N^{14}N-DNA, composed of one parental and one newly formed strand, after many more generations of DNA replication. In semi-conservative replication of DNA we find the explanation for the presence of parental DNA molecules in T2 phage progeny in the experiment conducted by Hershey and Chase (pp. 23–26), in which the original parental DNA was labelled with ^{32}P. Because of semi-conservative replication, the parental DNA is physically conserved and passed on to the phage progeny. It is possible that we ourselves still carry within us some of the actual DNA molecules provided by our parents at conception.

DNA replication in eukaryotes is also semi-conservative.

Many investigators have performed experiments demonstrating that semi-conservative replication occurs not only in prokaryotes like *E. coli* but in eukaryotic organisms as well. In 1958 J. Herbert Taylor and his co-workers reported one of the earliest experiments. To distinguish existing parental DNA strands from newly synthesized ones, Taylor utilized ³H-thymidine, a radioactive form of thymidine. This nucleoside is ultimately incorporated into DNA during replication to produce ³H-DNA. The presence of the ³H-DNA in a chromosome is detected by the use of autoradiography as we described in the account of *in situ* hybridization (pp. 85–86). The β particles emitted from the tritium when it decays produce silver grains in the photographic emulsion that covers the chromosomes, making it possible to determine whether a particular chromosome contains radioactive DNA.

In one set of experiments, Taylor and his collaborators allowed cells to replicate their DNA one time in the presence of ³H-thymidine (Figure 4-2). As each chromosome is replicated and forms two chromatids, the newly synthesized DNA is labelled with tritium. Silver grains are detected over both chromatids in each chromosome because every DNA double helix contains one strand of newly synthesized ³H-DNA and one non-radioactive parental strand (Figure 4-2). Taylor allowed these chromosomes to replicate their DNA a second time but in the absence of ³H-thymidine. The DNA synthesized during the second period of replication is not radioactive. The cells were also exposed to colchicine, a drug that prevents the formation of the spindle

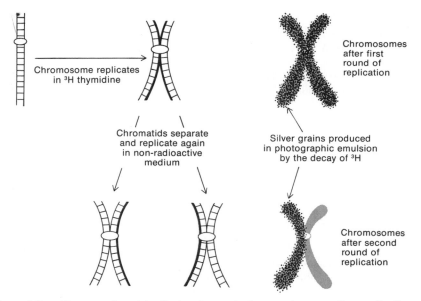

Figure 4-2   The experiment by Taylor demonstrating semiconservative replication of DNA in eukaryotes using autoradiography.

fibers necessary for separating newly replicated, paired chromosomes to opposite sides of a cell during nuclear division. In the presence of colchicine, the newly replicated chromatids, *sister chromatids,* of which each chromosome is composed are left lying side by side, still attached at their centromeres. After the second round of replication in the absence of ³H-thymidine, two kinds of sister chromatids are produced. Each pair of chromatids is composed of one non-radioactive chromatid and one radioactive chromatid. One chromatid is radioactive because it contains the DNA strand synthesized during the first replication of the DNA. The other chromatid is not radioactive because it contains a non-radioactive strand of DNA made during the second replication of the DNA, paired with the original non-radioactive parental strand. Because of semi-conservative replication, only one chromatid in each pair is radioactive. As an historical note, when Taylor and his co-workers first reported their results, the number of DNA double helices in a eukaryotic chromosome was not known. Thus, they could conclude only that chromosome replication was semi-conservative. Now because we know each chromosome contains a single double helix, we can conclude that the DNA itself is replicated semi-conservatively.

Autoradiography has one major drawback: sufficient numbers of β particles must enter the photographic emulsion to produce a clear autoradiograph, and in some cases it may be necessary to wait weeks or even months before developing the film. A new procedure, first reported by Samuel Latt in 1973, utilizes an analog of thymidine called 5-bromodeoxyuridine (BUDR) (Figure 4-3), which can substitute for thymidine in DNA. Tissue culture cells are grown for two generations in the presence of BUDR. The first replication of the DNA produces chromosomes with DNA composed of one strand containing thymidine and one newly formed strand containing BUDR. After the next round of semi-conservative replication, two types of DNA double helices are present, one with two strands containing BUDR, and one with a strand containing thymidine paired with a strand containing BUDR. Latt discovered a procedure using a fluorescent dye called Hoechst 33258 that binds to DNA for distinguishing chromosomes containing DNA with one or two strands of BUDR-DNA. Paul Perry and Sheldon Wolff have modified the procedures. Chromosomes with DNA molecules containing one strand of BUDR-DNA and one with normal DNA containing thymidine stain darkly

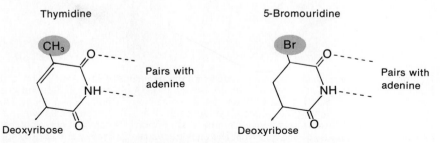

Figure 4-3   The structure of thymidine and its analog, 5-bromouridine.

when treated first with Hoechst 33258 and then with Giemsa (p. 44). A chromosome containing two strands of BUDR-DNA stains lightly. After one round of replication, all chromosomes stain darkly, but after the second round, this time in colchicine so that sister-chromatids remain paired, only half of the chromatids stain darkly (Figure 4-4). This procedure is called harlequin staining because the staining patterns of some of the paired chromatids resemble harlequin costumes.

Both in the early experiments by Taylor and his co-workers and in the recent ones by Latt, Perry, and Wolff, a second phenomenon of importance to genetics, *sister chromatid exchange,* was noted. In one photograph we see that in most cases one sister chromatid in a pair stains heavily, and the other does not. However, the other photograph shows pieces of staining and unstaining chromosomes that have exchanged places with each other. Breaks have occurred in the DNA that have allowed reciprocal exchanges between the paired sister chromatid arms. These sister strand exchanges were caused by treating the cells with cyclophosphamid, a chemical that causes mutation and cancer in test animals. We will discuss later (Chapter 14) the role of mutation as a cause of cancer.

Watson and Crick proposed that during replication the DNA unwinds progressively with the concomitant synthesis of a new strand of DNA.

Watson and Crick realized that one major problem was involved in the replication of double-helical DNA—namely, that for the two strands to be-

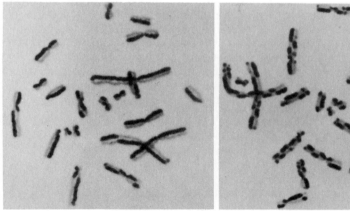

Figure 4-4   Harlequin chromosomes resulting from semiconservative replication and sister-strand exchanges. Chromatids with one strand of DNA containing BUDR and one with thymidine stain darkly; those with two strands containing thymidine stain lightly. (A) Chromosomes from cells grown without cyclophosphamid. (B) Chromosomes with numerous sister-strand exchanges from cells grown with cyclophosphamid, a chemical that causes cancer. Courtesy of Sheldon Wolff and Judy Bodycote, Laboratory of Radiobiology and Department of Anatomy, University of California, San Francisco.

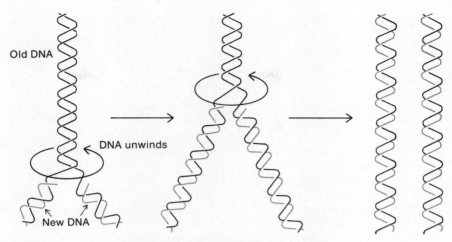

Figure 4-5   DNA progressively unwinds during replication.

come replicated, they must first unwind. It seemed unlikely that the DNA would completely unwind before synthesis of the complementary strands began. Watson and Crick proposed, therefore, that unwinding and synthesis of complementary strands occur simultaneously. Some of the DNA would be unwound, and, as soon as single-stranded regions became available, those regions would serve as templates for the synthesis of complementary DNA strands (Figure 4-5). The simultaneous unwinding of the DNA and the synthesis of the new complementary strands would progress along the length of a DNA molecule until the old molecule had been completely unwound, and the two new ones, containing one old and one new strand, were completed.

DNA replication is progressive.

We will now consider the evidence supporting Watson and Crick's proposal that DNA replication occurs in conjunction with the progressive unwinding of the double helix. So far, in discussing the experiments by Meselson and Stahl and by Taylor, we have only considered DNA molecules that are completely replicated. If we are to learn about the actual mechanism of replication, we must observe DNA during the replicative process. If the mechanism of replication is progressive, we expect to find DNA molecules with "forks" in them similar to the hypothetical replicating molecule shown in Figure 4-5, that is, single double helices should branch to form two double helices as unwinding and replication occurs. The first person to demonstrate the existence of a fork during DNA replication was John Cairns in a report published in 1962. *E. coli* cells were grown in a medium containing ^{3}H-thymidine long enough for the DNA to have been replicated one time and to be in the process of replicating a second time. This meant that the DNA with newly synthesized strands contained tritium. The ^{3}H-DNA was recovered

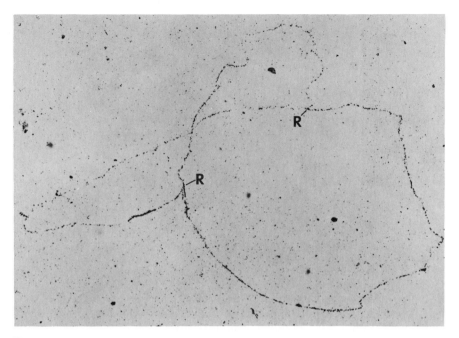

Figure 4-6   Replication forks (R) resulting from the progressive unwinding of DNA are seen by autoradiography of DNA labelled with ³H during replication. Courtesy of John Cairns, Imperial Cancer Fund, London. From Cairns, *Cold Spring Harbor Symposium on Quantitative Biology*, v. 28, p. 43, 1963, with permission.

intact from the cells and spread on a membrane filter over which a photographic emulsion was placed for autoradiography. The pattern of silver grains produced by the ³H-DNA was visible in the developed film with a microscope. A circular chromosome was revealed that, over part of its length, branched into two separate double helices (Figure 4-6). Cairns proposed that this configuration of the DNA resulted from the progressive unwinding and replication of the DNA in the manner suggested by Watson and Crick (Figure 4-5). The branch points, growing or *replication forks,* were sites where the DNA was unwinding and being replicated. Using autoradiography, similar forks have also been seen in DNA isolated from eukaryotic chromosomes that were undergoing replication.

When DNA in the process of replication is isolated and viewed with an electron microscope, structures similar to those seen by autoradiography are observed (Figure 4-7). Bubbles, which are believed to be the regions of the DNA that are undergoing replication, are visible. Such bubbles have been seen in DNA extracted during replication from both prokaryotes and eukaryotes and are called *eye forms* because their idealized shape resembles the outline of an eye.

DNA replication is usually bidirectional.

John Cairns and others replicated DNA from both prokaryotes and eukaryotes and identified structures that were compatible with a mechanism
of replication involving the simultaneous unwinding of a double helix with
immediate use of the separated strands as templates for the synthesis of
new DNA. Additional light on the process of replication was shed in the
laboratories of David Prescott and R. G. Wake who modified the Cairns
technique. Once again bacteria actively replicating their DNA were incubated
with ^{3}H-thymidine. The ^{3}H-thymidine was incorporated into the newly synthesized DNA. Initially ^{3}H-thymidine with low radioactivity was used and then
replaced after a few minutes with highly radioactive ^{3}H-thymidine. DNA synthesized during the first part of the experiment had low radioactivity, but
that synthesized during the last part of the experiment had high radioactivity.
If DNA unwinding and replication are progressive, we expect the DNA immediately adjacent to a growing fork to be highly radioactive because it was
synthesized last, but the DNA more distant from a replication fork would be
only slightly radioactive. In Figure 4-8, in which the results are schematically
depicted, we see that this is the case. But note that the DNA immediately
adjacent to *both* forks is highly radioactive, and that in the middle has low
radioactivity. DNA synthesis must be occurring at both forks. The replication

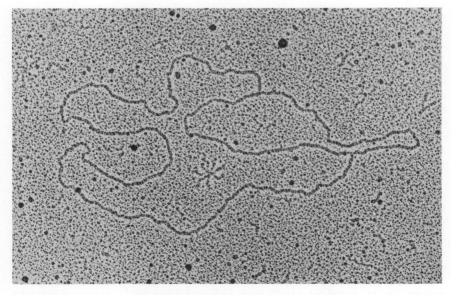

Figure 4-7   An "eye form" made during the replication of a small, circular chromosome
characteristic of prokaryotes. Courtesy of David Wolstenholme, Department of Biology,
University of Utah, Salt Lake City. From Wolstenholme, *Cold Spring Harbor Symposium
on Quantitative Biology*, v. 38, p. 267, 1973, with permission.

APPEARANCE OF
AUTORADIOGRAPH

INTERPRETATION

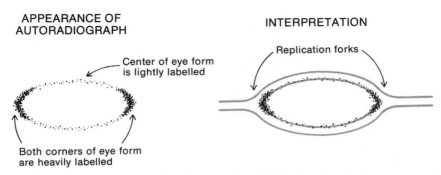

Center of eye form
is lightly labelled

Replication forks

Both corners of eye form
are heavily labelled

Figure 4-8   An autoradiographic demonstration of bidirectional replication.

of DNA starts at a particular point or *origin*, then moves in both directions away from the origin. Such *bidirectional* replication has been found in phages, bacteria, animal cells, and animal viruses. Unidirectional replication in which DNA synthesis occurs at only one of the two forks in an eye form is found in the phages P2 and 186, bacterial plasmids (for example, *ColE*1) and in animal viruses (for example, polyoma, herpes). A small proportion of DNA replication in animal cells may occur unidirectionally.

Replication initiates at a single origin in prokaryotic DNA but at multiple origins in eukaryotic DNA.

When replicating DNA is recovered from prokaryotes, a single eye form is visible because there is usually a single origin for the replication of DNA in prokaryotic chromosomes (Figure 4-6 and 4-7). Replication starts at this point and proceeds bidirectionally, each replication fork progressing with equal speed. In circular chromosomes, as in *E. coli*, termination of DNA replication occurs 180 degrees from the origin of replication, where the replication forks meet. The replication of the *E. coli* chromosome is characterized by a single *replication unit* or *replicon* containing a single origin and point of termination, 180 degrees apart.

In striking contrast, eukaryotes display numerous origins for DNA replication within a single chromosome and, thus, numerous replication units. A remarkable example (Figure 4-9) is seen in the replicating DNA recovered during early phases of the development of *Drosophila* where numerous eye forms are visible in a piece of DNA about 500,000 base pairs long or about one tenth the size of an *E. coli* chromosome. In eukaryotic chromosomes, as the multiple eye forms are extended during replication, they fuse to produce large looped structures. It is estimated that there are about 5,000 origins for DNA replication in the chromosomes of *D. melanogaster* or roughly one origin for every chromomere (p. 59). The rate of replication of DNA in *E. coli* is about 10^5 nucleotides per minute, but in many eukaryotes the rate is about 10^4 nucleotides per minute, the presence of nucleosomes being believed to slow replication. Nevertheless, because of the multiple

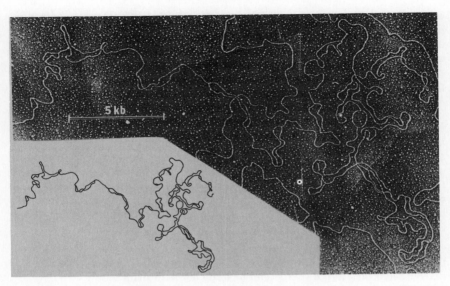

Figure 4-9   An electron micrograph of multiple eye forms in replicating DNA from *Drosophila* embryos. Courtesy of David Hogness, Department of Biochemistry, Stanford University School of Medicine, Stanford, California. From Kriegstein and Hogness, *Proceedings of the National Academy of Sciences, U.S.*, v. 71, p. 135, 1974.

origins for DNA replication, the entire genomes of some eukaryotes can be replicated faster than that of a bacterium. The entire DNA complement of *Drosophila melanogaster* can be replicated in two to three minutes; the process in *E. coli* takes about twenty minutes, even though *Drosophila* contains about twenty times more DNA than *E. coli*. In both organisms replication is rapid, and the rate at which replicating DNA must unwind is also rapid, occurring in *E. coli* at about 10,000 revolutions per minute.

THE ENZYMOLOGY OF DNA REPLICATION

We have now demonstrated that in cells DNA is replicated semi-conservatively in association with the progressive unwinding of the DNA double helix. The precise replication of DNA imposes specific requirements on an enzyme system responsible for the synthesis of DNA. One strand of DNA, acting as a template, directs the assembly of individual nucleotides into a new strand of DNA complementary to the first. Hence, an enzyme must exist that makes DNA out of nucleotide building blocks but does so only in the presence of existing DNA, which is to say, in the presence of a template. The product synthesized by this enzyme must be complementary to the existing strand. Such enzymes, DNA-dependent DNA *polymerases,* have now been isolated from a variety of organisms. The first such polymerase isolated and adequately characterized, DNA polymerase I, was extracted from *E. coli* cells by Arthur Kornberg and his associates. In our discussions of the enzymology of DNA replication, we will emphasize the characteristics of DNA polymerase I and other prokaryotic enzymes. Similar, though less well-characterized enzymes, are found in eukaryotes. As we shall see, the characteristics of the test-tube synthesis of DNA by polymerase I meet our expectations for an enzyme involved in DNA replication. We should keep in mind, however, that DNA replication, like the synthesis of any polymer, can be separated into three distinct phases—initiation (the beginning of synthesis of the polymer), elongation (the extension of an existing polymer), and termination (the completion of polymer synthesis). Although DNA polymerases are important in elongation, we will see that other enzymes are involved in DNA replication, particularly in initiation and termination. Let us begin with a consideration of elongation.

DNA-dependent DNA polymerases elongate DNA by adding deoxynucleotides to the free 3′ end of a nucleic acid primer.

The synthesis of DNA by all known DNA polymerases can be described by the following reaction.

$$n \begin{bmatrix} dATP \\ dGTP \\ dCTP \\ TTP \end{bmatrix} \xrightarrow{\text{DNA and DNA Polymerase}} DNA + 4n(P\text{-}P)$$

in which *n* moles of deoxyadenosine triphosphate (dATP), deoxyguanosinetriphosphate (dGTP), deoxycytidinetriphosphate (dCTP), and thymidinetriphosphate (TTP)—because thymidine is found only in DNA, not RNA, TTP is assumed to contain deoxyribose—are, in the presence of a DNA template, converted by DNA polymerase to DNA, with the release of 4*n* moles of pyrophosphate (P-P). The specific process by which a DNA polymer is elongated is depicted in Figure 4-10. In the presence of DNA polymerase, a deoxynucleotide triphosphate reacts with a free 3′ hydroxyl group in a terminal nucleotide of a polymer. The nucleotide is added to the 3′ end of the polymer by the enzyme with the simultaneous formation of a new phosphodiester bond and the release of pyrophosphate. The polymer has been elongated by one nucleotide. The added nucleotide with its free 3′ hydroxyl group is then ready for the addition of another nucleotide. The addition of nucleotides to the 3′ end can be repeated over and over, causing the polynucleotide chain to elongate. Because addition of nucleotides is always at the 3′ end, synthesis of a polynucleotide always starts at the 5′ end and then progresses to the 3′ end.

DNA polymerases, among which polymerase I is no exception, can only add nucleotides to the 3′ end of an existing polymer. Therefore, a *primer* molecule with a free 3′ end to which additional nucleotides can be added must be present. DNA polymerases will only use deoxyribonucleotide tri-

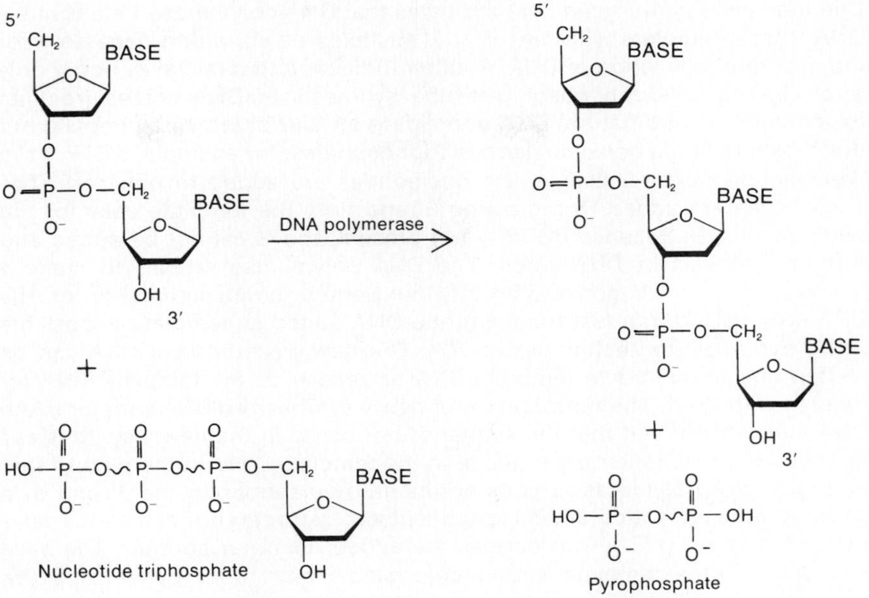

Figure 4-10   DNA synthesis. DNA polymerase adds a nucleotide to the 3' end of a polynucleotide.

phosphates for the synthesis of DNA. Deoxynucleotide monophosphates or diphosphates do not work. The addition of a nucleotide to a growing polymer is energetically favored. The bond between the two phosphate residues in the triphosphate, which is cleaved during the addition of a nucleotide to a 3' end, is a "high-energy" bond, containing more energy than the phosphodiester bond that is formed.

Synthesis of DNA by DNA polymerase utilizes existing DNA as a template.

We expect the newly synthesized DNA to be complementary to the existing DNA, which is necessary for further DNA synthesis. Therefore, the existing DNA should act as a template for the test-tube synthesis of DNA using DNA polymerase I. Early evidence from Kornberg's laboratory indicated that the DNA synthesized by DNA polymerase I had the same base composition as the DNA that was added as a template. We would expect this result if both strands of the template DNA were used for the synthesis of new DNA. Furthermore, when an artificial polymer composed only of deoxyadenosine (dA) and thymidine (T), a poly dAT, was used as a template, the product made by DNA polymerase I contained only dA and T and had no detectable dG or dC. This, of course, we expect if the poly dAT acts as a template because no bases complementary to G or C are present. The experiment has additional genetic importance because the absence of G and

C in the newly synthesized DNA indicates that DNA polymerase I makes new DNA from a template with few, if any, mistakes—a condition necessary for the accurate replication of DNA. Another indication that mistakes occur only rarely in the DNA-dependent test-tube synthesis of DNA comes from an experiment using a natural DNA containing all four bases as a template but omitting one of the deoxynucleotide triphosphates, for example, dGTP, from the reaction mixture. Only a few nucleotides are added to polymers. The nucleotides are added to a growing 3′ end until the template calls for the addition of a G. Because dGTP is not present, the G cannot be added and further synthesis of DNA stops. The DNA polymerase refuses to make a mistake. A final demonstration of the perfect complementarity of the DNA synthesized in a test tube with the DNA added as template is possible using hybridization techniques (p. 77). The newly synthesized DNA can be melted and renatured to template DNA as readily as the template DNA renatures with itself. The renaturation of newly synthesized DNA with template DNA makes it evident that the sequences of bases in the newly synthesized strands are complementary to those in the template DNA. In the picture that emerges DNA polymerase I adds additional nucleotides to the 3′ end of a growing polynucleotide chain, which is itself hydrogen bonded to the template (Figure 4-11). The nucleotides are added one after another. The base sequence of the template strand determines their order by utilizing the

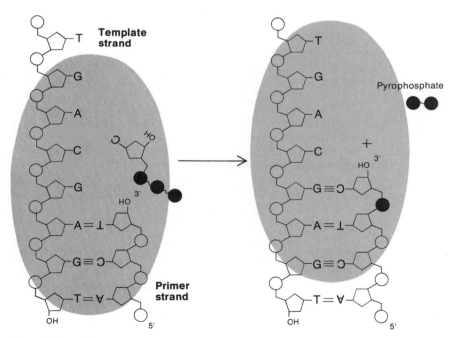

Figure 4-11   The template directed addition by base-pairing of a nucleotide to the 3′ end of an elongating strand of DNA.

specific pairing of complementary bases. The enzyme moves progressively along the template as it adds new nucleotides to the growing polynucleotide chain. We see a clear difference in function between the primer and the template. The primer provides a 3' end to which nucleotides are added; the template determines which nucleotides are added.

Prokaryotic DNA polymerases are self-correcting.

Any self-respecting, modern electric typewriter can correct its mistakes. If a wrong character is typed, a typist can back space the typewriter, erase the mistake, and then enter the correct character. DNA polymerases have been correcting their mistakes for millions of years. For in addition to adding nucleotides to 3' ends of primer strands, prokaryotic polymerases are also able to remove bases from 3' ends using an enzyme called a 3' exonuclease. A 3' exonuclease, like an endonuclease, can cleave phosphodiester bonds, however, it cleaves bonds only at the 3' end of a polynucleotide. Whether the polymerase uses its 3' exonuclease activity to remove a base or its polymerase activity to add one is determined by the condition of the nucleotide at the 3' end of the primer strand. If the last base on the primer strand is correctly paired with its template, then an additional base is added. If, however, the wrong base is present, having been erroneously added by the polymerase a moment earlier, and is mismatched with its complement, the base will be removed by the 3' exonuclease (Figure 4-12). The polymerase will then add the correct base and continue on its way. The mistake is corrected. The self-correcting activity provided by the 3' exonuclease of the DNA polymerase is a result of evolution that stresses the importance of error-free replication of DNA.

A second type of enzyme, DNA ligase, that is involved in DNA replication functions by sealing polynucleotides together.

In addition to DNA polymerase, which adds nucleotides to free 3' ends of deoxynucleotide polymers, another enzyme functions in the replication of DNA. This enzyme, discovered in 1967 by Martin Gelart and called *DNA ligase,* joins blocks of polynucleotides together and is involved in the completion or termination of polymer formation. An experiment reported by

Polymerase Adds   Removes   Adds
Wrong Base   Wrong Base   Correct Base

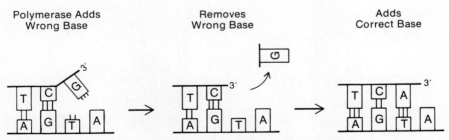

Figure 4-12   The self-correcting activity of DNA polymerase: the 3' exonuclease.

Mehran Goulian, Robert Sinsheimer, and Kornberg that utilizes single-stranded, circular DNA molecules from the phage ØX174 demonstrates the function of DNA ligase. DNA polymerase I was used *in vitro* (*in vitro* means "in glass," that is, in a test tube) to synthesize the DNA strand complementary to that of the phage DNA. The polymerase formed a complementary molecule by moving around the circular template until it encountered the 5' end of the strand it was synthesizing. Although this newly formed strand was complementary to the single-stranded template and contained all of the complementary bases, it was not a complete circle. DNA polymerase was unable to form the last phosphodiester bond. DNA ligase was then added. It joined the phosphate-bearing 5' end with the free 3' hydroxyl (Figure 4-13) to form a diester bond and complete the circle. Thus, while DNA polymerase is capable of elongating polynucleotides, DNA ligase is required to join the ends of two polynucleotides together, in this instance, during the termination of DNA replication.

DNA synthesis at a replication fork must occur in two directions.

We recall that the two strands in a double helix are antiparallel to each other. Newly synthesized strands must also be antiparallel to the template strands. Because DNA polymerase can only add nucleotides to 3' ends of primers at replication forks, the polymerization of new DNA must occur in opposite directions using the newly separated strands as templates. This condition creates a problem during the progressive unwinding and replication at the replication fork (Figure 4-14). Only one of the newly synthesized strands, the *continuous* or *leading* strand, is oriented so that its 3' end faces

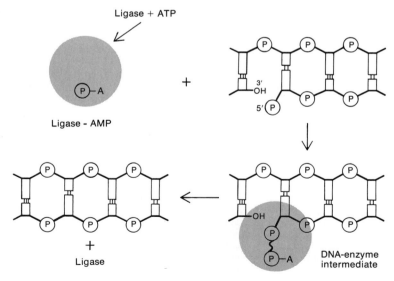

Figure 4-13   The mechanism of action of DNA ligase: Two adjacent polynucleotides are joined.

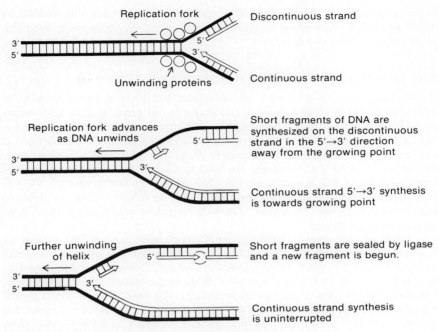

Figure 4-14   Polynucleotide synthesis at a replication fork.

towards the replication fork, thereby providing a primer for the polymerase and allowing continual addition of more nucleotides as the DNA unwinds. The other strand, the *discontinuous* or *lagging* strand, has its 5' end facing towards the replication fork and must be polymerized away from the replication fork.

Short DNA polymers called Okasaki fragments are made during replication.

The dilemma created by the necessity of having DNA synthesized in two directions at a replication fork is solved in the following way (Figure 4-14): Single-stranded templates are produced by the unwinding of the DNA at the replication fork. Synthesis of the continuous or leading strand occurs with unwinding and progresses towards the replication fork. After some unwinding and synthesis of the continuous strand has occurred, synthesis of the discontinuous or lagging strand starts near the replication fork and progresses away from the fork towards the 5' end of an already synthesized polymer. A short, separate polymer of DNA is created called, after its discoverers Reiji and Tuneko Okasaki, an *Okasaki fragment*. The fragment is elongated to fill the gap between the fork and the 5' end of the previously synthesized polymer (Figure 4-14). The life of an individual Okasaki fragment is short, as the 3' end of the fragment is soon joined to the 5' end of the previously synthesized polymer by DNA ligase to form a continuous strand. During their short lives the size of Okasaki fragments varies in different organisms, ranging from about 100 nucleotides in *Drosophila,* to 500 to 1,000 nucleotides in some phages, to over 1,000 nucleotides during the replication of DNA in bacteria.

RNA primers initiate synthesis of the discontinuous strand.

Although we have solved one problem, we have created another. If synthesis of the discontinuous strand starts at the growing fork and moves away from it, where is the primer necessary for elongation? The answer is that a primer composed of RNA, not DNA, is deposited immediately behind the replication fork by a DNA-dependent RNA polymerase. RNA polymerases, in contrast to DNA polymerases, are able to initiate and elongate a polynucleotide without the use of a primer. The primer in prokaryotes is 10–100 nucleotides long and provides the free 3' end to which a DNA polymerase can add nucleotides. Thus, each Okasaki fragment made has a short RNA sequence at its 5' end.

We might ask why an RNA primer is used? Why can't a DNA primer be used, or more generally, why can't DNA polymerases make polynucleotides without a primer if RNA polymerases can do so? The answer, according to Bruce Alberts, is that DNA-dependent RNA polymerases, unlike DNA polymerases, do not have 3' exonuclease activity. We have already noted (p. 30) that the hydrogen bonds between a single pair of complementary nucleotides are not sufficient to maintain paired bases. If, for the sake of argument, DNA polymerase did join two nucleotides together to start a polymer, they would be, at best, loosely paired with the template strand. The 3' exonuclease would think they were mispaired and act according to its function by cleaving the phosphodiester bond between them. DNA polymerase would destroy new phosphodiester bonds as fast as it makes them. RNA polymerase, however, because it has no 3' exonuclease, is able to initiate polymer synthesis without a primer. This analysis stresses again the importance of error-free replication of DNA. For the self-correcting activity of the DNA polymerase, even though necessitating the use of a separate RNA polymerase to initiate DNA synthesis, has been preserved during evolutionary history.

THE SEQUENCE OF EVENTS DURING DNA REPLICATION

We will describe DNA replication step by step, but we should keep in mind that in reality many of the steps occur simultaneously.

The DNA is unwound.

The first step in replication is to make single-stranded regions available as templates for the synthesis of the complementary polynucleotides by unwinding the DNA. This is accomplished by specific proteins called DNA unwinding proteins (Figure 4-14) that are attached to DNA at replication forks and unwind the DNA creating single-stranded regions. The internal unwinding of the double-helical DNA produces compensating twists and tension in the unwound double helix ahead of the replicating fork much like unbraiding a rope in its middle produces kinks in the unbraided portion of the rope. It appears that such tension is released through cuts in single strands in the DNA immediately ahead of the replication fork that allows the kinks to untwist. The cut DNA is then rejoined to its neighbors through phosphodiester bonds to restore the continuity of the double helix.

RNA primers are formed.

Synthesis of the continuous or leading strand occurs simultaneously with unwinding through the addition of nucleotides to the 3' end of the primer strand. Once the DNA is unwound and the continuous or leading strand synthesized, an RNA primer is made for synthesis of the discontinuous or lagging strand (Figure 4-15). The primer, consisting of as few as 10 ribonucleotides, enough to form a stable double helix composed of one strand of RNA and one of DNA, is then available for the addition of deoxynucleotides by DNA polymerase.

DNA is synthesized and the RNA primers are removed by DNA polymerases.

Once the primer is available, elongation of DNA by DNA polymerases occurs. In *E. coli*, a DNA-dependent DNA polymerase called polymerase III that is structurally different from DNA polymerase I adds deoxynucleotides to the available 3' end of the primer (Figure 4-15). After synthesizing a short polynucleotide strand, DNA polymerase I replaces DNA polymerase III, which continues to add deoxynucleotides, and eventually reaches the 5' end of the previously synthesized polynucleotide (Figure 4-15).

The previously synthesized polymer encountered by DNA polymerase I contains at its 5' end the RNA sequence used to prime synthesis of the previous DNA fragment. DNA polymerase I differs from DNA polymerase III in *E. coli* in that, in addition to the 3' exonuclease, it also contains a 5' exonuclease capable of cleaving phosphodiester bonds starting at the 5' end of a polynucleotide. When DNA polymerase I encounters the RNA at the 5' end of the existing polymer, it starts removing nucleotides. Hence, when the 5' end of an RNA primer is encountered, one end of DNA polymerase I is removing nucleotides while its other end is adding nucleotides to the growing Okasaki fragment. The exonuclease activity of the polymerase removes not only the RNA but also some DNA. Eventually the process of simultaneous removal and addition stops, and DNA polymerase drops off the job. The removal of the RNA primer is another example of protecting against possible errors during DNA synthesis. RNA polymerase, we have noted, contains no 3' exonuclease, and we suspect that the synthesis of the primer is more error prone than synthesis of DNA. The removal of the primer

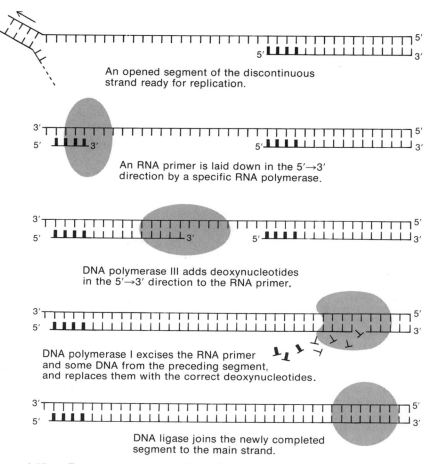

An opened segment of the discontinuous
strand ready for replication.

An RNA primer is laid down in the 5'→3'
direction by a specific RNA polymerase.

DNA polymerase III adds deoxynucleotides
in the 5'→3' direction to the RNA primer.

DNA polymerase I excises the RNA primer
and some DNA from the preceding segment,
and replaces them with the correct deoxynucleotides.

DNA ligase joins the newly completed
segment to the main strand.

Figure 4-15   The enzymology of the discontinuous synthesis of DNA.

and the replacement of the removed bases by a self-correcting DNA polymerase again assures error-free replication.

The fragments are sealed together by DNA ligase.

Once the RNA primer has been removed from the 5' end of an existing
polynucleotide, the 5' end of one polymer is ready to be joined to the 3' end
of the other. DNA ligase joins the two polynucleotides. The DNA behind the
replication fork has been replicated successfully and joined to previously
synthesized DNA. The process—unwinding, primer deposition, elongation,
and ligase sealing—is repeated over and over until an entire chromosome
is replicated.

Initiation of replication starts with RNA primers.

We now appreciate the role of RNA primers in the synthesis of the discontinuous strand of DNA. Without the presence of a primer, synthesis of the discontinuous strand would not be possible because initiation of synthesis would not take place. RNA primers are also used at the initiation of DNA replication prior to cell division. The initiation of DNA synthesis, whether at the single origin in prokaryotes or at the multiple origins in eukaryotes, could not occur without a primer. At an origin of replication, the two strands in the double helix become separated, creating a small bubble in the otherwise double-helical DNA (Figure 4-16); RNA primers are deposited on both of the separated strands. The synthesis of DNA can then start in the continuous direction by the addition of deoxynucleotides to free 3′ ends of both the primers. Synthesis in the discontinuous direction requires additional RNA primers to which deoxynucleotides can then be joined. After initiation, the progressive bidirectional elongation of the new DNA strands continues. The replication of the single-stranded chromosome of ØX174 also starts with the deposition of an RNA primer at a single point of the circular chromosome. DNA is then synthesized unidirectionally around the circular chromosome until the 5′ end of the RNA primer is reached. The 5′ exonuclease activity of the polymerase then removes the primer, and it is replaced with DNA. Finally, the complementary circle is closed through the action of DNA ligase. Further replication of the double-helical ØX174 DNA, called the replicating form or RF, is bidirectional.

Known mechanisms of DNA synthesis do not provide for the completion of replication of linear chromosomes.

The completion of the circular DNA strand complementary to the single-stranded ØX174 chromosome involves a joint effort of DNA polymerase and

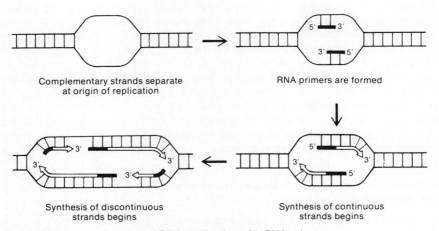

Complementary strands separate
at origin of replication

RNA primers are formed

Synthesis of discontinuous
strands begins

Synthesis of continuous
strands begins

Figure 4-16   The initiation of DNA replication with RNA primers.

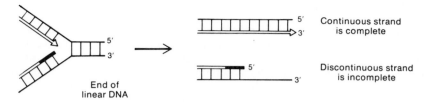

Figure 4-17   The dilemma of completely replicating a linear chromosome.

DNA ligase. The synthesis of DNA in a 5′ to 3′ direction around the circle necessarily results in the DNA polymerase encountering the 5′ end of the newly synthesized polynucleotide, with the last phosphodiester bond being formed by DNA ligase to complete the circle. In double-helical circular chromosomes in which DNA replication occurs either bidirectionally or unidirectionally, completion of replication poses no special problems. When replication is completed, 5′ ends encounter 3′ ends and will be sealed together by ligase. Thus, circular chromosomes are particularly suited for replication by the existing enzyme systems. Completion of replication in a linear DNA molecule, however, poses problems. Synthesis in the continuous direction can progress without difficulty to the end of a linear molecule (Figure 4-17); difficulties arise in the discontinuous direction. We could assume for want of anything better that synthesis of the RNA primer begins with the last nucleotide of the template (Figure 4-17). DNA polymerases could take over and elongate the polynucleotide with deoxynucleotides, which are then joined to the rest of the replicated DNA. Alas, this assumption leaves us with a short RNA segment on the end of our DNA. No opportunity exists for the RNA to be removed by the 5′ exonuclease activity of a DNA polymerase. If the RNA primer is left in place, is it not assumed to contain errors? Could it, in fact, serve as a template for a DNA-dependent DNA polymerase? Or are we left with a short unreplicated region? Regardless of what happens, having a circular chromosome is a real advantage during replication because of the apparent imperfections and limitations of the mechanisms of replicating DNA. From an evolutionary perspective we can see two possibilities. Circular chromosomes may be adaptations to the limitations in the mechanisms of replication. Alternatively, the mechanisms of replication may have evolved in systems with circular chromosomes and, hence, are not adapted to or adaptable to replication of linear chromosomes. It is likely that the replication machinery has not adapted to the presence of linear chromosomes; instead the chromosomes themselves may have evolved so as to accommodate the limitations of the replicating machinery.

Unreplicated ends of phage linear chromosomes are removed through concatemer formation.

During the replication of many linear phage chromosomes, for example T7 and T-even phages (phages T2 and T4), concatemers are formed. A *concatemer* is a molecule composed of a series of repeating units, like the links of a chain. For our purposes a concatemer is a chromosome composed of two or more linear repeats. We will describe a possible mechanism for the formation of concatemers during the replication of the T7 chromosome. The chromosomes of phage T7 and T-even phages are linear and are terminally redundant, that is, both ends or *termini* of a chromosome contain identical base sequences. Replication of the T7 chromosome (Figure 4-18) results in the production of two double helical molecules, each of which has one unreplicated single-stranded end. Because of the terminal redundancy, the single-stranded ends are complementary and can pair with each other. The two molecules can then be sealed together through the action of DNA ligase to produce a concatemer or dimer composed of two T7 chromosomes containing no single-stranded ends (Figure 4-18). The concatemer is still terminally redundant, having identical sequences at each end. Replication occurs again, with the production of two newly replicated dimer-sized molecules, each of which contains a single-stranded end. The single-stranded ends are complementary and can pair together to create a concatemer composed of four T7 chromosomes linked together in series. This concatemer can again replicate and fuse to produce one with 8 copies of the T7 chromosome, then 16, 32, and so on. The final concatemer produced after the fusion of the last pair of replicated products has no single-stranded ends. Concatemers are also produced during the replication of T-even phages, but the mechanism is not known and may differ from the one described here.

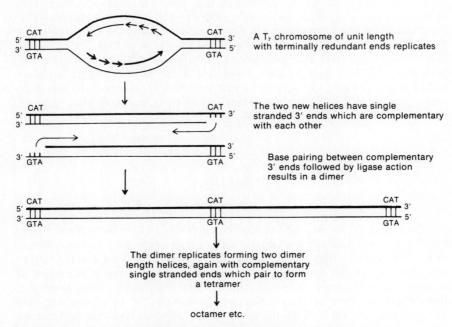

Figure 4-18   A model for concatemer formation in phage T7.

When new phages are produced, these large concatemers are cut up by endonucleases to produce chromosome-sized pieces to be packed into newly assembled phages.

In the human karyotype presented in Chapter 2 (Figure 2-7), we see that all 23 types of human chromosomes appear as rods and presumably contain linear DNA molecules. We are confident that concatemer formation, such as that described for the replication of the T7 chromosome, is not used to complete replication of eukaryotic nuclear DNA, the mechanism of which remains to be solved in the future.

Enzymes exist for the replication of chromosomes composed of RNA, notably an RNA-directed DNA polymerase found in RNA tumor viruses.

Enzymes for the replication of RNA chromosomes also exist. Some of these enzymes are called RNA-dependent RNA polymerases and produce exact copies of RNA chromosomes using RNA as a template. One form of enzyme deserves special mention because it is involved in the life cycle of animal viruses called RNA tumor or oncogenic viruses. Examples of RNA tumor viruses are the Rous sarcoma virus, which causes cancer in chickens and other animals, and the mammary tumor virus found in mice, which causes tumors in mouse mammary glands. During the replication of chromosomes in these viruses, DNA, instead of RNA, replicates are produced. Howard Temin originally suggested the possibility that DNA replicates were produced, and he and David Baltimore eventually demonstrated the presence of an RNA-dependent DNA polymerase, also called a *reverse transcriptase,* in Rous sarcoma virus (or, as Temin prefers to call it, an RNA-directed DNA polymerase because the enzyme is also capable of using DNA as a template). This enzyme plays an important role in the life cycle of RNA tumor viruses and in cancer in lower animals. No one has unequivocally demonstrated the presence of RNA tumor viruses in humans although there is growing evidence of the presence of a virus in human mammary tumors that has properties similar to the mouse mammary tumor virus and a human leukemia virus.

SUMMARY

DNA replication, as predicted by Watson and Crick, is semi-conservative, each strand in a double helix separating to act as a template for the synthesis of its complement. Replication occurs at the replication fork and is progressive, unwinding and synthesis of the complementary strands occurring simultaneously. Replication is also bidirectional, replication forks moving in both directions from an origin. Replication is mediated by several enzymes and proteins, including DNA-dependent DNA polymerases that add nucleotides to the 3′ end of a primer strand in conjunction with a template strand;

DNA ligase, which joins the 3' and 5' ends of blocks of nucleotides together; DNA unwinding proteins, which unwind the DNA at the replication fork; and DNA-dependent RNA polymerase, which provides an RNA primer for the initiation of replication and synthesis of Okasaki fragments during synthesis of the discontinuous strand. DNA polymerases, because of their 3' exonucleases, are self-correcting, removing errors occurring during replication, and assuring that the newly synthesized strand is an exact complement of its template. Thus, the mechanisms of replication satisfy the requirement for the error-free duplication of the genetic endowment.

5

The Maintenance of Genetic Continuity II: The Transmission of Chromosomes

All organisms, from the simplest to the most complex, have amazingly accurate mechanisms for replicating their DNA. Furthermore, the mechanisms of replication have in general been conserved during evolutionary history. All DNA is synthesized in a 5' to 3' direction, and all DNA-dependent DNA polymerases require primers to begin their work. In all organisms DNA is replicated partly in fragments, which are then stitched together through the intervention of DNA ligase. With the exception of some viruses, we find bidirectional DNA replication in all organisms, with two replication forks moving away from an origin of replication. The mechanisms for the successful replication of DNA must have evolved early, and they have been continually perfected during evolutionary history. It is not surprising, given the undeniable utility of these mechanisms, that the basic methods of replication remain essentially constant throughout the biological world.

Accurate replication of an organism's DNA is only, however, half the battle for genetic continuity. The replicated DNA sequences, embodying an organism's entire genetic endowment and so containing the instructions for making an entire new copy of an organism, must be accurately transmitted to the progeny, be they phages, new cells resulting from cell division, or baby elephants. The mechanisms for transmission of newly replicated chromosomes vary in the biological world. There are specialized mechanisms peculiar to specific types of phages, mechanisms that are similar in all bacteria, and still others that are similar in all eukaryotes. When they function correctly, all the mechanisms provide progeny with an exact and complete copy of the organism's genome and assure the maintenance of genetic continuity. As we shall see in the second part of this chapter, any failure of the mechanisms to transmit an exact complement of chromosomes—either too few or too many—can have disastrous consequences.

THE TRANSMISSION OF PHAGE AND BACTERIAL CHROMOSOMES

Transmission of phage chromosomes requires correct packaging of a chromosome into a phage head.

Mature phages contain a single chromosome of DNA or RNA that is enveloped by proteins. We know that the phage chromosome must be tightly coiled so as to be contained within the limited volume of the phage. Among the DNA phages, a distinction between those with single-stranded and double-stranded chromosomes can be made. Phages with single-stranded chromosomes have small genomes, 3,000 to 10,000 bases, and some have bodies that are filamentous (for example, M13 and f1). Others, such as ØX174, are icosahedral, that is, with twenty sides (Figure 5-1). Phages with double-stranded chromosomes have larger genomes, 30,000 to 150,000 base pairs, and have complex bodies—exemplified by λ and particularly by T2, with which Hershey and Chase worked—with a head, tail, and tail fibers (Figure 5-1). The proteins that are part of the structure of all phages are made in bacterial cells under the direction of the phage chromosome. (We will unravel in subsequent chapters just how this is done.) Following DNA replication, the phage proteins and nucleic acid assemble to produce new, mature phages. At the time of assembly, each complete phage must contain the entire complement of genetic information necessary for the production of yet more phages when it next has the opportunity to infect a bacterial cell. In the life cycles of phages, the form of the replicating DNA differs from the form in mature phages. The linear chromosomes of T2, T4, and T7 are replicated as concatemers, and the linear chromosome of phage λ is replicated as a circle. The single-stranded chromosome of ØX174 is replicated as a double-helical circle called a *replicating form* (*RF*). At the end of replication all of these molecules are converted to the form found in the mature phages. Packaging of phage chromosomes into phage heads requires a concatemer. Even circular chromosomes are converted into concatemers prior to phage assembly. Thus, the packaging of DNA into phage heads requires cutting the concatemeric DNA into appropriately sized pieces. The DNA is cut using either the *site specific* mechanism or the *headful* mechanism.

Linear T7 concatemers are cut into chromosomes of the correct size, using site specific cleavage.

During replication of T7 DNA (pp. 114–115), individual linear molecules are converted into concatemers, apparently through the pairing of single-stranded complementary ends. Mature phages contain only a single unit of the concatemer; thus, the concatemer must be cut back into pieces of constant size, each containing redundant ends. The means by which this is accomplished is not fully understood, but a mechanism has been suggested. We believe the concatemer is cut by an endonuclease, like a restriction endonuclease, at specific sites along the DNA. The cuts are made within the

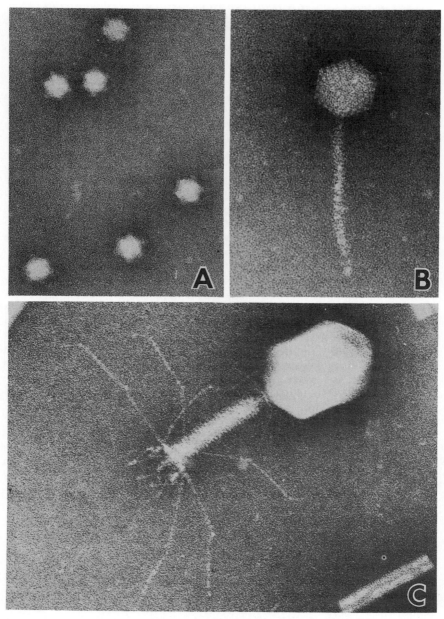

Figure 5-1   Electron micrographs of phages. (A) ØX174; (B) λ; (C) T4. Courtesy of Robley Williams, Virus Laboratory, University of California, Berkeley. From Williams and Fisher, *An Electron Micrographic Atlas of Viruses*, 1974, courtesy of Charles C Thomas, Publisher, Springfield, Illinois.

regions where the single-stranded ends paired to produce the concatemer (Figure 5-2). Single-stranded ends are generated with 5' ends. A 3' end is present in the complementary strand and is a primer allowing DNA polymerase to add nucleotides to form a complete double helix. A series of identical T7 chromosomes results, each with redundant ends. The chromosomes are then packaged into phage heads.

Linear T2 concatemers are cut, using the headful mechanism, into unit chromosomes that are terminally redundant and circularly permuted.

The linear chromosomes of T-even phages, of which phage T2 is an example, are also replicated as concatemers. The mechanism by which the T2 chromosome is packaged into a phage head differs from the one used by T7 and is called the headful mechanism. We believe that one end of the concatemer is packed into a phage until the correct amount of DNA fills the phage head. The DNA sticking out is then chopped off by an endonuclease. The concatemer is used repeatedly until the DNA is exhausted. Direct evidence for the headful mechanism is provided by mutant phages that lack endonuclease. The phages produced are normal except that long molecules of DNA extend from their tails. These DNA tails can be removed experimentally with DNase, after which the phages are able to infect bacteria. The headful mechanism has an interesting effect on the chromosomes that are packaged into phage heads. If we look at the structure of a concatemer, we see that the headful measuring system could start at one end of the concatemer, measure off a constant but correct amount of DNA, and produce a series of chromosomes, each having terminally redundant ends. As a result chromosomes with different combinations or permutations of sequences are produced, each containing at least one copy of every sequence but having different terminal duplications. For example, "A" is duplicated in the first

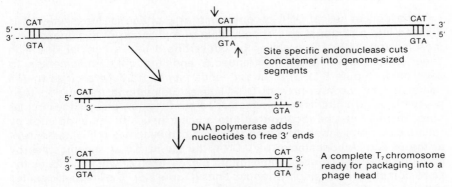

Figure 5-2   A possible mechanism for converting the T7 concatemer to genome-sized molecules for packaging into phages.

one, "B" in the second, "C" in the third, and so on, until a chromosome with "A" duplicated is produced again. Thus,

ABCDEFA/BCDEFAB/CDEFABC/DEFABCD/EFABCDE/FABCDEF/ABCDEFA

The same series of permutations would be produced if the basic repeat (ABCDEF) was present as a circle, and, starting with "A", was marked off every seven units. Thus, the different phage chromosomes produced are said to be circularly permuted.

Linear concatemers are produced from circular chromosomes by rolling circle replication.

In phages with circular chromosomes, such as the double-stranded λ and P2 and the replicating forms of the single-stranded ØX174 and fd, replication begins bidirectionally (see pp. 101–102). The method switches just before phage assembly from bidirectional replication to another method called rolling circle replication (Figure 5-3). In rolling circle replication one circular strand of the DNA is used continuously as a template to form a linear concatemer for the packaging of DNA into newly formed phages. An endonuclease cuts a phosphodiester bond between a specific pair of nucleotides in only one strand of the circular chromosome. The cut creates a 3′ end with a free hydroxyl. This 3′ end serves as the primer for the unidirectional synthesis of a strand of DNA, using the intact circular DNA as a template. As nucleotides are added to the elongating strand, the other strand of DNA, headed by a 5′ end, is extruded as a tail from the circle (Figure 5-4). The single-stranded tail can serve as a template for the synthesis of its complementary strand using RNA primers and the 5′ to 3′ growth of new DNA polymers. The DNA polymerase circumnavigates the circular chromosome several times and produces a tail that is a concatemer of the chromosome.

λ chromosomes with "sticky" ends are made by cutting a concatemer formed by rolling circle replication.

The double-helical DNA chromosome of λ phage is linear when it is injected into a bacterial cell. The problem of replication of this linear chromosome (pp. 113–114) is solved by converting it into a circular one. The linear chromosome has single-stranded 5′ ends that are complementary to each other (Figure 5-5). The "sticky" ends pair to produce a circle that is completed by the ever-present DNA ligase. Bidirectional replication then proceeds using a circular chromosome. To produce new phages the circular chromosome must be reconverted into a linear one with unpaired ends by rolling circle replication. The long tail produced becomes replicated except at the 5′ end. A concatemer of λ DNA, the tail is cut at specific sites by endonucleases into chromosome-length pieces, each of which contains the same complementary single-stranded ends (Figure 5-5). The linear molecules with their sticky ends are packaged into phage heads.

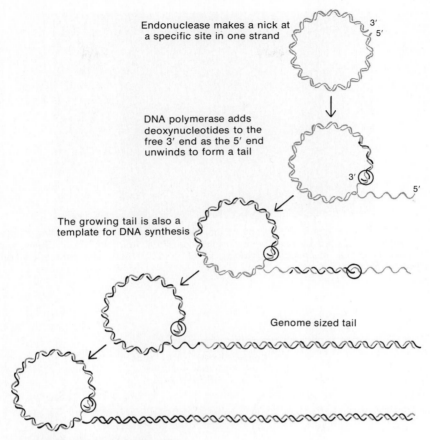

Endonuclease makes a nick at
a specific site in one strand

3'
5'

DNA polymerase adds
deoxynucleotides to the
free 3' end as the 5' end
unwinds to form a tail

3'
5'

The growing tail is also a
template for DNA synthesis

Genome sized tail

Synthesis continues around the
circle forming a concatemer

Figure 5-3   Rolling-circle replication. Linear concatemers are made in preparation for packaging of genome-sized molecules into phages.

The double-helical replicating form of ØX174 is converted into a single-stranded circle by rolling circle replication.

The mature phages ØX174 and f1 contain a circular chromosome composed of single-stranded DNA. For replication, the single-stranded circle is converted into a double-stranded replicating form (RF) that undergoes bi-directional replication (Figure 5-6). After replication is completed, a single-stranded circle must be formed from the RF. Only one of the strands of the RF, the + strand, not a mixture of + and − strands, is found in mature phages. A mechanism must exist for producing the appropriate single-stranded circles for inclusion in newly assembled phages. Production of the

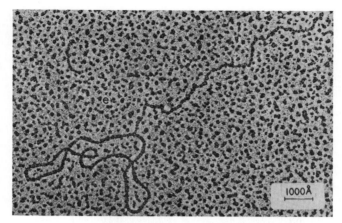

Figure 5-4   An electron micrograph of the ØX174 chromosome showing the single-stranded tail produced by rolling-circle replication. From Knippers, Razin, Davis, and Sinsheimer, *Journal of Molecular Biology*, v. 45, p. 237, 1969, with permission from Academic Press.

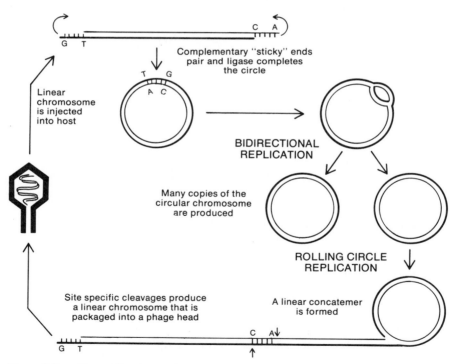

Figure 5-5   The replicative cycle of the λ chromosome.

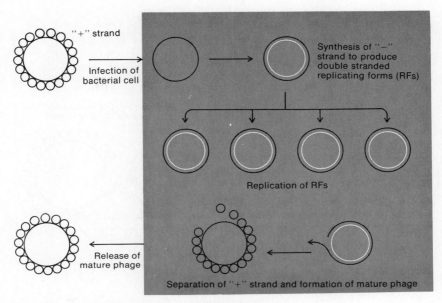

Figure 5-6   The replicative cycle of the ØX174 chromosome.

appropriate strand of DNA is a variation of rolling circle replication (Figure 5-7). A single-stranded tail is produced by rolling circle replication, except that the tail itself is not replicated but is covered with proteins, which maintain it as a single strand. At the 5′ end of the tail, a region contains tandem complementary sequences that pair to form a hairpin. Two such hairpins are produced in a single tail when replication progresses more than 360 degrees around the circular chromosome. The first hairpin is incomplete with a short unpaired region. The second hairpin is cut at a specific site by an endonuclease, freeing a tail fragment and creating another partial hairpin. The fragment that is released is a complete single-stranded ØX174 chromosome with a partial hairpin at the 5′ end and a sequence at the 3′ end that is complementary to the unpaired part of the hairpin. The complementary sequences pair to create a circle. The circle is single stranded except for the region of the hairpin where there is double helical DNA containing a short gap. The gap is closed by DNA ligase, producing a complete, covalently linked circle. The circular piece of DNA becomes part of a new phage.

Chromosomes are apparently transmitted during cell division in bacteria by attachment to the cell membrane.

Successful chromosome transmission in phages involves the packaging of newly formed chromosomes into phage heads constructed of proteins. Complete new phages are reconstructed after every generation of growth in a bacterial cell. Bacterial cells grow by cell division, in which an existing cell

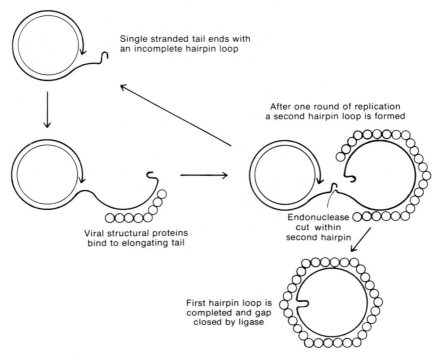

Single stranded tail ends with
an incomplete hairpin loop

After one round of replication
a second hairpin loop is formed

Viral structural proteins
bind to elongating tail

Endonuclease
cut within
second hairpin

First hairpin loop is
completed and gap
closed by ligase

Figure 5-7   Rolling circle replication of the ØX174 chromosome.

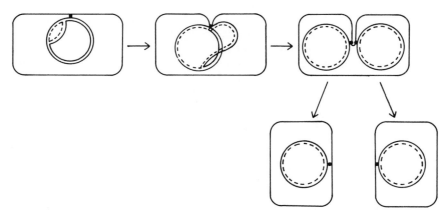

Figure 5-8   A scheme for chromosome replication and transmission during cell division
in bacteria.

divides to produce two new daughter cells. The daughter cells divide to produce 4 cells, the 4 divide to produce 8, then 16, and so on, in geometric growth. A colony or clone of cells is produced, all having the same genotype. At each division, one of the two newly replicated chromosomes, each a large circle of DNA, must be included in each daughter cell. Transmission of the separate copies of the newly replicated chromosomes may be accomplished by attachment of the chromosomes during replication to the cell membrane (Figure 5-8). When replication is completed, each newly replicated chromosome is attached to a separate place on the membrane. A membrane extends across the bacterial cell, dividing it in two. The membrane extends between the places where the chromosomes are attached (Figure 5-8). When the cell is divided by the membrane, one copy of the chromosome is on one side, the other copy on the other side of the membrane. One cell has become two, each with a copy of the chromosome.

THE TRANSMISSION OF EUKARYOTIC CHROMOSOMES.

In eukaryotes there are two types of nuclear divisions, mitosis and meiosis.

In eukaryotes, the type of nuclear division called mitosis produces daughter cells with chromosomal complements identical to those of the mother cell. Mitosis and cell division result in a population of descendents, each of which has the same genotype. The second type of nuclear division, meiosis, is required in the sexual reproductive cycle of higher organisms and assures the reduction in chromosome number from the diploid to the haploid state so that one of each type of homologous chromosome is present. In contrast to mitosis, the cells resulting from meiosis are not genetically identical. We shall describe the general mechanisms of mitosis first and then discuss meiosis.

The eukaryotic cell cycle is divided into four periods.

To reproduce themselves, eukaryotic cells pass through characteristic cycles of growth and division. Cells assimilate nutrients from their environment, increase in mass, and then divide. The cell cycle is divided into four periods based on the events occurring in the nucleus. The two most important periods are "S" (synthesis), a discrete stage when replication of the nuclear DNA occurs, and "M" (mitosis), when nuclear division occurs. Mitosis concludes with the separation of the newly replicated chromosomes and their inclusion into two newly formed nuclei. The two nuclei are separated into two new cells formed as a result of cell division (*cytokinesis*). In addition to these two periods are ones called G_1 (gap 1), which precedes synthesis, and G_2 (gap 2), which follows synthesis (Figure 5-9). M follows G_2 and the cycle is initiated once again upon completion of mitosis and cell division when the two daughter cells again enter G_1. During the collective

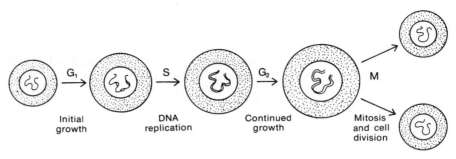

Figure 5-9   The eukaryotic cell cycle.

periods G_1, S, and G_2 (interphase), the chromosomes are uncondensed and are not distinguished as separate entities but instead appear as a formless collage. Because replication of DNA occurs during S, when the chromosomes condense after G_2 in preparation for mitosis, they appear as already duplicated structures.

Mitosis provides the two daughter cells with identical and complete sets of chromosomes.

Division of the nuclear DNA into separate chromosomes in eukaryotic cells places importance on the machinery that maintains the constancy of genetic information from one generation of cells to the next. To replicate a single bacterial chromosome and provide a copy of it to each of two daughter cells seems difficult enough. To replicate 46 chromosomes, as occurs in the diploid state in humans, and then deliver copies of each chromosome to the daughter cells seems impossibly complicated. As we will see the delivery is accomplished during mitosis in a highly organized and almost infallible manner. The replicated chromosomes are literally lined up and separated from each other, with one replicate of each delivered to each newly formed daughter cell.

Mitosis is divided into four cytologically distinct periods: prophase, metaphase, anaphase, and telophase.

During four active periods of mitosis—prophase, metaphase, anaphase, and telophase—the chromosomes undergo changes in appearance and distribution. These changes are explained diagramatically in Figure 5-10 and are shown in a series of micrographs in Figure 5-11.

Prophase. Outside of the nucleus in animal cells the centriole divides and the two daughter centrioles move apart. Within the nucleus the chromatin condenses to reveal individual chromosomes. Because each chromosome has already replicated, the content of DNA is twice that of the diploid cell, but the chromosome number remains unchanged because each chromosome is composed of two chromatids held together by a single centromere. The nucleolus begins to disappear along with the nuclear membrane.

   Metaphase. The centrioles are now 180° apart on opposite sides of the cell. The centrioles are the focus for the development of the spindle. The spindle is a many-fibered structure composed of microtubules that radiate from the centrioles towards the center of the cell where the chromosomes are located (Figure 5-12). Microtubules are polymers of a protein called tubulin. Actin, a contractile protein present in muscle cells, is also present in the spindle. The chromosomes move about within the spindle and eventually become arranged in a single plane, the metaphase or equatorial plate,

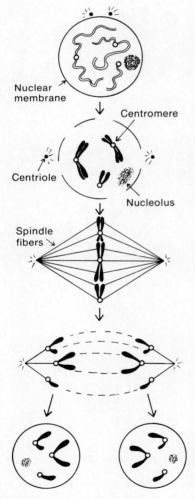

Early Prophase

Replicated chromosomes appear as two chromatids held together at the centromere. The centriole divides.

Late Prophase

The chromatids become short and thick. The centrioles move to opposite poles. The nucleolus and nuclear membrane disappear.

Metaphase

The paired chromatids align at the spindle equator. Spindle fibers attach to kinetochores within the centromeres.

Anaphase

Centromeres divide and chromosomes move to opposite poles by contraction of the spindle fibers. The shape of the moving chromosome is determined by the position of the centromere.

Telophase

Nuclear membranes form. Spindle fibers disappear. Each mitotic product has identical chromosome complements.

Figure 5-10   Mitosis.

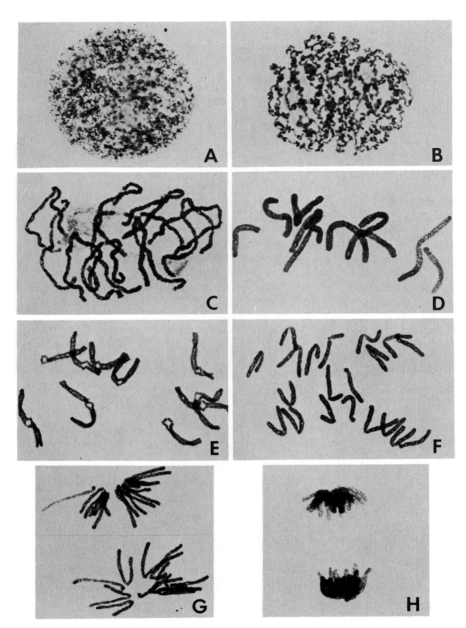

Figure 5-11   Stages of mitosis in the California coastal peony. Only the chromosomes are stained and can be seen. (A) Interphase; (B) Early prophase; (C) Late prophase; (D) Metaphase; (E) Early anaphase; (F) Anaphase; (G) Late anaphase; (H) Telophase. Material prepared and photographed by Spencer Brown and Marta Walters. Courtesy of Marta Walters, Botanical Garden, University of California, Santa Barbara.

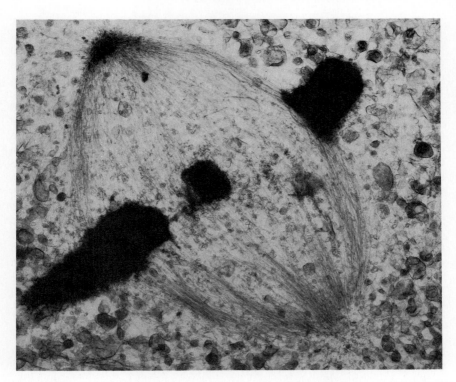

Figure 5-12   An electron micrograph of a cell in metaphase. The heavily stained, condensed chromosomes are on the equatorial plate. Microtubules radiate from the two centrioles to the chromosomes. Courtesy of J. Richard McIntosh, Department of Molecular and Cell Biology, University of Colorado, Boulder.

at the equator of the cell. The spindle fibers attach to the chromosomes through their kinetochores located in their centromeres. The replicated chromatids are ready to separate and to be pulled to opposite poles (pursuing the global analogy) of the cell by the spindle fibers.

Anaphase. The pairs of chromatids are separated from each other as complete chromosomes, each with its own centromere. The DNA content per chromosome set again equals the diploid amount. As the spindle fibers pull the chromosomes apart, the chromosomes form characteristic shapes. A metacentric chromosome with the centromere in the middle moves towards the pole as a "V"; an acrocentric chromosome with chromosome arms of unequal length moves as a "J"; a telocentric chromosome with the chromomere at one end moves as an "I". In this phase of mitosis the effects of colchicine, the drug that prevents microtubule and, thus, spindle formation, become most apparent: the chromatids, not being separated by spindle fibers, remain together in the center of the cell.

Telophase. The two groups of chromosomes are now at opposite poles. The spindle fibers disappear and the nuclear membrane reappears. A centriole is located immediately outside of each nuclear membrane in animal cells. As the nuclear membrane is completed, the chromosomes condense and take on the characteristics of the chromatin found during interphase. The nucleolus reappears.

All that remains to be done after mitosis is to separate the newly formed nuclei into daughter cells and complete the process of division. In animal cells this is accomplished by constriction of the cell in the middle until the daughter cells separate (Figure 5-13). The constriction results from the action of the contractile ring that circles the cell and then pinches it into two separate halves. In plant cells such constriction does not occur. Instead, a new cell membrane forms between the two separated nuclei by the coalescence of small membrane vesicles into a continuous sheet that divides the original cell into two. Cell division in yeast, the fungus used in baking bread and brewing beer and wine, differs from the process in both animal and plant cells. Yeast divides by budding. On the surface of a cell a small bud appears as a blister-shaped bulge. Mitosis occurs while the bud is still much smaller than the mother cell. Nevertheless, the bud receives a full complement of chromosomes. The contents of the bud are then separated from those of the original cell by formation of a membrane between the two cells. The bud separates and grows to equal the size of the mother cell.

Fusion of germ cells during sexual reproduction in eukaryotes requires reduction of the chromosome number from diploid to haploid.

The processes of natural selection favor species that conduct genetic experiments by mixing the hereditary endowments of different individuals together to create new genetic combinations. The mixing, a result of sexual reproduction, provides increased genetic variation upon which natural selection acts. In eukaryotes the mixing of hereditary endowments is accomplished by the fusion of germ cells or *gametes* (for example, eggs and sperm) with their entire sets of chromosomes. Two haploid cells, one from each parent, fuse to form a diploid cell, the zygote. If the diploid cell underwent

Animal cell

Plant cell

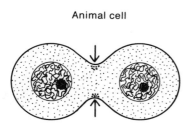 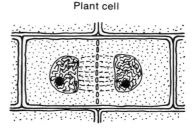

Daughter cells separate by constriction of a contractile ring of filaments

A cell plate is formed between daughter cells by coalescence of membrane vesicles

Figure 5-13   Cytokinesis in animal and plant cells.

numerous divisions to produce a multicellular organism that itself produced diploid gametes, problems would quickly arise. The diploid gametes would fuse to produce a tetraploid zygote, tetraploid gametes, an octaploid zygote, and so on. With each generation of sexual reproduction, the number of chromosomes would double, a geometric progression that, if other restraints did not intervene, would turn the entire universe into chromosomes. The problems of a geometric progression are obviated by meiosis, in which diploid cells, each containing two of every type of chromosome, are converted to haploid ones, containing one of every type of chromosome. Haploid gametes can then fuse to produce diploid zygotes, which can divide and produce more haploid gametes. Chromosome numbers are thereby maintained at constant levels.

The process that reduces chromosome number from diploid to haploid during meiosis is the one with the least likelihood of error.

The main function of meiosis is to reduce the diploid chromosome complement (containing two of every chromosome) to a haploid complement (containing one of every chromosome). We might try to solve a similar problem. Let us suppose we are given 40 objects of varying shapes. There are two of each shape present, and we are asked to separate the objects into two identical groups containing one of each object. To make it interesting, let us suppose that if we do the job, we live, but if we fail, we die. The best procedure is to first match every shape with its partner, side by side in a line, then separate the pairs into two groups, each group containing one of each object. Meiosis has to accomplish the same task, separating pairs of objects—homologous chromosomes—into two identical groups. And the procedure that separates homologous chromosomes is exactly the same as the one we used to separate the objects of various shapes. Indeed the stakes are the same. Failure in proper separation of homologous chromosomes will often result in death. In animals the organism formed from fusion of germ cells will die; in plants the cells, ova and pollen, produced by meiosis will die.

Meiosis occurs in two divisions: the first reduces chromosome number, the second separates replicated chromatids. Four haploid cells result.

We will discuss in detail chromosome behavior during meiosis. There are two meiotic divisions, meiosis I and II. In the first division, the replicated homologous chromosomes pair and then separate into two cells. This division actually accomplishes the reduction in chromosome number from diploid to haploid and is therefore called the reductional division. In the second division, called the equational division, the replicated chromatids in each of the two cells separate as in mitosis to produce four cells or meiotic products. As is true for mitosis, each meiotic division has four periods: prophase, metaphase, anaphase, and telophase. The prophase occurring during the first meiotic division, prophase I, is particularly long and is divided into several cytologically distinct stages. We will discuss the different stages of prophase I using a single pair of homologs as examples.

Meiosis I: Prophase I

Leptonema: The chromatin condenses during the leptotene stage to reveal the individual chromosomes. As in mitosis, condensation of the chromosomes is preceded by DNA replication, so that the chromosomes are already replicated and composed of two chromatids held together at the centromere. Nevertheless, the chromosomes often appear to be single at this stage because the sister chromatids are tightly paired. Because the chromosomes are newly replicated, the DNA content is twice that found in a diploid cell. The number of chromosomes, present as newly replicated chromatids, is, however, the same as in a diploid cell.

Zygonema: During the zygotene stage the separated homologous chromosomes pair with each other to form bivalents. (We should note one definition of homologous chromosomes is those chromosomes that pair during meiosis.) The pairing, *synapsis,* is so intimate that under a light microscope the number of chromosomes present often appears to be reduced by half and to equal the haploid number. Barring mutational differences, the homologous chromosomes, by definition, are identical. They contain identical DNA sequences, including the highly repetitive sequences located in the centric heterochromatin and the moderately repeated and single-copy sequences. Presumably the mechanism by which homologs "recognize" and then pair with each other is ultimately dependent on their common sequences, but we do not know whether recognition involves all types of sequences, or perhaps one type, for example, only the single-copy sequences. Nor is the molecular basis of synapsis understood. Nevertheless, the pairing of the homologous chromosomes, the objects with common shapes, completes the first essential step towards reducing the chromosome number from 2N to N.

Pachynema: The next stage during the first meiotic prophase, the pachytene stage, has special significance, not for the reduction of the chromosome number but for other genetic consequences of meiosis. It is during this stage that physical exchange of chromosome parts occurs between the homologous chromosomes. The exchange is somewhat similar to the sister chromatid exchanges we have discussed before (p. 98). In the pachytene stage the arms of the two pairs of sister chromatids separate from each other, and each sister chromatid arm becomes closely associated with an homologous partner. The group of four chromatids is known as a tetrad. Exchange of chromosome arms occurs between the homologous chromatids but not between the sister chromatids. This process, called crossing over or recombination, involves exact reciprocal exchanges between the chromosomal arms so that each recombined chromatid has the same amount of DNA as before. With electron microscopy we can identify a zipper-like structure called the synaptinemal complex that extends the length of the paired homologous chromatids (Figure 5-14). Formation of the synaptinemal complex is an essential part of recombination. The recombination process also involves DNA synthesis. Remembering that in higher animals as well as in many other organisms, each pair of homologs is composed of one chromosome from each parent, we realize that crossing over results in the creation of a chromosome that contains parts from both parents.

Diplonema: The partial separation of each pair of sister chromatids from their homologous counterparts marks the start of the diplotene stage. The chromatids are still held together, however, at the centromeres and at the points where recombinational exchanges between homologs have occurred. The point of attachment that results from crossing over, plus the extending chromosome arms, create an X-shaped structure called a *chiasma* (plural chiasmata) (from the Greek letter *chi* (χ) plus *asma*, meaning a crosspiece of wood) (Figure 5-15). Chiasmata are considered to be the cytological results of exchanges of chromosome arms.

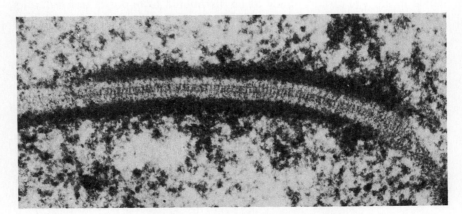

Figure 5-14   An electron micrograph of a synaptonemal complex between two paired chromosomes. Courtesy of Peter Moens, Department of Biology, York University, Downsview, Ontario.

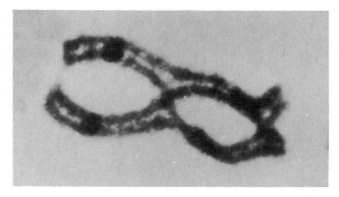

Figure 5-15   A light micrograph showing chiasmata in a tetrad during meiosis in salamander spermatocytes. Courtesy of James Kezer, Department of Biology, University of Oregon, Eugene. From "Meiosis in Salamander Spermatocytes" by J. Kezer in *The Mechanics of Inheritance* by F. W. Stahl, 1964, with permission from Prentice-Hall, Inc.

Diakinesis: The homologs are held together by chiasmata at their tips. The chromatids condense and appear as compact rods that are grouped as tetrads and distributed throughout the nucleus. The nucleolus has disappeared and the nuclear membrane starts to break down, signaling the end of the first meiotic prophase.

To summarize, during the first meiotic prophase replicated homologous chromosomes synapse, usually undergo recombination, and then condense as tetrads. Held together at the centromeres, pairs of sister chromatids in each tetrad are ready to be distributed to opposite poles during the remainder of the first meiotic division.

For the rest of our consideration of meiosis, instead of dealing primarily with a single pair of homologous chromosomes, we will trace the movement of three sets of homologs (Figure 5-16). We will also identify the origin, whether maternal or paternal, of each homolog. Photographs of chromosomes during several meiotic stages are presented in Figure 5-17.

Metaphase I. The nuclear membrane has broken down and the tetrads move to an equatorial plate within the matrix of the spindle. Pairs of homologs, each pair composed of two chromatids, are now aligned ready to separate from each other.

Anaphase I. The chromosomes in the tetrad disjoin so that pairs of sister chromatids, dyads, move to opposite poles. The maternal and the paternal homologs are thus separated (with the exception of regions that have been exchanged during crossing over). The number of chromosomes is reduced to the haploid number. The DNA content, however, is at the diploid level—each chromosome is still composed of two chromatids. Each pole receives a mixture of chromosomes of maternal and paternal origin, because the

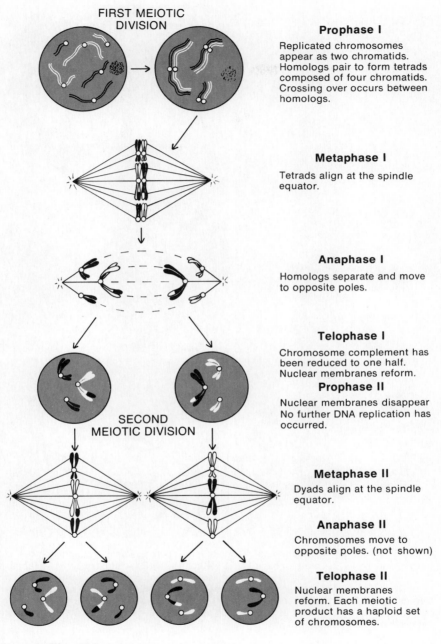

FIRST MEIOTIC DIVISION

Prophase I

Replicated chromosomes appear as two chromatids. Homologs pair to form tetrads composed of four chromatids. Crossing over occurs between homologs.

Metaphase I

Tetrads align at the spindle equator.

Anaphase I

Homologs separate and move to opposite poles.

Telophase I

Chromosome complement has been reduced to one half. Nuclear membranes reform.

Prophase II

Nuclear membranes disappear No further DNA replication has occurred.

SECOND MEIOTIC DIVISION

Metaphase II

Dyads align at the spindle equator.

Anaphase II

Chromosomes move to opposite poles. (not shown)

Telophase II

Nuclear membranes reform. Each meiotic product has a haploid set of chromosomes.

Figure 5-16   Meiosis.

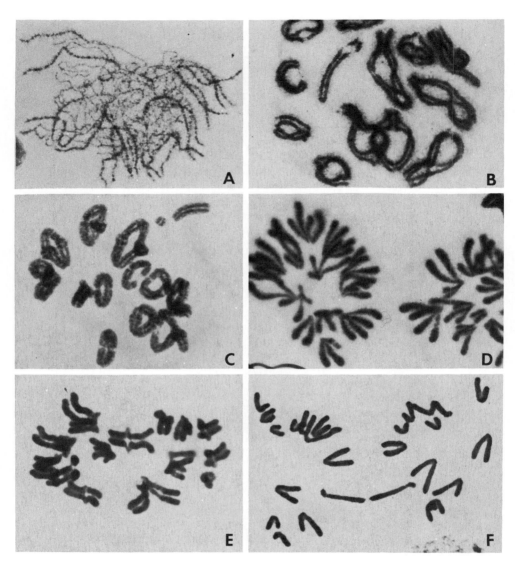

Figure 5-17   Light micrographs of stages of meiosis in salamander spermatocytes. (A) Early prophase I; (B) Late prophase I; (C) Metaphase I; (D) Anaphase I; (E) Metaphase II; (F) Anaphase II. Note that during anaphase I each chromosome is composed of two chromatids. In anaphase II each chromosome is a single entity. Courtesy of James Kezer, Department of Biology, University of Oregon, Eugene. (D) from "Meiosis in Salamander Spermatocytes" by J. Kezer, in *The Mechanics of Inheritance* by F. W. Stahl, 1964, with permission from Prentice-Hall, Inc.

homologs of paternal origin from one tetrad and those of maternal origin from another tetrad may move to the same pole. Thus, there is a mixing of chromosomes of paternal and maternal origin at each pole.

Telophase I. The dyads are at separate poles and new nuclear membranes form. A short interphase follows. This interphase is unusual; the chromosomes remain condensed and visible as distinct bodies and there is no further DNA synthesis. The second meiotic division, which operationally is identical to a mitotic division, now follows for both of the nuclei produced as a result of the first meiotic division.

Meiosis II

Prophase II. The nuclear membrane breaks down in both nuclei and the dyads are spaced well apart.

Metaphase II. The dyads move to a central plane as the spindle apparatus forms. The sister chromatids are now ready to separate from each other.

Anaphase II. The sister chromatids separate from each other as complete chromosomes that move to opposite poles. The DNA content now equals the haploid level, and the haploid number of chromosomes is present.

Telophase II. The chromosomes are grouped at opposite poles, and the nuclear membranes form around them, producing four nuclei. The reduction in the number of chromosomes is complete, with each of the four products of meiosis receiving one copy of each and every chromosome.

Replication, mitosis, meiosis, and fertilization are the basis for the constancy of karyotype in a species.

Each species has a characteristic karyotype. We can understand from the mechanisms of DNA replication, mitosis, meiosis, and fertilization how the karyotype is kept constant. The accuracy of DNA replication assures that sister chromatids have identical DNA sequences. All cells produced by mitosis will have the same chromosome complements as their ancestors. The meiotic divisions place one of each chromosome in each haploid product. When haploid nuclei fuse to produce a diploid nucleus in a zygote (a process called *syngamy*), the resulting diploid nucleus necessarily contains two of each chromosome. Thus, all of the events in the reproductive cycles of organisms constituting a given species assure that the karyotype remains constant.

The stages during the life cycle when meiosis and syngamy occur vary from species to species and determine whether an organism is mainly haploid, diploid, or both.

For completion of every reproductive sexual cycle in eukaryotes, at one stage the chromosome number must be reduced by meiosis and at another stage increased by fusion of germ cell nuclei (syngamy). Except for the fact

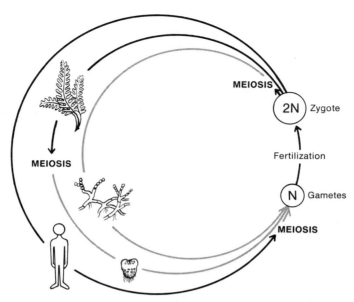

Figure 5-18   The positions of meiosis and fertilization in the life cycles of *Neurospora,* ferns, and humans.

that they alternate, there are no a priori restrictions on the timing of the two events, and a variety of situations exist (Figure 5-18).

1. Meiosis immediately follows syngamy. In some fungi, in molds such as *Neurospora* and in unicellular algae, for example, *Chlamydamonas,* meiosis occurs immediately following syngamy. In both *Neurospora* and *Chlamydamonas,* the vegetative or growth phase of the organism is haploid. In both examples, the meiotic products are contained within heavy walled cells that are resistant to unfavorable conditions, for example, desiccation. When environmental conditions are favorable for further growth, the cells germinate and undergo mitosis to produce large numbers of haploid cells.

2. Syngamy immediately follows meiosis. This condition is common in higher animals. The result of meiosis is the production of haploid cells, which differentiate without any further mitotic divisions to form gametes: sperm and ova. A sperm and ovum fuse to produce a diploid zygote. The diploid zygote then undergoes mitotic divisions to produce a multicellular organism composed of diploid cells. A line of cells within animals called germ line cells gives rise to the next generation of germ cells. In males the cells that give rise to meiotic cells are called spermatagonia (Figure 5-19). Spermatagonia undergo mitotic divisions to produce more cells some of which enter into meiosis. The meiotic cells are called primary spermatocytes during the first meiotic division and secondary spermatocytes during the second meiotic division. The products of the second division, spermatids,

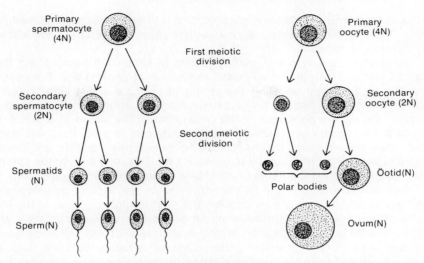

Figure 5-19   Meiosis and gametogenesis in mammals. (A) Meiosis in males leading to sperm formation. (B) Meiosis in females leading to egg and polar body formation.

differentiate into sperm. In females, premeiotic cells are called oogonia. Cells engaged in meiosis are respectively called primary and secondary oocytes during the first and second meiotic divisions. There is, however, an unequal distribution of cytoplasm during the cell divisions that accompany meiosis in females (Figure 5-19). After the first division, the secondary oocyte that is formed contains most of the cytoplasm, while its counterpart, a polar body, contains little cytoplasm. The oocyte and polar body divide again to produce four cells. Three polar bodies, containing little cytoplasm are present. One cell, the ootid, contains most of the cytoplasm and matures into the ovum. Only the ovum can be fertilized to produce a zygote, the polar bodies are non-functional in this regard.

    3. Syngamy and meiosis are separated during the life cycle. In some fungi and higher animals, meiosis and syngamy are immediately adjacent in the life cycle, so that organisms spend their growth phases almost exclusively as haploids or as diploids (Figure 5-18). We find that some fungi, for example, baker's yeast, *Saccharomyces cerevisiae,* and plants, particularly lower ones such as mosses, horse tails, and even ferns, spend part of their life cycles as haploids and part as diploids.

    In yeast, there are two mating types, *a* and *alpha* (α). Haploid *a* cells and α cells can fuse to produce diploid ones that will grow by mitotic budding as long as the medium is nutritionally suitable. When the nutrients are depleted, diploid cells undergo meiosis. Each cell produces four, thick-walled, haploid spores in a process called sporulation. Two of the spores are mating type *a* and two are α. When growth conditions are suitable the spores germinate and cells of opposite mating types fuse to produce diploid ones. If the *a* and α cells are separated, they will bud mitotically as haploid cells. Thus, an experimenter can manipulate yeast so that cells grow either as diploids or haploids.

In plants there is a biological program in which the haploid and the diploid states alternate, the so-called alternation of generations. For example, the croziers that uncoil in the spring to produce a fern with feathery leaves are diploid (Figure 5-18), constituting the sporophyte generation. Haploid spores are produced on the undersides of the leaves as a result of meiosis. The spores germinate and grow by mitosis to produce small, haploid, heart-shaped plants that constitute the gametophyte generation. These haploid forms then produce by mitosis, not meiosis, ova and motile sperm that fuse to produce the diploid sporophyte generation once again.

We find remnants of the alternation between the haploid gametophyte and the diploid sporophyte generation still exist in higher plants. In the life cycle of any flowering plant there are limited mitotic divisions immediately after meiosis. In the production of pollen grains, a diploid cell undergoes meiosis to produce spores that undergo mitotic divisions before a haploid nucleus, derived from the original nucleus of the pollen grain, fertilizes the egg nucleus to produce a diploid plant.

Non-disjunction of homologous chromosomes during meiosis can result in the presence or absence of a chromosome in the progeny.

We have noted frequently that the preservation of genetic information through accurate replication and transmission is essential. We have even suggested that the accurate disjunction of homologs during meiosis so as to produce haploid cells, each of which contains one of every chromosome, is essential for the well-being of the progeny. Indeed, correct disjunction can be a matter of life or death. There are rare abnormal conditions during meiosis that result in the production of meiotic products containing one chromosome too few and one too many. When such meiotic products undergo syngamy with a normal haploid nucleus, the zygote lacks one chromosome or has an extra chromosome, that is, $2N - 1$ or $2N + 1$ chromosomes. These conditions and others like them are examples of *aneuploidy*. Aneuploids are produced by the failure of homologous chromosomes to separate or disjoin during meiosis, a phenomenon called non-disjunction. An example of non-disjunction is diagrammed in Figure 5-20 for a cell containing 3 pairs of chromosomes in which one pair fails to disjoin during the first meiotic division. (Non-disjunction can also occur during the second division.) During the first division, instead of the paternal and maternal chromosomes separating to opposite poles, all four chromatids move to a single pole resulting in one nucleus lacking the chromosome and the other having four copies of the chromosome. During the second meiotic division of the second cell, the four chromatids separate, two going to each pole. Two of the four meiotic products have two copies of the chromosome, and the other two products have none. If these germ cells are involved in fusion with germ cells with normal haploid complements, then two types of aneu-

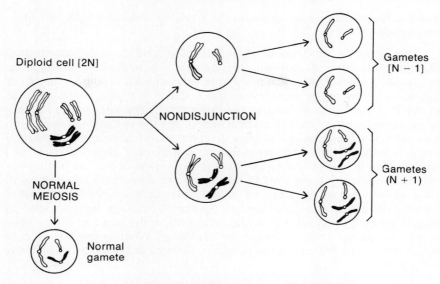

Figure 5-20   Primary non-disjunction during the first meiotic division.

ploid zygotes will be produced with chromosome constitutions of 2N + 1 or
2N − 1.

The presence of an extra copy of chromosome 21 in humans is highly deleterious, causing Down's syndrome.

An example of a deleterious effect of a single extra chromosome is
provided in humans by Down's syndrome (formerly called Mongolism because of a superficial similarity in facial characteristics of afflicted children
to orientals). Down's syndrome is characterized by mental retardation and
reduced life expectancy in addition to effects on facial characteristics. These
changes are the result of non-disjunction and of the presence of an extra
copy of chromosome 21 and, thus, Down's syndrome is also called Trisomy
21. Among human aneuploids, Trisomy 21 is unusual in that affected individuals are born alive. Aneuploidy for other non-sex chromosomes or *autosomes* (the X and Y sex chromosomes are exceptions to be discussed
shortly) generally results in death of the afflicted fetus before birth. For
example, about 50 per cent of spontaneously aborted fetuses have abnormal
karyotypes, including aneuploidy, while less than 1 per cent of children born
alive have abnormal karyotypes. No one really understands why an extra
copy of an autosome is deleterious. In general we believe that a genetically
successful individual must have the correct balance of gene products from
all of the genes present in the genome. The addition of an extra chromosome
and, hypothetically, the presence of the excess products derived from that
chromosome are believed to create a deleterious imbalance.

There is functional compensation for the X chromosome imbalance in males and females.

One common condition of chromosomal imbalance is not deleterious. Human males, after all, contain only a single X chromosome although females contain two. Yet both males and females contain the same number of autosomes. Why is the presence of the single X in males or the absence of the Y chromosome in females not deleterious? The answer lies in functional compensation, *dosage compensation,* for the imbalance. Mary Lyon was the first to suggest that dosage compensation for the extra X chromosome in female mammals involves turning one of the X chromosomes "off" so that it no longer produces genetic products. One of the two X chromosomes is randomly turned off during early cell divisions in development. Thus, in both mature females and mature males only one X chromosome is functionally active in a given cell. In *Drosophila* dosage compensation is accomplished by making the functional activity of the two X chromosomes in females equal to that of the single X in males. Whether the two X chromosomes are functionally turned down in females or turned up in males is not known.

Dosage compensation in humans involves the heterochromatization of one X chromosome in normal females to produce a Barr body.

An X chromosome in an XX female cell is turned off by condensing one of the X chromosomes during interphase. The condensed X chromosome is deeply staining as it is now heterochromatic and is found in interphase nuclei just inside the nuclear membrane. The heterochromatic X is called a Barr body after Murray Barr, who first observed it. The presence of a single Barr body in a cell reveals whether a particular cell has a karyotype with two X chromosomes. We believe heterochromatization, a result of tightly coiling the DNA into a hierarchy of solenoids (pp. 52–54), prevents the function of the genes on the X chromosome. The Y chromosome is already heterochromatic and functions in mammals only in the development of male sex characteristics. Presumably, the heterochromatic states of the Y and one of the X chromosomes in females prevent the genetic imbalance for sex chromosomes that occurs with an extra autosome, exemplified by Trisomy 21, in which the extra chromosome is not heterochromatic.

The effects of non-disjunction of sex chromosomes sometimes cause sterility but are less deleterious than those for autosomes.

Non-disjunction of sex chromosomes can also occur during meiosis. A variety of conditions with abnormal complements of sex chromosomes has been identified in humans (Table 5-1). Three comparatively common conditions are XXY males, XO females, and XYY males.

XXY males. Males with two X chromosomes and a single Y chromosome are born at a frequency of about 0.13 per cent. The presence of two X

Table 5-1   Abnormal Sex-Chromosome Complements in Humans

Condition	Frequency at Birth	Sex	Characteristics	Number of Barr Bodies
XO (Turner's)	0.04%	Female	Sterile, short stature, broad neck, reduced intelligence	0
XYY	0.15%	Male	Tall, on average lower intelligence	0
XXY (Klinefelter's)	0.13%	Male	Sterile, some breast development, on average lower intelligence	1
XXX	0.16%	Female	Some fertility, on average lower intelligence	2
XXXX	<0.03%	Female	Fertile	3

Data on incidence of aneuploidy from Eric Engel, in *The Metabolic Basis of Inherited Disease,* John B. Stanbury, James B. Wyngaarden, and Donald S. Fredrickson, Eds., McGraw-Hill Book Company, 1972.

chromosomes, one of which becomes heterochromatic so that these males have a single Barr body in each nucleus, results in a series of conditions called Klinefelter's syndrome after Harry Klinefelter, who first described them. XXY males have normal life expectancy but, on the average, have reduced intelligence, are sterile, and have some breast development. The XXY condition can result from non-disjunction in the male or female parent. Normally, during meiosis in males (Figure 5-21), the X and Y chromosomes pair and then disjoin. Non-disjunction during the first meiotic division pro-

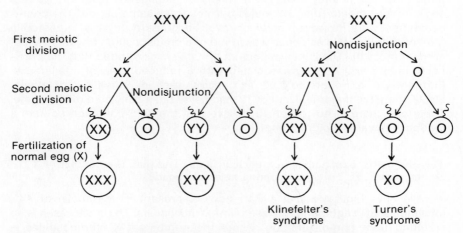

Figure 5-21   Non-disjunction of X and Y chromosomes during meiosis in humans.

duces sperm containing an X and a Y chromosome. If an XY-carrying sperm fertilizes a normal X-carrying ovum, an XXY zygote will result. Non-disjunction can occur in the female to produce an XX-carrying ovum. If that ovum is fertilized by a normal Y-carrying sperm, an XXY zygote will result.

XO females. Females with a single X chromosome are born at a frequency of 0.04 per cent in humans, much lower than the frequency of the XXY constitution. The presence of a single X chromosome results in a set of conditions called Turner's Syndrome after Henry Turner, who first described them. Because only a single X chromosome is present, these females lack a Barr body. Females with an XO karyotype have normal life expectancy and, on the average, reduced intelligence, are somewhat shorter than average, have a webbed or broad neck, and are sterile, having no or only rudimentary ovaries. XO females can result from non-disjunction in male or female parents. An X-carrying sperm can fertilize an ovum with no X or an ovum with an X can be fertilized by a sperm lacking a sex chromosome (Figure 5-21). Parenthetically, we can note that fertilization of an ovum lacking an X chromosome by a sperm carrying a Y chromosome results in early death of the zygote because no X chromosome is present.

XYY males. Males with an XYY constitution are born at a frequency of about 0.15 per cent. XYY males can be produced by non-disjunction of the Y chromatids during the second meiotic division (Figure 5-21). XYY males are taller than average, become sexually mature at an early age, and are completely fertile. A controversy rages about behavioral abnormalities among XYY males. High frequencies of XYY males have been found in penal and mental institutions, and there has been a suggestion that XYY males are predisposed towards violent crimes. A recent evaluation, however, of the XYY condition by a group of 12 Danish and American investigators leads to a different conclusion. These investigators attempted to identify every XYY male born in Denmark over a three-year period. They found XYY individuals performed significantly lower on a variety of intelligence tests and had an arrest record greater than XY males born over the same period. The crimes committed were, however, petty or involved property (that is, automobile theft or burglary). No evidence was found of a predisposition towards violent crime. In fact the Danish investigators tentatively concluded that the criminality record was directly associated with a reduced level of intelligence. Furthermore, although only 12 XYY males were identified out of almost 29,000 men, the majority, 7 out of 12, had never been convicted of any crime. Therefore, there is no reason to believe that the XYY condition confers a tendency towards violent crime.

The number of Barr bodies in a nucleus is one less than the number of X chromosomes, all but one becoming heterochromatic.

We have indicated that a Barr body is present in the nuclei of XXY individuals although none is present in XO individuals. There are cases, also resulting from non-disjunction, where the number of X chromosomes is greater than two. For example, a triple-X constitution can result from the fertilization of an ovum carrying two X chromosomes by an X-carrying sperm, and an individual with four X chromosomes can be produced by the union of two gametes each carrying two X chromosomes. In all situations, all but one of the X chromosomes becomes heterochromatic (Table 5-1). Thus, individuals with three X chromosomes have two Barr bodies and those with four X chromosomes, three Barr bodies. In all cases, then, there is but one functional X chromosome present. In contrast to Trisomy 21, we find that individuals with three or even four X chromosomes are reasonably normal. Indeed those females with four X chromosomes are fertile and have normal intelligence.

Fertile individuals with multiple X or Y chromosomes produce a high frequency of aneuploid offspring.

The incidence of non-disjunction during meiosis of normal diploid cells, that is, primary non-disjunction, is low. We have seen that the frequency of producing XO females or XXY males is at most about one in 1,000. But if an XYY male or a fertile XXX female has offspring, there will be a high frequency of aneuploidy in their progeny (Figure 5-22). In the case of the XYY male, during meiosis the two replicated Y chromosomes and the X chromosome pair so that two X chromatids are paired with two pairs of Y chromatids. During the first meiotic division the X's usually go to one pole and the two pairs of Y's to the other. The second meiotic division produces two sperm with single X's and two with two Y's. If a YY-carrying sperm fertilizes an X-carrying ovum, another XYY individual will be produced. Alternatively, although less frequently, during the first meiotic division a pair of Y chromatids may travel to one pole with the X chromatids so that, ultimately, two XY-carrying sperm will be produced along with two Y-carrying sperm. If an XY-carrying sperm fertilizes a normal ovum, an XXY zygote results. Half of

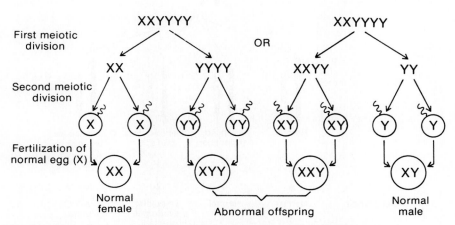

Figure 5-22   Secondary non-disjunction of X and Y chromosomes in humans.

the possible meiotic products carry an extra Y chromosome, and, thus, half the progeny produced could be aneuploids. The production of aneuploid germ cells by individuals who are already aneuploid involves a second form of non-disjunction, secondary non-disjunction, which, as we have seen, is a common event.

The partial loss or addition of a chromosome is also deleterious.

It is also possible to have an excess or deficiency of parts of chromosomes. This condition results from translocations in which part of an arm from one chromosome is broken from its normal position and attached to another, non-homologous chromosome. Translocations can be reciprocal with two arms of each of two non-homologous chromosomes changing positions or non-reciprocal with an arm being removed from one chromosome and attached to another. A case of non-reciprocal translocation in humans demonstrates the possible consequences of translocations. Part of an arm of chromosome 5 has become attached to chromosome 13 (Figure 5-23). The karyotype is atypical in that chromosome 13 carrying the translocated piece is now longer than normal, and chromosome 5 is shorter than normal. The condition is called a balanced translocation for, despite the aberration in the karyotype, the individual is phenotypically normal in as much as the normal complement of chromosomal arms is present. When

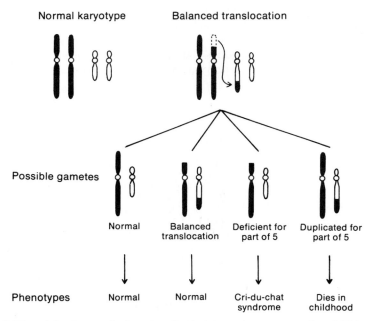

Figure 5-23   Inheritance of a translocation between chromosomes 5 and 13 leading to the cri-du-chat syndrome in humans.

meiosis occurs in this individual, four types of germ cells are produced as we can see in Figure 5-23. One type is normal, with normal chromosomes 5 and 13. A second type is balanced, containing a deficient chromosome 5 and a chromosome 13 carrying an arm of chromosome 5. If either of these gametes fuses with a gamete with a normal karyotype, a phenotypically normal individual will result. In addition, two abnormal types of germ cells are produced that upon fertilization will result in karyotypes with one or three copies of part of one arm of chromosome 5. When one copy is present, fertilization involves a gamete with a normal chromosome 13 but a deficient chromosome 5. When three copies are present, fertilization occurs with a gamete in which chromosome 5 is normal but chromosome 13 carries the additional piece of chromosome 5. In either case the zygote formed has an abnormal karyotype, and the individual so affected, although born alive, soon dies. In the case where one copy of the arm of chromosome 5 is present, there is severe retardation and an abnormal behavior pattern characterized by frequent cat-like crying. Hence the condition is called the "cri du chat" (cry of the cat) syndrome.

The karyotype of humans can be determined before birth by a process called amniocentesis.

During gestation in humans, the amniotic fluid that bathes the fetus in the uterus contains cells that originate from the fetus. It is possible to recover a sample of amniotic fluid and the cells in it with a syringe without damage to the fetus, grow these cells in tissue culture, and determine the karyotype of the fetus. It would be expensive and, because the procedure is somewhat delicate, inadvisable to determine the karyotypes of all fetuses before birth. There are some situations, however, when amniocentesis is indicated.

The frequency of giving birth to children with Trisomy 21 increases dramatically in women over the age of forty.

Trisomy 21 has a significant socioeconomic impact on society. Families with a child suffering from Down's syndrome understandably are exposed to a particular travail and stress that families with normal children escape. Down's syndrome is by no means rare. It is estimated that 10 to 20 per cent of all patients in institutions for the mentally retarded have Trisomy 21. In the United States this amounts to over 100,000 individuals for whom continuing care must be provided beyond the capability of the ordinary family. Furthermore, individuals afflicted with Trisomy 21 are extremely limited in their abilities to participate in and enjoy the activities that characterize human existence. The frequency of giving birth to children affected by Trisomy 21 increases markedly in women over 40 (Figure 5-24), a 50-fold increase compared to the incidence among women under 30. There is also a slight increase in frequency if the father is over 50.

It is reasonable to use amniocentesis to determine for pregnant women over 40 whether the karyotype of the fetus is normal. If Trisomy 21 is detected, the fetus, assuming the parents are willing and in agreement, can

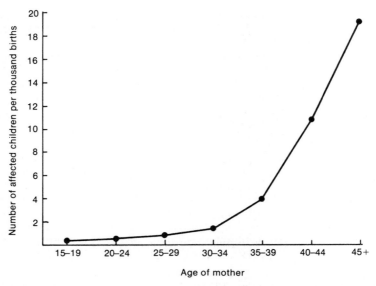

Figure 5-24   Effect of age of the mother on the incidence of trisomy 21 in humans. After Lejeune, *Progress in Medical Genetics,* v. 3, pp. 144–177, 1964.

then be aborted. Amniocentesis is also indicated for pregnancies following birth of an earlier child with an abnormal karyotype causing a deleterious condition, for example, that producing the cry of the cat syndrome.

SUMMARY

We have described several varieties of chromosome transmission—those found in phages, bacteria, and in many eukaryotes. Concatemers form in phages, both to eliminate unreplicated ends in linear chromosomes and to convert circular double-helical chromosomes into single-stranded ones ($\emptyset$X174) or into linear ones with single-stranded ends (λ) for packaging into phage heads. During cell division in bacteria the newly replicated chromosomes are transmitted by attachment to the cell membrane. Mitosis and meiosis in eukaryotes are the ultimate mechanisms for chromosome transfer. In mitosis the replicated chromosomes are aligned at the equator of the cell, and one of each pair of chromatids moves to opposite poles. In the first meiotic division the homologous chromosomes pair before being mustered at the cell equator, and homologs, consisting of paired chromatids, separate to opposite poles causing the reduction in chromosome number. During the second meiotic division the replicated chromatids are separated. Each meiotic product receives one of each chromosome.

The necessity of correct transmission is highlighted by non-disjunction. Aneuploid zygotes, with one too few or one too many autosomes, are inviable. In mammals death usually occurs before birth. Aneuploidy for sex chromosomes is less deleterious because of the formation of Barr bodies in which all but one X chromosome becomes heterochromatic.

In general, however, we see that the mechanisms forged by evolution—DNA replication and chromosome transmission—for the successful propagation of genetic endowments operate with few errors.

6

Mendelian Genetics

We have discussed the replication of entire chromosomes and their transmission from parent to offspring. We will now consider the transmission of limited parts of chromosomes with particular nucleotide sequences. The enormous length of double-helical molecules and the astronomical number of possible nucleotide sequences confer on DNA the ability to encode all of the information necessary for the structure and function of complex organisms. The study of how a particular unique nucleotide sequence is transmitted from parent to offspring, in the midst of millions of others, seems formidable. Nevertheless, we already know in general how a particular sequence is transmitted. In eukaryotes we know how nuclear chromosomes are delivered to daughter cells during mitosis and how homologous and nonhomologous chromosomes are transmitted through meiotic divisions. Therefore, we have general expectations for the patterns of transmission of nucleotide sequences. The sequences will be carried along on the chromosomes of which they are integral parts. What we need, however, is a means for determining which chromosome embodies a particular sequence, where on the chromosome it is located, and into which zygote it is ultimately delivered during sexual reproduction.

Genetic analysis provides a means for investigating the inheritance of specific DNA sequences.

Possibly, if we utilized highly complicated chemical techniques, we could follow the movement of particular base sequences through mitosis and meiosis and, upon fertilization, into newly created zygotes. Perhaps we could monitor the transmission of a particular base sequence, even in a unineme chromosome, by using some form of technologically advanced *in situ* hybridization and by making substantial investments of time and, no doubt, money as well. We can do such an analysis, however, if some specific DNA sequences control specific traits, that is, are genes, and if we study the inheritance of variant forms of those traits. All people do not have brown eyes, nor Roman noses. All mice do not have greyish brown hair nor all flies straight wings, and not all yeasts can ferment sugar to produce alcohol.

152

Many such variant attributes, differing from those normally present, are inherited, offspring having the same characteristics as their parents. When a variant attribute *is* inherited, we can assume that the difference between the variant and normal attribute is the result of a difference in the content of genetic information, in the base sequences of DNA. From analysis of the phenotypes of the offspring produced by mating or crossing variant-attributed organisms with normal ones, we can determine whether a particular phenotype is the result of a single difference in a DNA sequence or the result of several differences. How we are able to do this is the subject of this and the following four chapters. The great power of the genetic analyses that we will describe lies in their enabling us to draw conclusions about the transmission and even the function of DNA molecules simply by observing the phenotypes of offspring produced in particular crosses. Indeed, the study of the inheritance of a phenotype is in reality a study of the inheritance of a DNA sequence.

Genetically homogeneous, true-breeding strains must be established before making crosses between strains.

The basic techniques necessary for following the transmission of specific genes through meiosis from parent to offspring were established about one hundred and twenty years ago by the Augustinian monk, Gregor Mendel. Mendel appreciated, as we should, that if any conclusions are to be drawn from crossing phenotypically different organisms, it is first necessary to establish a genetic basis for the difference. Not all differences in phenotypes are the result of differences in genotypes. Certain types of baldness in men are inherited. But baldness can also result from shaving heads. Mendel realized that before studying the types of offspring produced by crossing phenotypically different strains, it is necessary to begin with true-breeding individuals. For example, peas with tall plants constitute a true-breeding strain if, when crossed with each other, they produce only tall offspring generation after generation. True-breeding strains of short plants can also be established. Because the plants breed true, we can deduce that the phenotypic condition is inherited and that the plants are genetically homogeneous, that is, the tall strain, for example, contains only genetic information for the production of more tall plants and lacks the information for the production of short ones.

By crossing strains differing for a single phenotypic trait, Mendel deduced that the traits resulted from single genetic factors.

Mendel made a series of crosses using the garden pea, *Pisum sativum,* between pairs of strains that differed by only a single inherited characteristic or trait, such as between strains of tall versus short plants, round versus wrinkled seeds, or yellow versus green cotyledons. To make the crosses he removed the anthers, the male flower parts, from plants of one strain (so that the peas could not self-fertilize) and transferred pollen from a second strain to the pistils, the female flower parts, of the first strain (Figure 6-1).

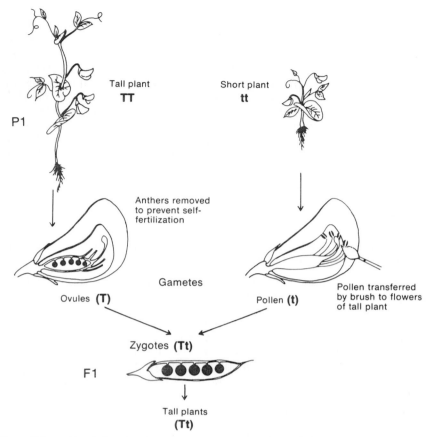

Figure 6-1   Mendel's cross between tall and short pea plants to produce F_1 offspring that are all tall.

These parent plants, abbreviated P_1, produced progeny, called the first filial generation and abbreviated F_1, all of which resembled only one of the two parental strains. For example, in the cross of tall with short plants, all of the progeny were tall. The trait that is expressed in the F_1 is called the *dominant* trait, and the one that is not expressed is the *recessive* trait. Mendel then crossed the F_1 plants with themselves by simply allowing the F_1 plants to self-fertilize. He recovered the next generation of progeny, the second filial or F_2 generation, and recorded the phenotype of each plant. He found many plants with the dominant trait and others with the recessive trait. Moreover, he recognized that the numbers of progeny present closely approximated 75 per cent dominant ones and 25 per cent recessive ones, a 3:1 ratio of dominant to recessive plants, in crosses involving each pair of traits (Table 6-1).

Table 6-1   Mendel's F_2 Results for Crosses Involving One Pair of Traits

Characteristic	Dominant		Recessive		Total
	Number	Per Cent	Number	Per Cent	
Form of Seed	5,474	(74.7)	1,850	(25.3)	7,324
Color of Cotyledons	6,022	(75.1)	2,001	(24.9)	8,023
Color of Seed Coats	705	(75.9)	224	(24.1)	929
Form of Pod	882	(74.7)	299	(25.3)	1,181
Color of Pod	428	(73.8)	152	(26.2)	580
Position of Flowers	651	(75.9)	207	(24.1)	858
Length of Stem	787	(74.0)	277	(26.0)	1,064
Total	14,949	(74.9)	5,010	(25.1)	19,959

These observations were particularly important for they allowed Mendel to deduce basic characteristics of gene transmission in eukaryotes. Because each trait is inherited, Mendel deduced that hereditary factors exist that cause each of the two phenotypic traits. A factor for each parental trait had to be present in the F_1 plants even though these plants resembled only one parent, because the F_1 plants subsequently produced both types of plants. Furthermore, the factors are discrete entities whose inherent characteristics are preserved in the F_1 plants even in the presence of an alternative factor. Finally, for reasons we will explain shortly, Mendel concluded that each plant carried two such factors.

This conclusion stems from the following analysis: We assume in each cross that one parent plant has two dominant factors (for example, TT) and one has two recessive factors (for example, tt, (Figure 6-1). Each parent produces germ cells, each carrying one factor, that fuse to produce the F_1 offspring, all of which have one recessive and one dominant factor and have the phenotypic characteristics of the dominant trait. The germ cells produced by each F_1 plant carry either a dominant or a recessive factor but not both. Thus, the factors are *segregated* from each other (Figure 6-2). Each F_1 plant produces germ cells carrying each factor in equal numbers, half (0.5) of the germ cells having the dominant factor and the other half (0.5) having the recessive factor. The F_2 offspring are produced by the random fusion of these germ cells (Figure 6-2). We can calculate the frequency at which we expect F_2 progeny with the dominant or the recessive trait to be produced. The expected frequency of the joint occurrence of two independent events is the product of their separate probabilities. For the Mendelian factors, in the F_2 we expect plants with two dominant factors to be produced at a frequency of 25 per cent (0.5×0.5). Those with two recessive factors should also occur at a 25 per cent frequency (Figure 6-2). Offspring with one dominant and one recessive factor should be present at a frequency of 50 per cent, a male germ cell carrying a dominant factor fusing with a female one with a recessive factor at a frequency of 25 per cent and a female one with a dominant factor fusing with a male one with a recessive factor at a frequency of 50 per cent (Figure 6-2). Of the F_2 progeny 75 per cent will have at least one dominant factor and will express the dominant trait. Only 25 per cent of the progeny will have two recessive

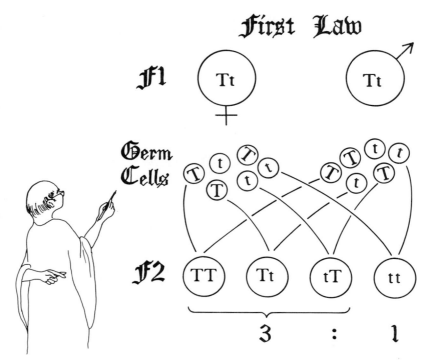

Figure 6-2   Segregation of Mendelian factors (alleles) in F_1 plants and the production of a 3:1 phenotypic ratio in the F_2

factors and express the recessive trait. Because Mendel's results closely approximated the frequencies expected assuming each plant carried two hereditary factors, Mendel deduced that each plant carried two such factors for each trait.

The pattern of inheritance of a single pair of traits results from the distribution of homologous chromosomes during meiosis.

We can understand the physical basis for Mendel's results, the Mendelian Law of Segregation, by taking a short cut through several decades of genetical research. The inheritance of the *albino* trait in mice and of its normal counterpart provides a suitable example. We assume that the difference in phenotype between the *albino* strain, lacking pigment, and the normal one, with pigment, results from a difference in a unique base sequence in the two types of mice. This unique sequence constitutes a gene and must be located at a specific site, a *locus*, on a particular chromosome. Each strain carries an alternative form of the gene, an *allele*. One form of the gene, the common or normal allele (also called the wild-type allele) is present in mice with typical brownish-grey coat color and brown eyes. Mice of the other strain with pigmentless fur and eyes have another form of the gene, the *albino* allele. The mice are diploid, with two of each chromosome. Each strain has two copies of one allele and breeds true. Furthermore, each copy of an allele must be located in the same position on each of two homologous chromosomes. By convention genes in higher eukaryotes are named after the mutant or rare allele. The genotype of the *albino* strain is abbreviated *alb/alb* and that of the normal strain, *alb⁺/alb⁺* or simply +/+ with the superscript "+" designating the common or wild-type allele of the gene. When a diploid organism carries two identical alleles, such as *alb/alb*, it is said to be a *homozygote* or to be homozygous for an allele.

Mice from one strain are mated with those from the other (Figure 6-3) (a *monohybrid* cross because differences in only one gene are involved). Each type of parent produces only one kind of germ cell, each one carrying either the *alb* allele or the *alb⁺* allele. The F_1 offspring can only have one possible genotype, *alb/alb⁺*, and must have two different forms of the same gene, two different alleles, and are said to be *heterozygous*. The phenotype of the F_1 offspring is not, however, certain, but depends on the expression of the traits as described later (pp. 161–163). In this case the expression of the normal allele is dominant, and the F_1 heterozygous mice all have normal coat and eye color.

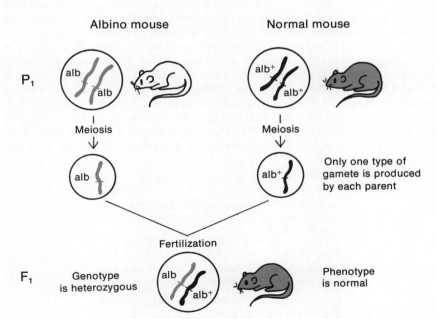

Figure 6-3   A cross between normal and albino mice. The chromosomal location of genes and the processes of meiosis and fertilization produce F_1 heterozygotes.

F₁ individuals, heterozygous for a single pair of alleles, should produce two types of germ cells in equal numbers.

During meiosis in the F_1 offspring, the homologous chromosomes carrying the two different alleles synapse, disjoin, and proceed through meiosis. Germ cells are produced, each of which carries one copy of each homologous chromosome and one copy of one of the two alleles, either *alb* or *alb⁺* (Figure 6-4). If meiosis operates in a normal manner, each germ cell receives one and only one allele. According to Mendel's terminology, the alleles have segregated. In male mice, each meiosis leads to the formation of four sperm, two (0.5) of which carry the *alb* allele and two (0.5) the *alb⁺* allele. In female mice, only one of the four haploid nuclei produced as a result of meiosis will end up in an ovum, as three nuclei will be included in polar bodies (Figure 5-19). By chance, half of the ova produced will receive a homolog with the *alb* allele and half will receive a homolog with the *alb⁺* allele. Thus, as a result of meiosis in the F_1 mice, germ cells with chromosomes carrying the *alb* or the *alb⁺* allele are produced in equal numbers (Figure 6-4).

For monohybrid crosses the expected frequency of the three possible F_2 genotypes is determined by assuming the two types of F_1 germ cells randomly fuse to produce the F_2 progeny.

We can predict the results of a cross between F_1 *alb*/*alb⁺* heterozygotes. Each F_2 offspring will have one of three possible genotypes: heterozygous (*alb*/*alb⁺*), dominant homozygous (*alb⁺*/*alb⁺*), or recessive homozygous (*alb*/*alb*). We assume that the fusion of any two germ cells to produce a zygote is a random process. The frequency with which a particular genotype is produced in the F_2 generation is the probability of any two germ cells fusing at random and is determined conveniently using a device called a Punnett Square (Table 6-2) that was introduced by an early twentieth century geneticist, Reginald Punnett. We know that half of the sperm and ova produced carry the *alb* allele and that half carry the *alb⁺* allele. From the Punnett Square we see that the probability of forming a zygote that is *alb*/*alb* is 0.25. Half of the time a zygote will receive an *alb* allele from its mother, and half

Table 6-2   Frequencies of Expected F_2 Genotypes for Crosses Involving One Pair of Alleles

F_1 gametes	0.5 alb⁺	0.5 alb
0.5 alb⁺	alb⁺/alb⁺	alb⁺/alb
0.5 alb	alb/alb⁺	**alb/alb**

Each gamete combination in the F_2 is produced in a frequency of 0.25 (0.5 × 0.5).

The genotype in which the recessive phenotype is expressed is shown in bold type.

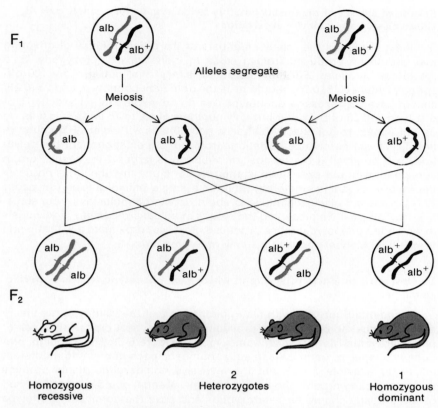

Figure 6-4   The chromosomal basis of segregation. The disjunction of homologous chromosomes carrying alternative alleles and the random fusion of gametes produce the Mendelian F_2 genotypic proportions.

of these, 0.25 (0.5 × 0.5) will also receive an *alb* allele from its father. The probability of forming an *alb⁺*/*alb⁺* zygote is also 0.25. The probability of forming a heterozygous zygote is 0.5. The paternal germ cell could be *alb* and the maternal one *alb⁺*, or the maternal germ cell could be *alb⁺* and the paternal one *alb*. The expected genotypic frequencies for *alb*/*alb*:*alb*/*alb⁺*: *alb⁺*/*alb⁺* are 0.25:0.5:0.25 or 1:2:1. We can also determine the expected phenotypic frequencies. The homozygotes for the dominant allele and the heterozygotes will be phenotypically indistinguishable and different from the homozygotes for the recessive allele. We do not, however, know which of the offspring with the dominant trait are homozygous and which are heterozygous. Nevertheless, 0.75 of the offspring will have the dominant *alb⁺* phenotype and 0.25 the recessive *alb* phenotype, a ratio of 3:1 (Figure 6-4). That, of course, is the ratio obtained by Mendel in his crosses.

Results of crosses of organisms possibly heterozygous for a single pair of alleles closely approximate expectations.

We can predict the relative numbers of the different kinds of offspring that should be produced from a cross of individuals heterozygous for a single pair of alleles. For dominance, we expect that out of every 100 offspring produced 75 (0.75) should have the dominant phenotype and 25 (0.25) should have the recessive phenotype. We do not expect to get exactly 75 of one type and 25 of the other but only numbers close to 75 and 25, just as we do not expect to get exactly 50 heads and 50 tails with every 100 coins we flip. There will be chance deviations from our expectations, but the deviations will be small if the number of offspring is large. If the results are in agreement with our expectations, then we conclude that the cross *probably* did involve organisms heterozygous for a single pair of alleles (on pages 171–175 we will discuss limitations about drawing conclusions from statistics). In Table 6-1 we presented the results of actual crosses of F_2 individuals in which the phenotypic ratios produced closely approximate a 3:1 ratio and from which Mendel deduced his Law of Segregation.

The test cross produces offspring in which each phenotype reflects a specific genotype.

Mendel did not stop with his F_2 plants. To test the validity of his theory of inheritance, he developed a procedure called a "test cross," in which F_1 plants are mated to recessive homozygotes (Figure 6-5). The F_1 plants produce two types of germ cells in equal numbers, each containing a different allele, for example T or t, but the recessive homozygous plants produce germ cells carrying only the recessive allele. Mendel could predict that half of the progeny should be heterozygous and have the dominant phenotype and the other half should be homozygous and have the recessive phenotype (Figure 6-5). His results were in good agreement with his expectations and confirmed his conclusions about the inheritance of single traits. The test cross has an advantage over the cross of F_1 organisms with themselves. In

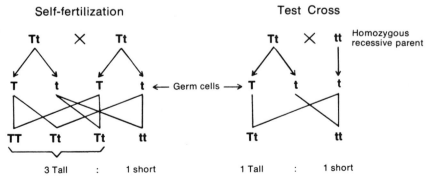

Figure 6-5   Comparison of progeny ratios from self-fertilization and test crosses.

the F_2 the genotype of an organism with a dominant phenotype is uncertain. It can be either homozygous or heterozygous. In test cross progeny each phenotype is associated with a specific genotype, a phenotypically dominant organism being heterozygous, and a phenotypically recessive one being homozygous. Thus, there is no uncertainty about the genotype of a particular organism.

In the light of present knowledge about meiosis and on the basis of our assumptions, Mendel's results are easily understood. The Mendelian factors are carried on chromosomes. Because in diploid organisms each chromosome is present in two copies, there are two copies of each of Mendel's factors. The pattern of inheritance of the factors stems from the distribution of chromosomes during meiosis. It is important to remember that Mendel, however, did not have the help of any knowledge of DNA, its distribution in chromosomes, or the behavior of chromosomes during meiosis. On the basis of frequencies of particular phenotypes resulting from carefully controlled crosses alone, he was able to deduce basic laws of inheritance. Mendel's work serves vividly to demonstrate the power of genetic analysis in elucidating cellular mechanisms, in this case, the mechanisms of gene transmission. At a time when genetic research is becoming the province of those who do DNA sequence determination, it is good to pause and remember that there are still many problems in biology, such as the genetic basis of behavior, that are amenable to the techniques of genetic analysis but are not elucidated by determining DNA sequences. There is more to genetics—and to life—than the order of nucleotides in a chromosome.

Dominance relationships between pairs of alleles range from complete dominance to co-dominance.

In addition to the situation in which the expression of one allele is completely dominant over that of another, we find others in which the phenotype in the F_1 is intermediate between that of the two parental traits and still others in which both traits are expressed. In all of these cases the organism is genetically heterozygous for one pair of alleles. Therefore, we realize that these dominance relationships result from different phenotypic expressions of the alleles. Furthermore, as we will now discuss, whether the expression of one allele is completely dominant, intermediate, or equal to that of the other depends to some degree on the manner in which the phenotype is characterized.

1. **The heterozygous offspring are phenotypically identical to one of the parents (dominance, recessiveness).** When *albino* mice are crossed with wild-type mice, the F_1 offspring are all phenotypically normal and look like the wild-type parents (Figure 6-3). Yet we know that the F_1 offspring must be heterozygous, having one *alb* allele and one *alb+* allele. The presence of the single *alb+* allele is sufficient in this case to produce a wild-type phenotype. When we evaluate phenotypes by the general appearance of the organisms, we often observe conditions of dominance and recessiveness for such traits as the presence or absence of pigmentation in mammals, or the shape of wings or eyes in *Drosophila*. In humans, inherited diseases such as hemo-

philia (loss of blood-clotting ability), Tay-Sachs (a fatal disease causing mental and motor retardation and, typically, death before the age of five), and the Lesch-Nyhan Syndrome (a fatal disease producing mental retardation and self-destructive behavior) are all recessive traits inasmuch as heterozygotes are not afflicted with the disease.

2. **The heterozygous offspring are intermediate in phenotype between the parents (incomplete dominance or semidominance).** The phenotype of the F_1 offspring may be between those of the parents. For example, in a cross of red snapdragons with white ones, the F_1 heterozygotes are all pink. In the F_2 generation the homozygous red, heterozygous pink, and homozygous white snapdragons are produced in proportions of 1:2:1 reflecting the 1:2:1 genotypic ratios. Because neither allele is dominant in a heterozygote, this condition is characterized as having incomplete dominance. When we evaluate phenotypes only by the general appearance of the organism, we find that examples of incomplete dominance are rare compared to complete dominance. But if we extend our analysis to the biochemical level, we find incomplete dominance is common. For example, many abnormal phenotypes are associated with the loss of an enzyme found in normal individuals. Normally, each gene copy in a diploid cell is responsible for the production of half of the amount of an enzyme. Enzymes are catalysts, and often more enzyme is present than is needed and half the amount of enzyme is usually sufficient for normal physical appearance. In humans individuals heterozygous for deleterious traits usually appear physically normal, but because they have only one normal allele they may have only half the normal amount of a specific enzyme. The inherited recessive human disease we just mentioned, Tay-Sachs disease, is characterized by the absence of hexosaminidase A, an enzyme that cleaves amino sugars from components to which the sugars are attached. People who are heterozygous, having one normal allele and one mutant allele, and are *carriers* of the disease allele, do not suffer from Tay-Sachs disease, but, on the average, have only half as much hexosaminidase A as people who are homozygous for the + allele. By the criterion of manifestation of a disease, the Tay-Sachs allele is recessive. But by the criterion of the quantity of enzyme present, the heterozygotes are intermediate between the respective homozygotes in the same sense that pink snapdragons are intermediate between red and white ones. Thus, although heterozygotes may appear to be identical to homozygotes, they differ biochemically.

3. **The heterozygous offspring have characteristics of both parents (co-dominance).** Antigens are inherited in a co-dominant manner. Antigens are substances, often proteins to which sugars are attached, that have the ability to elicit the formation of antibodies when injected into a vertebrate. Antibodies, or immunoglobulins, are proteins found in the serum that recognize a specific antigen, bind to it, and can cause its precipitation. Normally antibodies are formed against foreign substances, such as viruses and bacteria that cause diseases, as part of the defense mechanisms against infection. In humans, because of the practices of blood transfusions, organ transplants, and skin grafts, antibodies are also produced against antigens found in one individual but absent in another. Differences between individuals in

the proteins carried on the surfaces of all types of cells are particularly important in determining the acceptability of a skin or organ transplant or blood transfusion. The cell surface proteins are antigenic and, for transfusion or transplantation to be successful, recipients must have the same or nearly the same cell surface antigens as the donors. If the antigens differ, then antibodies will be made against transplanted tissue and will lead to its destruction, or in the case of transfusions, antibodies will clump the transfused red blood cells and interfere with circulation. The cell surface antigens of red blood cells are very well characterized. We have identified about 21 human blood group antigen systems, such as the ABO, the MN, and the rh blood groups.

People with blood group type A have the A antigen, people with B, the B antigen, and those with type O have neither A nor B antigens. Type A individuals can be homozygous for the A allele, or be heterozygous, carrying A and O alleles. Type B individuals can be either homozygous B, or heterozygous B/O. An O individual must be homozygous for the O allele, while an AB individual is an A/B heterozygote (Table 6-3). Parents, one of whom is homozygous for the A allele and the other for the B allele will have children who are exclusively AB. Because the children will have both A and B antigens, each allele is expressed phenotypically and the expression is characterized as being co-dominant. AB individuals are called "universal acceptors" and can receive transfusions from A individuals, B individuals, or O individuals; none of the donor cells will have antigens foreign to the recipient. In contrast, individuals with blood group O can receive transfusions of blood cells only from type O individuals. People with the type O blood group are, however, "universal donors," because their blood, lacking A and B antigens, can be used for transfusion of A, B, AB, and O individuals.

In matings between homozygous strains differing for two pairs of alleles the F_1 progeny must be heterozygous for both pairs of alleles.

We now turn our attention to the inheritance of two pairs of alleles carried on different, that is, non-homologous, chromosomes. Mendel anticipated us by making crosses between strains that differed for two traits, for

Table 6-3   The ABO Red Cell Antigens: Genetics and Transfusion Compatibilities

Donors		Recipient Phenotypes			
Genotype	Phenotype	A	B	AB	O
AA or AO	A	+	Clumps	+	Clumps
BB or BO	B	Clumps	+	+	Clumps
AB	AB	Clumps	Clumps	+	Clumps
OO	O	+	+	+	+

+ indicates that transfusion is acceptable.

"Clumps" indicates that donor cells are clumped (agglutinated) by antibody made against red cell antigen not present in recipient.

example, tall plants with round peas crossed with short plants with wrinkled peas. We can predict the kinds of numerical results Mendel obtained, however, by determining the ratios at which we expect progeny expressing the traits to be produced. Once again, we start with genetically true-breeding strains. *Drosophila melanogaster,* the fruit or vinegar fly, is extensively used in genetical research because of its small size and ease of culture in the laboratory and because its short generation time (the time required to go from one generation of mature adults to the next) of about 10 days at 25° C permits the rapid recovery of several generations of offspring. We can establish a stock of flies in which the wings are *vestigial* (*vg*) and the eyes have a normal shape (*ey$^+$*) and another stock in which the wings have normal shape (*vg$^+$*) but the flies are *eyeless* (*ey*) or have eyes greatly reduced in size. Insects have compound eyes composed of many clusters of cells. In *ey* flies all or most of the cell clusters are missing (Figure 6-6). Both stocks of flies breed true and are double homozygotes: *vg/vg*; *ey$^+$/ey$^+$* and *vg$^+$/vg$^+$*; *ey/ey*. When flies from the two strains are crossed with each other (a *dihybrid* cross because two genes are involved), the offspring must be heterozygous for both pairs of alleles, *vg/vg$^+$* and *ey/ey$^+$* (Figure 6-7). The F_1 offspring all appear normal because the expression of both *vg* and *ey* is recessive to the + alleles.

F_1 double heterozygotes should produce four types of germ cells in equal numbers.

We have specifically chosen an example in which each pair of alleles is carried on a different chromosome. In order to establish the types of germ cells that the F_1 flies will produce, we need to consider the movement of two pairs of chromosomes through meiosis. During the first meiotic division each pair of homologs synapse and line up at the equatorial plate (Figure 6-7). Then the homologous chromosomes and the alleles carried on them separate. In one case we see that the chromosome carrying the *vg* allele moves to the left and the chromosome carrying the *vg$^+$* allele moves to the right. In the other case the chromosomes carrying the two alleles move in reverse directions. The homologs carrying the *ey* and the *ey$^+$* alleles also move to opposite poles. Four possible type of germ cells are produced: *vg, ey*; *vg, ey$^+$*; *vg$^+$, ey*; *vg$^+$, ey$^+$*. We assume that each pair of homologs is oriented during the first meiotic division independently of the other pair, that is, there is an equal likelihood of the paternal homolog of one pair moving to one pole with either the maternal or paternal homolog of the other pair. Thus, the four types of germ cells should be produced in equal frequencies in proportions of 1:1:1:1 (Figure 6-7).

For dihybrid crosses the expected frequencies of the nine F_2 genotypes are determined by assuming the four types of F_1 germ cells fuse randomly to produce the F_2 progeny.

The four types of germ cells produced by F_1 flies can unite in 16 possible combinations, as we can see by using a Punnett Square (Table 6-4). Assuming

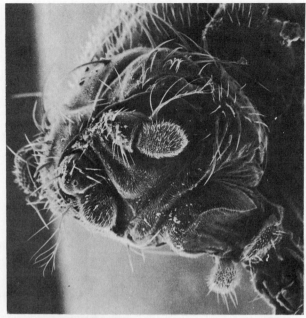

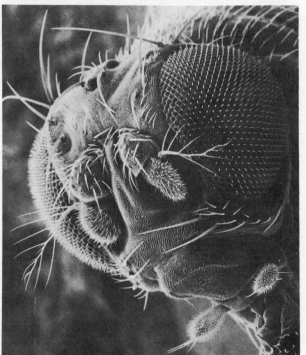

Figure 6-6   Scanning electron micrographs of normal and mutant adult *Drosophila melanogaster* with (A) normal compound eyes and (B) without eyes. Courtesy of Larry Salkoff, Department of Biology, Yale University, New Haven, Connecticut.

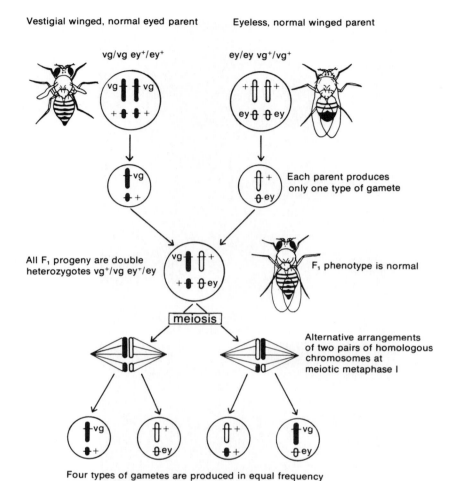

Vestigial winged, normal eyed parent   Eyeless, normal winged parent

vg/vg ey⁺/ey⁺   ey/ey vg⁺/vg⁺

Each parent produces
only one type of gamete

All F₁ progeny are double
heterozygotes vg⁺/vg ey⁺/ey   F₁ phenotype is normal

meiosis

Alternative arrangements
of two pairs of homologous
chromosomes at
meiotic metaphase I

Four types of gametes are produced in equal frequency

Figure 6-7   The chromosomal basis of independent assortment of two pairs of alleles is demonstrated by mating F_1 progeny.

that the union of germ cells is random, the resulting nine genotypes are produced in the following proportions: 1:2:1:2:4:2:1:2:1, which is the polynomial expansion of the F_2 genotypic ratios of a monohybrid cross (1:2:1 × 1:2:1). Offspring homozygous for two pairs of alleles are each represented once in the Punnett Square. Those homozygous for one pair of alleles and heterozygous for the other pair are represented twice. The largest single class, represented four times, is composed of offspring heterozygous for both pairs of alleles. The phenotypic proportions are 9:3:3:1, which is an expansion of the 3:1 phenotypic ratio of the monohybrid cross (3:1 × 3:1). The offspring expressing both dominant traits are most frequent ($\frac{9}{16}$), those expressing one dominant and one recessive trait are less frequent ($\frac{3}{16}$ each) and those with both recessive traits are least frequent ($\frac{1}{16}$). The phenotypic frequencies in the F_2 change markedly for dihybrid crosses if one or both of the traits has incomplete dominance or is co-dominant. If one trait is incompletely dominant and the other is dominant, then the phenotypic proportions in the F_2 are 3:6:3:1:2:1 (1:2:1 × 3:1) and if both of the traits are incompletely dominant, then the phenotypic ratios are the same as the genotypic ones (1:2:1:2:4:2:1:2:1).

The test cross simplifies the analysis of the meiotic products produced by F_1 organisms heterozygous for two pairs of alleles.

As we noted before, we cannot tell from the phenotype of an F_2 organism whether it is heterozygous or homozygous for the dominant allele. This uncertainty is eliminated in a test cross by mating F_1 heterozygotes to organisms homozygous for the recessive allele. The phenotypes of the progeny then reflect directly the genotypes of the germ cells produced by the F_1 generation. For dihybrid crosses, the complex genotypic proportions produced by mating F_1 double heterozygotes with themselves are greatly simplified by mating F_1 heterozygotes with organisms homozygous for both recessive alleles in a test cross (Figure 6-8). The F_1 organisms should produce four kinds of germ cells in equal frequency. The doubly homozygous parent produces only one type of germ cell carrying both recessive alleles, for example, *vg* and *ey*. Thus, the offspring produced have only four possible genotypes (Figure 6-8) instead of 16, each of which is reflected in a single phenotype. Also with incomplete dominance or co-dominance, only four phenotypic classes are produced.

Results from crosses involving organisms possibly heterozygous for two pairs of alleles can closely approximate expectations.

We have specific expectations for the results of crossing F_1 individuals that are possibly heterozygous for two pairs of alleles carried on non-homologous chromosomes. We find in many crosses that the results obtained are in accord with our expectations (Table 6-5), namely in $F_1 \times F_1$ crosses the classes of progeny are produced in approximately 9:3:3:1 phenotypic ratios and in test crosses in approximately 1:1:1:1 proportions. From such results, we can conclude that the two phenotypic traits are probably controlled by two genes carried on non-homologous chromosomes.

In the crosses we have just considered, we have duplicated some of Mendel's experiments using both $F_1 \times F_1$ crosses and test crosses. Mendel analyzed the numerical results of his dihybrid crosses and deduced that factors for one pair of traits travelled from parent to offspring independently of the factors for the second pair of traits and, therefore, that different pairs of factors were assorted independently of each other during inheritance. We now realize that the basis for the "law" of *independent assortment* depends on the location of two pairs of alleles (or factors) on two pairs of non-homologous chromosomes and results from the independent orientation of each pair of chromosomes on the equatorial plate during meiosis.

Table 6-4   Frequencies of Expected F_2 Genotypes for Crosses Involving Two Pairs of Independently Assorting Alleles

	Sperm			
F_1 Gametes	vg$^+$ ey$^+$ 0.25	vg$^+$ ey 0.25	vg ey$^+$ 0.25	vg ey 0.25
vg$^+$ ey$^+$ 0.25	vg$^+$/vg$^+$ ey$^+$/ey$^+$	vg$^+$/vg$^+$ ey$^+$/ey	vg$^+$/vg ey$^+$/ey$^+$	vg$^+$/vg ey$^+$/ey
vg$^+$ ey 0.25	vg$^+$/vg$^+$ ey$^+$/ey	vg$^+$/vg$^+$ **ey/ey**	vg$^+$/vg ey$^+$/ey	vg$^+$/vg **ey/ey**
vg ey$^+$ 0.25	vg$^+$/vg ey$^+$/ey$^+$	vg$^+$/vg ey$^+$/ey	**vg/vg** ey$^+$/ey	**vg/vg** ey$^+$/ey
vg ey 0.25	vg$^+$/vg ey$^+$/ey	vg$^+$/vg **ey/ey**	**vg/vg** ey$^+$/ey	**vg/vg** **ey/ey**

Eggs (label spanning the left side of the rows)

Each gamete combination is produced in a frequency of 0.0625 (0.25 × 0.25). Genotypes in which a recessive phenotype is expressed are shown in bold type.

We can predict the frequency of recovery of genotypic and phenotypic classes for any number of independently assorting pairs of alleles.

We have seen for two specific cases, involving one pair of alleles and two pairs of alleles carried on non-homologous chromosomes, that the frequencies at which the genotypic and phenotypic classes are produced result from the movement of chromosomes during meiosis. Because we understand the mechanics of meiosis, we can generalize about the transmission of any number of allele pairs. Given that maternal and paternal homologs of all pairs of chromosomes orient independently of each other during meiosis, it follows that any number of allele pairs will be transmitted independently of each other as long as each pair resides on a different pair of homologous chromosomes. Independent assortment may occur for trihybrid crosses involving three pairs of alleles, for pentahybrid crosses involving 5 pairs of alleles, and in humans, where there are 23 pairs of chromosomes, independent assortment may occur for 23 pairs of alleles. The possible combinations of genotypes and phenotypes expected from these higher order crosses are summarized in Table 6-6 for both $F_1 \times F_1$ matings and test crosses. The ratios are determined in each case by an expansion of the expected genotypic and phenotypic frequencies as a function of the number, n, of pairs of alleles involved. The general solution for genotypic ratios in $F_1 \times F_1$ crosses is $(1:2:1)^n$ and for phenotypic ratios $(3:1)^n$ where there is complete dominance. The ratios of genotypic classes produced in test crosses are much less complex, being $(1:1)^n$. The number of phenotypic classes in test crosses is the same as in $F_1 \times F_1$ crosses, but the proportions of the phenotypic classes, 1:1:1:1: . . . :1, are far less cumbersome. The value of the test cross in simplifying the patterns of inheritance cannot be overly emphasized. For example, in an $F_1 \times F_1$ cross involving heterozygosity for three allele pairs, 64 possible genotypic combinations result in the F_2 in

Table 6-5   Summary of F_2 and Test Cross Results for Crosses
Involving Two Different Traits

Experiment	F_2 Progeny				Test Cross Progeny			
	AB	Ab	aB	ab	AB	Ab	aB	ab
1.	315	101	108	32	31	27	26	26
2.	138	65	60	38	24	25	22	27
Total	453	166	168	70	55	52	48	53

Data are from Mendel's experiments. Uppercase denotes dominant phenotype; lowercase, recessive phenotype.

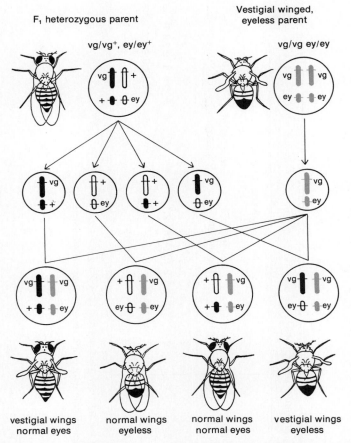

Figure 6-8   The chromosomal basis for independent assortment of two pairs of alleles
is demonstrated in a test cross.

Table 6-6   The Relationship Between the Number of Allele Pairs in a Cross and the Genotypic and Phenotypic Classes in F_2 and in a Test Cross

Number of Allele Pairs	F_2 Expectations				Test Cross Expectations	
	Number of Genotypes	Genotypic Proportions	Number of Phenotypes	Phenotypic Proportions	Number of Genotypes and Phenotypes	Genotypic and Phenotypic Proportions
1	3	1:2:1	2	3:1	2	1:1
2	9	1:2:1:2:4:2:1:2:1	4	9:3:3:1	4	1:1:1:1
3	27	*	8	27:9:9:9:3:3:3:1	8	1:1:1:1:1:1:1:1
n	3^n	$(1:2:1)^n$	2^n	$(3:1)^n$	2^n	$(1:1)^n$

* 1:2:1:2:4:2:1:2:1:2:4:2:4:8:4:2:4:2:1:2:1:2:4:2:1:2:1.
Complete dominance and recessiveness is assumed.

variable frequencies (Table 6-6), while in a test cross only 8 possible combinations exist with expected proportions of 1:1:1:1:1:1:1:1.

Probability is an intrinsic element of genetics.

We realize from the preceding discussion of inheritance of different pairs of alleles that genetic crosses involve a distinct element of random chance. While meiosis is a precise mechanism, the determination of which allele actually gets into the ovum instead of a polar body, and the allelic nature of the sperm which actually fertilizes the egg are random processes. The frequencies that we calculate with Punnett Squares are the expected occurrence of the different genotypes in an infinite number of offspring, assuming segregation and independent assortment. When there are finite numbers of offspring, as in the data of the previous examples, the actual number of individuals with a particular genotype is generally close to but not exactly the same as the expected number. The decision we must make is whether differences between the numbers we expect and the numbers we actually observe can be attributed purely to chance or whether the differences are so large that we should conclude that our expectations are wrong.

To make such a decision we need some feeling for the degree to which the results of a random process are likely to deviate from their expectation. Are there limits by which chance deviations can be reasonably expected to be bound? Consider flipping a coin ten times. We expect to get five heads and five tails, but would not be surprised by 3, 4, 5, 6, or 7 heads. It is possible that we could get ten heads and no tails—the probability is $(\frac{1}{2})^{10} = \frac{1}{1024}$. The probability is so small, however, that we do not reasonably expect such an outcome, and if it were to occur we would probably check to see whether we had been using a two-headed coin. Now suppose we flip a coin one hundred times. We would not be surprised by a result between 40 and 60 heads. Thirty heads out of one hundred flips (30 per cent heads) would seem somewhat far from the expected number of fifty. In our first example, however, three heads out of ten flips (also 30 per cent) seemed reasonable. These two examples should convince us that there are limits, caused by probability, on chance deviations and that the limits depend on the size of the sample. In a large sample we expect the results to be proportionately closer to the expectation than they are in a small sample.

The χ^2 Goodness of Fit Test is a statistical method for comparing results with expectations.

Determination of what constitutes a reasonable deviation from expectation is a matter of probability theory. For a given set of results a precise question we can ask is: if our expectations are correct, what is the probability of getting results that deviate from their expected numbers by as much as, or more than, our results do? This question can be answered in several ways. A procedure that is often used to evaluate results is known as the chi-square (χ^2) test. It is based on the squares of the deviations from the expected numbers for each class of offspring. Application of the test is easily described in tabular form with a particular example. Table 6-7 contains data obtained by Carl Correns in 1900 when, during his rediscovery of Mendel's work, he repeated Mendel's cross between strains of garden peas with yellow and green seeds.

The column labelled *Obs.* gives the number of seeds produced or *observed* in each class. The column labelled *Exp.* gives the *expected* number of seeds in each class assuming a 3:1 ratio in a total of 1,847 seeds. Deviations between observed and expected numbers are given for each class in the column labelled *Dev.* In the last column the number, $(Dev.)^2/Exp.$ is calculated for each class. The sum for all classes, of this last column, in this example two, gives a number denoted by χ^2 and called the chi-square value. For this example χ^2 is 0.221. Clearly, the larger the deviations are in comparison to the expected numbers, the larger the χ^2 value will be. Conversely, the closer the results fit the expectations, the smaller the χ^2 value will be. It is important to note that the computations are all done with actual or expected numbers and not with percentages or proportions. Also, as presented here, the test is not always reliable if the expected number of one or more classes is small, for example, less than 5.

It is possible to determine the probability of obtaining a χ^2 value that is equal to, or larger than, any number obtained. Table 6-8 gives the χ^2 values that are associated with various different probabilities. Before we can use the table, however, we must learn to take into account a quantity known as *degrees of freedom.* It will be easier to grasp the meaning of this important quantity if we first consider another example. In Table 6-9, results from Mendel's two-factor cross of yellow versus green plants and round versus wrinkled seeds are given.

In this case there are four classes of offspring in the F_2 generation. The expected numbers are based on a total of 556 offspring and the genetic hypothesis of 9:3:3:1 phenotypic ratios. For each class we calculate $(Dev.)^2/Exp.$ and then we add the four numbers to get a χ^2 value of 0.470. The χ^2 value in this example is twice as large as that in the previous example, but there are four classes of offspring that contribute to the χ^2 value, whereas there were only two classes in the previous example. The concept of degrees of freedom takes into account the difference in the numbers of classes between the two examples. As the name implies, degrees of freedom are related to the number of different quantities that can independently deviate from their expected numbers. The expected number in each class is based on the total number of progeny. The sum of deviations, as seen in Tables

Table 6-7 χ^2 Analysis of Results in F_2 of a Cross by Correns of Strains of Peas with Yellow and Green Seeds

Class	Obs.	Exp.	Dev.	$\dfrac{(Dev.)^2}{Exp.}$
Yellow seeds	1,394	1,385.25	8.75	0.055
Green seeds	453	461.75	−8.75	0.166
Total	1,847	1,847	0	0.221

Expected numbers are based on an assumed ratio of 3:1.

Table 6-8 χ^2 Values for 1 to 10 Degrees of Freedom That Are Associated With Various Probabilities

Degrees of Freedom	Probabilities									
	.95	.90	.70	.50	.30	.20	.10	.05	.01	.001
1	.004	.016	.15	.46	1.07	1.64	2.71	3.84	6.64	10.83
2	.10	.21	.71	1.39	2.41	3.22	4.61	5.99	9.21	13.82
3	.35	.58	1.42	2.37	3.67	4.64	6.25	7.82	11.35	16.27
4	.71	1.06	2.20	3.36	4.88	5.99	7.78	9.49	13.28	18.47
5	1.15	1.61	3.00	4.35	6.06	7.29	9.24	11.07	15.09	20.52
6	1.64	2.20	3.83	5.35	7.23	8.56	10.65	12.59	16.81	22.46
7	2.17	2.83	4.67	6.35	8.38	9.80	12.02	14.07	18.48	24.32
8	2.73	3.49	5.53	7.34	9.52	11.03	13.36	15.51	20.09	26.13
9	3.33	4.17	6.39	8.34	10.66	12.24	14.68	16.92	21.67	27.88
10	3.94	4.87	7.27	9.34	11.78	13.44	15.99	18.31	23.21	29.59

← accept | reject →

at .05 level

For each value in the table, the associated probability gives the likelihood of obtaining a χ^2 with the given degrees of freedom that is as large or larger than the value in the table.

Taken from Table IV of Fisher & Yates: *Statistical Tables for Biological, Agricultural and Medical Research,* published by Longman Group Ltd. London, (previously published by Oliver & Boyd Ltd. Edinburgh) and by permission of the authors and publishers.

6-7 and 6-9, always equals zero and implies that the deviations are not all independent of each other. Consequently the number of degrees of freedom must be less than the number of classes. As long as nothing in the *observed* data other than the total sample size is used to determine the expected numbers of offspring in each class, the degrees of freedom are one less than the number of classes. In the first example, there is one degree of freedom; in the second, three degrees of freedom.

Now, knowing the degrees of freedom for our two examples, we can use Table 6-8 to make probability statements about the χ^2 values. We will consider Correns's data first. Looking at the row for one degree of freedom, we see that our χ^2 value of 0.221 falls between 0.15 and 0.46. Looking next at the probabilities at the top of the table, we see that these numbers correspond respectively to probabilities of 0.7 and 0.5. This means that more than

Table 6-9 χ^2 Analysis of Mendel's F_2 Results for a Dihybrid Cross

Class	Obs.	Exp.	Dev.	$\dfrac{(Dev.)^2}{Exp.}$
Round and Yellow	315	312.75	2.25	0.016
Round and Green	108	104.25	3.75	0.135
Wrinkled and Yellow	101	104.25	−3.25	0.101
Wrinkled and Green	32	34.75	−2.75	0.218
Total	556	556	0	$\chi^2 = 0.470$

Expected numbers are based on assumed proportions of 9:3:3:1.

50 per cent of the time but less than 70 per cent of the time, we can expect deviations by chance alone that produce χ^2 values as large as, or larger than, the χ^2 value obtained by using Correns's data. Because larger deviations could occur over half of the time, we can take the data to be consistent with the expected 3:1 phenotypic ratio. Turning to the second example, which gives Mendel's data, there are three degrees of freedom. The χ^2 value of 0.470 falls between the table entries of 0.35 and 0.58 with corresponding probabilities of 0.95 and 0.90. In other words, Mendel's data fit the expectations well, very well in fact. In fewer than one experiment out of ten can we hope to get results that are this close to expectations.

Decisions that we make from χ^2 tests are based on probability.

We should note from these examples that using the χ^2 test does not tell us in any absolute way whether our data fit our expectations. Rather, we simply learn with what probability chance deviations will be as great as or greater than the deviations we obtained. When the deviations are so great that it is highly unlikely that they result from chance, we can reasonably conclude that our expectations about the mode of gene transmission were wrong. However, if it is reasonably probable that deviations from our expectations as great as those observed would have occurred by chance alone, our data are not inconsistent with our expectations, but our expectations are not necessarily correct. We can only say that the deviations *may* be merely the result of chance and the data, therefore, are *consistent* with our predictions; we cannot exclude the possibility that our expectations are wrong, but only by a small amount, or are correct, but for the wrong reason, and cannot, therefore, assert that we have *proved* that our expectations were, in fact, based on a correct hypothesis.

Consistency is not proof that our expectations are correct but inconsistency is strong evidence that they are wrong. Thus, we are really testing for inconsistency. Decisions made from statistical tests are normally based upon probability levels at which we accept that the data are inconsistent with expectations. The most commonly used criterion for inconsistency is the so-called 5 per cent level. If the deviations are so large that they would occur by chance less than one time in twenty, we have reasonable confidence that they are not caused by chance and, therefore, that our expectations are wrong. Applied to the χ^2 probability table (Table 6-8), this criterion dictates that when a χ^2 value is greater than the number in the column with probability 0.05, we reject our expectations as incorrect. In such cases we say that the data are significantly different from our expectations at the 0.05 level. When, as in our two examples, the χ^2 value is less than the number at the 0.05 level, we conclude that the results are consistent with our expectations.

Let us consider an additional example. In Table 6-10 data are given for the phenotypes of offspring produced in a test cross that involves two different genes affecting corn seeds and the χ^2 value is calculated. The expected numbers are based on 1:1:1:1 phenotypic ratios arising from two pairs of alleles that assort randomly during meiosis. Even without statistical

Table 6-10 χ^2 Analysis of a Dihybrid Test Cross in Maize

Class	Obs.	Exp.	Dev.	$\dfrac{(Dev.)^2}{Exp.}$
Colored and Waxy	2,542	1,677	865	446.2
Colorless and Non-waxy	2,710	1,677	1,033	636.3
Colored and Non-waxy	717	1,677	−960	549.5
Colorless and Waxy	739	1,677	−938	524.7
Total	6,708	6,708	0	$\chi^2 = 2,156.7$

Expected numbers are based on assumed proportions of 1:1:1:1.

analysis we realize that the results differ from our expectations and the χ^2 value is enormous. There are three degrees of freedom. From the χ^2 table (Table 6-8) we see that a χ^2 value as large as 16.27 will occur only 1 time in 1000 ($p = 0.001$). Our value is much, much larger. Our expectations are unequivocally wrong. Either our assistant botched the cross or we are not dealing with two genes that assort randomly in accordance with Mendel's second law. We have more genetics to do with this cross and we will return to it in Chapter 8.

Mendel's results were apparently too good to be true.

In Table 6-1 we presented a series of Gregor Mendel's crosses in which the results were in accord with a 3:1 phenotypic ratio expected for segregation of one pair of alleles. We have already conducted a χ^2 test on another of Mendel's crosses (Table 6-9) and found that the probability of as good or better results is less than one time in ten. The χ^2 values for the results presented in Table 6-10 are equally wonderful, all indicating that by chance alone we would expect the results of each experiment less than one time in ten. Mendel was either exceedingly lucky, inspired, or had extraterrestrial help. In an article published in 1936, entitled "Has Mendel's Work Been Rediscovered?," Ronald A. Fisher questioned the authenticity of Mendel's results. Fisher, a noted English statistician and one of the founders of population genetics, pointed out that for all of Mendel's experiments the probability of getting as great or greater deviations from expectations was 0.99993. That is to say, by chance alone, we would obtain results as good as, or better than, Mendel's less than one in 10,000 times. Fisher further noted that in some experiments described by Mendel the results did not, for probabilistic reasons, fit expectations. Fisher seems to suggest that Mendel, after obtaining good fits in his early experiments, adjusted the results from his later experiments to fit his expectations. In all likelihood some sort of adjustment took place. In the view of many contemporary geneticists, among them George Beadle, whom we will meet in Chapter 11, this adjustment probably did not involve willful changes from the actual results in the numbers of offspring reported. Rather, it seems likely that Mendel simply collected data until the numbers were in good accord with his expectations.

Such a practice is, of course, statistically invalid in as much as it will bias data towards expectations. Mendel, with or without biased data, correctly deduced basic mechanisms of inheritance.

The analysis of family pedigrees is used to determine the pattern of gene transmission in humans.

In the crosses we have analysed, the determination of the pattern of gene transmission has depended upon the creation of strains that breed true, mating of these strains under controlled conditions, and analysis of large numbers of offspring from these matings. Following these procedures we are able to draw conclusions about the genetic transmission of a particular phenotypic trait. Each step of these procedures, although readily carried out with some plants and animals, cannot, for what are, we hope, obvious reasons, be carried out in humans. Although the mechanisms of gene transmission are essentially the same in humans as in other animals, the analysis of inheritance in humans requires very different approaches. We will consider one approach, which uses human tissue culture cells, in Chapter 9. Another falls into place here.

The approach that we will consider at this juncture takes advantage of two conditions. First, although one cannot decree the type of genetic cross necessary for investigating the inheritance of a particular variant trait, the enormous size of the human population makes it likely that the cross has occurred and that the results are available for study. Because many variant traits are deleterious and come to the attention of physicians, it is possible to find records of the birth of children with variant traits in the files of hospitals and public health organizations. Second, although specific crosses are not intentionally made, we can often reconstruct the genetic history that resulted in the birth of a child with a variant condition. This endeavor is aided by conversations with affected individuals, their immediate families, or both (an approach useless with *Drosophila* and garden peas). The information we obtain from such sources deals primarily with the inheritance of traits that come to the attention of medical practitioners and authorities and are, therefore, likely to be deleterious.

In the analysis of human gene transmission, one often starts with the identification of an affected individual, the *propositus,* followed by the construction of a genetic history or pedigree of the family of the propositus. Figure 6-9 shows a hypothetical example of a pedigree for the inheritance of Tay-Sachs disease. By convention circles denote females and squares denote males. The children produced by a couple are indicated by a line drawn back to another line connecting the couple. Affected individuals are represented by solid squares or circles and the propositus is indicated by an arrow. Finally, each generation is displayed on a different level.

The characteristics of the pedigree that suggest Tay-Sachs disease is inherited as a recessive allele are the existence of two affected offspring in one family, the presence of the propositus in a second family, and the absence of affected individuals among the children of the grandparents (generation II) or among the grandparents themselves (generation I). Finding

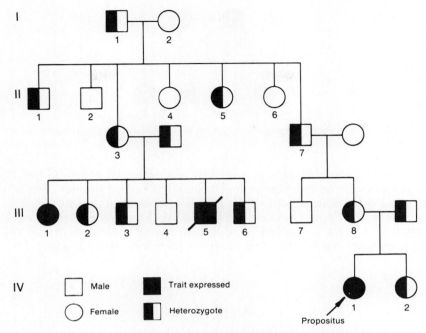

Figure 6-9   Hypothetical human pedigree for the inheritance of a recessive trait.

the disease in two related families eliminates the possibility that it arose as a rare mutation in the propositus or the other afflicted children. It is possible to identify carriers of the recessive allele for Tay-Sachs disease by measuring the level of hexosaminidase A (p. 162). The individuals in the pedigree who are demonstrably carriers of the disease are identified (by shading half of a square or circle) and confirm the allele's recessive nature. In other genetic diseases, the identification of heterozygous carriers may not be possible yet, and an unequivocal demonstration that a condition is inherited as a recessive may require the analysis of several independent pedigrees similar to the one shown in Figure 6-9. Figure 6-10 gives another pedigree to show the inheritance of a dominant allele causing dwarfism. In this pedigree, two individuals, both of whom are affected, produce both normal and affected children. Such a result could not occur if both parents were homozygous for a recessive trait or if one were homozygous for a dominant trait because all of the offspring should have the abnormal phenotype. Thus, we can conclude that both parents were heterozygous for a dominant allele.

The examples we have discussed demonstrate that patterns of inheritance of deleterious human genetic diseases can be ascertained by the analysis of family pedigrees. In addition to demonstrating that the patterns of inheritance in humans are the same as those found in other higher organisms, these analyses have medical value in allowing an informed physician to give knowledgeable advice to parents who have given birth to children

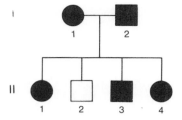

Figure 6-10   Hypothetical human pedigree for the inheritance of a dominant trait.

with known genetic defects. In particular, the physician is able to provide the parents with an estimate of the likelihood of having another child with the same deleterious condition. For example, when both parents are heterozygous for the Tay-Sachs allele, the physician will be able to tell the parents that there is a one in four likelihood of their having another child with Tay-Sachs disease, that is, a child homozygous for the recessive allele.

SUMMARY

In the beginning there was Mendel, thinking his lonely thoughts alone. And he said, "Let there be peas," and there were peas and it was good. And he put the peas in the garden saying unto them "Increase and multiply, segregate and assort yourselves independently," and they did and it was good. And now it came to pass that when Mendel gathered up his peas, he divided them into round and wrinkled, and called the round dominant and the wrinkled recessive, and it was good.

But now Mendel saw that there were four hundred and fifty round peas and a hundred and two wrinkled ones; this was not good. For the law stateth that there should be three round for every wrinkled. And Mendel said unto himself "Gott in Himmel, an enemy has done this, he has sown bad peas in my garden in the cover of night." And Mendel smote the table in righteous wrath, saying "Depart from me, you cursed and evil peas, into the outer darkness where thou shalt be devoured by the rats and the mice," and lo, it was done and there remained three hundred round peas and one hundred wrinkled peas, and it was good. It was very, very good. And Mendel published.

"Let There Be Peas," by Gregory G. Doyle, Department of Genetics, University of Missouri, Columbia, Missouri.

Genes Carried on a Single Chromosome: The Genetics of Lambda Phage

The independent movement of non-homologous chromosomes during meiosis, as revealed by the pattern of inheritance of pairs of alleles, provides a picture of the effects of meoisis. Sets of chromosomes of maternal and paternal origin carrying, in all likelihood, alternative forms of some genes are randomly assorted among germ cells. In organisms with large numbers of chromosomes, the germ cells resulting from meiosis will only rarely carry the chromosomes contributed by a single parent. We can calculate the likelihood that a germ cell will receive chromosomes from only one parent. The probability is 0.5^n, where n is the haploid number of chromosomes, that all of the chromosomes will originate from one parent, and 2×0.5^n that they will all be of maternal *or* paternal origin. In *Drosophila melanogaster* with a haploid set of just four chromosomes; on average, $\frac{1}{8}$ (0.125 or 2×0.5^4) of the germ cells are expected to have chromosomes all of maternal or paternal origin. In humans, with a haploid set of 23 chromosomes, the probability is $2 \times (\frac{1}{2})^{23}$, about one chance in five million. Thus, meoisis not only causes reduction in chromosome number from the diploid to the haploid state but also assures that the chromosome complement will be scrambled with the creation of new genetic constitutions consisting of groupings of alleles different from the groupings in the parents. Meiosis plays the role of the dealer in genetic blackjack and creates new combinations of alleles. Some of the new combinations may be particularly adaptive in a given environment, thereby providing new evolutionary opportunities, new lines of victors.

Recombination between homologous chromosomes is another means for varying genetic constitutions.

The likelihood of receiving a set of chromosomes all of paternal or of maternal origin is even lower than we calculated. We recall that during meiosis there is recombination between homologs, creating chromosomes partly of paternal and partly of maternal origin. In our hypothetical view of a chromosome, we proposed that each chromosome contains various genes distributed along its length. We expect that because of mutation different homologous chromosomes will have different sets of alleles. Recombination between homologous chromosomes will also create new allelic combinations that provide any species with new evolutionary opportunities and with increased flexibility to adapt to the selective pressures that drive evolution. We should not be surprised then to find that recombination in one form or another exists in most organisms from the simplest bacteriophages to the most complex eukaryotes. Although at this moment no one has demonstrated exactly how recombination occurs, enough information exists to formulate reasonable models of the mechanisms of recombination. All of these models descend from those Robin Holliday and Harold Whitehouse first proposed in the early 1960's. Figure 7-1 presents a synthetic model based in part on one formulated by George Radding and Mathew Meselson and in part on one formulated by David Dressler. Mechanisms of recombination may differ somewhat between prokaryotes and eukaryotes and our model is by no means absolute. It will serve, however, as a framework for our consideration of recombination from a genetic perspective and, we hope, will make it easier to understand the effects of recombination on gene transmission. Note that enzymes, some of which are involved in DNA replication, mediate recombination.

Recombination is initiated by the formation of duplexes between homologous regions of DNA.

In our discussion of meiosis we saw that synapsis brings two homologous chromosomes together in such a manner that the sequences of one chromosome are apparently exactly juxtaposed to the sequences of the other chromosome. In all models of recombination we assume that exact pairing of homologous regions occurs; like sequences on the two chromosomes are lined up in register, side by side. Recombination is initiated between the closely opposed homologs when an endonuclease makes a nick in one of the two strands of the double-helical DNA of one homolog by cleaving an ester bond and creating a free 3' end (Figure 7-1(1)). The endonuclease may cleave at specific regions or sequences. For now we will assume that the endonuclease attack can occur between any two bases with equal likelihood and that recombination can be initiated anywhere along the entire DNA molecule comprising a chromosome. The free 3' end acts as a primer and the unbroken strand as a template for the synthesis of DNA by DNA polymerase. As DNA synthesis proceeds, the nicked strand, headed by its phosphate-bearing 5' end is peeled away by the newly synthesized polynucleotide in a manner similar to what occurs during rolling circle replication (p. 123, Figure 5-3). This single-stranded molecule pairs with its complementary region on the homologous chromosome, forming a double helical region called a duplex and displacing the strand with which its complement was paired (Figure 7-1(2)). There are often minor differences in base sequences between homologous chromosomes. Possibly one base pair out of one hundred differs, for example, being an A:T pair on one chromosome and a G:C pair on the other. Thus, the two strands in the duplex are usually not perfectly complementary and the duplex is referred to as a *heteroduplex*.

Following heteroduplex formation covalently linked bridges between the two homologous chromosomes are established.

The strand of DNA that is displaced when the heteroduplex forms is now removed and ultimately degraded to mononucleotides, first through endonuclease, then exonuclease, attack. A free 3' end and a 5' end are created on either side of the heteroduplex. The polynucleotide from the lower chromosome in the heteroduplex now has its 5' end sealed by DNA ligase to the neighboring 3' end from the upper chromosome. A covalent bridge between the two homologs is established (Figure 7-1(3)). The heteroduplex lengthens. A single-stranded DNA molecule, headed by a 5' phosphate transfers from the upper chromosome and is covalently linked to the free 3' end available on the lower chromosome, again through the action of DNA ligase. As Figure 7-1(4) shows, two asymmetrical heteroduplexes form, one being longer than the other. There is variation in the sizes relative to each other of the two heteroduplexes. During recombination in yeast, for example, it appears that the second of the heteroduplexes formed is virtually non-existent. In *Ascobolus,* another fungus, it appears that the second heteroduplex formed is present but is shorter than the first. We have arbitrarily depicted this condition in Figure 7-1.

The covalently linked chromosomes undergo isomerization.

The two homologs held together by covalent bridges now undergo isomerization by the rotation of the bottom chromosome 180 degrees with respect to the top chromosome. The structure created, as shown in Figure 7-1(6), resembles a large X but no longer has two strands crossing each other. Such X-shaped structures, *chi-forms,* have been seen in electron micrographs of phage chromosomes undergoing recombination and provide evidence that the kinds of mechanisms we have described have biological reality (Figure 7-2).

The bridges are broken through endonuclease attack creating either two recombinant or non-recombinant molecules.

The joined homologs are now separated from each other by endonuclease attack on either pair of opposite strands (Figure 7-1(7)). Following the endonuclease cleavages, the chromosomes separate. Each chromosome contains a small gap in one strand of DNA, which is closed by the action of the ever-present ligase. Which pair of strands is cleaved determines whether the molecules are recombinant or not. In Figure 7-1 the ends of the original chromosomes have been marked, one *A, B* and the other *a, b*. For recombination to occur, the resulting chromosomes should be *A, b* and *a, B*. If the bottom and the top pair of strands are cleaved (Figure 7-1), two recombinant chromosomes are created. If the two strands on the sides are cleaved, the chromosomes are not recombinant. The ends of the chromosomes are still the same as they were originally. However, in the short regions where het-

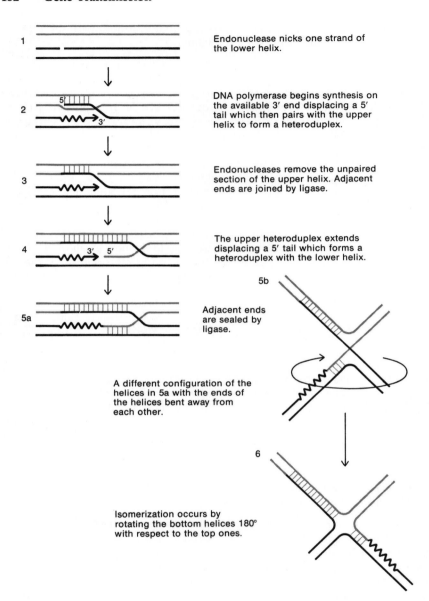

Figure 7-1   A postulated mechanism of recombination between homologous chromosomes. After Meselsohn and Radding, *Proceedings of the National Academy of Sciences, U.S.,* v. 72, p. 358, 1975, and Potter and Dressler, *Proceedings of the National Academy of Sciences, U.S.,* v. 73, p. 3000, 1976.

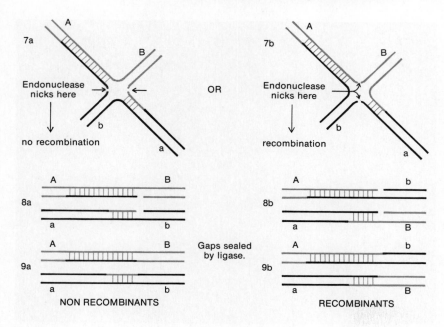

NON RECOMBINANTS   RECOMBINANTS

eroduplexes were formed, single strands of DNA have been exchanged between the two homologous chromosomes.

Recombination can be detected genetically, allowing determination of the order in which genes are located on a chromosome.

As with random assortment of chromosomes during meiosis, the presence of recombination can be detected by observing the offspring of crosses between stocks carrying appropriate genetic differences. The detection of recombination is illustrated in Figure 7-3, where allele differences at two different loci, a and b, on two homologous chromosomes are indicated. In the original chromosomes obtained from the parents, the *parental* chromosomes, alleles a^+ and b^+ were on one homolog, and a^- and b^- were on the other. In the absence of recombination, the original parental allele configurations a^-b^- and a^+b^+ are retained. As a result of reciprocal recombination, the recombinant chromosomes a^+b^- and a^-b^+ are produced. In an appropriate cross, for example, one involving organisms with a haploid chromosome number of one where one parent was genotypically a^-b^- and the other a^+b^+, the production of progeny that are phenotypically a^-b^+ and a^+b^- indicates that recombination has occurred. We shall see that analysis of the numbers of the different progeny produced allows us not only to detect whether recombination has occurred but also to determine the order in which a set of genes is distributed on a chromosome and to estimate the distances between the genes. In short, we will be able to make a genetic map of the genes on the chromosome.

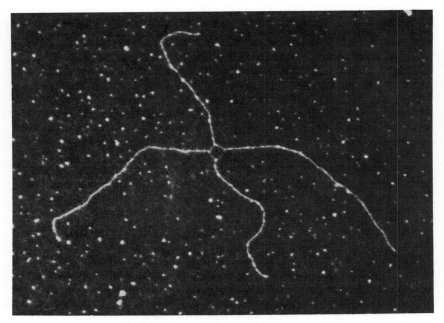

Figure 7-2   An electron micrograph of phage DNA undergoing recombination. Courtesy of David Dressler and Huntington Potter, Biology Laboratories, Harvard University, Cambridge, Mass. From Potter and Dressler, *Proceedings of the National Academy of Sciences, U.S.*, v. 74, p. 4168, 1977.

Because recombination between loci may be infrequent (sometimes less than 1 per cent of the progeny may be recombinant), the construction of genetic maps is far more demanding of organisms used for genetic analysis than the demonstration of segregation or random assortment. Only a small number of organisms—probably not many more than twenty and most of these phages and bacteria—have been extensively mapped. These organisms, by and large, share a common set of properties that make them ideal for genetic mapping: short generation times; the production of large numbers of offspring—hundreds, thousands, even millions in some cases, thereby eliminating most mammals; genetic variants that are readily identified and maintainable; easy propagation under laboratory conditions; and, very important, the presence of sex, that is, a mechanism that allows the admixture of different genotypes. In eukaryotes extensive genetic mapping has been done mostly for fungi, in particular the baker's yeast, *Saccharomyces cerevisiae,* and the pink bread mold, *Neurospora crassa,* but also *Aspergillus, Ascobolus,* and others. Among multicellular animals, *Drosophila melanogaster* has the best defined genetic map (it is possible to grow one million flies!). There is rapidly increasing knowledge of the genetics of a small, soil-living nematode *Coenorhabditis elegans.* Among mammals the

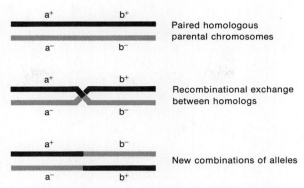

Figure 7-3   The effects of recombination between two pairs of linked alleles.

genetic map of the mouse is best characterized. The genetically best characterized higher plant is *Zea mays* (each kernel of corn is one offspring), but only about 50 genes have been mapped in the organism that started it all, the garden pea. Several bacteria are, by comparison to eukaryotes, extremely well characterized genetically, including *Escherichia coli, Bacillus subtilis,* and *Salmonella typhimurium.* The methods of genetic mapping in bacteria are unique to these organisms, and we will discuss them in Chapter 10. Extensive genetic maps have also been constructed of the chromosomes of phages. We will use one phage in particular, phage lambda (λ), as a model for the study of recombination.

The characteristics of the life cycle of λ phage are particularly suitable for genetic study.

Phage λ has a basic structure similar to that of phage T2 (Figure 5-1), with a single linear chromosome of DNA enclosed in a protein head from which a tail extends. The DNA is injected through the tail when the bacterial host, *E. coli,* is infected. The single λ chromosome of DNA contains about 46,500 nucleotide pairs. It is linear in the phage, but (as we noted on p. 122) upon entering the bacterial cell it becomes circular through the pairing of single-stranded complementary ("sticky") ends. It is in the circular form that replication of λ DNA occurs, with rolling circle replication generating linear chromosomes for packaging into new phages (p. 124, Figure 5-6). In contrast to T2, the λ chromosome, after infecting the bacterial host, can enter one of two different life cycles. In one cycle, the *lysogenic* cycle of growth, the phage DNA is integrated by recombination into the bacterial chromosome. It is then called a prophage and is replicated along with the bacterial chromosome (Figure 7-4). Following division of the bacterial host cell, each bacterial daughter cell contains within its chromosome a λ prophage. A bacterium harboring a prophage is called a lysogen (a generator of lysis) because the integration process can be reversed by, for example, exposing

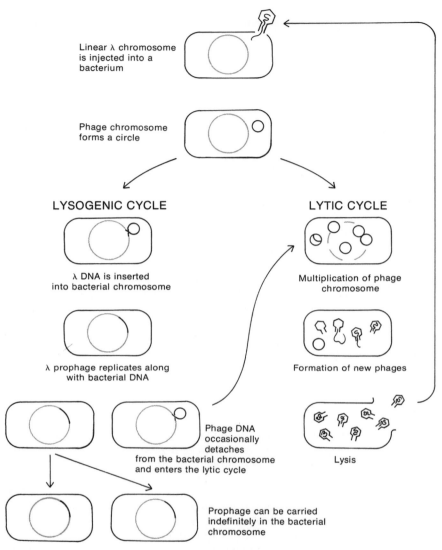

Linear λ chromosome
is injected into a
bacterium

Phage chromosome
forms a circle

LYSOGENIC CYCLE

LYTIC CYCLE

λ DNA is inserted
into bacterial chromosome

Multiplication of phage
chromosome

λ prophage replicates along
with bacterial DNA

Formation of new phages

Phage DNA
occasionally
detaches
from the bacterial chromosome
and enters the lytic cycle

Lysis

Prophage can be carried
indefinitely in the bacterial
chromosome

Figure 7-4   The lytic and lysogenic cycles of the growth of a temperate phage.

the cells to UV light. Such exposure causes the prophage to separate from
the bacterial chromosome and resume its circular structure (Figure 7-4). The
circular chromosome can then replicate rapidly, leading to the production
of new phages and the lysis of the cell.

The rapid replication of the phage chromosome, the *lytic,* or *productive,*
cycle of growth, is the second life cycle in which the λ chromosome can

participate. The phage chromosome can also enter directly into the lytic cycle immediately following infection of a bacterial cell. Normally, the two cycles are mutually exclusive. A lysogen with a λ prophage cannot be infected by another λ phage, that is, it has immunity to further infection, though the cell can be infected by another, unrelated phage. Phages such as λ, which can reproduce by either lysogenic or lytic cycles of growth, are called *temperate* phages. Those such as T2 that can reproduce only by lytic growth cycles are called *virulent* phages. The dual life cycle of temperate phages, of which λ is a very well-studied example, makes them particularly useful to geneticists because the genetic characterizations of the phage chromosome can be assayed in the lysogenic cycle by studying the properties of the bacteria or in the lytic cycle by studying the properties of the phages. One of the properties of phages that is often studied is the lysis of host cells.

Inherited differences in plaque morphology and formation have been used for studying the genetic structure of the λ chromosome.

When fewer than 100 phages are distributed on the surface of solid medium in a Petri dish on which a thick layer of bacteria is present—a so-called bacterial lawn—each phage serves as the focus for a series of infections. The phage infects a bacterial cell, lysis of the cell occurs with the release of progeny descended from that single phage, and the progeny infect neighboring bacterial cells. The process of infection and lysis continues with the progressive destruction of bacterial cells and the creation of a transparent circular area, called a *plaque,* in the bacterial lawn (Figure 7-5). Each plaque is the result of lysis of bacteria by phages, all of which are descended from the original infecting phage and, hence, all of which have the same genotype. The plaques produced by normal temperate phages such as lambda have a characteristic cloudiness that is distinct from the transparent plaques produced by virulent phages such as T2. The cloudiness stems from the presence of lysogens, bacteria-carrying prophages, that cannot be infected and lysed by other phages. Inherited variants of λ exist that alter the appearance or formation of the plaques. Inasmuch as each plaque is the product of phages descended from a single progenitor, the properties of plaques can be used as phenotypic traits for the study of phage genetics. Descriptions of variants in plaque morphology or formation follow:

Plaque size mutants. Wild-type phage infect, reproduce, and lyse bacteria at a characteristic rate producing plaques for a given period of time of uniform size. Mutants that are defective in any of these processes may produce plaques that are distinctly smaller than normal. One such variant in lambda is called *small* (*s*), another, unrelated one that produces even smaller plaques, is called *minute* (*mi*) (Figure 7-5).

Plaque clearness mutants. Normal lambda produces plaques that are cloudy because of the presence of lysogens. Some strains cannot enter into the lysogenic cycle and therefore produce transparent or almost transparent plaques (Figure 7-5). *Clear I* (*cI*) is one such mutant that produces a trans-

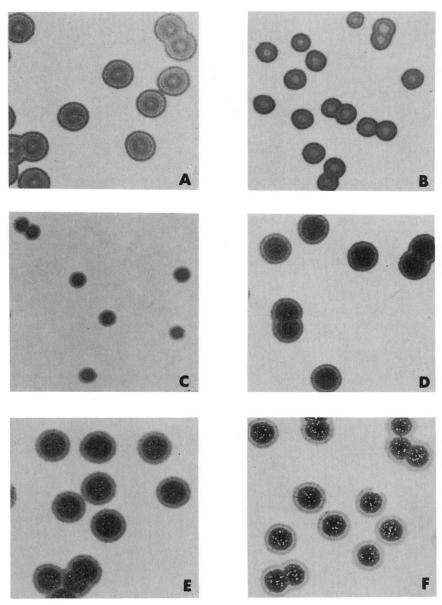

Figure 7-5   Normal and mutant plaques produced in a lawn of *Escherichia coli* by phage λ. (A) normal; (B) *small*; (C) *minute*; (D) *clear II* (*cII*); (E) *clear III* (*cIII*); (F *clear I* (*cI*). Courtesy of A. Dale Kaiser, Department of Biochemistry, Stanford University School of Medicine, Stanford, California. From Kaiser, *Virology*, v. 1, p. 424, 1955, with permission from Academic Press.

parent plaque. Others, called *clear II* and *clear III* (*cII* and *cIII*) produce nearly transparent plaques that are distinguishable from each other and from *cI*.

Virulent mutants. The mutants mentioned above affect plaque appearance or morphology and are analogous to mutants that cause morphological alterations in plants or animals (such as round versus wrinkled peas). Another class of mutants, *virulent mutants,* are so named because they are able to infect and kill lysogens and hence able to overcome the normal immunity of the lysogen.

Lethal mutants. A lethal mutant is one in which viable progeny are not produced. Lethal mutants are common and extensively used in genetic studies of all organisms (Tay-Sachs disease is a recessive lethal). Among phage a lethal mutant is one in which no phage progeny and hence no plaques are produced. How are lethal mutants identified and maintained if they produce no progeny? In diploid organisms, recessive lethals, for example, the Tay-Sachs allele, can be maintained in a heterozygous state. In λ, because there is only a single chromosome per phage, heterozygotes cannot normally exist. There are, however, two major ways available for the detection and maintenance of lethal mutations in λ. One way involves the use of *conditional* lethal mutations in which, by definition, a particular mutant is viable under one set of conditions, the *permissive* conditions, but is not viable under another set of conditions, the *restrictive* conditions. For example, the mutant may be a temperature-sensitive lethal, growing and producing lysis of bacteria at 25° C., the permissive temperature but will be a lethal mutant at 37° C., the restrictive temperature. Conditional lethal mutants or genetic constitutions are found in all organisms including humans. Insulin-dependent diabetes, for example, in the absence of insulin therapy is lethal. The second procedure for the maintenance and identification of lethals is peculiar to temperate phages that can enter either lytic or lysogenic growth cycles. Lethal mutations that prevent lysis of the host cell can be maintained in the prophage state because the inability to cause cell lysis need not prevent the replication of the prophage in the bacterial chromosome. The presence of the prophage can be demonstrated by the fact that the bacteria are immune to infection and lysis by normal lambda phages. Such lethals, called viability mutations by λ geneticists, are designated by upper-case letters (A, B, C, etc.). These lethals can be studied genetically by releasing the prophage from the host chromosome and simultaneously infecting the bacteria with viable phages. Chromosomes carrying lethal alleles can then be included in phage bodies synthesized under the direction of a normal chromosome. Phages with chromosomes carrying lethals will be released along with those carrying normal chromosomes.

We see that the variety of variants available with discernible phenotypes makes λ suitable for studies on gene transmission. In our discussions of gene transmission in λ we will use the same genetic nomenclature we used in Chapter 6. Mutant alleles will be designated by abbreviations, for example, *mi* for *minute*, and the normal allele will be designated *mi$^+$* or simply $+$. The position on the chromosome where the *mi* gene is located is the *mi* locus.

We can detect lambda phage progeny with recombinant genetic constitutions following simultaneous infection of bacteria with two genetically distinct strains of phage.

Reciprocal recombination, that is, reciprocal exchanges between DNA molecules of the kind presented in Figure 7-1, occurs during the replication of lambda DNA. To detect recombination genetically, single bacterial cells are simultaneously infected by phages with two distinct but complementary genetic constitutions, bringing two DNA molecules with differences in some base sequences together into one cell. In practice, a tube containing liquid medium in which *E. Coli* cells are growing in suspension is inoculated simultaneously with phages of two different genotypes (Figure 7-6). One phage strain might be genotypically *cI⁺* and *mi* (*cloudy* and *minute*) and the other, *cI* and *mi⁺* (*clear* and *large*). High concentrations of phages are used to increase the likelihood that a single bacterium will have DNA injected into it by two phages. The DNA's replicate within the same cell during which time recombination can occur. The bacteria lyse and the phage progeny are released into the culture medium. The medium is then diluted and distributed onto a bacterial lawn in a Petri dish so that there are about 50 phages per dish. The plates are incubated to allow the phage to grow and produce plaques. When the plates are examined we typically find that most of the plaques are like those produced by the two parent strains, that is, *large* and *clear* or *minute* and *cloudy*. Some plaques, however, have appearances characteristic of both parents. They are either *minute* and *clear,* resulting from the growth of phages that are genotypically *mi* and *cI*, or they are *large* and *cloudy,* resulting from the growth of phages that are genotypically *mi⁺* and *cI⁺*. Thus, some of the phages have genotypes in which an allele of one gene was derived from one phage parent and that of another gene from the other phage parent. Given that lambda has a single chromosome, we can conclude that recombination has occurred between the *mi* and *cI* loci during the replication of the two types of genotypically distinct phage chromosomes within single bacterial cells. Furthermore, we can surmise that recombination between DNA molecules during the replication of lambda chromosomes is occurring all of the time but has been detected in this case through the use of chromosomes that have genetic differences that produce demonstrable phenotypic alterations.

The analysis of the types of progeny produced as a result of recombination provides a means for determining the order of gene loci on chromosomes.

The analysis of the numbers of recombinant and non-recombinant progeny provides a basis for determining the order in which genes are distributed on chromosomes. It is important to keep in mind that, according to the model of recombination which we presented in Figure 7-1, recombinational exchanges can be initiated with equal likelihood at any point between homologous chromosomes. It is also important to keep in mind that recombination is a relatively rare event, the recovery of recombinant progeny in the

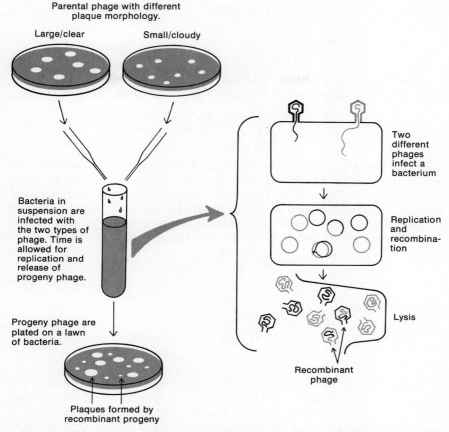

Figure 7-6   The experimental procedure for studying recombination between phage chromosomes.

offspring of a cross being low compared to the recovery of non-recombinant progeny. For example, for loci that are close together, 1 recombinant may be produced for every 100 progeny. If the loci are farther apart, the frequency of the recombinant progeny should increase because the opportunity for recombination between the loci has increased. Five recombinant progeny might be produced in every 100 offspring. Furthermore, knowing the frequency at which recombination occurs, say between loci A and B, between A and C, and between B and C, we can construct a genetic map of the relative positions of the different loci. The analytic process is similar to that used to determine the order of three stations along a monorail. If a train travels at a constant speed and takes 20 minutes to go from station X to Y, 30 minutes from station X to Z, and 50 minutes from Y to Z, we know that station X must be in the middle, located 20 minutes from Y and 30 minutes from Z.

Genetic distance between two loci located along the single track of a DNA molecule is measured by recombination frequency in which

$$\text{per cent recombination} = \frac{\text{Number of recombinant progeny}}{\substack{\text{Total of all progeny} \\ \text{(recombinant and non-recombinant)}}} \times 100$$

Note that both reciprocal types of recombinant progeny—for example, Ab and aB—are added together to determine the number of recombinants between A and B. Thus, if we find that there is 5 per cent recombination between loci A and B; and 20 per cent between A and C; and 15 per cent between B and C, then we know that locus B is in the middle, separated by 5 per cent recombination from locus A and 15 per cent from locus C.

```
A              B                              C
├──────────────┼──────────────────────────────┤
   5 per cent        15 per cent
```

GENETIC MAPPING WITH TWO-FACTOR CROSSES

In the preceding discussion we determined the order of three loci from data on the recombination frequency between each of the three possible pairs of loci. In practice we estimate the recombination frequency between two loci by making crosses between strains of phages with allelic differences at two genes, a "two-factor" cross. The earliest genetic maps constructed from data obtained by the simultaneous infection of bacteria with two genetically different strains of phage lambda were published by Francois Jacob and Elie Wollman in 1954 and by A. Dale Kaiser in 1955. We will use some of Kaiser's results shown in Table 7-1 to construct a genetic map for the *s* (*small*), *mi* (*minute*), and *cII* (*clear II*) loci. Several aspects of the results are worthy of emphasis.

1. From any cross, we recover progeny phages representing all four possible genotypic combinations: two parental types and two recombinant types. The parental types are those progeny with the same genotype as either parent and the recombinant types are those progeny with genotypes composed of one allele of one gene from one parent and one allele of the other gene from the other parent.

2. The frequencies of production of progeny with the four different possible genotypic combinations are far from equal (we do not need a χ^2 test). We recover the parental types at higher frequency than the recombinant types. Thus, the different allelic combinations are not randomly produced as

Table 7-1   Results from Two-factor Crosses in Bacteriophage λ

Parents	Non-Recombinant (parental)		Recombinant		Total Progeny	Per cent Recombination
1. $cII + \times cII^+mi$	$cII\ mi^+$	$cII^+\ mi$	$cII^+\ mi^+$	$cII\ mi$		
	5,162	6,510	311	341	12,324	5.3
2. $s\ cII^+ \times s^+\ cII$	$s\ cII^+$	$s^+\ cII$	$s^+\ cII^+$	$s\ cII$		
	7,101	5,851	145	169	13,266	2.4
3. $s\ mi \times s^+\ mi^+$	$s\ mi$	$s^+\ mi^+$	$s^+\ mi$	$s\ mi^+$		
	13,253	13,083	1,155	1,024	28,515	7.6

From A. Dale Kaiser, *Virology*, 1:424 (1955).

might occur *if* λ had several chromosomes and the three loci were all located on different chromosomes. In that case, following Mendelian random assortment, we would expect 1:1:1:1 proportions from a two-factor cross.

3. The unequal recovery of the different allele combinations is not a result of pronounced lethality of any of the alleles. We see that each allele is recovered in approximately equal frequency in any cross. These data do not support this contention particularly well, but $s = s^+ = mi^+ = mi$ in the third cross.

4. We interpret the disproportionately high frequency of production of the parental, non-recombinant types as a result of the two loci in question being inherited together on the same chromosome and of recombination between homologous chromosomes occurring infrequently between the two loci. Thus, most progeny phages have chromosomes genetically identical to one or the other type of parental phage. Only a few progeny phages have chromosomes that because of recombination are genetically recombinant. Such disproportionate recovery of allelic combinations—frequent parental types, infrequent recombinant types—demonstrates the absence of random assortment and reveals the presence of joint transmission of alleles from parent to offspring, called *linkage*.

5. The two recombinant allele combinations are recovered in approximately equal frequency as we expect if recombination is reciprocal. In cross 1, there are 311 + + progeny and 341 *cII mi* progeny (a 1:1.1 ratio); in cross 2, 145 + + and 169 *s cII* progeny (a 1:1.2 ratio); in cross 3, 1024 *s* ± and 1155 ± *mi* progeny (a ratio of 1:1.1). Parental types are also recovered in approximately equal frequencies. It is worth noting that it does not matter what the original allele combinations in the parents are as long as each parental strain has a different allele at each locus.

We calculate recombination frequencies by using the formula on p. 192 (total recombinants divided by total offspring multiplied by 100 per cent). For cross 1 there are 652 recombinant progeny recovered (341 + 311) along with 11,672 progeny with the parental allele configurations (5, 162 + 6,510) for a grand total of 12,324. Recombination frequency is 652/12,324 × 100 per cent = 5.3 per cent. Thus, there is 5.3 per cent recombination between the *cII* locus and the *mi* locus. Similarly, we can calculate that there is 2.4

per cent recombination between the *cII* and the *s* loci and 7.6 per cent recombination between the *s* and *mi* loci. We can construct only one linear genetic map consistent with the recombination data.

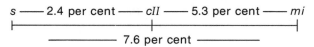

If we try to construct genetic maps with *s* or *mi* in the middle, they will not work. Incidentally, putting the *s* locus on the right and the *mi* locus on the left does not count as a different genetic map; we are simply viewing the same map from the other side. The frequencies do not quite sum up (2.4 per cent + 5.3 per cent ≠ 7.6 per cent) but are remarkably close. Recombination, unlike a monorail train traveling at constant speed, is a probabilistic phenomenon. There is a given probability that recombination will occur between two loci, but, as in random assortment, there is statistical error. We can see from this exercise in genetic cartography that it is possible to construct genetic maps specifying the order of three loci on a chromosome by using three two-factor crosses. It is preferable, however, to construct genetic maps using three-factor crosses (three allele pairs at a time) because a genetic map can be constructed from a single cross.

GENETIC MAPPING WITH THREE-FACTOR CROSSES

In crosses involving three loci on the same chromosome, progeny with the parental allele configuration are expected in the highest frequency and those resulting from exchange of the middle allele pair only, in the lowest frequency.

The possible chromosome configurations produced from three-factor crosses involving differences at three loci on the same chromosome are pictured in Figure 7-7. We expect four pairs of complementary products. The first pair of products, constituting the non-recombinant or parental class, consists of chromosomes with allele configurations identical to the parental ones. Because recombination is an infrequent event, progeny with the parental allele configuration are expected in highest frequency. The second and third classes of products result from recombination between the middle locus and one of the two outside loci. Progeny with either of these allele configurations are recovered at a lower frequency than the parental class; the frequency at which they are present is determined by the frequency of recombination between the middle and each of the flanking loci. The fourth class of products results from double recombination, two simultaneous exchanges of the homologous chromosomes between the middle and each of the outside loci.

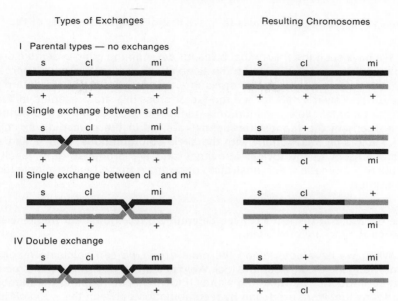

Figure 7-7   The effects of recombination between three pairs of linked alleles.

Let us assume, for now, that during double recombination, the two exchanges occur independently of each other. The presence of one recombinational exchange between the middle locus and one of the flanking loci does not influence the probability of an exchange occurring between the middle locus and the other flanking locus. In the previous example, we determined that the recombination frequency between s and cII is about 2.3 per cent, and between cII and mi is 5.3 per cent. If recombinational exchanges in two adjacent regions are independent, then the expected frequency of double recombination is the product of the recombination frequencies between the two pairs of loci, that is, 2.3 per cent × 5.3 per cent (0.023 × 0.053), which equals 0.12 per cent (0.0012). We see that the expected frequency of double recombination is substantially smaller than the frequency of either single recombination. Thus, in a three-factor cross, we expect progeny produced as a result of double recombination to be present in the lowest frequency. When we have identified the parental allele constitution (that of the progeny present in the highest frequency) and the allele constitution produced by double recombination (that of the progeny class present in the lowest frequency), we can determine the order of the three loci. The middle locus is identified by observing which single pair of parental alleles had to be exchanged to produce the allele combination present in lowest frequency. For example, in Figure 7-7, the parental configurations are $s\ cII\ mi$ and $+ + +$; the double recombinant configurations are $s + mi$ and $+\ cII\ +$, which are produced by exchanging the middle alleles, cII and cII^+.

In analyzing a three-factor cross in λ, we first determine the order of the loci.

We have seen the theoretical basis for determining gene order in a cross involving three genes carried on the same chromosome. When we approach actual data from a three-factor cross in λ, our first concern is to determine gene order. Kaiser's data for a three-factor cross in λ are presented in Table 7-2. The parental allele constitutions—identified because they are present in the progeny in the highest frequency—are, as in the preceding example, + + + and s cII mi. Progeny with the allele configurations + cII + and s + mi are present in the lowest frequency and are produced as a result of double recombination, demonstrating once again that cII is the middle locus of the three.

Once the order of the loci has been determined, recombination frequencies between the loci are calculated.

When we have determined the gene order, we can calculate recombination frequencies between the loci. We can start by determining the recombination frequency between the s and cII loci. We add together the numbers of progeny that were produced by recombination between the s and cII loci. The allelic configurations s + + and + cII mi were produced from the parental allele configurations + + + and s cII mi by recombination between the s and cII loci. The progeny produced as a result of double recombination, s + mi and + cII +, also resulted, in part, from recombination between the s and cII loci. We determine the total number of progeny produced because of recombination between the s and cII loci by adding the number of double recombinants to the appropriate number of single recombinants as follows.

			Number	Frequency
Single recombinant class	s + +		30	
	+ cII mi		32	2.95%
Double recombinant class	s + mi		5	
	+ cII +		13	0.85%
Total recombinants			80	3.8%
Total progeny			2,091	

We determine the recombination frequency between the s and cII loci by dividing 80, the total number of recombinants, by 2,091, the total number of progeny, with the result that the recombination frequency is 3.8 per cent from this experiment.

We can calculate the recombination frequency between the cII and the mi loci in the same manner. The progeny classes + + mi and s cII + were created as a result of single recombination between the cII and the mi loci. Recombination between these two loci also occurred to produce progeny with chromosomes that result from double recombination. Thus, we determine the total number of recombinant progeny produced as a result of recombination between the cII and the mi loci by adding the number of

Table 7-2   Results from a Three-factor Cross in Bacteriophage λ

Phenotypes	Number of Progeny
Parental Class:	
$s^+ cII^+ mi^+$	975
$s\ cII\ mi$	924
Single Recombinant Classes:	
$s\ cII^+ mi^+$	30
$s^+ cII\ mi$	32
$s\ cII\ mi$	61
$s^+ cII^+ mi$	51
Double Recombinant Class:	
$s\ cII^+ mi$	5
$s^+ cII\ mi^+$	13
Total Progeny	2,091

From A. Dale Kaiser, *Virology,* 1: 424 (1955).

double recombinants to the number of appropriate single recombinants as follows.

				Number	Frequency
Single recombinant class	s	cII	+	61	
	+	+	mi	51	5.35%
Double recombinant class	+	cII	+	13	
	s	+	mi	5	0.85%
Total recombinants				130	6.2%
Total progeny				2,091	

The recombination frequency between *cII* and *mi* loci is 6.2 per cent, determined by dividing 130, the total number of recombinants, by 2,091, the total number of progeny.

We know the order of the three loci and the recombination frequencies between the middle locus and each of the flanking loci. We can construct the following genetic map:

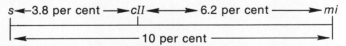

The recombination frequency between *s* and *mi* is determined by adding the recombination frequencies between *s* and *cII* and between *cII* and *mi* together. This map is similar to the one produced from two-factor crosses (p. 194) but differs in the frequencies of recombination between the loci. The differences in results call attention to the fact that estimations of recombination frequency vary from experiment to experiment and are not precise. The order of the three loci is, however, the same in the two experiments.

The frequency of double recombination in phages is often greater than expected.

Is the assumption that one recombinational exchange on a chromosome is independent of another correct? The frequency of double recombination occurring in the three-factor cross we have been laboring over is greater than expected. The recombination frequencies in the two intervals are 3.8 per cent and 6.2 per cent. The expected frequency of double recombination is 0.24 per cent ($0.062 \times 0.038 = 0.0024$). The observed frequency of double recombination is 0.85 per cent, roughly three times greater than expected. Levels of double recombination greater than expected are commonly found in phage crosses when the loci involved are genetically fairly close together (for example, with less than 5 per cent recombination occurring between them). The higher than expected frequency of double recombination indicates that recombinational exchanges are not independent of each other. The presence of one exchange increases the probability of a second exchange occurring within the same region of the DNA as the first one. The phenomenon of one recombinational exchange influencing the likelihood of a nearby second recombinational exchange is called *interference.* There are two types of interference, which for historical reasons are peculiarly named. The first type discovered, *positive interference* or simply interference, denotes a reduction in the frequency of double recombination over small recombinational distances; one recombinational exchange interferes with the likelihood that a second one will take place. Positive interference is commonly found in eukaryotes. The type of interference found in phages, which increases the frequency of double recombination, is called *negative interference* (clearly a double negative). The magnitude of interference is expressed quantitatively by a relationship called the *coefficient of coincidence.* The coefficient of coincidence is simply

$$\frac{\text{Observed frequency of double recombination}}{\text{Expected frequency of double recombination}}.$$

When the coefficient of coincidence equals 1, there is no interference and recombinational exchanges are independent of each other. For values less than 1, there is positive interference; one recombinational exchange prevents the occurrence of a second one nearby. For values greater than 1, there is negative interference; one recombinational exchange increases the likelihood of a second one nearby.

Three-factor crosses call attention to the existence of possible systematic errors in determination of recombination frequencies between any two loci.

In the preceding example, the recombination frequency between the two outside or flanking loci, *s* and *mi*, was determined by adding together the frequencies between *s* and *cII* and between *cII* and *mi*. For either of the two intervals the frequency of recombination was determined by adding the frequency of the single recombinants and the frequency of the double recombinants. Thus, we have the following:

Recombination frequency s to *cII* = single recombination frequency (s to *cII*) + double recombination frequency.

Recombination frequency *cII* to *mi* = single recombination frequency (*cII* to *mi*) + double recombination frequency.

Recombination frequency between s and *mi* (the two outside loci) = single recombination frequency (s to *cII*, 2.95 per cent) + single recombination frequency (*cII* to *mi*, 5.35 per cent) + *twice* the double recombination frequency (2 × 0.85 per cent) = 10.0 per cent.

It is not surprising that when we determine the total recombination frequency between the outside loci, it is necessary to add in twice the double recombination frequency. After all, each double recombination involves two recombinational exchanges, and, if we wish to know the total recombination frequency, we must count both of these exchanges. Using the data from Table 7-2, let us calculate the recombination frequency between the two outside loci. At the same time, let us assume that there is no allelic difference at the middle locus, *cII*, as in a two-factor cross between s and *mi*. We can no longer detect double recombination because double recombination does not result in a change of the allele configurations of the parental chromosomes. We can only detect single recombinations or, more correctly, any odd number (1, 3, 5, etc.) of recombinations between the two loci. The recombination frequency between the s and *mi* loci (without detecting double recombinants) is determined as follows:

				Number	Frequency
Single Recombinants	s	+	+	30	2.95%
	+	*cII*	*m*	32	
Double Recombinants	+	*cII*	+	13	0.85%
NOT DETECTED	s	+	*mi*	5	
Single Recombinants	s	*cII*	+	61	5.35%
	+	+	*mi*	51	
Double Recombinants	+	*cII*	+	13	0.85%
NOT DETECTED	s	+	*mi*	5	
Total Detected Recombinants				174	8.3%
Total Progeny				2,091	

Because the double recombinants are not detected, the recombination frequency between s and *mi* is only 8.3 per cent, instead of 10 per cent. It is 1.7 per cent, twice the frequency of double recombination, lower than when double recombination is detected. As a generality, then, we realize that because of the presence of undetected double recombination, recombination frequencies *determined* between any two loci are or may be underestimates of the actual recombination frequency.

The unit of genetic distance on a chromosome, the map unit, is defined as equaling 1 per cent of actual recombination frequency.

Geneticists have defined the unit of recombinational distance, the map unit or the *centimorgan,* as equaling 1 per cent of actual recombination. We have assumed that the probability of recombination is constant for a given length of DNA. If the assumption is correct, map distances are correlated with physical distances. However, we have also seen that our determination of the actual amount of recombination between two loci is made difficult by the presence of undetected double recombination. Only when there is little or no double recombination does the frequency of observed recombination closely mirror the frequency of actual recombination and directly provide a good measure of map distance. When there is substantial undetected double recombination, the frequency of observed recombination is an underestimate of actual recombination.

We can estimate the magnitude of the error in determining actual recombination frequency between two loci. In the example on p. 199 we determined that the observed recombination frequency between *s* and *mi* was 8.3 per cent instead of 10 per cent when double recombination was not considered. The difference, 1.8 per cent, is an 18 per cent underestimation of the actual recombination frequency (1.8 per cent/10 per cent × 100). For a map maker that could be considered a serious error. Imagine the results if a cartographer said that a mountain on an approach to a landing field is 820 meters high when in fact it is 1,000.

In the case of recombination frequencies we have expectations, based on the assumption that recombinational exchanges are independent of each other, for the amount of double recombination that is taking place. If the assumption is correct, when single recombination frequencies are low, there should be little error caused by the presence of undetected double recombination. For example, for two adjacent intervals, each with 0.5 per cent recombination frequency, we expect a frequency of double recombination of 0.0025 per cent (0.005 × 0.005 = 0.000025 or 0.0025 per cent). In the absence of allele differences at the middle locus, we expect the detected recombination frequency between the two outside loci to be 1 per cent − 2 × 0.0025 per cent, or 0.995 per cent, an error of only 0.5 per cent (0.005/1 × 100 = 0.5 per cent). In contrast, if the recombination frequency in each of the adjacent intervals is 5 per cent, then the expected frequency of double recombination is 0.25 per cent. In the absence of an allelic difference at the middle locus, we expect the detected frequency of recombination to be 10 per cent − 2 × 0.25 per cent or 9.5 per cent, an error of 5 per cent (0.5/10 = 5 per cent). From these two examples we can see that when recombination frequency between loci is low, 1 per cent or lower, there will be little error in the determination of actual recombination frequency because there is little undetected double recombination.

There is a theoretical relationship between map distance and observed recombination frequency between two loci with observed recombination frequency being limited to 50 per cent.

In general, for any two loci only odd numbered recombinational exchanges (1, 3, 5, etc.) result in observed recombination of parental allele constitutions in the progeny. Double recombination and all even-numbered exchanges (2, 4, 6, etc.) are not detected. There may be repeated recombinational exchanges between loci. In 1917, long before the first recombinational studies with phages, J. B. S. Haldane determined the effect of repeated recombinational exchanges on the relationship of observed recombinational frequency and actual recombinational frequency, that is, map distance, between any two loci. Haldane realized that the probability (P) of detectable recombination (odd-numbered exchanges) between two loci is related to the actual amount of recombination (d) between the loci (odd plus even numbered exchanges) by the following mathematical series:

$$P = \left(\frac{d}{1!} + \frac{d^3}{3!} + \frac{d^5}{5!} + ...\right) e^{-d} = 0.5(1 - e^{-2d}).$$

where e is the base for natural logarithms and the symbol ! denotes "factorial," which means that all positive integers less than or equal to the number in question are multiplied together. For example, 4! is $4 \times 3 \times 2 \times 1 = 24$. Three points may be made about Haldane's equation. First, when the distance between two loci is zero, $P = 0$ $(1 - e^{-2d}) = 1 - e^0 = 1 - 1 = 0)$. Second, for small values of d, that is, short map distances (about 1 to 5 per cent actual recombination), the equation simplifies to $P = d$, or observed recombination frequency equals map distance. Third, for long distances, that is, large values of d, e^{-2d} approaches zero and $P = 0.5$. These relationships are seen clearly when Haldane's equation is plotted graphically (Figure 7-8). The 50 per cent limit of observed recombination frequency indicates that as loci become farther and farther apart, the recombinational frequency is essentially constant and we cannot estimate accurately whether loci are 100 or 200 map units apart. It is also important to realize that when recombination frequency between two loci approaches 50 per cent, recombinant types and parental types are recovered in equal numbers, that is, in 1:1:1:1 proportions; and the loci, although located on the same chromosome, will, by definition, be assorting independently. Thus, the presence of independent assortment does not assure that loci are carried on different chromosomes.

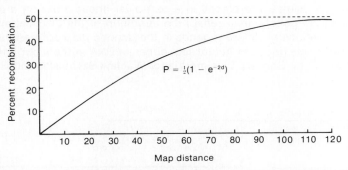

Figure 7-8   The relationship between map distance (actual recombination frequency) and the percentage of detected recombination.

An extensive genetic map of the λ chromosome has been developed.

We now know that it is possible to determine the order of three loci on a chromosome and to estimate the actual recombinational distance between the loci. If such a map can be constructed for three loci, it can be constructed for any number of loci. Even so, we realize that under some conditions observed recombinational frequencies are poor estimates of map distance because of the possible presence of undetected double recombination. Thus, to obtain an accurate genetic map of an entire chromosome in which distances between different loci are expressed in map distances, it is necessary to construct the map bit by bit, using loci that are comparatively close together. In phages, because of negative interference, such an approach also has its pitfalls, and it is necessary to correct for the effects of negative interference. Figure 7-9 shows an extensive map of the λ chromosome, encompassing numerous loci, developed by summing up distances between neighboring loci and correcting for negative interference. A linear map is depicted reflecting the linear nature of the chromosome found in mature phages.

Genetic maps of phage chromosomes other than λ have also been constructed and have many properties similar to the λ map. A particularly interesting one is of phage T4. The map is circular despite the fact that the T4 chromosome is linear, but we will recall that the T4 chromosome is circularly permuted as a result of concatamer formation and the headful mechanism of packing the chromosome into the phage head (p. 121). Thus, any locus on the T4 chromosome is flanked at least in some chromosomes by loci on both sides. Thus, for any locus on the map, other loci are found on both sides. The result is that there is no end to the T4 genetic map. A line without an end is a circle. A genetic map without an end is a circular map.

The genetic map of the λ chromosome has been related to the physical map of the chromosome.

We might well question the validity of the procedures for constructing genetic maps of chromosomes in which different loci, identified through the use of variant alleles, are placed in a particular linear order on a chromosome. Does the genetic map reflect the condition of the physical map, that is, do the positions of different genes in the genetic map correspond to the positions of these genes as actual sequences of DNA in the λ chromosome? Work reported by Barbara Westmoreland, Waclaw Szybalski, and Hans Ris

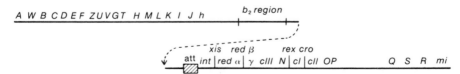

Figure 7-9   The linear genetic map of the λ chromosome. The chromosome is arbitrarily separated into two segments.

in the late 1960's and by Norman Davidson and his co-workers in the early and middle 1970's indicates that the genetic map corresponds accurately to the physical map. The basis for comparison of the physical and genetic maps depended on the use of λ chromosomes in which physical aberrations were present, for example, the presence of an extra sequence of DNA or the absence of a region of the DNA normally present. It is possible to locate such aberrations genetically in the λ chromosome. We know the recombination frequency between two loci in the normal chromosome. We then study a chromosome with a deficiency in which much of the DNA between these two loci is now absent. The two loci are physically closer together and as a result the map distance between them is reduced. Thus, we can determine the genetic positions of several deficiencies by identifying where, in a chromosome, recombinational distance between two loci is reduced. Similarly, we can identify insertions of additional DNA by determining where, in the abnormal chromosome, map distance has increased between two loci.

Knowledge of the physical location of aberrations in the λ chromosome derives from creating duplexes between aberrant DNA and normal DNA.

Let us consider two strains of λ, one of which has a normal chromosome and one of which has a deletion of about 5,000 base pairs called b2, in part of its DNA. We can prepare DNA from both strains, denature it, and, using density gradient centrifugations under denaturing conditions (so the strands do not renature), separate the complementary strands from each other and recover them. Single strands from the aberrant DNA and their complements from the normal DNA are available. They can be renatured with each other to form a double helix. We can anticipate the physical appearance of such double-helical molecules. Those regions of the DNA strand with the deficiency, which are complementary to the normal strand, will pair with the normal strand. However, the region in the normal strand that does *not* have a counterpart in the DNA that is complementary with the deficiency cannot renature and will remain as a single stranded blister on the side of the double-helical DNA. The blister can be detected with electron microscopy (Figure 7-10). The physical position of the blister, that is, the position of the

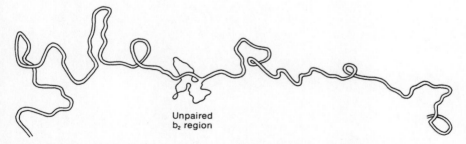

Unpaired
b₂ region

Figure 7-10   A representation of the duplex formed between complementary strands of a normal λ DNA molecule and a molecule with the b₂ deficiency. After Westmoreland, Szybalski, and Ris, *Science*, v. 163, p. 1343, 1969.

deficiency, can then be determined by measuring its position from each end of the λ chromosome. Westmoreland and her colleages found the b2 deficiency to reside 44.3 per cent of the length of the λ chromosome from one end. This corresponds precisely with the position of the b2 deletion on the λ genetic map (Figure 7-9). Other aberrations have been physically placed along the λ chromosome. Their positions correspond to the genetic positions of the aberrations. Thus, we can conclude that the positions of the genes on the genetic map accurately reflect their actual physical positions on the λ chromosome.

SUMMARY

Recombination between homologous chromosomes occurs in organisms from phages to bacteria. Some of the enzymes active in DNA replication, namely, DNA polymerases and ligases, mediate recombination. Exchanges between homologous chromosomes can occur at any point along the length of paired chromosomes. Recombination creates chromosomes with genetic constitutions different from those found in the parents. For loci that are physically close, recombination is rare, and recombinant allele combinations are recovered less frequently than parental ones. Such deviations from random assortment constitute linkage and reflect the physical location of the loci on the same chromosome. By analysis of the frequencies at which progeny with non-recombinant and recombinant chromosomes are produced we can determine the positions of genes along a chromosome and thereby construct a genetic map. The units of distance on the map, map units, reflect the actual amount of recombination that occurs. The genetic map of λ corresponds closely to its physical map.

Linkage in Eukaryotes

The procedures for construction of a phage genetic map are basic and are used for genetic cartography of most organisms. We introduce chromosomes with allele differences at two or more loci into common cells; in phage by mixed, simultaneous infection with two strains; in eukaryotes by crossing two genetically different strains to produce an F_1. We then determine the recombinational products by examining the phenotypes of subsequent offspring. In phage these are simply the progeny; in eukaryotes, we mate the F_1 organisms to an appropriate strain and examine the offspring of that cross. We have used phage to introduce the methods of making genetic maps because each phage has only one chromosome and we do not have to distinguish genes carried on different chromosomes that randomly assort from those carried on the same chromosome and that may be linked. Furthermore, in phage there are no problems associated with diploidy such as the expression of dominant and recessive alleles, which might make it difficult to determine the genotypes of the offspring from their phenotypes.

Historically, however, the techniques for making genetic maps were first developed using eukaryotes. Following the rediscovery of Mendel's work in 1900 by Carl Correns and Hugo de Vries, a view called the "chromosomal theory of heredity" developed, holding that Mendelian factors, the genes, were carried on chromosomes. The laboratory of Thomas Hunt Morgan was particularly involved with efforts to substantiate the theory. Morgan and his associates at Columbia University in New York City first took advantage of *Drosophila melanogaster* as a tool for genetical research. It was Morgan's view that several Mendelian factors could be located on the same chromosome and therefore inherited as a "block" of linked traits rather than as individual elements. Alfred H. Sturtevant, who as an undergraduate worked in Morgan's laboratory, investigated six such linked traits and through analysis similar to our analysis of the three traits in λ was able to construct the first genetic map, which was published in 1913. Sturtevant's map had the genes for the six traits distributed in a linear array with different recombinational distances between them and was an important contribution to substantiating the chromosomal theory of heredity. On the one hand, the linear array of the elements in the map correlated nicely with the linear structure

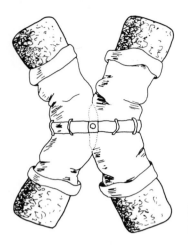

Jeans on chromosomes. After the sculpture by Robert Varner, Department of Genetics, University of California, Berkeley.

of eukaryotic chromosomes. On the other hand, different Mendelian traits residing at specific points on the genetic map like beads on a string supported the concept of the particulate nature of the Mendelian factors and demonstrated that a given chromosome contained an array of these factors. Thus, the conclusions deduced from formal genetic analysis by Sturtevant in the early 1900's indicating that genes occupy limited, but specific, regions of chromosomes, anticipated our view that genes are particular DNA sequences occupying particular places in chromosomal DNA. The work by Sturtevant and other members of Morgan's laboratory developed the basic procedures for genetic mapping in eukaryotes. In any cross in eukaryotes involving several genes, it is necessary first to demonstrate whether the genes are linked or assort independently and, second, for any linked genes, to determine their order and the recombinational frequencies between them.

The test cross is the method of choice for the study of transmission of several genes.

An advantage of phage in addition to the presence of the single chromosome is that all genetic traits are expressed phenotypically because, normally, each gene is present only once per genome. Many eukaryotes are diploid in their dominant growth form and the determination of the genotype from the phenotype may be difficult because of the presence of dominant alleles. As we noted on p. 160, we avoid the difficulties involving dominance by using a test cross in which F_1 organisms, heterozygous for all different allele pairs, are crossed to organisms homozygous for all of the recessive alleles. In the resulting progeny the contribution of the homozygous recessive parent can be ignored in as much as the phenotypes directly reflect the genetic constitutions of the germ cells contributed by the F_1 heterozygous parent. Thus, the recombinational exchanges, which occurred during meiosis in the F_1 parents, can be evaluated directly. In a three-factor cross in a eukaryote the first step is to produce F_1 organisms heterozygous for the allele pairs at all three loci and then cross the F_1 organisms to organisms homozygous for the three recessive alleles.

The analysis of a three-factor cross in maize.

We will analyse a three-factor cross in a higher diploid organism using an example from maize genetics. To the casual observer maize may not seem to be a particularly suitable organism for genetic studies. It does not have the small size or short generation time of *Drosophila,* not to mention phage. Maize has, however, many desired properties. The haploid chromosome number is 10 and the chromosomes are readily seen by light microscopy. There is a reservoir of mutant genes in maize and anyone who has had the happy fortune to drive across Iowa in the summer knows that maize plants can be produced in enormous numbers. Of particular importance, however, is the nature of the mature seeds or kernels. Every ear of corn contains several hundred seeds, the embryoes in which are each the result of fusion of a single female germ cell (ovum) and a single male cell (pollen nucleus). Many of the known mutant traits affect the color, shape, or composition of the kernels themselves (Figure 8-1). In some instances, simply by observing 10–20 ears of corn and noting the phenotypes of each kernel, we obtain enough data to draw conclusions about the transmission of a particular set of genes.

A cross involving three genes controlling the phenotypes of the kernels in maize was reported in 1922 by Claude B. Hutchison and provides an excellent example of the processes involved in the analysis of a three-factor cross in a eukaryote. The genes Hutchison studied were named after characteristics of their rare alleles: *Colored (Cl)*, a dominant allele causing a part of the seed, the aleurone, to be colored; *shrunken (sh)*, a recessive allele reducing the endosperm, the nutritional layer of the seed; and *Waxy (Wx)*, a dominant allele causing a waxy appearance of the endosperm. We shall refer to the recessive alleles of these three genes as *cl*, *sh*, and *wx* and the dominant ones as *Cl*, *Sh*, and *Wx*. Hutchison crossed two strains to produce

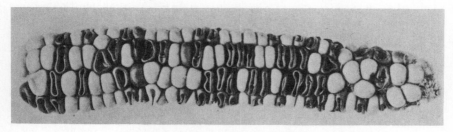

Figure 8-1   Two pairs of traits expressed in kernels are displayed in a single ear of corn. Most kernels are *shrunken* and *colored* or *full* and *colorless*. A few recombinant kernels are *shrunken* and *colorless* or *full* and *colored*. From Hutchison, *Cornell University Agricultural Experiment Station Memoir 60*, 1922.

Table 8-1   Results From a Three-factor Cross in Maize

Phenotype	Number
Cl sh wx	116
cl Sh Wx	113
Cl Sh wx	601
cl sh Wx	626
Cl Sh Wx	2,538
cl sh wx	2,708
Cl sh Wx	4
cl Sh wx	2
Total Number of Progeny	6,708

the triple heterozygote F_1 (*Cl/cl*, *Sh/sh*, *Wx/wx*) and then crossed the F_1 plants in a test cross to plants homozygous for all three pairs of recessive alleles (*cl/cl*, *sh/sh*, *wx/wx*). The numbers of offspring (that is, kernels on about 30 ears of corn) that he obtained from the test cross are presented in Table 8-1. For convenience we have listed the progeny in Table 8-1 in complementary pairs, which, as we can see, are produced in approximately equal numbers. Keep in mind that although we are observing the phenotypes of the kernels, we know the genotypes of the germ cells produced by the F_1 organisms because we have conducted a test cross.

The first step in the analysis of a three-factor cross in a higher eukaryote is to determine the mode of gene transmission. Is there linkage? Is there independent assortment?

We know from the classic Mendelian results that allele pairs may be carried on different chromosomes and may assort independently during meiosis. We also know from our consideration of gene transmission in phage that single chromosomes carry numerous genes and, therefore, that many genes will be inherited together in a linked fashion. Because most higher eukaryotes have several chromosomes, both random assortment and linkage are possible. For a given three-factor cross all three loci may be linked, two may be linked, or none. These possibilities must be distinguished. With some experience in analysing three-factor crosses, it will often be possible to make the distinction at a glance, inasmuch as each condition will result in different phenotypic proportions. At first, however, it is prudent to examine systematically the inheritance of two pairs of alleles at a time and determine if they assort independently or are linked. Let us look first at the *Cl* and *cl* allele pair and their association with the *Sh* and *sh* allele pair. From Hutchison's data (Table 8-1) we can construct the following:

Cl together with *Sh*:	2,538 + 601 = 3,139	(46.8%)
Cl together with *sh*:	116 + 4 = 120	(1.8%)
cl together with *Sh*:	113 + 2 = 115	(1.7%)
cl together with *sh*:	2,708 + 626 = 3,334	(49.7%)
Total Progeny	6,708	(100%)

If the allele pairs at the *Colored* and *shrunken* loci were assorting independently, we would have expected 1:1:1:1 proportions of progeny for this cross. The ratios differ substantially from 1:1:1:1 and we can conclude, even without the benefit of a χ^2 test, that the two loci are linked.

We now consider the *Colored* and the *Waxy* loci and determine the frequencies of the different allele combinations resulting from the test cross:

Cl together with *Wx*:	2,538 +	4 = 2,542	(37.9%)
Cl together with *wx*:	116 + 601 =	717	(10.7%)
cl together with *wx*:	2,708 +	2 = 2,710	(40.4%)
cl together with *Wx*:	113 + 626 =	739	(11.0%)
Total progeny		6,708	(100%)

The *Colored* and the *Waxy* loci also do not assort independently. We can conclude that all three loci are linked. In this example, there is no need to determine the frequencies of association of the alleles at the *shrunken* locus with those at the *Waxy* locus. Because they are both linked to the *Colored* locus, they must also be linked to each other.

We can determine whether the loci are all linked, all are unlinked, or only two are linked by inspecting the ratios of all types of progeny in the test cross.

We can draw conclusions about the mode of transmission of genes in a three-factor cross by inspection of the proportions of the eight classes of progeny produced by the test cross. We already know that for random assortment we expect all progeny classes to be recovered in equal frequency; therefore, the progeny classes should be in 1:1:1:1:1:1:1:1 proportions. If all three genes are linked, we expect the parental allele configurations to be recovered at the highest frequency, those resulting from double recombination at lowest frequency, and the two progeny classes resulting from single recombination at intermediate frequencies (x and y). The frequency of the double recombination class should be approximately equal to the product ($x \times y$) of the frequencies of the single recombinant classes if there is no interference. Setting the recovery of the parental classes equal to 1, then the progeny classes should be recovered in 1:1:x:x:y:y:xy:xy: proportions where x and y are both less than 1. Finally, if two loci are linked and one is unlinked, we expect the parental allele combinations for the linked loci to be present in higher frequency than those produced by recombination. The unlinked pair of alleles should randomly assort with respect to the parental and the recombinant allele configurations. We can set the frequency at which the parental allele configuration of the linked genes is recovered with either of the alleles of the unlinked gene equal to 1. We can call z the frequency at which the recombinant allele configuration of the linked genes is recovered with either of the alleles of the unlinked gene. We realize that z is less than one. The ratio at which the progeny are recovered is then 1:1:1:1:z:z:z:z. These relationships are summarized in Table 8-2.

Table 8-2   Expected Proportions of Phenotypes for
Three-factor Test Crosses Involving 0, 2, and 3 Linked
Genes

Linkage Relationships	Expected Proportions of Offspring
1. No linked genes:	1:1:1:1:1:1:1:1
2. Two linked genes, one unlinked:	1:1:1:1:z:z:z:z
3. Three linked genes:	1:1:x:x:y:y:xy:xy

The values of x, y, and z are all between 0 and 1; z is dependent on
the frequency of recombination between the two linked genes; x
and y are dependent on the frequency of recombination of the
center gene and each flanking gene, and the product xy is the
expected frequency of double recombination, assuming no interference.

**Once we establish that all or two loci are linked, we determine
recombination frequency between linked loci.**

To construct a genetic map for three linked loci, in this case *Colored,
shrunken,* and *Waxy*, we proceed as we did for the three-factor cross in
lambda. We identify the parental class, which is present in highest frequency,
and determine which locus is in the middle by comparing the allele configurations in the parental classes to those in the double recombinant class,
which is present at lowest frequency. The recombination frequencies between the two outer loci and the middle locus are then determined. The data
are presented as follows.

Genotypes and Numbers of Progeny				Recombination Configurations
A. *Cl Sh Wx*	2,538			—Cl——Sh——Wx—
cl sh wx	2,708	5,246	(78.2%)	—cl——sh——wx—
B. *Cl Sh wx*	601			—Cl——Sh⟍ ⟋Wx—
cl sh Wx	626	1,227	(18.3%)	—cl——sh⟋⟍wx—
C. *Cl sh wx*	116			—Cl⟍ ⟋Sh——Wx—
cl Sh Wx	113	229	(3.4%)	—cl⟋⟍sh——wx—
D. *Cl sh Wx*	4			—Cl⟍ ⟋Sh⟍ ⟋Wx—
cl Sh wx	2	6	(0.1%)	—cl⟋⟍sh⟋⟍wx—
Total Progeny		6,708	(100.0%)	

Class A is present in the highest frequency. Therefore, the allelic configurations of the chromosomes of the parents that were crossed to produce
the F_1 were *Cl Sh Wx* on one chromosome and *cl sh wx* on the other
chromosome. Class D is present in the lowest frequency, with the allelic
configurations differing from the parental ones by exchange of alleles at the
Sh locus. Therefore, the exchange between chromosomes of the *Sh* and the
sh alleles resulted from double recombination and the *Sh* locus is in the
middle.

We can now estimate map distances. The map distance between the *Cl* and the *Sh* loci is estimated by determining the recombination frequency between these two loci. In class C and in the double recombination class, D, the configurations of the alleles at the *Cl* and *Sh* loci have recombined with respect to the allele configurations of the parental chromosomes. The recombination frequency between the *Cl* and *Sh* loci is the sum of the numbers of progeny in classes C and D divided by the total number of progeny × 100 and is 3.5 per cent (116 + 113 + 4 + 2 = 235; 235/6708 = 3.5 per cent). To determine the recombination frequency between the *Sh* and *Wx* loci, we add together the progeny classes produced as a result of recombination between these two loci during meiosis in the F_1. Classes B and D are produced as a result of recombination between the *Sh* and *Wx* loci and the recombination frequency is 18.4 per cent (601 + 626 + 4 + 2 = 1233; 1233/6708 = 18.4 per cent). Because we know the recombinational frequencies between the middle locus and each of the two flanking loci, we can produce the following map:

$$\underset{\overline{\hspace{1.2cm}3.5\hspace{2cm}18.4\hspace{1.2cm}}}{\text{Cl}\qquad\text{Sh}\qquad\qquad\qquad\qquad\text{Wx}}.$$

The frequency of double recombination in higher organisms is frequently lower than expected.

Assuming that recombinational exchanges between homologs are independent of each other, we know that the expected frequency of double recombination is the product of the recombination frequencies within each interval between the loci. In the preceding example the expected frequency of double recombination is 0.035 × 0.184 = 0.0064. The observed frequency of double recombination is 0.0009 (6/6708). The coefficient of coincidence, the observed frequency of double recombination divided by the expected frequency of double recombination, is 0.14 (0.0009/0.0064) and is substantially below 1, indicating positive interference. As we noted, in Chapter 7, positive interference is characteristic of recombination in eukaryotes. Because of the reduction in the frequency of double recombination, the 1:1 relationship between map distance and recombination frequency (Figure 7-8) is maintained over longer map distances than it would be if there were no interference. In this example from maize the observed recombination frequencies are accurate estimates of map distance.

The ability to recover the products of a single meiosis is particularly useful for the study and analysis of recombination.

In the recombinational studies using phage and maize, we have looked at the average results obtained with a large number of progeny and, therefore, from a large number of matings or meiotic divisions. In some organisms, such as yeasts (*Saccharomyces*), molds (*Neurospora, Aspergillus*), unicellular algae (*Chlamydamonas*), and some mosses, we can recover the four haploid products from a single meiosis and determine the genotypes from the phenotypes of each cell. The genotypes of each set óf meiotic products can be analysed in a process called *tetrad analysis*, and conclusions can be drawn about processes occurring during meiosis that could not be deduced readily, if at all, by looking at the total yield of products from a large number of meioses. Examples of such analyses will be provided shortly, but first let us briefly describe the life cycle of one organism well-suited for such analyses, the pink bread mold *Neurospora crassa.*

In *Neurospora* the meiotic products are produced in a linear order corresponding to the meiotic divisions.

The life cycle of *N. crassa* is presented in Figure 8-2, and shows that the dominant life form of the organism consists of branched filaments called hyphae, the total mass of which is called a mycelium. Hyphae are produced by rapid, vegetative (mitotic) growth, filaments growing at up to 5 mm per hour at 30° C. The hyphae contain haploid nuclei. Asexual spores, cells with heavy walls that resist drying out called *conidia*, are produced by mitoses when conditions become unsuitable for the continued growth of the hyphae. When conditions are again suitable for growth these spores germinate to produce new filamentous masses. *N. crassa* also undergoes sexual reproduction. The mold consists of two mating types called *A* and *a*, which are morphologically indistinguishable; both can form female reproductive structures called *protoperithecia* or fruiting bodies. Each fruiting body consists of a ball of cells inside of which is a coiled structure called the ascogonium. Protruding from the fruiting body, waving around in the wind so to speak, is a finely branched structure, the trichogyne. Sexual reproduction is initiated when a spore or a filament from the opposite mating type comes into contact with the trichogyne. A haploid nucleus from the fertilizing cell travels down the trichogyne and into the ascogonium. Eventually a cell, the ascus initial, is formed. It contains two nuclei, one of which is derived from each of the mating types. The two nuclei fuse to produce a diploid nucleus and cell. This diploid nucleus is the only one that exists during the life cycle of *Neurospora*. The diploid cell immediately enters into meiosis, the two meiotic divisions producing four cells. A single mitotic division follows, converting each meiotic product into a pair of genetically identical cells called a spore-pair. The three divisions all occur within the ascus, an elongated cell derived from the ascus initial. The divisions occur along the length of the ascus in a manner that assures that the resulting four pairs of spores, the ascospores, are distributed within the ascus in the same order in which they are produced during meiosis (Figure 8-2).

Analysis of a cross in *N. crassa* involving a single pair of alleles provides genetic evidence that meiotic recombination occurs when four chromatids are present.

A cross in *N. crassa* involving a single pair of alleles reveals the meiotic stage in which recombination occurs. A particularly useful allelic difference is at the *lysine-5* locus (one of the loci in *N. crassa* involved in the synthesis

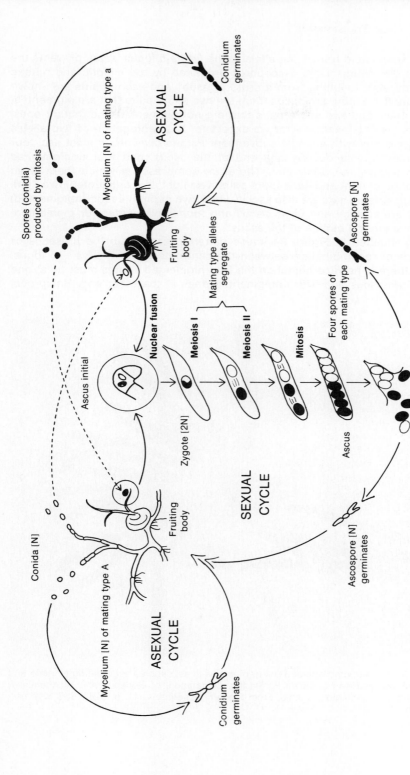

Figure 8-2   The life cycle of *Neurospora crassa*. The products of a single meiosis and one mitosis are aligned within an ascus.

of the amino acid lysine). An allele at the *lysine-5* locus, *asco*, prevents the normal darkening of the ascospores produced by the meiotic and mitotic divisions. Asci resulting from a cross of *asco*⁺ and *asco* strains are shown in Figure 8-3. Within any ascus there are two spore pairs that are pigmented, the + ones, and two spore pairs that are not pigmented, the mutant ones. Such a result is exactly what we expect for the segregation of two alleles during meiosis with a single subsequent mitotic division. In most asci four pigmented spores occupy one end of the ascus, and four unpigmented spores occupy the other end. There are some asci in which there is an alternation of light and dark spore pairs: two light, two pigmented, two light, two pigmented. There are also asci in which two pigmented (or unpigmented) spores are at each end of an ascus and four unpigmented (or pigmented) spores are in the middle of the ascus.

As depicted in Figure 8-4 these patterns result from the presence or absence of recombination between the centromere and the *lys − 5* locus. When there is no recombination the four pigmented spores must be at one end of an ascus and four unpigmented ones at the other end. Reciprocal

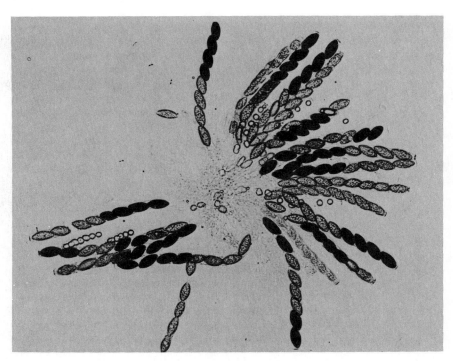

Figure 8-3   A micrograph of a group of asci with ascospores segregating for *asco* and *asco*⁺. Photo courtesy of David R. Stadler, Department of Genetics and Biochemistry, University of Washington, Seattle. From Stadler, *Genetics,* v. 41, p. 528, 1956, with permission.

Segregation without recombination

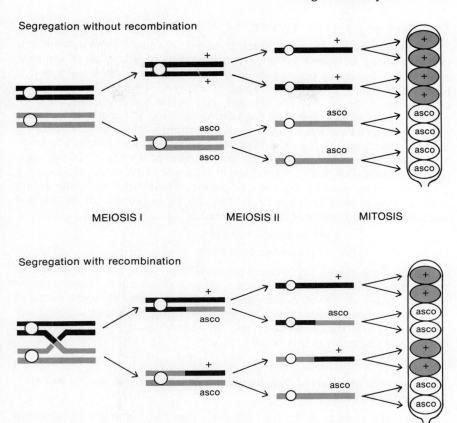

MEIOSIS I   MEIOSIS II   MITOSIS

Segregation with recombination

Figure 8-4   The arrangement of *asco* and *asco*⁺ spores as a result of segregation with and without recombination.

recombination between the centromere and the *lys* − 5 locus, during a stage of meiosis when four chromatids are present, produces an alternating pattern of spores within the ascus (*asco*⁺, *asco*, *asco*⁺, *asco*). By changing the orientation of the paired chromatids during the second meiotic division, recombination produces a spore pattern with pigmented spores at both ends and unpigmented ones in the middle (*asco*⁺, *asco*, *asco*, *asco*⁺). If recombination did not occur when four chromatids were present, the alternating pattern of spore pairs could not be produced. The presence of the alternating spore pattern, while not eliminating the possibility that recombination would occur when only two chromosomes were present, demonstrates that recombination *does* occur when four chromatids are present. This bears out our claim, made in describing meiosis (pp. 134–135), that recombination occurs in tetrads during the first meiotic prophase.

Tetrad analysis allows the mapping of a chromosome's centromere.

In the preceding example, recombination between the centromere and the *lys-5* locus produced the alternating patterns of wild type and mutant spores. The frequency of production of the alternating spore patterns is a measure of the recombination frequency between the *lys-5* locus and the centromere. If the centromere and the *lys-5* locus are physically close together, recombination between the centromere and the locus will be rare and the production of the alternating spore pattern will also be rare. Ignoring double recombination, the recombination frequency is the number of recombinant tetrads divided by 2 times the total number of tetrads × 100 per cent. (A tetrad constitutes the four products from a single meiosis, in this case a set of four spore pairs.) If there is a single recombinational exchange during meiosis, two recombinant chromosomes will be produced per tetrad, and the recombination frequency is 50 per cent or

$$\frac{1 \text{ recombinant tetrad}}{2 \times \text{total number of tetrads}} \times 100 \text{ per cent,}$$

in this case one tetrad. For n recombinant tetrads out of a total of N tetrads, the recombination frequency equals $n/2N \times 100$ per cent. Thus, recombination frequencies can be determined from the numbers of recombinant and non-recombinant tetrads. It is also possible, using more complicated calculations, to take the presence of double recombination into consideration, but we will not do so in this book.

The frequency of observed recombination between two loci during meiosis is limited to 50 per cent.

We have already noted (pp. 200–201) that the frequency of observed recombination between two loci is limited to 50 per cent. We can demonstrate this limitation during meiosis by another approach. Let us imagine four chromatids lined up side by side with one recombinational exchange between two non-sister chromatids, that is, between homologs. If no further recombination were to occur, the frequency of recombination from this single exchange would be 50 per cent, two of the chromosomes would be non-recombinant with respect to allelic markers, and two of the chromosomes would be recombinant. Thus, if recombination is to exceed 50 per cent, a second recombinational exchange must occur. We assume that the second recombinational exchange between any two non-sister chromatids has an equal likelihood of involving any two of the non-sister chromatids. The second recombinational exchange can produce four types of double recombinations: two chromatid doubles, two kinds of three chromatid doubles, and four chromatid doubles.

Figure 8-5 displays the four types of double recombinational exchanges and their effects on the alignment of the alleles. The two chromatid double recombination involves a second exchange with the same pair of chromatids involved in the first exchange. The effect genetically is that all four of the chromatids appear non-recombinant as far as the allele alignment is con-

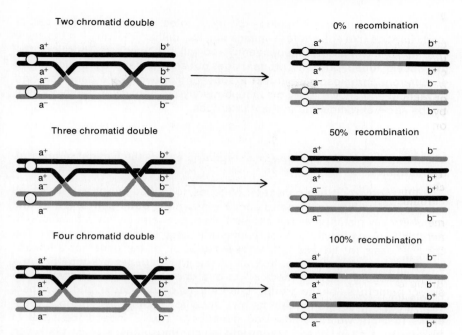

Figure 8-5   Double recombination at the four-chromatid stage during meiosis limits the frequency of observed recombination to 50 per cent.

cerned. Three chromatid double recombination involves one chromatid involved in the first recombinational exchange plus a non-sister chromatid not involved in that exchange—a total of three chromatids in the double recombination. The three chromatid double (Figure 8-5) produces two chromosomes that appear recombinant and two that do not. There are two types of three chromatid exchanges, one involving chromatids 1, 2, and 3 and the other involving chromatids 2, 3, and 4. The two types of three chromatid exchanges produce four recombinant and four non-recombinant chromosomes. The last type of double recombination, the four chromatid double, produces four products, all of which are recombinant for their marker alleles. If we sum up the possible products, we find that eight products appear non-recombinant and that eight are recombinant, that is, 50 per cent appear recombinant. Thus, assuming that recombinational exchanges are random, we have another demonstration that observed recombination between two loci is limited to a maximum frequency of 50 per cent.

This demonstration directs our attention to other matters. First, the kinds of double recombinations (two chromatid doubles, three chromatid doubles, and four chromatid doubles) are expected in a ratio of 1:2:1 if recombination occurs when four chromatids are present. Use of multiple markers and tetrad analysis demonstrates this condition and confirms that recombination occurs during meiosis when four chromatids are present. Second, the 50 per cent limitation on recombination—we have noted it before, but it is particularly important in eukaryotes—means that loci distantly located on the same chromosome will assort independently and will not be linked. Therefore, the presence of independent assortment does not assure that two loci are carried on different chromosomes. In fact it appears that two of the genes studied by Mendel and shown to assort independently are actually distantly located on the same chromosome in the garden pea.

The number of linkage groups in an organism equals the haploid number of chromosomes.

A *linkage group* is a collection of genes, identified by variant alleles, all members of which are linked to one or more of the other genes. Genes that are physically distant may not appear linked, but can be shown to belong to the same linkage group by their common linkage to intervening genes. The physical basis for linkage is the presence of all of the genes of a linkage group on one chromosome. In λ all of the known genes are together in a single linkage group and there is but one chromosome. In eukaryotes, we expect numerous linkage groups, one for each chromosome, and we expect each linkage group to contain its own specific set of genes. In those organisms that are well characterized genetically, this is indeed the case. In maize there are ten linkage groups and ten chromosomes; in *N. crassa* seven linkage groups and seven chromosomes. The yeast *S. cervisiae* has eighteen linkage groups, but because the chromosomes are so small, there are no certain estimates of chromosome number (the average haploid number, however, is between 17 and 19). Among higher animals, if we count the X but not the Y, the number of linkage groups in *D. melanogaster* equals the haploid chromosome number, four. The mouse, *Mus musculus,* has 20 linkage groups and 20 chromosomes. In humans, genes have been placed on all 23 chromosomes, but the techniques used did not involve the establishment of 23 different linkage groups through recombinational analysis (see Chapter 9).

Particular linkage groups can be associated with particular chromosomes: sex linkage.

If the genes of a particular linkage group owe their linkage to common residence on a particular chromosome, then it should be possible to identify which chromosome carries a particular linkage group. An early well-documented case of such an identification occurred in *Drosophila* with the assignment of genes to the X chromosome by Morgan and his co-workers. The assignment depended on the "peculiar" behavior of certain pairs of alleles during transmission. A good example of sex linkage in *Drosophila melanogaster* is provided using the recessive trait *white* (*w*), a mutant condition that results in the total absence of pigment from the normally brownish-red eyes of the fly. When males with white eyes are crossed to females homozygous for w^+, the following results are produced:

F_1: All males and females have normal eyes as would be expected for a cross involving dominant and recessive alleles.

F_2: Half of the males have white eyes, half have normal eyes. All of the females have normal eyes.

When the original cross is made reciprocally, that is, females with white eyes are crossed with males with normal eyes, the following results are produced:

F_1: All the males have white eyes; all the females have normal eyes.

F_2: Half of the males and half of the females have white eyes; half have normal eyes.

Morgan explained these "peculiar" results by deducing that the *white* gene is on the X chromosome and that the Y chromosome does not carry a copy of the *w* gene. The patterns of inheritance are then understood (Figure 8-6). In the first cross, the males carry the *w* allele and have white eyes because there is no *w* gene on the Y chromosome. The condition in males in which there is an allele on the X and none on the Y is referred to as the *hemizygous* condition. The females are homozygous for the w^+ allele. In the F_1 the males receive their X chromosomes only from their mothers and are w^+ hemizygotes with brownish-red eyes. The F_1 females receive one X chromosome carrying the *w* allele from their fathers and one X chromosome carrying the w^+ allele from their mothers. Hence, all the F_1 females are heterozygotes, w/w^+, and have brownish-red eyes. In the F_2, half of the males receive a *w* allele and half a w^+ allele from their mothers. Hence, half the males have white eyes and half have brownish-red eyes. All of the F_2 females receive a w^+ allele from their fathers and must have normal-colored eyes, half being heterozygous, w/w^+, and half being homozygous, w^+/w^+.

In the reciprocal cross, the F_1 males receive X chromosomes carrying the *w* allele from their mothers, are hemizygous *w*, and have white eyes. The F_1 females receive an X chromosome from their fathers carrying the w^+ allele, an X from their mothers carrying the *w* allele, and are all heterozygotes (w/w^+) with brownish-red eyes. When the F_1 flies are mated, half of the ova produced in the females carry an X chromosome with the w^+ allele and half an X chromosome with the *w* allele. If these ova are fertilized by sperm carrying Y chromosomes, then half of the males are w^+ hemizygotes with normal eyes and half are *w* hemizygotes with white eyes. All of the sperm carrying X chromosomes carry the *w* allele, so that the females produced are half w/w homozygotes with white eyes and half are w/w^+ heterozygotes with brownish-red eyes. The correlation between the pattern of inheritance of the w^+ and *w* alleles and that of the X chromosomes indicates that the *white* gene is carried on the X chromosome. Furthermore, all genes in the same linkage group as *white* have the same pattern of inheritance. Thus, this particular linkage group can be assigned with confidence to the X chromosome, thereby showing sex linkage.

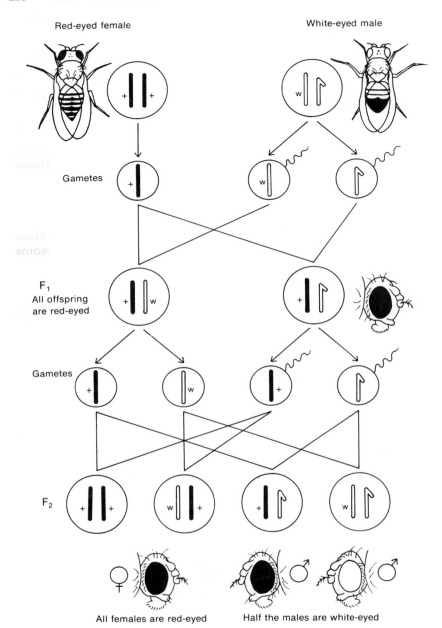

Figure 8-6   Sex-linked inheritance of *white* eye color in *Drosophila*.

In human pedigrees the appearance of a recessive trait every other generation in males indicates sex linkage.

In the first cross described on page 219, starting with mutant males and normal females, we saw that the F_1 progeny were all normal, but that in the F_2, half of the male progeny were mutant and had white eyes. Human pedigrees in which a mutant trait of a male parent is absent in his children but reappears in male grandchildren indicates sex linkage and is depicted in Figure 8-7. There are numerous sex-linked traits in humans, including red/green color blindness (which Sturtevant, the *Drosophila* geneticist, had), a form of hemophilia (abnormal bleeding because of reduced clotting of the blood), and the Lesch-Nyhan syndrome (p. 162). Queen Victoria, as shown in Figure 8-7, carried an allele for hemophilia on one of her X chromosomes and passed it to many of her descendents. One of Victoria's granddaughters, Alix, married Tsar Nicholas II and the allele eventually found its way into their son Alexander. Some believe that the preoccupation of Nicholas with his son's hemophilia distracted his attention from political problems and significantly contributed to the Russian Revolution and the overthrow of the Imperial government.

The presence of non-disjunction of X chromosomes was first deduced from genetic experiments in *Drosophila*.

Non-disjunction (pp. 142–143), we recall, is the failure of homologs during the first meiotic division or of sister chromatids during the second meiotic division to disjoin. Non-disjunction leads to aneuploidy, producing karyotypes with an extra chromosome or a deficiency of a single chromo-

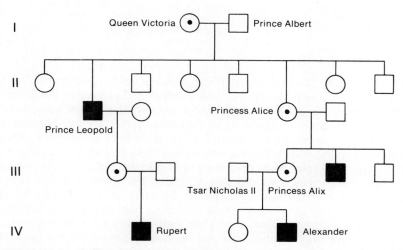

Figure 8-7   A pedigree of some of the descendants of Queen Victoria demonstrating the sex-linked inheritance of hemophilia. Black squares represent affected males; circles with black centers, carrier females.

some (2N + 1 or 2N − 1). It is interesting, not only as a piece of the history of genetics but also as another demonstration of the power of genetic analysis, that non-disjunction of X chromosomes was first deduced in *Drosophila* on the basis of genetic evidence and then confirmed by cytological observations. Evidence for non-disjunction of the X chromosomes in *Drosophila* females was first obtained by Thomas Hunt Morgan in a cross involving the *white* locus. Morgan crossed males with white eyes with females with red eyes. In the F_1 progeny 1,237 males and females had red eyes; 3 males had white eyes. With the exception of these three males the results are consistent with the inheritance of a sex-linked trait (Figure 8-6). Morgan initially attributed the presence of these males to new mutations. Several years later, Calvin Bridges, a graduate student working in Morgan's laboratory, proposed a correct explanation based on primary non-disjunction of the X chromosomes in females for the appearance of the exceptional offspring. We can see the effects of non-disjunction of the X chromosomes in females in a cross between males with normal eyes and females with white eyes. We designate the chromosomes carrying the w allele, X^w and those carrying the + allele, X^+. Primary non-disjunction in the females produces ova that carry two X chromosomes and some that carry none. These ova produce the following types of offspring when fertilized by X or Y carrying sperm:

Genotypes produced by primary non-disjunction

		Female Germ Cells	
		X^wX^w	O
Male	X^+	$X^wX^wX^+$ females that usually die	X^+O males with normal eyes
Germ			
Cells	Y	X^wX^w females with white eyes	YO dies

Bridges correctly concluded that YO zygotes die before adulthood. The XXX female offspring—initially called "super females" because of the presence of the extra X chromosome, now called metafemales—have poor viability. There is nothing super about these particular females—they had the wild type eye color expected for females even when they did survive. Bridges proposed that the male offspring with normal eyes had an XO karyotype and received an X^+ chromosome from their fathers instead of an X^w chromosome from their mothers as is normal. The XO males are, however, sterile because they lack the Y chromosome necessary for male fertility. Finally, Bridges proposed that the white-eyed female offspring had XXY karyotypes having received two X chromosomes from their mothers and a Y from their fathers. We should note that the chromosomal sex determination mechanisms differ between *Drosophila* and mammals. In mammals XXY is male, in flies, female; and in mammals XO is female, in flies, male, as long as the autosomal complements are diploid. In *Drosophila,* XXY females are completely fertile and because they already carry an extra sex chromosome, they undergo frequent secondary non-disjunction. Secondary non-disjunction produces large numbers of XX- and particularly XY-carrying ova, as well as ones carrying X and Y chromosomes. When Bridges mated XXY females with normal males, offspring were produced in the following proportions:

Secondary non-disjunction in XXY female *Drosophila*

		Female Germ Cells			
		X^wX^w (2)	X^wY (23)	X^w (23)	Y (2)
Male	X(1)	X^wX^wX dies	X^wXY female	XX^w female	XY male with normal eyes
Germ					
Cells	Y(1)	X^wX^wY female with white eyes	X^wYY male	X^wY male	YY dies

Numbers in parentheses indicate the relative numbers of male and female germ cells produced.

Now, large numbers of offspring had phenotypes different from those expected for normal sex linkage. Namely, males with normal eyes had presumably received their X chromosomes from their fathers and white-eyed females were presumed to have received two X chromosomes, each carrying a *w* allele from their mothers. These white-eyed females, according to Bridges' explanation, had an XXY karyotype. Light microscopic observations of chromosomes from these females revealed an XXY karyotype and confirmed Bridges' proposals. The abnormal inheritance of the *w* alleles in these crosses resulted from the abnormal transmission of the X chromosomes themselves and provided strong support for the chromosomal theory of inheritance that Morgan and his colleagues were espousing.

Permanent transmission of paternal X chromosomes to sons and maternal X chromosomes to daughters occurs when two X chromosomes are attached to a single centromere.

Lilian Morgan (whose husband, incidently, was Thomas Hunt Morgan) extended the work on the X chromosome. She discovered a pattern of inheritance of sex-linked genes that was even more extreme than the one caused by non-disjunction. In a cross of white-eyed females (*w/w*) with normal males with brownish-red eyes, all of the males in the F_1 had wild-type, brownish-red eyes and all the females had white eyes. Lilian Morgan deduced and subsequently demonstrated that the two X chromosomes in the female had become attached and shared a single centromere, producing an attached-X stock. During meiosis in these females (Figure 8-8), two types of germ cells are produced, those with the attached-X chromosome and those lacking any sex chromosome. Mating with a normal male produces XXY females with white eyes and XO sterile males with brownish-red eyes. Crossing the XXY females with XY males (Figure 8-8), produces a stable line

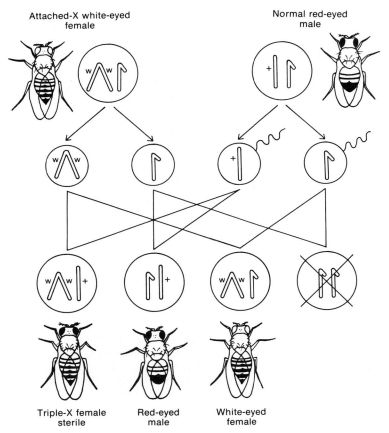

Figure 8-8   Permanent maternal inheritance of attached-X chromosomes in *Drosophila*.

in which the male progeny are XY, receiving a Y from their mothers and an X from their fathers, and the female progeny are XXY, receiving the attached-X from their mothers and the Y from their fathers.

Chromosomal abnormalities allow assignment of linkage groups to specific autosomes.

The pattern of inheritance of X-linked genes, corresponding to the transmission of X chromosomes, allows the assignment of one linkage group to a particular chromosome. Hence, sex-linked genes are readily identifiable in any organism with sex chromosomes. Indeed, in humans, until recently the only genes that had been assigned to a particular chromosome were those carried on the X. In *Drosophila* the assignment of genes to the X chromosome was confirmed through the abnormal transmission of X chromosomes, which correlated with the abnormal pattern of inheritance of the chromosomes, for example, attached-X chromosomes. The deduced existence of abnormal X chromosomes was confirmed by light microscopy. We can also assign linkage groups to specific autosomes, that is, non-sex chromosomes by utilizing chromosomal abnormalities. Two abnormalities are particularly useful: *translocations* and *inversions.* In a translocation, as discussed before (pp. 148–149), part of an arm of one chromosome is broken from its normal position and attached to another chromosome. Translocations are reciprocal or non-reciprocal, that is, two arms can be exchanged or one arm can be added to an otherwise complete non-homologous chromosome. The chromosomal rearrangement in the *cri du chat* syndrome is a non-reciprocal translocation (Figure 5-23). An inversion is a rearrangement in which a chromosomal segment is inverted with respect to its normal condition. The word "inrevsion" contains an inversion of three letters. The inheritance of translocated chromosomes and chromosomes with inversions differs from normal and allows the detection of translocations and inversions genetically. In addition, the translocated and inverted chromosomes can be detected by light microscopy. By combining the genetic and microscopic observations, linkage groups can be assigned to particular chromosomes.

Translocations cause aberrant chromosomal transmission, with only part of the meiotic products resulting in viable offspring.

X-rays cause breaks in chromosomal arms that can heal in new configurations and cause translocations. We will consider the effects of a non-reciprocal translocation on the inheritance of two linkage groups. The following hypothetical chromosomes are involved:

A B C   and   L M N

where the letters designate sequences specific to each of the chromosomes, which, we can see, are non-homologous. A non-reciprocal translocation produces the following:

A B C N   and   L M.

When the stock is homozygous for the translocation, that is, when both copies of the two chromosomes in a diploid cell have the translocation condition, the germ cells contain all of the DNA sequences from A to C and from L to N and produce viable offspring. When we cross the translocated strain to a normal one, a translocation heterozygote results, that is, an organism that carries one copy of each normal chromosome and one copy of each translocated chromosome (Figure 8-9). The translocation heterozygote is viable and "balanced," having two copies of all of the sequences. During meiotic prophase I in the translocation heterozygote, the homologous regions of the chromosomes synapse, but their new physical relationships lead to unusual pairing configurations that can be visualized by light microscopy (Figure 8-9). When meiosis is complete, two types of germ cells are produced from normal disjunctions, half having an abnormal sequence content and half having a complete sequence content (Figure 8-9).

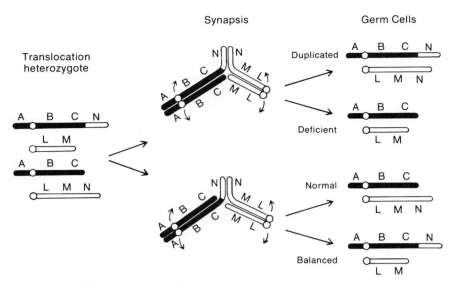

Figure 8-9   Effects of reciprocal translocation of chromosomal segments.

Recalling the *cri du chat* syndrome (p. 148), we remember that the presence of an extra chromosomal arm or the absence of a chromosomal arm is lethal. In the example now before us, we see that two sets of meiotic products contain complete sequences and will produce viable zygotes and two sets have abnormal sequences and will produce inviable offspring. In the absence of the translocation, genes on the ABC chromosome are not linked with those in region N of the LMN chromosome, and their alleles assort independently. However, because of the translocation we now find that genes in the N region—being part of the same chromosome—are linked to those in the ABC chromosome. We can deduce that the chromosomal segment carrying genes in the N region has been translocated to the chromosome carrying the ABC linkage group. When we view chromosomes from the translocated stock by light microscopy, we see that one chromosome is shorter than normal (the one that donated part of a chromosome arm) and another is longer (the recipient). This condition indicates that the normal counterpart of the short chromosome carries linkage group LMN, and the normal counterpart of the longer chromosome carries linkage group ABC. We also know that genes in the N region, the one determined to be translocated, reside in the translocated segment. Thus, we are able to assign two linkage groups to two specific chromosomes, and by an extension of these methods, we could assign all linkage groups to specific chromosomes.

The recovery of recombinant progeny from inversion heterozygotes is suppressed.

Inversions also affect the pattern of gene transmission and the appearance of the karyotype and can be used to assign linkage groups to specific

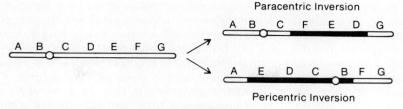

Paracentric Inversion

A B C F E D G

A E D C B F G

Pericentric Inversion

Figure 8-10   Paracentric and pericentric inversions.

chromosomes. There are two types of inversions, *paracentric* inversions and *pericentric* inversions. Paracentric inversions are separate from the centromere. Pericentric inversions include the centromere within the inverted segment (Figure 8-10). When a strain carrying a chromosome with an inversion is crossed to a normal strain the progeny are inversion heterozygotes. The chromosome carrying the inversion can be identified microscopically when it is synapsed with the normal chromosome, either during meiosis or, in the case of *Drosophila,* using polytene chromosomes, because these chromosomes are composed of paired homologs that have undergone endoreplication. The homologous sequences of the chromosomes synapse. For the chromosomal regions separate from the inverted segment, the pairing is normal. However, for the inverted segment to pair with its homologous region in the normal chromosome a segment must twist around in a loop (Figure 8-11). This inversion loop is readily identified microscopically (Figure 8-11).

In inversion heterozygotes, recombination within the boundaries of the inversion produces abnormal chromosomes. The effect of recombination depends on the nature of the inversion.

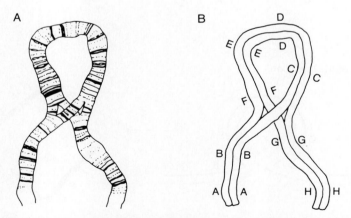

Figure 8-11   A loop in the salivary chromosomes of an inversion heterozygote of *Drosophila.* (A) Drawing of a cytological preparation. After Dobzhanzky and Socolov *Journal of Heredity,* v. 30, 9, 1939. (B) Pairing of homologous regions.

Paracentric inversions. In chromosomes heterozygous for a paracentric inversion, recombination within the inversion loop creates one chromosome containing two centromeres (a dicentric) and another one lacking a centromere (an acentric) (Figure 8-12). During the subsequent meiotic division, a bridge forms as the two centromeres move to opposite poles and eventually breaks, leaving each meiotic product with a piece of chromosome. The acentric fragment moves to neither pole and is lost. As a result of the recombination the products of meiosis lack their normal sequence content and any zygotes formed are lethal. Thus, the only offspring are those that carry chromosomes that did not recombine within the inversion loop.

Pericentric inversions. In chromosomes heterozygous for a pericentric inversion, recombination within the inversion loop creates two chromosomes, each of which has a centromere but lacks the sequences from one end of the chromosome and is duplicated for those at the other end (Figure 8-12). Germ cells carrying these chromosomes are deficient for some sequences and duplicated for others. If these cells fertilize a normal germ cell, then the zygote produced has an abnormal sequence content and is inviable. Again, only cells carrying chromosomes that have not undergone recombination within the inverted segment will be recovered.

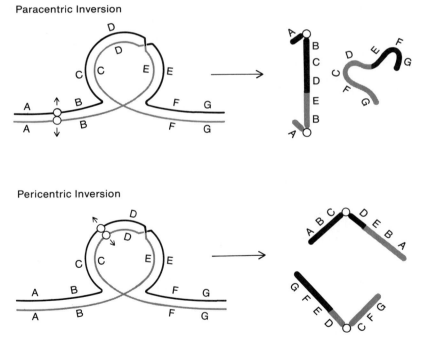

Figure 8-12   The consequences of recombination within paracentric and pericentric inversion loops.

Both types of inversions prevent the recovery of progeny that are recombinant for the allelic markers carried on the inverted chromosomes. The absence of observed recombination identifies the linkage group carried by the inverted chromosome. When this information is combined with microscopic identification of the inverted chromosome, a linkage group can be assigned to a particular chromosome.

We can relate genetic maps to physical maps in eukaryotes using chromosomal aberrations.

In Chapter 7, we discussed the procedures used to relate the genetic map to the physical map of the λ chromosome. The comparison relied on the use of chromosomal abnormalities that could be detected by viewing hybrid DNA molecules with an electron microscope. We can also relate the genetic maps and the physical map of chromosomes in eukaryotes through the use of chromosomal abnormalities. Because of the polytene chromosomes and the high level of resolution they afford, *Drosophila melanogaster* is an organism particularly suited for relating physical and genetic maps. The polytene chromosomes are produced by replication of the DNA without any separation of the newly replicated strands. At the same time, the homologous chromosomes pair in somatic cells. If one views a polytene chromosome in which one chromosome is "standard" and the other chromosome is abnormal in its structure, it is possible to detect the precise position of an abnormality. For example, a deficiency involving a few hundred thousand nucleotide pairs can be located in a polytene chromosome preparation from an organism that is heterozygous for the deficient chromosome and a normal chromosome. The two chromosomes are paired throughout their entire length except for the deficient region with which the corresponding sequences on the normal chromosome cannot pair. The normal chromosome bulges out opposite the deficiency (Figure 8-13a) unless the deficiency is located at one of the tips of a chromosome, in which case an unpaired region is present (Figure 8-13b). Deficiencies are also genetic markers, typically being recessive lethals or causing visible phenotypic alterations, and can be mapped genetically. Their physical positions can be determined by light microscopy, and we can compare genetic and physical locations of the

A

B

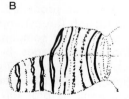

Figure 8-13   Positions of deficiencies are readily demonstrated in polytene chromosomes of deficiency heterozygotes. (A) Internal deficiency. After Bridges, Skoog, and Lee, *Genetics*, v. 21, 788–795, 1936. (B) Terminal deficiency. After Demerec and Hoover, *Journal of Heredity*, v. 27, p. 206, 1936.

deficiencies. Such a comparison of the X chromosome of *D. melanogaster* is depicted in Figure 8-14. We see that the order of genes on the genetic map is the same as their physical distribution on the chromosome. The physical and genetic spacing between the loci differ because the frequency of recombination per unit chromosome length is higher near the middle of the chromosome than near the telomeric tip or near the centromere.

Particular genes can be associated with single polytene chromosome bands in *Drosophila*.

The use of deficiencies in *Drosophila* can be extended to allow a precise locating of a particular gene, for example, the locus of the X-linked gene *facet* (*fa*). The compound eyes of insects are composed of large numbers of individual facets (Figure 6-6). The *fa* allele causes abnormalities in the structure of the facets, giving the eye an overall rough appearance. The *fa* allele is recessive and in the hemizygous form results in males with rough eyes. If a *fa* allele is heterozygous with a deficiency that includes the *fa* locus, the fly also has rough eyes. We can obtain a collection of deficiencies, all of which result in rough eyes when made heterozygous with a *fa* allele. These deficiencies will be of varying sizes and have different chromosomal boundaries, but all will have in common the absence of the chromosomal region where the *fa* gene is normally located. By examining a series of chromosomes with deficiencies by light microscopy, we can identify one band or chromomere that is absent in all of the deficiencies (Figure 8-15). In normal chromosomes, the *fa* locus must reside at or near this band. In the case of the *fa* gene it has been found to reside in or near band 3C7 on the X chromosome. Other genes have been similarly placed to specific bands, for example, *white* to 3C2.

Genetic maps of eukaryotic chromosomes are much less complete than those of phages.

The ultimate result of mapping procedures and the procedures relating specific chromosomes to specific linkage groups is a detailed genetic map of an organism. As we have noted before, in eukaryotes extensive genetic maps have been produced for only a handful of organisms. These include yeast (although the cytology is deficient), *Neurospora, Drosophila,* maize, and mice. We can compare the completeness of eukaryotic genetic maps with those from phages. In λ, which has a chromosome with about 46,500 nucleotide pairs, some 40 genes have been identified and mapped, about one gene per 1000 base pairs. On the X chromosome of *Drosophila melanogaster,* which is probably the best mapped chromosome of any higher eukaryote, some 500 genes have been identified and mapped, about one per 70,000 base pairs. Either the genetic map is incomplete and there is much work to do, or the genetic organization of genes in *Drosophila* in particular and eukaryotes in general differs from phage. In fact, both are probably true!

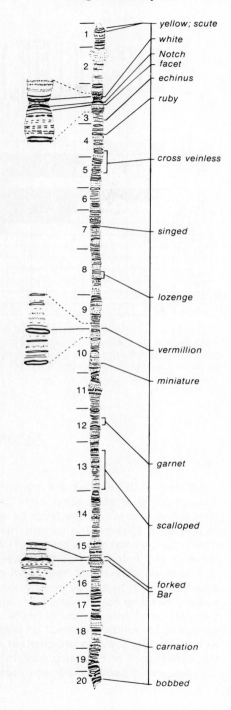

Figure 8-14   The genetic map (based on recombination frequencies) of the X chromosome of *Drosophila melanogaster* is compared with the cytological map (based on the positions of chromosomal aberrations).

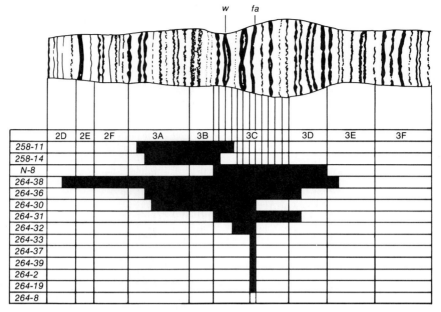

Figure 8-15   The *facet* gene is placed by deficiency mapping to a single band (or interband) in the polytene chromosomes of *Drosophila melanogaster*. The only band absent in all constitutions that, when heterozygous with a recessive *facet* allele, expresses the *facet* phenotype is 3C7. After Slizynska, *Genetics,* v. 23, pp. 291–299, 1938.

SUMMARY

Genes in eukaryotes are transmitted from parent to offspring either independently of each other or in linked fashion. Genetic maps are constructed by analysis of the types of test cross progeny in a manner similar to that used to construct genetic maps in phages. Recombination occurs when four chromatids are present. The frequency of detected recombination between two loci is limited to 50 per cent. The number of linkage groups should equal the haploid number of chromosomes. Linkage groups can be associated with particular chromosomes, the assignment to the X chromosome depending on the pattern of inheritance of a linkage group paralleling the pattern of inheritance of an X chromosome. Assignment of linkage groups to autosomes depends on the use of chromosomal aberrations such as translocations and inversions that affect both patterns of inheritance and the appearance of chromosomes. In *Drosophila melanogaster,* using deficiencies, genes can be assigned to specific bands in polytene chromosomes.

Non-nuclear Inheritance
and Somatic Cell Genetics

NON-NUCLEAR INHERITANCE

The genetic maps that show that genes are distributed in a linear array are consistent with our notion that genes are distributed at different places along a DNA molecule and DNA molecules carry genetic information. DNA is not, however, limited to the nucleus of eukaryotic cells but is also found in other organelles, notably in mitochondria, the centers of energy production, and in plant chloroplasts, the centers of photosynthesis. The presence of DNA in mitochondria and chloroplasts makes it likely that these organelles carry genetic information but is not proof. We have already established many of the attributes and conditions that allow us to view nuclear DNA as a carrier of genetic information. We can do no less for organelle DNA. We must demonstrate that there is constancy in the organelle chromosome constitution in the same sense that there is constancy in the nuclear chromosomal karyotype, that the DNA is faithfully replicated, and, finally, that replicated copies of the DNA are included in successive generations of organelles by demonstrating that the organelles reproduce themselves and are not created *de novo* under the complete direction of the nuclear genome. We must also demonstrate that the inheritance of some traits is associated with inheritance of particular types of mitochondria or chloroplasts. We will briefly establish for mitochondrial and chloroplast genomes what we have already established for the nuclear genome. We will show that mitochondria and chloroplasts satisfy the minimal requirements to be carriers of genetic information. They contain chromosomes that are constant in size and nucleotide sequence, possess mechanisms for the replication of their chromosomes, reproduce themselves, and, finally, carry genes specifying some phenotypes.

Mitochondria and chloroplasts contain circular chromosomes of double-helical DNA.

Each species has one or a few forms of mitochondrial or chloroplast chromosomes. The chromosomes of mitochondria and chloroplasts, like those of bacteria, are composed of double-helical DNA and are circular. They contain no histones and, thus, no nucleosomes. The size of the mitochondrial or chloroplast genome varies from species to species but is essentially constant within the mitochondria of a particular species. Mitochondria of higher animals usually have smaller chromosomes than those of all fungi and plants (Table 9-1). The mitochondrial chromosome of humans, for example, contains only about 14,000 nucleotide pairs, fewer than that of *Drosophila* (18,000 nucleotide pairs) or fungi (yeast, 75,000; *Neurospora crassa,* 60,000). Despite their larger size, the mitochondrial chromosomes of lower eukaryotes contain about the same amount of single-copy DNA as those of higher eukaryotes, the additional DNA resulting from the presence of short, highly repeated sequences. In general, the size of the mitochondrial chromosome is similar to that of a bacterial plasmid. Chloroplast genomes are about as large as those of mitochondria. Not only are mitochondrial and chloroplast chromosomes of constant size but they are also of constant sequence in a particular species. (We ignore inherited variation, which, relative to nuclear DNA, is substantial.) The constant sequence of organelle DNA is demonstrated by digestion with restriction endonucleases (pp. 70–71) and the production of restriction maps. Because restriction enzymes cleave at specific sequences, the production of restriction maps with organelle DNA implies that the sequences are essentially

Table 9-1   Properties of Nuclear, Mitochondrial, and Chloroplast DNA's.

	Nuclear DNA		Mitochondrial DNA		Chloroplast DNA	
	Per cent GC	Genome Size	Per cent GC	Genome Size	Per cent GC	Genome Size
ANIMALS						
Human	41	2.75×10^9	41	14×10^3	—	—
Chick	43	1.2×10^9	49	15×10^3	—	—
Sea Urchin	39	7.2×10^8	45	13×10^3	—	—
D. melanogaster	42	1.75×10^8	22	18×10^3	—	—
FUNGI						
N. crassa	53	4.3×10^7	42	60×10^3	—	—
S. cerevisiae	44	1.75×10^7	23	75×10^3	—	—
PLANTS						
C. reinhardi	64	1×10^8	71	2×10^5	36	30×10^3
Maize	49	6.6×10^9	48	2.2×10^5	39	13×10^3
Tobacco	38	1.1×10^9	51	2×10^5	47	15×10^3
Sweet Pea	35	9.4×10^9	38	9×10^4	46	12×10^3

Genome size in base pairs. *C. reinhardi* = *Chlamydomonas reinhardi*. Each mitochondrion is believed to contain about 5 chromosomes and each chloroplast about 20 chromosomes. Genome size of mitochondrial and chloroplast DNA is assumed to equal the size of one chromosome except in Maize mitochondria where several different chromosomes are believed to be present.

Data mainly compiled from "Chloroplast DNA:Physical and Genetic Studies" by Ruth Sager and Gladys Schlanger and "Mitochondrial DNA" by Margit M. K. Nass, in *Handbook of Genetics,* Robert C. King, Ed., v. 5, Molecular Genetics, p. 371 and p. 477, Plenum Press, 1976.

constant. The constancy in mitochondrial or chloroplast sequences in a species implies that the organelle chromosomes replicate. The size of any mitochondrial or chloroplast genome is substantially smaller than the nuclear genome of the same species. Also, the base compositions of the mitochondrial or chloroplast DNA usually differ from those of nuclear DNA. Thus, in three fundamental characteristics—base composition, size and structure—mitochondrial or chloroplast DNA differs strikingly from nuclear DNA.

Mitochondrial and chloroplast DNAs are replicated semiconservatively by polymerases that are unique to the organelles.

The DNA composing the chromosomes of mitochondria and chloroplasts is replicated semiconservatively through mechanisms similar to those we have already described. At least in mice, the replication of the mitochondrial chromosome proceeds unidirectionally from one replication origin (The eye form in Figure 4-7 actually shows the replication origin in the mitochondrial chromosome of the mouse.) This replication is under the direction of DNA polymerases found exclusively in mitochondria. The polymerases differ from nuclear polymerases. Thus, at least some of the enzymes involved in mitochondrial DNA replication are distinct from their nuclear counterparts.

Mitochondria and chloroplasts reproduce themselves.

Mitochondria and chloroplasts grow and divide. In an electron micrographic study, Irene Manton demonstrated in the 1950's that these organelles divide to produce daughter organelles. Manton utilized the unicellular alga, *Micromonas,* which contains a single mitochondrion and a single chloroplast. She demonstrated that the mitochondrion and chloroplast divided in synchrony with the nucleus and that the daughter organelles were equally distributed to the two daughter cells.

In 1963, David Luck drew the same conclusion after a detailed study of mitochondrial division in *N. crassa* using density labelling procedures. Luck discovered that when a strain of *Neurospora* was grown in a medium rich in a lipid precursor and low in protein, the mitochondria produced were less dense than those found when the *Neurospora* were grown in a medium rich in protein and low in lipid precursor. When *Neurospora* that had been grown in the lipid-rich, low-protein medium (and, therefore, had low-density mitochondria) were transferred to high-protein medium for one generation of cell growth, the number of mitochondria doubled and all were of a density intermediate between the low-density and high-density ones. The interpretation of Luck's experiments (Figure 9-1) is that during growth the mitochondria assimilate different amounts of lipid. The light mitochondria assimilate less lipid after transfer to the protein-rich medium and, therefore, become more dense. When the mitochondria divide, each daughter mitochondrion has a density intermediate between the light and the heavy ones. If new mitochondria were being made independently of the old ones, then two types of mitochondria should be present, half of them heavy, half light.

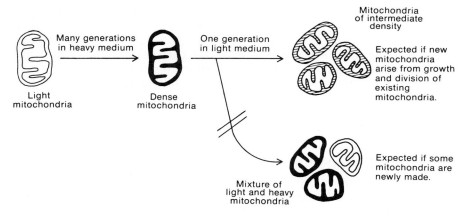

Figure 9-1   The experiment by Luck demonstrating that mitochondria arise by division of existing mitochondria.

Finding only mitochondria with intermediate density, Luck concluded that all mitochondria derived from existing ones. We can draw a similar conclusion about chloroplasts.

Because female gametes are rich in cytoplasm and male gametes contain little, we expect traits specified by organelle genomes to be inherited through the maternal parent.

The properties necessary for genetic information to be carried in mitochondria or chloroplasts are present: The organelles contain chromosomes of constant physical size and nucleotide sequence, the chromosomes are replicated semiconservatively, and the organelles themselves undergo division and are transferred to daughter cells. We now have the problem of identifying traits specified by mitochondrial or chloroplast genomes. We expect such traits to be inherited independently of the nuclear chromosomes and to be transmitted in the cytoplasm. The ovum of a higher plant or animal contains a large amount of cytoplasm that includes, in animals, mitochondria, and in plants, chloroplasts and mitochondria. The male gamete, in contrast, is deficient in cytoplasm. At the time of fertilization the male gamete usually contributes no mitochondria or chloroplasts to the zygote. We inherit our mitochondria from our mothers, and we expect that traits that are encoded in the mitochondrial or chloroplast chromosomes to be transmitted through maternal parents. Even in lower eukaryotes, for example, in fungi, we find that one germ cell contributes both nucleus and cytoplasm to the zygote, and the other germ cell contributes only a nucleus. In *Neurospora crassa* where there are two mating types, *A* and *a* (p. 212), the contribution of the conidium (spore) is limited to the nucleus, and the contribution of the ascogonium within the fruiting body includes both nucleus and cytoplasm, and, of course, the mitochondria.

Non-nuclear or cytoplasmic inheritance is demonstrated by showing that the contribution from the maternal parent determines the genotype of the offspring.

The mutant trait *poky* in *N. crassa* provides a classic example of maternal inheritance and, therefore, presumably of inheritance of traits carried in the mitochondrial genome. Mary and Herschel Mitchell described the *poky* trait in 1952. The mutant is phenotypically characterized by hyphae that grow much more slowly than their wild-type counterparts. The non-nuclear or maternal inheritance of *poky* is demonstrated (Figure 9-2) by reciprocal crosses in which either the "male" or "female" parent is *poky*:

poky "females" × *poky*⁺ "males" produce ascospores, all of which germinate to form *poky* hyphae.

poky⁺ "females" × *poky* "males" produce ascospores all of which germinate to form normal hyphae.

There are two striking differences between these results and the results from *Neurospora* that we discussed earlier. First, there is no 1:1 segregation of the *poky* and *poky*⁺ spores as we expect for traits carried on nuclear chromosomes, for example, the mating type alleles, *A* and *a* (Figure 9-2). Because the *poky* and *poky*⁺ traits do not segregate, we know that the *poky* gene is not linked to any nuclear chromosome (all of the nuclear chromosomes, of course, segregate during meiosis). Second, the genotype of the female parent determines the genotype of the offspring. As long as the female gamete is *poky*, all of the offspring will be *poky*. This condition passes from generation to generation as is characteristic of an inherited trait.

The *poky* phenotype is associated with mitochondrial abnormalities.

We are justified in concluding that the *poky* condition is not transmitted by the nuclear genome. But can we conclude that it is transmitted by the mitochondrial genome? Support comes from several observations on the properties of mitochondria in *poky* mutants. All mitochondria contain ribo-

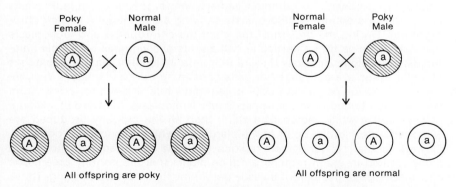

Figure 9-2   The absence of normal segregation in reciprocal matings involving *poky* and *poky*⁺ alleles in *Neurospora*.

somes, particles on which protein synthesis occurs (pp. 320–322) that are distinct from the ribosomes found in the cytoplasm. David Luck and his co-workers have demonstrated that the ribosomes of *poky* mutants are abnormal, perhaps because of a single abnormal component. The abnormality of mitochondrial structure coupled with the maternal inheritance of the trait provides a strong but circumstantial argument for the conclusion that a mutant gene in the mitochondrial chromosome causes the *poky* trait.

Mitochondrial mutants characterized by slow growth are also found in yeast.

In yeast, there are mutants called *petites,* characterized phenotypically by slow growth, that are inherited in a non-nuclear fashion. Like *poky, petites* have abnormal mitochondria in which components in the electron transport chain are missing as are components of the ribosomes (just as in *poky*). The base compositions of the *petite* mitochondrial DNAs differ from those of the wild type. (The GC composition of the mitochondrial DNA of some *petites* can be higher or lower than in wild type.) Regions of the normal mitochondrial chromosome are absent in *petites.* The observation that the mitochondrial DNA is abnormal in these cytoplasmic mutants compels us to conclude that the *petite* phenotype is inherited through the mitochondrial genome. As in *Neurospora,* there are mitochondrial components in yeast that are specified by genes in the mitochondrial chromosome.

Mutant traits transmitted by chloroplasts have also been identified.

Many plants such as wandering jews (*Zebrina pendula*) and four o'clocks (*Mirabilis jalapa*) have variegated leaves in which parts of the leaves, often stripes, are white and other parts are green (Figure 9-3). There are two populations of plastids present in these plants: normal green chloroplasts containing chlorophyll and abnormal colorless ones, *leukoplasts,* in which chlorophyll is absent. The simplest explanation for the nature of leukoplasts is that they are abnormal chloroplasts carrying a mutant gene in their chromosomes. During the growth of the plant the chloroplasts and leukoplasts may segregate so that a particular cell and its descendents have only chloroplasts or only leukoplasts (Figure 9-3). In a variegated plant, the patches of white tissue are composed of cells containing only leukoplasts. We presume that leaves containing both green and white tissue have arisen from cells that contained both chloroplasts and leukoplasts. In four o'clocks, if a flower is on a white region of a stem, then all the progeny produced are white regardless of the color of the plant from which the pollen is obtained. Flowers located in variegated regions give rise to variegated plants and those located in green regions produce plants that are entirely green. The characteristics of the maternal tissue determine those of the offspring. These characteristics are inherited generation after generation through the maternal cytoplasm.

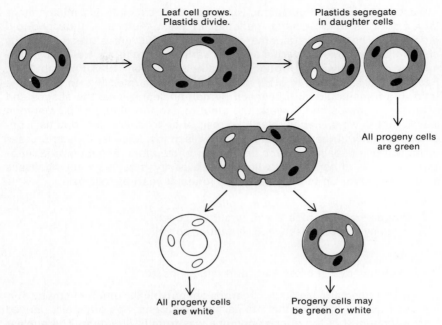

Leaf cell grows.
Plastids divide.

Plastids segregate
in daughter cells

All progeny cells
are green

All progeny cells
are white

Progeny cells may
be green or white

Figure 9-3   The basis for variegation of white and green tissue in four o'clocks.

The components of mitochondria and chloroplasts are specified by genes in both the nuclear and organelle genomes.

In the examples we have just discussed, we learned that there are genes separate from those in the nuclei. The genes are carried in the chromosomes of either chloroplasts or mitochondria and specify components in these organelles. The total inventory of mitochondrial and chloroplast components, however, is specified by genes in both the nuclear and organelle genomes. Two examples will suffice. In mitochondria, one of the major components of the electron transport chain is cytochrome C. In yeast, the gene for cytochrome C is located on chromosome 10 in the nuclear genome. In chloroplasts the first enzyme involved in the fixation of CO_2 during photosynthesis, ribulose bis-phosphate decarboxylase, contains two parts. One of these parts, the small component, is specified by a gene in the chloroplast chromosome; the other, the large component, is specified by a nuclear gene. This enzyme, incidently, is the major protein found in chloroplasts and in plant tissues, where it constitutes about 50 per cent of the protein. It is, therefore, the most prevalent protein in the world.

Non-nuclear genes control a wide range of phenotypic characteristics.

The amount of DNA in the nuclear genomes of eukaryotes is several orders of magnitude larger than that present in the mitochondrial or chloroplast genomes. The human nuclear genome, for example, is 2 million times larger than its mitochondrial genome; in yeast there is a thousand-fold difference. The nuclear genome of maize is 20,000 times greater than its chloroplast genome. Nevertheless, the variety of traits controlled by non-nuclear genes is wide. Through extensive study from 1923 through 1948, Paul Michaelis, a German plant geneticist, demonstrated that in plants the following traits could be influenced by non-nuclear genes: the shapes and positions of leaves and flowers, the amount and the nature of pigments in leaves and flowers, viability, fertility, resistance to chemicals, optimal conditions for growth, height, and others. In his study, Michaelis made crosses using two plant species, *Epilobium luteum* and *Epilobium hirsutum* (*Epilobium* is a genus of herbs with willow-like leaves and small flowers). Michaelis made crosses for up to 25 generations (one per year) as follows:

1. *E. luteum* ova × *E. hirsutum* pollen.
2. Progeny ova of cross 1 × *E. hirsutum* pollen.
3. Progeny ova of cross 2 × *E. hirsutum* pollen.
.
.
.
25. Progeny ova of cross 24 × *E. hirsutum* pollen.

He identified and isolated numerous phenotypically different strains during the crosses. In all of the strains, the cytoplasm was originally derived from *E. luteum,* but the nuclear genome was from *E. hirsutum.* The nuclear genomes in the different strains became identical because the progeny in each cross were always mated to a single true-breeding strain of *E. hirsutum.* Thus, any inherited differences between the strains had to result from differences in non-nuclear genes. We can conclude that the contribution of the non-nuclear genome to the phenotype, although secondary to the contribution of the nuclear genome, is substantial. There are no animal cells without mitochondria and no independently viable higher plants without chloroplasts. The point is simple: we should not be nuclear chauvinists.

Chromosomes of chloroplasts, as well as chromosomes of mitochondria, recombine. In *Chlamydomonas* the genetic map of the chloroplast chromosome is circular.

Ruth Sager and her collaborators in a study utilizing mutants carried by chloroplast chromosomes in *Chlamydomonas reinhardi* provided a particularly interesting demonstration of the recombination that occurs between organelle chromosomes. *Chlamydomonas* is a haploid unicellular alga that swims using two flagellae. As in yeast and *Neurospora,* there are two mating types in *Chlamydomonas,* called *mating type+, mt+* and *mating type −, mt−,* each of which is determined by an allele carried on a nuclear chromosome. Mating occurs (Figure 9-4) when two cells of the opposite mating type fuse to form a diploid zygote, which then develops into a heavy-walled cyst. The cyst germinates after 10 days to produce four or eight haploid cells as a result either of meiosis or of meiosis followed by a single mitotic division. Because the mating type alleles segregate, half of the cells are *mt−* and half

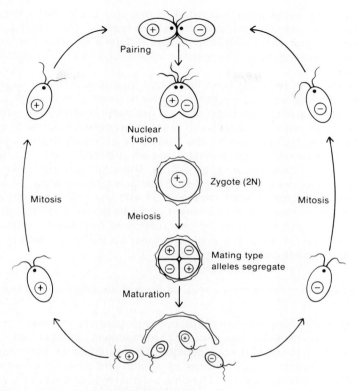

Figure 9-4   The life cycle of *Chlamydomonas*.

are *mt*⁺. There are traits inherited in a non-nuclear fashion and carried on the chloroplast chromosome. An example of such a trait is a mutant that renders cells resistant to the antibiotic erythromycin, which normally kills *Chlamydomonas* cells. Sensitive cells are *ery*ˢ and resistant cells are *ery*ʳ. In a cross of *mt*⁺, *ery*ʳ × *mt*⁻, *ery*ˢ the offspring are usually all *ery*ʳ. In the reciprocal cross, *mt*⁺, *ery*ˢ × *mt*⁻, *ery*ʳ, the offspring are usually all sensitive to erythromycin.

The pattern of inheritance of resistance to erythromycin is typical of non-nuclear genes. The *uniparental* pattern of inheritance occurs in *Chlamydomonas* despite contributions by both mating types of equal amounts of cytoplasm to the zygote and the absence of any evidence of degradation of the chloroplasts from the *mt*⁻ cells. Some mechanism must assure that chromosomes of chloroplasts from *mt*⁺ cells are preferentially recovered in the meiotic products. There is, however, a low level of biparental inheritance of non-nuclear traits in *Chlamydomonas,* with about 5 per cent of the offspring receiving traits (for example, *ery*ʳ and *ery*ˢ) from both parents. We interpret biparental inheritance as resulting from the presence in the progeny of two types of chloroplasts or chloroplast chromosomes, one from each parent. During subsequent mitotic divisions the allelic traits segregate, daughter cells receiving only one of the two types of chloroplasts. This is reminiscent of the variegation that occurs in four o'clocks when leaf cells receive only leukoplasts or chloroplasts instead of a mixture of chloroplasts and leukoplasts.

In some cells resulting from biparental inheritance, stable segregants are produced that carry allele constitutions originating from both parents. We conclude that the production of these cells with recombinant genotypes results from recombination between chloroplast chromosomes. Sager and her colleagues, by determining the frequency at which stable lines with varying recombinant genetic constitutions are produced, constructed a genetic map of the chloroplast chromosome. There are two particularly interesting aspects of the map (Figure 9-5). It contains a hypothetical attachment site by which the chromosome is associated with an unidentified chloroplast structure. We may recall the attachment of the bacterial chromosome to the cell membrane and its role in the segregation of replicated chromosomes during cell division. Also, the map is circular as we would expect for the circular chloroplast chromosome.

Chloroplasts and mitochondria appear to be evolutionarily related to bacteria.

There are numerous similarities between bacteria, mitochondria, and chloroplasts. All three contain circular chromosomes composed of "naked" DNA lacking histones and nucleosomes. The structure of the ribosomes, sites of protein synthesis, of all three are similar and differ from those found in the cytoplasm of eukaryotes. The DNA polymerases of bacteria, and some mitochondria are sensitive to the drug rifampicin, which binds to the polymerases and prevents the replication of DNA. Eukaryotic nuclear polymerases are insensitive to rifampicin.

We can guess that the evolutionary origin of mitochondria, for example, resulted from the infection of a eukaryotic cell by a bacterial ancestor. The descendents of the original infective bacterium and infected cell have evolved to be mutually dependent, the mitochondria performing functions necessary to the well-being of the cell (the production of energy in the form of ATP) and the cell performing functions necessary for the mitochondria (providing some of the structural components). An animal cell could not live without its mitochondria (animal cells die when they are deprived of oxygen) nor can a mitochondrion exist and reproduce outside of the cell (it simply cannot make all of its structural components).

If organelles evolved from infecting bacteria, it would not be too surprising to find repetitions of such an evolutionary event. Perhaps we can identify traits inherited in a non-nuclear fashion through agents that are, in a sense, infecting the cells and may be engaged in the first stages of evolving into new intracellular organelles. We will look at two examples, one involving a bacterium whose origin was initially external to the cell in which it is now harbored, the other involving an RNA cancer virus, mouse mammary tumor virus, whose evolutionary origin may have been within the cell itself.

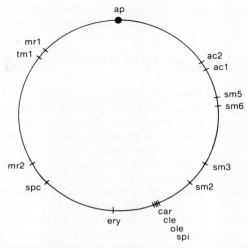

Figure 9-5   A circular genetic map of the chromosome of *Chlamydomonas*. After Singer, Sager, and Ramanis, *Genetics*, v. 83, p. 341, 1976.

Maternally inherited abnormal sex-ratios in *Drosophila* offspring are caused by the presence of a bacterium.

We expect the segregation of sex chromosomes and the random union of gametes to produce a 1:1 sex ratio in many higher organisms. There are examples in several species of *Drosophila,* for example, *D. prosaltans, D. bifasciata,* and *D. borealis,* found in the wild (that is, in the jungle or bush as opposed to laboratories) where offspring are almost exclusively female, a condition inherited in a non-nuclear fashion through females. Samples of cytoplasm can be taken from eggs laid by females producing only female offspring and can then be injected into females that produce both male and female progeny, for example, a laboratory strain of *D. melanogaster.* After an appropriate incubation period, the injected females will start producing only or mainly female offspring.

A microscopic analysis of the blood of larvae in a strain with abnormal sex-ratios reveals the presence of a small spirocheate bacterium resembling members of the genus *Treponema* (*T. pallidum* is the spirocheate that causes syphilis). Growth of these bacteria is retarded by penicillin, and it is possible to "cure" the abnormal sex-ratio condition. Successful growth of the bacteria that cause the abnormal sex-ratio condition is also dependent on the genetic constitution of the host. Some strains of *D. melanogaster* will not allow growth of the spirochetes, indicating that some genetic constitutions confer resistance to the bacterium. In those strains and species where spirochete infection occurs, the flies may contribute substances necessary for the growth of the bacteria. The bacterium, by greatly increasing the frequency of female offspring, may be helping the flies. In *Drosophila* males will mate with several females. The excess females may produce more offspring, increasing the size of the population carrying the spirochete. If so, we may be witnessing the evolution of a new cellular "organelle."

Maternally inherited mammary tumors in mice are caused by mammary tumor virus (MTV).

The Roscoe B. Jackson Laboratory in Bar Harbor, Maine, has long specialized in studying the genetics of mice. One of the early projects there was the establishment of a number of different strains of mice with all members in each strain being as genetically identical as possible—in essence, an endless number of identical twins. The availability of these strains, some of which are highly susceptible to particular diseases, has allowed the identification of conditions that would have been difficult to identify using wild mice. The female mice in some of the strains, for example, C3H, have a high level of mammary tumors, and those in other strains, for example, C57 Black 6, have low frequencies. When females from a strain with a high tumor incidence are mated with males from a strain with a low incidence, the female offspring have a high frequency of mammary tumors. When the reciprocal cross is made, female offspring have a low frequency of tumors. The results are much in accord with standard cytoplasmic inheritance, but one additional experiment demonstrates that the inheritance is unusual. When the female offspring from a high-incidence strain are nursed by mothers from a low strain, the offspring have a low incidence of mammary tumors. Conversely, when newborn females from a low strain are suckled by mothers from the strain with a high incidence of mammary tumors, tumors develop with increased frequency. There is a factor that causes mammary tumors in the milk of mothers from the high-incidence strain. The factor is an RNA-containing virus, mammary tumor virus (MTV), and is passed to offspring generation after generation through the milk of lactating mothers. Exposure to the virus is required for the production of mammary tumors. Nevertheless, susceptibility to tumor induction varies from strain to strain because of genetic differences among the strains.

The action of mouse mammary tumor virus depends on the integration of a DNA transcript into the mouse nuclear DNA.

MTV has a structure characteristic of other RNA tumor viruses (Figure 9-6) and contains two copies of RNA chromosomes, each with about 9,000 nucleotides. The two single-stranded chromosomes are partially paired with each other through a short region of complementarity. Upon infecting a cell, the RNA chromosomes serve as templates for synthesis of DNA by reverse transcriptase (p. 116), a normal component of the virus. A 9,000-base-pair double-helical copy is produced and takes on a circular form. The circular DNA then integrates into a specific site in a host chromosome by a mechanism assumed to be similar to the one by which λ integrates into the *E. coli* chromosome (Figure 7-3). When integrated, the MTV DNA, a *provirus,* serves as a template for the production of new complementary RNA molecules, which, in the presence of virus proteins, form new viruses. The progeny

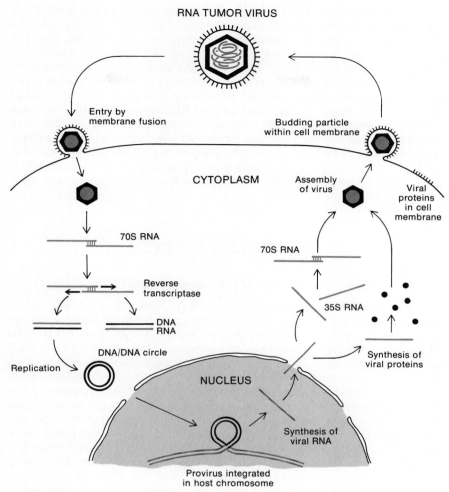

Figure 9-6   The life cycle of an RNA tumor virus such as mammary tumor virus.

viruses are then budded from the cell within vesicles derived from the cell membrane. The host cell is not, however, destroyed. Integration of the MTV DNA into a chromosome is essential for the conversion of normal mammary cells into malignant ones, a process called *oncogenic transformation* or just transformation (but distinct from genetic transformation in Pneumococcus, p. 20). The means by which a normal cell is transformed into a cancer cell by MTV is not understood. In the mouse, the virus does not immediately produce a tumor. Tumors usually occur in "old age," after the mouse has given birth to several offspring. In addition to its transfer in milk, it is also possible for MTV to be transferred through germ cells as a provirus.

Mouse mammary tumor viruses were demonstrated before 1960. Demonstration of similar viruses in humans has been slow and difficult, but there is recent and increasing evidence for their existence. For example, proteins, presumably viral proteins, highly similar to mouse mammary tumor virus proteins, were found in human breast cancers by workers in the laboratory of Sol Spiegelman in 1978. The proteins, providing evidence for the presence of what is presumed to be a human mammary tumor virus, were found in 51 of 131 different malignant human mammary tumors. It is possible that these mammary tumors are, at least in some instances, "inherited" in the same manner as those in mice. We must stress, however, that there is little direct evidence for such inheritance. Also, in contrast to the inbred strains of mice, which are genetically susceptible to MTV, the average woman undoubtedly has genetic resistance to tumor induction by the hypothetical human MTV.

SOMATIC CELL GENETICS

Mitotic or somatic cells are also used to investigate inheritance.

The gene transmission systems we have considered so far involve meiotic cells primarily. Even the studies on non-nuclear inheritance that we discussed in the first part of this chapter depended mostly on the analysis of meiotic products, for example, the absence of segregation of the *poky* allele and its normal counterpart. Genetic events grouped under the heading "somatic cell genetics" occur in and between cells that are dividing mitotically, so that new gene combinations result. Recombination between homologous chromosomes of mitotically dividing cells that are heterozygous for a pair of alleles can produce daughter cells that differ in their genetic constitutions. We will also discuss conditions under which mitotically dividing cells fuse with other cells, resulting in a mixture of two chromosomal complements. In some cases, during subsequent mitotic divisions following fusion, chromosomes originating from one cell or the other are lost. Stable lines of cells can be established with varying chromosomal complements. We shall be concerned in this section, on the one hand, with recombination that occurs within somatic cells in an organism, and, on the other hand, with genetic exchanges that occur in artificially fused hybrid cells grown in culture.

Recombination between homologus chromosomes in mitotically dividing cells can produce daughter cells with different genotypes.

The processes of mitosis assure that all cells descended from a single zygote are genetically identical and constitute a clone. A low frequency of spontaneous recombination does, however, occur in mitotically dividing cells. The level of recombination is increased by x-radiation of somatic tissues. X-rays cause breaks in chromosome arms and produce translocations (p. 225). The x-ray-induced recombinational exchanges between homologous chromosomes in mitotically dividing cells also involve breakage and reunion of chromosome arms to produce recombination. The x-ray-induced recombinational exchanges are similar to the recombinational exchanges during meiosis in that they are reciprocal. It is possible that somatic recombination occurs during the cell cycle in G_1 before DNA replication. We shall assume here, however, that somatic recombination occurs after DNA replication in G_2 when four chromatids are present.

The detection of recombination in somatic cells in whole multicellular organisms requires the use of autonomous traits.

To detect somatic recombination we again require the use of appropriate genetic markers. The requirements are more stringent than in meiotic cells because the phenotypes of individual cells must be *autonomous,* that is, be expressed in the midst of cells with other genotypes and phenotypes. Early examples of somatic recombination were reported in 1936 by Curt Stern using *Drosophila melanogaster.* The steps involved in the detection of somatic recombination in *Drosophila* are illustrated using females heterozygous for the X-linked alleles w and w^+. Because w is recessive, the eye tissue of the heterozygous females has the normal brownish-red color. During cell division rare recombinational exchanges may occur between one member of each pair of replicated X chromatids at a location between the centromere and the w locus (Figure 9-7). During the subsequent mitotic division the replicated sister chromatids move to opposite poles. Depending on the orientation of the centromeres of the recombined and unrecombined chromatids, either recombinant or non-recombinant daughter cells will be produced. With one orientation we can see (Figure 9-7) that one daughter cell receives two copies of the w allele and the other cell receives two copies of the w^+ allele. Both cells are now homozygous instead of heterozygous. Following further mitotic divisions, two patches of tissue will be produced, one containing w/w cells and the other, w^+/w^+ cells. The two types of homozygous tissue are surrounded by heterozygous cells. If the recombination occurred in cells that are going to develop into eye tissue, the w/w tissue can be detected in the adult fly as a patch of white tissue surrounded by brownish-red tissue. The expression of the w trait is autonomous. The w^+/w^+ tissue cannot be detected because it is phenotypically indistinguishable from the heterozygous tissue.

It is possible, however, to choose genetic constitutions in which we can detect both populations of homozygous cells resulting from reciprocal somatic recombination. Allelic differences at two neighboring, linked loci can demonstrate that the production of the reciprocal recombination classes results from somatic recombination. At the X-linked *yellow* (y) locus of *Drosophila,* the recessive y allele causes a yellow body color distinct from the normal grey color caused by the y^+ allele. At another X-linked locus near to y, the *singed* (sn) allele causes the bristles on the body of the fly to appear to be singed, and the dominant sn^+ allele produces normal bristles. Recombination can occur between the centromere and the two loci (Figure 9-8). During the subsequent mitotic division, daughter cells with two different genotypes can be produced, one sn^+/sn^+, y/y and the other, sn/sn, y^+/y^+.

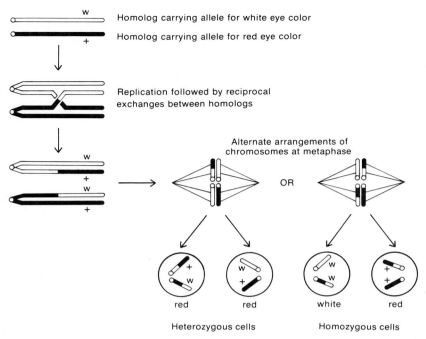

Figure 9-7   Somatic recombination. Recombination in cells heterozygous for a single pair of alleles produces homozygous white cells in the midst of heterozygous red cells.

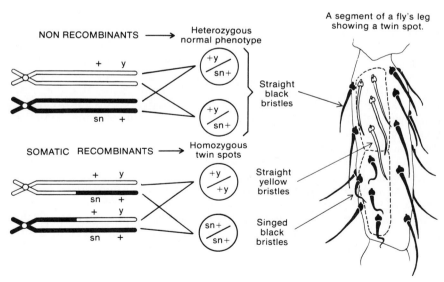

Figure 9-8   The production of a twin spot resulting from recombination in somatic cells heterozygous for two pairs of alleles carried on the same chromosome.

Further mitotic division of the daughter cells produces two patches of tissue, twin spots, that are genotypically different. A twin spot can be identified in the adult fly by the presence of a tissue patch with *singed* bristles and normal body color next to one with *yellow* color and normal bristles (Figure 9-8). The recovery of twin spots demonstrates that somatic recombination can be reciprocal.

Somatic cells can be fused to produce new genetic constitutions.

We will change our focus now and consider genetic manipulations in tissue culture cells that occur both between and within cells. We are all aware that germ cells fuse to produce zygotes. Fusion can also occur between somatic or mitotic cells to produce by parasexual means (para means "aside from") new cells containing the sum of the nuclear genomes of both original cells. Fusion of tissue culture cells is greatly increased by the use of inactivated viruses such as Sendaii virus or by chemical agents such as polyethylene glycol. The original fused cells contain two nuclei (Figure 9-9),

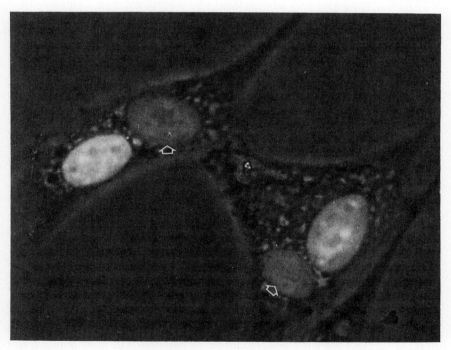

Figure 9-9   A photo of a binucleate cell resulting from cell fusion. Prior to fusion one group of cells was stained with Hoechst 33258, the fluorescent dye that binds to DNA. In the fused cells one nucleus does not fluoresce brightly (arrow). Photo courtesy of Woodring Wright, Department of Cell Biology, University of Texas Health Science Center, Dallas. From Wright, *Experimental Cell Research*, v. 112, p. 395, 1978, with permission from Academic Press.

but, following cell division, the chromosome complements of the two cells are located together in a single nucleus. Cells derived from fusion can continue to divide. During subsequent divisions, there is a tendency for various chromosomes to be lost, so that the population of cells descended from the original fused cells is not genetically homogeneous. There is substantial variation in karyotypes from cell to cell. Cell fusion is not limited to cells of the same species but occurs between cells of different genera (for example, mouse and man) and even different orders (mammals and birds). In addition to fusions of cells from vertebrates, there has been fusion of fungal, plant, and insect cells. For cell fusion to be useful in genetic studies, for example, in gene mapping, several criteria must be met. There must be a series of genetic markers with identifiable phenotypes. There must also be a means to recover and distinguish the fused cells from unfused ones. Finally, there must be a means by which the karyotypes of the fused cells can be related to the phenotypes of the cells.

Gene mapping studies in humans have been greatly aided by somatic cell genetics.

In addition to the sex chromosomes, there are 22 pairs of autosomes in the human genome. Yet before 1968, the only human genes that had been assigned to a particular chromosome were those carried on the X chromosome. The assignment of genes to the X chromosome was made possible by the unique patterns of inheritance of sex-linked genes (pp. 18–22). No such unique patterns exist for autosomal genes and although the study of pedigrees (pp. 177–178) established that some genes were linked (for example, the nail-patella syndrome (causing abnormalities in bone and nail structure) and the ABO blood group loci were known to be linked), no one knew on which human autosome any gene was carried. In 1968, by comparing the pattern of inheritance of an aberrant form of Chromosome 1 with the inheritance of Duffy blood group antigens (the Duffy blood groups, like the ABO system, are antigens found on the surface of red blood cells), Roger Donahue and his collaborators established that the Duffy blood group locus was on chromosome 1. Delay in the first assignment of a locus to a specific autosome until 1968 is ample testament to the difficulties of gene mapping in humans through pedigree analysis. But only nine years after the assignment of the first gene to a specific autosome, at least one gene had been localized to every chromosome in the human genome. The rapid advancement in mapping the human genome was made possible, in part, by intensified analysis of human pedigrees on the part of human geneticists. The techniques for cell fusion and genetic analysis to which we previously referred have opened up new possibilities for the cartography of the human genome. Indeed the field has advanced so fast that the term somatic cell genetics is used by many persons to refer exclusively to the analysis of human and mammalian genomes through the use of fused tissue culture cells. In particular, the use of human–mouse and human–hamster cell hybrids has proved to be important in the mapping of the human genome. These hybrids are especially useful, as we will document presently, because of the preferential loss of human chromosomes from the hybrid cells and the establishment of hybrid cell lines carrying only one or a few human chromosomes.

Mouse–human or hamster–human hybrid cells are selectively recovered using the HAT (hypoxanthine-aminopterin-thymidine) technique.

One of the first requirements for using cell fusion and the resulting hybrid cells for genetic analysis is the necessity for preferential recovery of the appropriate cell hybrids free from unfused cells. A particularly powerful technique, which goes under the acronym "HAT" for hypoxanthine-aminopterin-thymidine, is often used for recovery of hybrids. The HAT technique, originally devised by Waclaw Szybalski and his colleagues in 1962 and modified two years later by John Littlefield, uses conditions under which only hybrid cells are capable of synthesizing DNA and therefore of increasing in number by cell division. Non-hybrid cells cannot synthesize DNA and are not recovered.

The synthesis of DNA depends on the presence of purine and pyrimidine triphosphates that are formed from their respective monophosphates. The purine and pyrimidine monophosphates can be formed in two ways—by their synthesis from precursors and by salvage pathways in which enzymes reutilize purine and pyrimidines produced by breakdown of polynucleotides. In the presence of the drug aminopterin, the synthesis of new purines and pyrimidines is inhibited so that cells are completely dependent on the salvage pathways for the formation of purine and pyrimidine triphosphates and the synthesis of DNA. The HAT technique makes use of aminopterin to inhibit the new synthesis of purines and pyrimidines from precursors and also uses two different cell lines, one lacking an enzyme involved in salvage of purines and the other lacking an enzyme involved in salvage of pyrimidines.

As an example (Figure 9-10) of the HAT technique, we could use a mouse cell line homozygous for an allele that eliminates the presence of hypoxanthine phosphoribosyl transferase (HPRT), an enzyme involved in the salvage

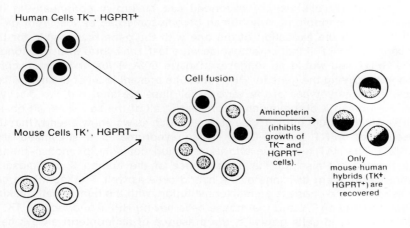

Figure 9-10   The *HAT* technique for recovering fused (hybrid) cells.

of purines. The mouse cells have the enzyme thymidine kinase (TK), which converts the pyrimidine thymidine to thymidine monophosphate. The human cells have an inherited deficiency in TK but contain HPRT. Thus, for different reasons, neither the human nor the mouse cells can grow in the presence of aminopterin, the human cells being unable to make pyrimidine monophosphates, the mouse cells, purine monophosphates. Hybrid cells formed by the fusion of human and mouse cells are, however, able to grow in the presence of aminopterin. The hybrid cells have HPRT because of the presence of the human chromosomes and are able to salvage purines and make purine monophosphates. These cells also have TK because of the presence of the mouse chromosomes and are able to make pyrimidine monophosphates. Thus, the hybrid cells are the only ones that survive in the presence of aminopterin.

Mouse–human hybrid cells preferentially lose human chromosomes.

One of the characteristics of mouse–human hybrid cells is that as the cells divide there is preferential loss for unknown reasons of human chromosomes. Ultimately, a mixture of cells is produced containing a full complement of mouse chromosomes and a few human chromosomes. Individual cells can be isolated that give rise to cell lines with stable karyotypes, some of which contain different complements of human chromosomes. The human chromosomes that are present in a particular cell line can be identified by a Giemsa staining technique (p. 44) because human chromosomes stain differently from mouse chromosomes (Figure 9-11). The result is the establishment of a series of mouse–human hybrid cell lines, each with a known limited complement of human chromosomes.

We can assign a gene to a specific chromosome by comparing the human chromosomal complements of several cell lines with the presence or absence of a phenotypic trait.

The hybrid cells we just described can be grown continually in the presence of aminopterin. Any human chromosome can be lost from these cell lines with the exception of the one with the gene responsible for the presence of HPRT. If that chromosome were lost, the hybrid would no longer have HPRT and would no longer synthesize DNA. Thus, the human chromosome carrying the gene for HPRT must be present in all of the hybrid cell lines. If the karyotypes of the various cell lines are analysed, we discover that the human X chromosome is present in all of the cell lines, an observation which should not surprise us in as much as we know already that the absence of HPRT in the Lysch-Nyhan syndrome is a sex-linked trait (p. 221). The HAT technique affords selection not only for mouse–human cell hybrids but also for the maintenance of the human X chromosome during subsequent divisions of the hybrid cells as well. The technique can be used with the opposite genetic constitution, with the human cells lacking HPRT and having TK and the mouse cells having HPRT and lacking TK. In that case, hybrid cells grown in the presence of aminopterin always have

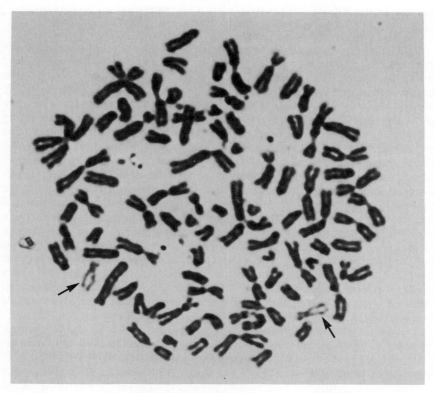

Figure 9-11   The chromosomal complement of a mouse–human cell hybrid in which only one human chromosome remains (arrow). Photo courtesy of David Cox, Department of Pediatrics, University of California, San Francisco.

human chromosome 17, which we can conclude is the one carrying the gene allowing the presence of TK.

In the examples we have described, the chromosomal locations of the genes necessary for the presence of TK and HPRT were identified by the selective recovery of hybrid cell lines carrying the human chromosomes on which the two genes are located. The use of the HAT technique is not limited to these two enzymes. It can be modified to identify the chromosomes carrying numerous other genes, using both selective and non-selective procedures. The assignment of a gene for a non-selected trait, that of the gene for lactate dehydrogenase A (LDH_A), to human chromosome 11 was reported in 1972. Human LDH_A differs from its counterpart in mouse cells and can be identified in mouse–human cell hybrids. With the HAT technique, mouse–human hybrids were created, and lines of cells carrying different subsets of human chromosomes were identified. The lines were then inspected for the presence of human LDH_A. The human enzyme was always found in cells that contained human chromosome 11 but not in cells lacking this chromosome.

The approximate positions of different genes on human chromosomes have been determined using translocations and somatic cell techniques.

Once a set of genes has been assigned to a common chromosome the next step is to determine the order of the genes and their locations along the chromosome. Knowing that particular markers are located on the same chromosome, one may search for family pedigrees with appropriate genetic constitutions for recombinational analysis. From these analyses, which are somewhat similar to the ones we have already discussed for determining gene order on chromosomes, a linkage map can be derived. Translocations are also used to map genes to limited regions of chromosomes. We recall, for example, that in the *cri du chat* syndrome (Figure 5-23), a piece of chromosome 5 has been translocated to chromosome 13. If a cell culture were established carrying the *cri du chat* translocation, it would be possible to determine whether a particular gene already assigned to chromosome 5 is on the translocated or untranslocated part of the chromosome. If the gene is on the translocated part, the short translocated part will always be present in human–mouse cell hybrids carrying the gene. Furthermore, the Giemsa banding technique (p. 44), which allows the delineation of different regions on human chromosomes, can be used to determine the positions where breaks occurred to produce translocations. By relating the presence or absence of a trait to the position of a break in a translocation, it is possible to assign a gene to a limited region of a human chromosome.

Geneticists have assigned over 125 genes to specific human autosomes, in addition to those assigned to the X chromosome (Table 9-2). Since the first gene assignment to chromosome 1 in 1968, innovations of somatic cell genetics coupled with the intensified study of human pedigrees has made for substantial progress. The human genetic map is on its way to respectability. We estimate, however, that there may be as many as 50,000 genes in

Table 9-2   Human Chromosome Gene Assignments.

Chromosome 1	*Chromosome* 2	*Chromosome* 5
Amylase	Acid phosphatase-1	Diphtheria toxin
Adenylate kinase-2	Aryl hydrocarbon	sensitivity
Cataracts	hydroxylase	Hexosaminidase-B
Elliptocytosis-1	Interferon-1	Interferon-2
Enolase-1	Isocitrate dehydrogenase	
Fumarate hydratase-1 and	-1	*Chromosome* 6
-2	Malate dehydrogenase-1	Complement component
α-Fucosidase		-2, -4 and -8
Duffy blood group		Chido blood group
Guanylate kinase-1 and -2	*Chromosome* 3	Glyoxylase I
Phosphoglucomutase-1	Aconitase (mitochondrial)	Human lymphocyte
6-Phosphogluconate	Galactose-1-phosphate	antigens
dehydrogenase	uridyl-transferase	Malic enzyme-1
Rhesus (rh) blood group	Herpes virus sensitivity	Olivopontocerebellar
5S RNA genes		atrophy I
Scianna blood group	*Chromosome* 4	P blood group
Uridine monophosphate	Group-specific	Pepsinogen
kinase	component	Phosphoglucomutase-3
	Phosphoglucomutase-2	Rodgers blood group

Chromosome 7
β-Glucuronidase
Hydroxyacyl-CoA
    dehydrogenase
Malate dehydrogenase
    (mitochondrial)

Chromosome 8
Glutathione reductase

Chromosome 9
ABO blood groups
Aconitase (soluble form)
Adenylate kinase-1 and -3
Arginosuccinate
    synthetase
Nail-patella syndrome

Chromosome 10
Adenosine kinase
Glutamate oxaloacetic
    transaminase-1
Glutamate semialdehyde
    synthetase
Hexokinase-1
Inorganic
    pyrophosphatase

Chromosome 11
Acid phosphatase-2
Lethal antigen (3 loci)
Esterase-A4
Hemoglobin, β chain
Lactate dehydrogenase-A

Chromosome 12
Citrate synthase
    (mitochondrial)
Enolase-2
Glyceraldehyde-3-
    phosphate
    dehydrogenase
Lactate dehydrogenase B
Peptidase B

Chromosome 13
Esterase D
Ribosomal RNA genes

Chromosome 14
Nucleoside
    phosphorylase
Ribosomal RNA genes
Tryptophanyl-tRNA
    synthetase
Immunoglobulin heavy
    chains

Chromosome 15
β-2-Microglobulin
Hexosaminidase A
Isocitrate dehydrogenase
    (mitochondrial)
Manosephosphate
    isomerase
Pyruvate kinase-3
Ribosomal RNA genes

Chromosome 16
Adenine
    phosphoribosyltransfer-
    ase
α Haptoglobin
Hemoglobin, α chain
Lecithin-cholesterol acyl-
    transferase
Thymidine kinase
    (mitochondrial)

Chromosome 17
Galactokinase
Thymidine kinase
    (soluble)

Chromosome 18
Peptidase A

Chromosome 19
Peptidase D
Phosphohexose
    isomerase
Polio sensitivity

Chromosome 20
Adenosine deaminase
Desmosterol-to-
    cholesterol enzyme
Inosine triphosphatase

Chromosome 21
Antiviral protein
Ribosomal RNA genes
Superoxide dismutase-1

Chromosome 22
β-Galactosidase
Ribosomal RNA genes

X Chromosome
Color blindness
α-Galactosidase (Fabry
    disease)
Glucose-6-phosphate
    dehydrogenase
Hemophilia
Hypoxanthine
    phophoribosyl-
    transferase (Lesch-
    Nyhan syndrome)
Phosphoglycerate kinase

Y Chromosome
Y cell surface antigen
Testis determining factor

Modified from McKusick and Ruddle, *Science*, v. 196, pp. 390–405, 1977.

the human genome, and we have a long way to go before we have accurately placed a substantial fraction of our own genes in their nucleotidic pigeon holes.

Plants, because of the developmental totipotency of their cells, offer unique opportunities for somatic cell genetics.

One of the major differences between plants and animals is the means by which the germ cells are derived during development. In animals, cells that develop into eggs and sperm are set aside early in development as a separate population of diploid cells or *germ-line* cells. Only those cells are destined to give rise to gametes and only those cells must retain and be able to call into service all of the genes necessary for the development of an entire animal. The rest, the somatic cells, become progressively restricted in their developmental potential and normally lose the capacity to produce an entire organism. The restriction in developmental potential of animal cells has importance for medicine. For example, kidney cells when grown in culture, at best, produce more kidney cells. We have not yet found the way to have kidney cells form new kidneys or, for that matter, to produce any other organ. The situation is unfortunate, for if it were otherwise the disaster of kidney failure would not confront us—we could grow the appropriate replacement part in organ culture.

In plants, germ line cells are not set aside as a separate cell population; flowers develop from somatic tissues that are not specialized. Thus, any cell lineage can give rise to a flower and must retain access to the full repertoire of its genetic constitution. Plant somatic cells do not have the limitations of animal cells. In a classic experiment, Frederick Steward was able to demonstrate in 1958 that a single carrot cell divided to produce a mass of cells that when challenged with the appropriate milieu of plant hormones could produce both roots and shoots. Flowers developed on the shoots and produced seed from which completely normal carrots were grown. Thus, individual carrot cells have complete developmental potential; they are totipotent.

The totipotency of plant cells, including those grown in culture, provides unique opportunities for somatic genetics. For it is theoretically possible to fuse cells from two different plants and have the fused cells retain their developmental potential so that new plants, which might otherwise be impossible to produce, grow from the fused cells. Exotic possibilities come to mind—maize plants that harbor nitrogen-fixing bacteria in root nodules, an ability normally reserved to legumes, tomatoes that grow on trees, or apples the size of watermelons that grow on vines. But let us see what has actually been achieved so far.

Hybrid tobacco plants have been produced from fused somatic cells.

Two tobacco species, *Nicotiana glauca* (2n = 24) and *Nicotiana langsdorffii* (2n = 18) can be crossed sexually to produce a hybrid plant (2n = 42). Peter Carlson, Harold Smith, and Rosemarie Dearing, reported in 1972 that they were able to produce the same hybrid by fusing somatic cells. Cells from the two species were stripped of their cell walls and fused. Cells from the two original species were unable to grow in the absence of plant hormones, but some of the fused cells could grow to produce tissue masses. These cell masses were then induced, also using plant hormones, to form shoots and leaves and, ultimately, flowers. The appearance and the structure of the leaves were indistinguishable from those of the sexually produced hybrid plants. The flowers produced seeds that germinated to produce plants identical with those produced sexually. Thus, fused plant somatic cells can grow and differentiate to produce a hybrid plant.

Cultures of haploid plant cells can be established and used to recover new mutants that, following reconstitution of the plant, can be sexually inherited.

We noted before that in higher plants there are remnants of the alternation of generation between the haploid gametophyte stage and the diploid sporophyte (pp. 140–142). Both male and female haploid cells resulting from meiosis undergo limited mitotic divisions in anticipation of cell fusion to produce new diploid plants. These mitotic cells can be used to establish haploid cell cultures. Haploid cell cultures have a major advantage over diploid ones. Because only one copy of most genes is present, recessive mutants will be expressed. Thus, it is much easier to induce and recover new mutants in haploid cells than in diploid ones.

The work of two German plant geneticists, Andreas Muller and Reinhard Grafe, provided in 1978 an example of mutant induction and recovery in haploid plant cell cultures of *Nicotiana tobaccum* (Figure 9-12). Normal plant cells can use nitrate as a source of nitrogen for the synthesis of nitrogen containing compounds such as amino acids. The first step in this utilization requires the conversion of nitrate to nitrite by the enzyme nitrate reductase. Nitrate reductase also converts chlorate to chlorite, a reaction that results in the death of the cells. Plant cells grow in the absence of nitrate if amino acids are added to the culture medium. By exposing haploid *N. tobaccum* cells to mutagens and then growing them in the presence of chlorate and amino acids, Muller and Grafe were able to recover mutant cells that lacked nitrate reductase, such cells being the only ones able to

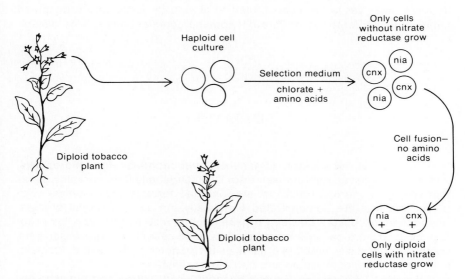

Figure 9-12   A technique for creating plants with new genetic constitutions. Diploid plants are reconstituted from fused haploid cells that carry specific, selected genetic constitutions.

live. The presence of nitrate reductase requires two normal genes. The loss of nitrate reductase can result from mutation in either one of the two genes to produce alleles called *cnx* and *nia*. In a later report published in 1978, Kristina Glimelius and Tage Eriksson in Sweden, working with Grafe and Muller, were able to recover fused diploid cells by mixing *cnx* cells with *nia* cells (Figure 9–12). The fused cells were recovered because they could grow using nitrate as a nitrogen source although neither type of the unfused mutant cells was able to do so. In the fused cells, each of the two genomes supplied the function that the other lacked, much like cell fusion involving cells lacking HPRT and TK.

Previously diploid cell cultures of *N. tobaccum* have been reconstituted to produce fertile plants and Grafe and Muller and their Swedish co-workers reported in 1979 the formation of plants from their hybrid cells. The importance of their results lies in demonstrating the utility of recovering mutants from haploid cell cultures and the possibility of growing normal diploid plants from fused cells. Although the particular mutants with which Grafe and Muller work are of no immediate agricultural significance, other mutants may be extremely important. Plants, too, are victims of bacterial, fungal, and viral diseases. It is slow and tedious to produce plant strains that are resistant to diseases through standard breeding techniques. However, in cell cultures, those cells that are resistant to particular pathogens may be rapidly recovered because the cells will continue to divide. Indeed, a group of researchers in Minnesota, headed by C. Edward Green, have obtained corn plants resistant to the fungus *Helminthosporium maydis,* which causes corn leaf blight, by reconstituting plants from cells resistant to the fungal toxin. Unfortunately, many plants that are agriculturally important, such as wheat and rice, have yet to be reconstituted from cultured cells, and widespread improvement of crop plants through somatic genetics may still be distant.

SUMMARY

Non-nuclear inheritance. Organelles, mitochondria and chloroplasts, contain small circular chromosomes of DNA. Organelle DNA contains genes that specify numerous traits that are inherited through the maternal cytoplasm and that are demonstrated genetically through reciprocal crosses. Recombination occurs between organelle chromosomes. Organelles probably evolved from bacteria infecting eukaryotic cells. Infectious organisms, bacteria and viruses, emulate organelles by producing phenotypic effects in eukaryotes that are inherited in a non-nuclear fashion.

Somatic Cell Genetics. Genetic changes occur within and between mitotically dividing cells and produce cells with new genotypes. Within heterozygous cells, recombination can occur so that each daughter cell is homozygous for one of the alleles. Both plant and animal cells can be fused to produce cells containing two nuclear complements. Fused cells can be selectively recovered, for example, using the HAT technique. Somatic cell hybridization between human and mouse tissue culture cells has greatly facilitated mapping of the human genome. Plant cell cultures can be induced to form normal plants. Selection for desirable genetic traits in plant cell cultures, followed by production of fertile plants from these cultures, may provide important opportunities for improvement of crop plants.

10

Bacterial Genetics

During the early years of genetics from 1900, when Carl Correns and Hugo De Vries rediscovered and repeated Mendel's work, through the 1930's, the emphasis was on multicellular organisms. Studies using primarily *Drosophila,* mice, and maize highlighted the role of meiosis in the transfer of genes from parent to offspring. Early studies with fungi such as *Neurospora* also concentrated on the transmission of traits through meiotic cells. Little interest was shown in bacteria because it was believed that the bacterial life cycle involved only asexual reproduction, each cell division producing two daughter cells with identical genetic constitutions. There seemed to be no possibility of employing mutant traits to investigate the nature of bacterial chromosomes and gene transmission. Then, in the 1940's all that changed. Sex was discovered in *Escherichia coli*!

To be precise, Joshua Lederberg and Edward L. Tatum reported in the Cold Spring Harbor Symposium of 1946 that when they mixed two strains of *E. coli* together, each mutant for three different traits, they recovered cells with new genetic constitutions, including some which were wild type for all six traits. Evidently chromosomes from the two strains were coming together and recombining to produce bacteria that carried chromosomes with wild-type alleles at all six loci. A transformation mechanism was ruled out because the bacteria had to make physical contact with each other in order to produce the recombinant progeny. Physical contact strongly implied that some kind of mating occurred between the two strains.

We now recognize three major mechanisms by which transfer of genetic information from one bacterium to another occurs: *Transformation* (pp. 20–23), involving the cellular uptake of DNA molecules from the culture medium; *Transduction,* involving the transfer by phages of pieces of bacterial chromosomes from one bacterium to another; and *Conjugation,* the process Lederberg and Tatum discovered, involving the transfer of part of a bacterial chromosome from one cell to another via a tube linking them. In the first two of these processes the donor bacteria is destroyed by lysis before the DNA or infecting phages are released into the medium. In conjugation, both donor and recipient cells can survive the mating. In this chapter we review the mechanisms of gene transfer in bacteria, starting with a brief discussion

260

of bacterial phenotypes and selective techniques used to recover cells with recombinant genetic constitutions.

The ability to grow under different culture conditions is the major phenotypic criterion used in bacterial crosses.

The presence of a bacterium is conveniently determined by placing it on a solid nutrient medium in a Petri dish. The bacterium goes through repeated divisions to produce a clone or colony composed of thousands of descendents, each with an identical genotype. The colonies, unlike the individual bacterium, are readily visible to the naked eye. Usually all colonies look alike (the rough and smooth forms of Pneumococcus being somewhat exceptional), which limits the opportunity to identify mutant bacteria by morphological criteria. The major phenotypic criterion is simply the ability of the bacterium to grow at all. Four major types of mutational change have been extensively used by geneticists to analyze bacterial gene transmission.

1. Sensitivity or resistance to drugs. There are a large number of chemical compounds that prevent the growth of bacterial cells; one example is sodium azide, which poisons the electron transport chain and blocks normal respiration. Antibiotics are produced by other organisms, particularly fungi and bacteria; penicillin and streptomycin come to mind. Many bacterial strains are unable to grow when antibiotics are added to the culture medium. Such strains are designated as *sensitive,* for example, streptomycin-sensitive, abbreviated *strep*s. Other strains will grow and are designated *resistant,* for example, streptomycin-resistant, abbreviated *strep*r. Bacterial strains are also sensitive or resistant to infection by phages.

2. Dependence on specific nutrients in the culture medium. Most normal bacteria can grow on very simple culture media, supplied only with a carbon source, such as a sugar; a source of nitrogen, such as nitrate; and various inorganic constituents, Na^+, K^+, PO_4^{--}, to name but a few. The list is small because bacteria can manufacture all of the other compounds required for growth (amino acids, purines, pyrimidines, vitamins, etc.) from these simple compounds. Some bacterial mutants are incapable of manufacturing all of their required compounds and can only grow when these compounds are provided in their growth medium. We call such strains *auxotrophs.* A strain that requires the amino acid leucine is a leucine auxotroph, *leucine*⁻, or *leu,*⁻ and is distinguished from the normal strain, *leucine*⁺ or *leu*⁺.

3. Inability to utilize specific nutrients. Some mutant strains are limited in the selection of basic nutrients that will promote growth. For example, standard strains are able to utilize a variety of different sugars, including the disaccharide lactose, as a carbon source, but some mutant strains lack the ability to use lactose and are designated *lactose*⁻ or *lac*⁻. Other strains are unable to use the monosaccharide galactose as a carbon source and are designated *galactose*⁻ or *gal*⁻. Normal strains are *lac*⁺ and *gal*⁺.

4. Conditional lethal mutations. Strains that carry conditional lethals, like similar phage strains (p. 189), are unable to grow under certain *restrictive* conditions (for instance, at a temperature of 37° C) but can grow under other, *permissive* conditions (for instance at 25° C).

These mutants all differ from each other, but all four classes of mutants are characterized by the presence or absence of bacterial colonies, that is, by the ability or the inability to grow.

Recombinants are recovered using selective techniques.

In our consideration of somatic cell genetics we discussed a selective procedure, the HAT technique, for the recovery of cells that had undergone fusion. Selective techniques are also commonly used in bacterial genetics to identify recombinant types that are produced as rarely as one in a million from crosses between two different strains.

We can illustrate selective techniques by describing in greater detail Lederberg and Tatum's experiments revealing recombination in *E. coli.* Recall that the two strains they used were each mutant for three different characteristics. One was a triple auxotroph for the amino acids leucine (*leu⁻*) and threonine (*thr⁻*) and the vitamin thiamine (*thi⁻*). In other words, this strain required all three of these substances for its growth. The other strain was triply auxotrophic for the amino acids phenylalanine (*phe⁻*) and cysteine (*cys⁻*) and the vitamin biotin (*bio⁻*). Lederberg and Tatum mixed the two strains together in a medium that contained all of the necessary supplements for the growth of both strains, but then placed or *plated* the cells on a solid medium which contained none of the growth requirements (Figure 10-1). On this medium they were able to recover *prototrophs,* that is, cells with the same nutritional requirements as wild type cells, in this case *thr⁺, leu⁺, thi⁺,*

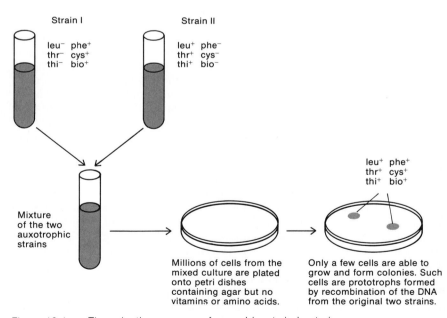

Figure 10-1   The selective recovery of recombinants in bacteria.

bio⁺, *phe⁺*, and *cys⁺*, because the recombinant cells could grow without supplements. The recombinant cells growing in an unsupplemented medium were easily selected from non-recombinant cells. Although these recombinant cells were extremely rare, fewer than one in a thousand, they were rapidly identified through the selective technique.

TRANSFORMATION

The first mechanism of genetic transfer found in bacteria is transformation. Recently transformation has also been reported in eukaryotes, including yeast, *Drosophila,* and in mammalian tissue culture cells. In bacteria, transformation occurs in several different species, such as *Hemophilus influenzae, Bacillus subtilis, Pseudomonas aeruginosa,* and *Escherichia coli,* as well as in Pneumococcus, where it initially provided evidence that DNA carried genetic information (pp. 20–23). Transformation is caused by DNA from one strain (the *donor* strain) being incorporated into the chromosome of a second strain (the *recipient* strain). Donor DNA need not be isolated and purified by the experimenter, because transformation occurs when cells are grown together in the same culture vessel or if dead cells are simply mixed with live ones as in Griffith's original experiments. There are several steps leading to the inclusion of the donor DNA into the chromosome of a recipient Pneumococcus cell (Figure 10-2).

1. The recipient cells take up double-helical DNA (single-stranded DNA cannot be taken up), but only a small percentage of the cells in a population is competent to take up the DNA.

2. During uptake one strand of the double helical DNA is digested by nucleases, leaving the other strand intact.

3. The remaining single strand forms a heteroduplex (p. 180) by base pairing with its complement in the chromosome of the recipient cell.

4. The single-stranded loop of DNA displaced by the heteroduplex is excised, leaving the donor DNA in its place.

5. Covalent bonds form through the action of DNA polymerases and ligases to seal the donor DNA to the host chromosome. The insertion of homologous donor DNA into the host chromosome is very efficient. About 50 per cent of cells that have taken up DNA will incorporate the donor DNA into their chromosomes.

This sequence of events creates a heteroduplex region in the host chromosome, in which one strand of the duplex was originally present in the chromosome of the recipient cell and the other strand came from the donor cell. Because the donor and the recipient cell are genetically different, the two strands will not be perfectly complementary to each other (Figure 10-2). For purposes of illustration, we may suppose that one strand has a small deletion, a few missing base pairs, that makes itself known in a mutant phenotype. When the chromosome is replicated, each strand of the DNA will

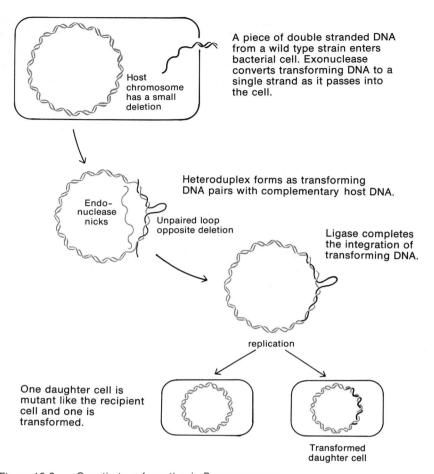

A piece of double stranded DNA from a wild type strain enters bacterial cell. Exonuclease converts transforming DNA to a single strand as it passes into the cell.

Host chromosome has a small deletion

Heteroduplex forms as transforming DNA pairs with complementary host DNA.

Endo-nuclease nicks

Unpaired loop opposite deletion

Ligase completes the integration of transforming DNA.

replication

One daughter cell is mutant like the recipient cell and one is transformed.

Transformed daughter cell

Figure 10-2   Genetic transformation in *Pneumococcus*.

serve as a template for the synthesis of its complement. The strand with the small deletion will produce a double helix, both strands of which contain the deletion. The normal strand will produce a normal complement and a normal double helix. Upon cell division, one daughter cell will be normal; and one will be mutant; one will be as before, the other will be transformed.

Transformation can be used to construct genetic maps in bacteria: mapping by co-transformation.

Let us imagine that mutant alleles for two different genes are carried together in the same bacterial strain and that the genes are physically distant from each other, say, diametrically opposite one another in the circular chromosome. During isolation of the donor DNA, the DNA is sheared, and the two genetic traits end up on physically separate pieces of DNA. Because transformation is a rare event, owing to the inefficiency of the uptake of the DNA, a recipient cell transformed for one trait is unlikely to be transformed for the second trait. The recipient cell will be transformed for both traits only if two separate transformation processes occur simultaneously. In contrast, let us imagine that the two genes are side by side in the bacterial chromosome. During DNA isolation the two genes are likely to remain together in the same piece of DNA. Therefore, when the recipient cell is transformed for one trait, a good possibility exists that it will be transformed for the second trait at the same time, in other words, that there will be *co-transformation.* We deduce that co-transformation occurs if the frequency of simultaneous transformation for both traits is greater than expected for two individual transformations. For example, if transformation for each trait occurs at a frequency of 0.01, we expect the frequency of simultaneous transformation for both traits to be about 0.0001 (0.01 × 0.01) with independent transformation and significantly greater with co-transformation.

Actual demonstration of co-transformation is somewhat more complicated. Because only a few cells in a population can take up DNA, it is possible to add DNA in such excess that every competent cell becomes transformed for both traits, giving the appearance of co-transformation. To demonstrate true co-transformation, recipient cells are exposed to a wide range of concentrations of DNA, carrying appropriate allelic markers for the two genes. As the concentration of DNA is increased, the frequency of transformation increases until a plateau is reached when all of the recipient cells that are capable of being transformed, the competent cells, are transformed (Figure 10-3). If the two traits under study are co-transformed, that

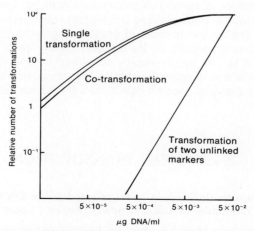

Figure 10-3   The frequency of transformation for one and two traits as a function of the concentration of transforming DNA. After Goodgal, *Journal of General Physiology,* v. 45, p. 205, 1961.

is, carried on the same piece of DNA, the frequency of simultaneous transformation for both traits increases linearly and parallels the increase in transformation for a single trait. However, if the two traits are carried on separate pieces of DNA, the frequency of simultaneous transformation increases exponentially, not linearly, because the frequency of simultaneous transformation is the product of the frequencies of two independent transformation processes (Figure 10-3).

TRANSDUCTION

Transduction depends on the lytic or lysogenic cycle of phage growth.

In transduction, recipient cells are genotypically converted by DNA transferred from donor bacteria within the body of a phage. The mechanisms of transduction depend on how phages grow in bacterial cells. As we noted earlier (Figure 7-3), there are two alternative life cycles by which phages reproduce, the lysogenic cycle and the lytic cycle. To review briefly, in the lysogenic cycle the phage chromosome instead of being rapidly replicated inserts itself into the bacterial chromosome and is then replicated and passed along with the bacterial chromosome to daughter cells. The prophage, however, can detach from the bacterial chromosome and enter the lytic cycle of growth, characterized by the production of large numbers of phages and lysis of the bacterial cell. Phages that can enter both lysogenic and lytic cycles are called *temperate,* while those that can only enter the lytic cycle of growth are called *virulent* phages. Transduction is mediated only by temperate phages. There are two types of transduction, generalized (unrestricted) transduction, occurring in association with the lytic cycle of temperate phage growth and specialized (restricted) transduction, occurring in association with the lysogenic cycle of growth. In generalized transduction, a recipient cell can apparently be transduced for any gene from the donor chromosome, that is, can be genetically converted by DNA from any part of the donor chromosome. In specialized transduction, a recipient cell is transduced for only one or two specific genes from the donor cell.

GENERALIZED TRANSDUCTION

In generalized transduction, pieces of fragmented donor DNA are accidentally and randomly included in newly formed phages.

Norton Zinder and Joshua Lederberg first described generalized transduction in 1952. Some temperate phages that participate in generalized

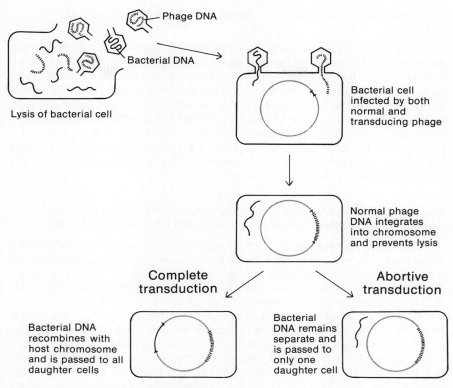

Figure 10-4   Generalized transduction.

transduction are P22 in *Salmonella typhimurium,* P1 in *E. coli*, SP10 and PBS1 in *B. subtilis,* and F116 in *Pseudomonas aeruginosa.* Upon entering the lytic cycle of growth, the phage chromosome is rapidly replicated, and the bacterial chromosome is fragmented by endonuclease attack (Figure 10-4). On completion of replication, phage proteins are synthesized and new phages assembled by packaging DNA into phage heads. The mechanisms of packaging are not, however, foolproof and pieces of bacterial DNA can substitute for phage DNA at a low frequency (about one in 100,000). Thus, some phages, containing bacterial DNA, not phage DNA are produced. Apparently, any region of the bacterial chromosome can be included in a phage, but the size of the piece is small, about 1 per cent of the size of the bacterial chromosome (40,000–90,000 base pairs), the actual size being determined by the amount of DNA normally packaged into the phage head. On lysis, rare phages carrying bacterial DNA and others carrying phage chromosomes are released into the culture medium.

Phage-transferred bacterial DNA cannot cause lysis of the recipient bacterium.

A phage-carrying bacteria DNA retains the capacity to adsorb on the surface of its bacterial host and to inject bacterial DNA into a recipient cell. The bacterial DNA fragment does not carry genes specifying on its own autonomous replication nor does it have information for the synthesis of phage proteins. Thus, a cell infected only with bacterial DNA cannot undergo lysis. That cell, however, is a target for normal phages that might destroy it. Such destruction is prevented when a bacterial cell that is a recipient of a fragment of bacterial DNA also receives a phage chromosome that enters the lysogenic cycle of growth. The presence of the lysogen confers immunity on the cell and prevents its destruction.

The bacterial DNA recombines with the chromosome of the recipient cell (complete transduction) or remains unintegrated (abortive transduction).

Two possible fates await the fragment of bacterial DNA. It can become integrated into the chromosomes of the recipient cell by pairing with its homologous region, and then, through recombination, can effect an inherited change in the genetic constitution of the recipient bacterium (Figure 10-4). This process is called *complete transduction.* Alternatively, in a process called *abortive transduction,* the donor DNA may remain unintegrated. In the integrated state the donor DNA is replicated with the bacterial chromosome. In the unintegrated state the donor DNA is not replicated, remaining as a single double helix in the host cell. In the absence of integration, following cell division, only one of two daughter cells receives the donor DNA fragment. If other divisions follow, only a single cell in a colony would harbor the unintegrated fragment.

To identify transduced cells, selective techniques are used so that only prototrophic cells can grow. For example, a *leu⁻* cell can be transduced to a *leu⁺* cell. In complete transduction cell division occurs rapidly in a medium lacking leucine to produce a normal-sized colony. In abortive transduction, the one cell containing the chromosomal fragment is able to grow and divide. Only one of the two daughter cells, however, is able to continue to grow and divide. Thus, instead of geometric growth (the number of cells doubling at each division), there is arithmetic growth (the number of cells increasing by one after each division). The result is that abortive transduction produces a microcolony, which is identifiable and readily distinguished from a normal-sized colony resulting from complete transduction. Abortive transduction is the usual result of transferring a chromosomal fragment via a phage into a recipient cell; it occurs ten to twenty times more frequently than complete transduction.

In complete transduction, genetic maps can be constructed using an analysis similar to that for transformation.

In complete transduction the size of the integrated piece of DNA is about 1 per cent of the size of the bacterial genome. Thus, during generalized transduction, it is unlikely that two randomly chosen genes will be carried by the same piece of DNA. Only those genes that are close together in the bacterial chromosome will be introduced into a recipient cell in a single DNA fragment by a single phage; that is, only closely linked genes will undergo *co-transduction* (similar to *co-transformation*). The frequency of transduction is very low; only one cell out of 10^5 to 10^7 is transduced for a single trait. If two traits were specified by genes distantly located from each other on a bacterial chromosome, it is highly unlikely that a single cell would be transduced for both genes. Only one simultaneous transductant would be expected for every 10^{10} to 10^{14} cells, about one million times less frequently than the transduction of a single trait. Reversing the logic, we can conjecture that two genes are physically close together on a bacterial chromosome if they co-transduce, that is, produce simultaneous transductions at frequencies much higher than expected for two independent events.

We can determine the order of genes that co-transduce.

The frequency at which complete generalized transduction occurs is very low; nevertheless, it is the commonly utilized procedure for determining the order of closely linked genes in a bacterium. For not only can transduction be used to determine which genes are close together, it also can be extended to order closely linked genes by using two- and three-factor crosses. In *E. coli* the two genes *thr* and *leu* are linked to *ara* (*ara* controls the ability to metabolize the five-carbon sugar arabinose). The following, with the genes in the correct order, are two paired double helices of DNA, one a fragment brought in by a phage (*leu+*, *ara−*, *thr+*) and the other the homologous region of the chromosome of the recipient cell (*leu−*, *ara+*, *thr−*). Allelic differences exist for each of the three genes.

Donor DNA

— *leu+*— *ara−*— *thr+*—

— *leu−*— *ara+*— *thr−*—

Bacterial Chromosome

For any part of the DNA of the donor chromosome to be integrated into the chromosome of the recipient cell at least two recombinational exchanges must occur. If only a single recombinational exchange occurred the host chromosome would lose both its circular nature and its ability to replicate properly (Figure 10-5). Any single pair of alleles or all three pairs of alleles can exchange positions as a result of two recombinational exchanges:

—leu+   ara−   thr+____   leu+__ ara− __ thr+

—leu−   ara+   thr−____   leu−__ ara+ __ thr−

Bacterial Chromosome   Bacterial Chromosome

Similarly, the middle pair of alleles along with either pair of flanking alleles can also exchange through double recombination:

Bacterial Chromosome   Bacterial Chromosome

However, in order to have only the two outside pairs of alleles exchange positions, four recombinational exchanges are required:

Bacterial Chromosome

Consequently, we expect the *leu⁺*, *ara⁺*, *thr⁺* class of recombinants to be present at a substantially lower frequency than the other recombinational classes. By determining the recombinational class present at the lowest frequency, we can determine which of the three genes occupies the center position.

We should keep in mind though that through selective procedures all transductants are recovered as prototrophs. One approach is to recover prototrophs for only one of the traits, for example, leucine and then determine what the genetic constitutions are at the *ara* and *thr* loci. Possible results from such an analysis are:

Genotype	Recombination Pattern	Number
leu⁺ ara⁻ thr⁺		30
leu⁺ ara⁻ thr⁻		750
leu⁺ ara⁺ thr⁻		250
leu⁺ ara⁺ thr⁺		0

There are four general points that can be made about the data. First, all recovered transductants are, of course, *leu⁺*. Second, cells that were transduced for *leu⁺* and *ara⁻* are more numerous than any of the others. We interpret this to mean that the *leu* and *ara* loci are so close together that they are usually contained in the same piece of transducing DNA and are usually not separated from each other during recombination with the host chromosome. Third, the *leu⁺*, *ara⁻*, *thr⁺* class is present at low frequency, indicating that the *thr* locus is substantially farther from the *leu* locus than is the ara locus. Finally, the absence of the *leu⁺*, *ara⁺*, *thr⁺* class results mainly from the requirement that four recombinational exchanges be made to produce the *leu⁺*, *ara⁺*, *thr⁺* combination and indicates that the *ara* locus is between *leu* and *thr*.

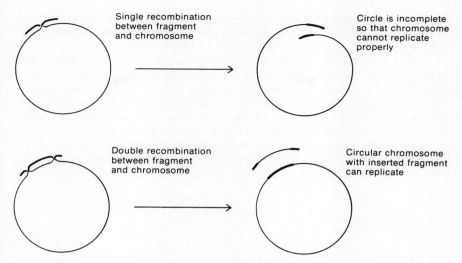

Single recombination between fragment and chromosome

Circle is incomplete so that chromosome cannot replicate properly

Double recombination between fragment and chromosome

Circular chromosome with inserted fragment can replicate

Figure 10-5   Successful integration of transducing DNA during generalized transduction requires double recombination.

SPECIALIZED TRANSDUCTION

In generalized transduction, any region of the bacterial chromosome can be transduced. In specialized or restricted transduction, however, only one or two specific genes are transduced, the inclusion of bacterial genes occurring as an error when a prophage detaches from its specific site in a bacterial chromosome. The means by which bacterial genes are included in the phage chromosome is illustrated by the prime example of a specialized transducing phage, the phage λ.

There is a single specific site in the normal *E. coli* chromosome where the λ chromosome integrates.

In the lysogenic cycle, the circular form of the λ chromosome integrates by recombination into a specific region of the *E. coli* chromosome, the λ attachment site, located between the *galactose* and *biotin* loci. The insertion is always at the same place when normal bacteria and normal phage are used. Corresponding to the λ attachment site is a similar region on the λ chromosome simply called the attachment site (abbreviated *att*). The base sequence of the two attachment regions have been determined by Arthur Landry and Wilma Ross to consist of identical sequences of 15 nucleotide pairs:

5' . . . GCTTTTTTATAGTAA . . . 3'
3' . . . CGAAAAAATATCATT . . . 5'

The base sequences on either side of the *att* regions of both chromosomes have been determined and have been found to differ from each other. Thus, the region of homology is limited to only 15 nucleotide pairs.

There is a single, reciprocal recombinational exchange between the circular λ and bacterial chromosomes, resulting in the insertion of the λ chromosome into the bacterial chromosome between recombined attachment sites (Figure 10-6). The inserted λ prophage can now be replicated with the bacterial chromosome and passed to subsequent generations of cells. The integration process can be reversed and the phage excised from the bacterial chromosome by a single reciprocal recombinational exchange involving the *att* sites once again (Figure 10-6). The excised DNA then enters the lytic cycle and causes lysis of the bacterial cell with the production of

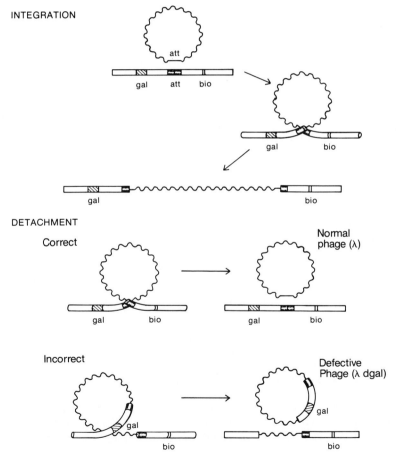

Figure 10-6   Integration and detachment of the λ chromosome from the host chromosome.

phage progeny. The precise recombinational process must differ from the one described earlier (Figure 7-1), in which there is extensive homology between recombining chromosomes and the formation of heteroduplexes about 1,000 nucleotide pairs long. In the process by which λ integrates into a bacterial chromosome and others like it, long heteroduplex regions cannot be formed.

Errors during the detachment of the λ chromosome result in segments of bacterial DNA replacing segments of λ DNA.

Excision of the λ chromosome from the *E. coli* chromosome sometimes occurs abnormally. The inserted λ prophage is flanked on one side by the *galactose* (*gal*) locus and on the other side by the *biotin* (*bio*) locus. In rare instances when the prophage is induced to detach from the *E. coli* chromosome, it carries the *gal* or the *bio* region (never both) in place of some of its own DNA. A phage chromosome carrying the *gal* or *bio* region results from misplaced recombination between regions outside of the *att* sites (Figure 10-6). This phage chromosome is deficient in some of its own sequences, and resultingly defective, but contains in their place some bacterial sequences. The defective chromosome carries either the *gal* region and is denoted λ *dgal* (*d* for defective) or the *bio* region and is denoted λ *dbio*. Such abnormal chromosomes are produced at a frequency of only about one per million. The defective chromosomes are able to replicate, and produce phages that are released into the medium.

Defective λ chromosomes are inserted, along with chromosomes of normal phages, into chromosomes of recipient bacteria.

Defective λ phages can inject their DNA into recipient *E. coli* cells in the same manner as normal λ chromosomes, but they cannot establish themselves as lysogens. Successful insertion into the bacterial chromosome depends on the prior insertion of a normal λ chromosome from a second phage into the bacterial chromosome. The defective λ chromosome can insert itself into the *E. coli* chromosome via the *gal* (or *bio* region, which it shares with the bacterial chromosome (Figure 10-7). The recombination results in a bacterial chromosome carrying a normal λ prophage and a defective prophage containing a *gal* (or *bio*) region. Because the bacterium already carried a *gal* locus in its chromosome, the resulting chromosome is now diploid for the *gal* locus, but haploid for the remainder of the bacterial genome. If the donor strain were *gal*⁺ (able to utilize galactose as a carbon source) and the recipient strain were *gal*⁻ (unable to utilize galactose), the recipient cell is heterozygous and has a *gal*⁺ phenotype. If the bacterial cells were plated on a medium in which galactose is the only carbon source, the transduced cells would be able to grow and could be identified. Specialized transduction cannot be used to map the positions of a large number of different genes. It can, however, be used for studying recombination within a limited region of a chromosome, for example, within a gene (Chapter 15).

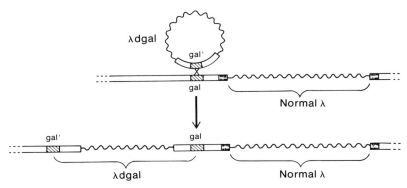

Figure 10-7   Specialized transduction by λ *dgal.*

High-frequency specialized transduction occurs when transduced bacteria serve as donors in transduction experiments.

A bacterial clone descended from a cell transduced by a defective λ phage is composed of cells, each of which has a normal λ prophage and a defective prophage carrying a bacterial gene. When these lysogens are induced to enter the lytic cycle, both types of λ chromosome are replicated, the normal and the defective one. Upon lysis, a large number of phages are produced with defective chromosomes carrying a bacterial gene. Because of the large numbers of defective phages here, in contrast to the situation described earlier where they were rare, a large number of transductants are produced.

CONJUGATION

To geneticists, conjugation, the third mechanism of gene transfer in bacteria, has other uses than do transduction and transformation, which are particularly suited for mapping studies over short distances. Conjugation is best suited for mapping studies involving long distances. Thus, when conjugation is available, it is used to construct a gross chromosomal map. Transduction might be used for filling in the details. Conjugation is not, however, found in all bacteria, necessitating the use of transduction or transformation for general mapping, a slow process at best. Conjugation is known to occur naturally in only a few bacterial species, *Escherichia coli, Pseudomonas aeruginosa,* and *Vibrio comma.* During conjugation, part but usually not all of the genetic complement of one bacterium is transferred to a second bacterium. As in transformation and transduction, a donor strain and a recipient strain participate in the genetic exchange. In contrast to syngamy in eukaryotes, bacterial conjugation involves neither the complete fusion of two cells nor an equal genetic contribution from each parent cell to the progeny derived from the cross.

Two types of cells are involved in conjugation.

The ability to initiate conjugation requires the presence of a plasmid called a transfer or fertility (*F*) factor. The *F* factor in *E. coli* is a small plasmid, a circular double helical chromosome composed of DNA and containing 100,000 nucleotide pairs (about 2 to 3 per cent of the size of the bacterial chromosome). The *F* factor undergoes replication autonomously, with its replication coordinated with cell division so that only one copy of the *F* plasmid is present per cell. Thus, a clone of cells carrying the *F* factor, F^+ cells, can be established. Cells without *F* factors are designated F^-. Electron microscopy can distinguish F^- cells from F^+ cells because the F^+ cells have filamentous structures, *pilli*, extending from their surfaces (Figure 10-8). When F^+ and F^- cells are mixed together, pairs of F^+ and F^- cells become attached via *conjugation tubes*, which are believed to be derived from the pilli of the F^+ cells.

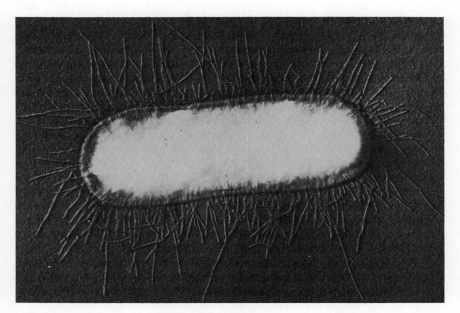

Figure 10-8   A bacterium covered with pili (also called fimbriae). Photo courtesy of J. P. Duguid, Bacteriology Department, Queen's College, Dundee, Scotland. From Duguid, Anderson, and Campbell, *Journal of Pathology and Bacteriology,* v. 92, p. 107, 1966, with permission.

F⁺ cells transfer a copy of the *F* factor to *F⁻* cells through the conjugation tube.

Once a conjugation tube is established between an *F⁺* and *F⁻* cell, the *F* factor undergoes rolling circle replication (Figure 10-9). The single-stranded DNA molecule spun off from the circular *F* element passes through the conjugation tube connecting the two bacteria. After entering the *F⁻* cell, the single-stranded *F* DNA is a template for the synthesis of its complement and resumes a circular configuration. The conjugated cells separate into what are termed exconjugants. The original *F⁻* cell has been converted to an *F⁺* and upon cell division will give rise to more *F⁺* cells. The original *F⁺* cell remains *F⁺*. Additional conjugations will continue until all of the *F⁻* cells in the culture flask have been converted to *F⁺*. Because the effect of mixing *F⁺* and *F⁻* cells together is to convert all of the cells to the *F⁺* condition, the *F* plasmid should be considered an infectious agent.

Among *F⁺* cells are a few cells with the *F* factor integrated into the bacterial chromosome so that transfer of the chromosome occurs during conjugation.

The *F* element does not always exist as an independent circular plasmid. It can integrate into the bacterial chromosome in a manner similar to that in which the circular λ chromosome inserts into the *E. coli* chromosome. There are usually a few cells (about 0.1 per cent) in a large population of *F⁺* cells wherein the *F* factor has been inserted into the *E. coli* chromosome. In this position, the *F* factor undergoes replication along with the bacterial chromosome (much like a prophage) and strains of cells with integrated *F* factors can be established. Cells in which the *F* factor is part of the nuclear chromosome are called *Hfr* (high-frequency recombination) cells. *Hfr* cells, like *F⁺* cells, have pilli on their surfaces and are able to conjugate with *F⁻* cells (Figure 10-10). In response to conjugation, rolling circle replication is initiated within the boundaries of the integrated *F* DNA. A single strand of DNA

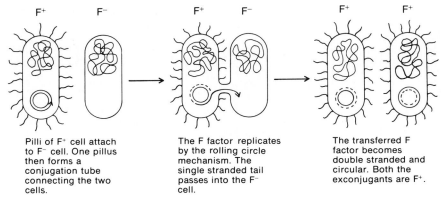

F⁺ F⁻	F⁺ F⁻	F⁺ F⁺

Pilli of F⁺ cell attach to F⁻ cell. One pillus then forms a conjugation tube connecting the two cells.

The F factor replicates by the rolling circle mechanism. The single stranded tail passes into the F⁻ cell.

The transferred F factor becomes double stranded and circular. Both the exconjugants are F⁺.

Figure 10-9   Transfer of the *F* factor during conjugation between *F⁺* and *F⁻* cells.

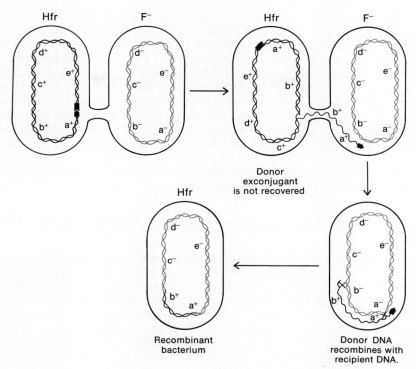

Figure 10-10   Transfer of the bacterial chromosome during conjugation between *Hfr* and *F⁻* cells.

is transferred from the *Hfr* cell through the conjugation tube into the *F⁻* cell. The transferred DNA has *F* factor DNA at its 5′ end, but is composed mostly of DNA derived from the bacterial chromosome. Thus, chromosomal genes are transferred from one *E. coli* cell to another. The conjugal bliss is short-lived, alas, and the two cells usually separate before transfer of much of the *Hfr* chromosome. The *F⁻* cell receives only part of the bacterial chromosome and part of the *F* factor, the remainder of the F factor being left behind in the donor cell, and remains *F⁻*. The portion of the chromosomal DNA that is transferred can undergo recombination with the chromosome of the recipient cell and produce new genetic combinations.

Different Hfr strains can be established and used to construct genetic maps of the E. coli chromosome.

Because, in an *Hfr* strain, the integrated *F* factor is replicated along with the nuclear chromosome, it is possible to establish a clone of *Hfr* cells. Unlike λ, which inserts into a single site on the *E. coli* chromosome, *F* factors integrate into the bacterial chromosome in a large number of sites. Among

Hfr strains of independent origin, we expect many to have the *F* factor inserted at different locations in the bacterial chromosome. Because transfer of the nuclear chromosome from the *Hfr* to the *F⁻* cell is initiated within the *F* factor, different *Hfr* strains start transfer of the *E. coli* chromosome at different positions and in different directions, that is, clockwise or counter-clockwise. By comparing the order in which different marker alleles are transferred by different *Hfr* strains, it is possible to construct a general map of the *E. coli* chromosome. In the following sections we discuss four major factors that must be considered in using *Hfr* × *F⁻* crosses to construct genetic maps: (1) the selective recovery of recombinant cells, (2) the rate of transfer of the chromosome from the donor to the recipient cell, (3) recombination of the donor DNA with the chromosome of the recipient cell, (4) the order of gene transfer in different *Hfr* strains.

Selective techniques are used to recover recombinant progeny in *Hfr* × *F⁻* crosses.

In Lederberg and Tatum's original experiments, they identified recombinants by recovering prototrophs that could grow in the absence of supplements. Selective techniques are still standardly used to recover recombinants in bacterial conjugation experiments. We mate bacteria that differ in multiple traits. Some of these traits, the selected traits, serve to recover recombinant cells (as in the generalized transduction experiments, p. 270). The rest of the traits, the marker traits, serve only to construct genetic maps. To illustrate, consider that an *Hfr* and an *F⁻* strain with the following genotypes are mated:

Hfr: thr⁺, leu⁺, strep^s, azide^r, T_1^r, gal⁺, lac⁻

F⁻: thr⁻, leu⁻, strep^r, azide^s, T_1^s, gal⁻, lac⁺.

The cells have alternative allele constitutions for all of the genes under study. We can use *strep^r* from the *F⁻* strain and *thr⁺* and *leu⁺* from the *Hfr* strain as selected traits. Cells growing in the absence of threonine and leucine and the presence of streptomycin are recovered and are all *thr⁺*, *leu⁺* and *strep^r*. These cells are all derived by recombination between the two strains. The recombinant cells are recovered and examined to determine whether they are recombinant for the traits at the *azide*, T_1, *gal*, and *lac* loci. Because all of the recombinant cells must carry the *strep^r* allele, derived from the *F⁻* cell, we can determine nothing about the transfer of the *strep^s* allele from the *F⁺* cell.

Interrupted mating experiments reveal gene order in *Hfr* × *F⁻* matings.

Starting with part of the integrated *F* factor, there is progressive transfer of the chromosome from an *Hfr* cell to an *F⁻* cell. The entire transfer requires about 90 minutes at 37° C, and the rate of transfer is apparently constant. Hence 25 per cent of the chromosome is transferred after 22.5 minutes, 50 per cent after 45 minutes, and so on. A gene near the head, the 5′ end, of the transferred single strand will enter the F^- cell before a gene in the middle or at the end of the transferred DNA. The relative positions of genes are determined by interrupting the mating at different times after the initiation of conjugation. This is done by separating pairs of conjugating cells in a blender. If the DNA carrying a particular allele has already entered the F^- cell, it can undergo recombination with the chromosome of the F^- cell and is identifiable in the recovered recombinant cells.

By determining from the start of conjugation the time of entry of each allele from an *Hfr* cell into a recovered recombinant, we can determine the gene order on the *E. coli* chromosome. Among recombinant cells that are *leu+*, *thr+*, *strep*r, we find azide resistant cells 9 minutes after conjugation starts, T_1 resistant cells after 10 minutes, *lac+* cells after 17 minutes, and *gal+* cells after 23 minutes (Figure 10-11). From these times of entry, we deduce the following order of the genes on the *E. coli* chromosome: *az*, T_1, *lac*, *gal*. In the case of the *thr+* and *leu+* alleles derived from the *Hfr* strain, we find the first *thr+*, *leu+*, *strep*r cells about eight minutes after conjugation starts. Therefore, the *threonine* and *leucine* loci are near the head end of the transferred chromosome segment in this cross. It is important that the selected traits derived from the *Hfr* strain be transferred before the marker traits when we use the time of entry technique. If the selected traits were transferred last, all of the recombinant bacteria would already have received the marker traits and their time of entry could not be determined.

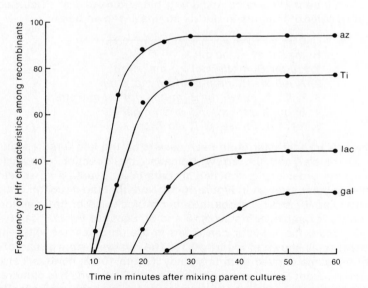

Figure 10-11   Time and efficiency of transfer of unselected donor alleles at the *az*, *T1*, *lac*, and *gal* loci among *F- thr+*, *leu+*, *strep*r recombinants in an interrupted mating experiment. After Jacob and Wollman, *Sexuality and the Genetics of Bacteria*, New York: Academic Press, 1961.

The frequency of recovery of marker traits depends on their proximity to the selected traits transferred from the *Hfr* cell.

Examining the frequency at which the different marker traits are recovered in *thr*+, *leu*+, *strep*r recombinants reveals that those alleles transferred early during the mating are recovered at higher frequency than those transferred later (Figure 10-11). Differences in the frequency of recovery vary with the physical distances between the marker traits and the selected traits and are reflected in the differences in the time of transfer of the traits. Because all recovered cells must be *thr*+ and *leu*+, the region of the *Hfr* chromosome carrying the two genes must recombine with the chromosome of the *F*⁻ cell. Those genes that are physically close to the *threonine* and *leucine* loci can be integrated into the *F*⁻ chromosome by the same recombinational exchanges that insert the *leu*+ and *thr*+ alleles. If the genes are distant from the selected ones, it is less likely that the same recombinational exchanges will insert both selected and marker alleles into the chromosome of the *F*⁻ cell. Two pairs of recombinational exchanges might be needed, and the frequency of recovery of the distant genes in the recombinant cells is reduced.

Use of *Hfr* strains with *F* factors inserted at different locations reveals the genetically circular nature of the *E. coli* chromosome.

In 1958 Francois Jacob and Elie Wollman reported a study of matings using six different *Hfr* strains, including the strain we just discussed. The order of transfer of markers in the six strains is shown below:

$$0 \longleftarrow\rule{6cm}{0.4pt}$$

H. *thr leu az T₁ lac gal λ*
1. *leu thr B₁ meth mtol xyl mal strep*
2. *T₁ leu thr B₁ meth mtol xyl mal strep*
3. *T₆ lac T₁ az leu thr B₁ meth mtol xyl mal strep*
4. *B₁ meth mtol xyl mal strep λ gal*
5. *meth B₁ thr leu az T₁ lac T₆ gal λ.*

The first trait transferred in each case is on the left, 0 representing the origin of transfer and the arrow indicating the direction of transfer. By matching the order of the transfer of traits in an overlapping scheme, we deduce a circular array as in Figure 10-12. Note that the direction of transfer in a particular *Hfr* strain can be in either a clockwise or counterclockwise direction (for example, compare *Hfr* 3 and 5). Because transfer can occur in either direction, the *F* factor can insert in the chromosome with either of two polarities as shown in Figure 10-13. The relatively simple map of the *E. coli* chromosome derived from this study (Figure 10-12) contrasts strikingly with a more recent *E. coli* genetic map (Figure 10-14), which is cluttered with over 400 genes placed there by investigators utilizing conjugation techniques to determine the approximate location of genes and transduction techniques to determine the positions of closely linked genes.

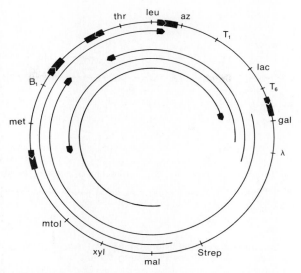

Figure 10-12   Diagram showing how the circular genetic map of *Escherichia coli* is deduced from the order of marker transfer in different *Hfr* strains.

F factors carrying chromosomal genes can detach from the bacterial chromosome.

Like λ prophages, integrated *F* DNA can become detached from the bacterial chromosome and return to its former life as a small, circular, autonomously replicating entity. Detachment involves a single reciprocal exchange that produces a circular *F* factor and leaves the circular bacterial chromosome intact. Also as in λ prophages, the detachment of the F factor can occur with the inclusion of an adjoining region of the bacterial chromosome. The *F* factor, now designated *F'* (F prime), carries genes typically present in the bacterial chromosome although that chromosome is correspondingly deficient in those genes. For example, the *lac* locus can be included in the *F'* factor (now referred to as *F' lac*) and be absent from the nuclear chromosome. Nevertheless, the bacterium contains a complete genetic complement, which is shared between the *F'* and the nuclear chromosome.

Nuclear genes carried by an F' factor are transferred during conjugation with F⁻ cells.

F' cells can undergo conjugation and transfer their *F'* factor to an *F⁻* cell. A nuclear gene, which is physically part of the *F'* element, is also transferred. The process is referred to as *F-duction* or *sexduction*. Because the *F'* factor replicates autonomously, the *F⁻* cell becomes *F'* and establishes

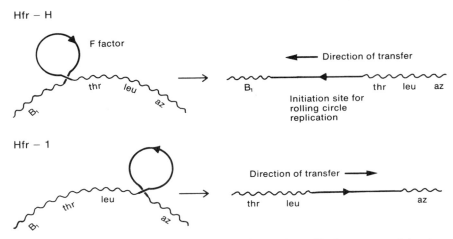

Figure 10-13   The orientation of the inserted *F* factor in an *Hfr* chromosome determines the direction of transfer of the chromosome during conjugation.

a new line of *F'* cells. Typically, the recipient cell has its own copy of the region of the bacterial chromosome carried by the *F'* element. The resulting cell line will be partially diploid, having two copies of one or a few genes and one copy of all of the others.

There are repeated sequences in bacterial chromosomes: insertion sequence elements (IS elements).

In our consideration of the renaturation of *E. coli* DNA (p. 79), we noted that the kinetics of renaturation indicate that the genome of the bacterium is composed of only single-copy sequences. Even so, the sensitivity of the technique could not eliminate the possible presence of low levels of repeated sequences. Indeed, repeated sequences *are* present in bacterial chromosomes. The types of repeated sequences we wish to introduce at this point, because they are involved in the integration of *F* into the bacterial chromosome, are called insertion sequence elements or IS elements. IS elements contain no genes specifying known products and are characteristically shorter than 2,000 base pairs. In *E. coli* a series of four prominent IS elements are known. They have different sequences and are identified as IS1, IS2, IS3, and IS4. The number of copies of different IS elements present in the chromosome ranges from about five to thirty. IS1 contains about 800 base pairs and about 8 copies are present in the *E. coli* chromosome. IS2 is about 1,300 base pairs long and is present in *E. coli* in about 5 copies. The total proportion of the genome devoted to IS elements is less than 0.3 per cent. The termini of IS elements are redundant, the repeats being inverted with respect to each other, so that complementary sequences are located on the same strand. Upon denaturation and renaturation they will pair with each other to form a "lollipop" configuration (Figure 10-15). The positions

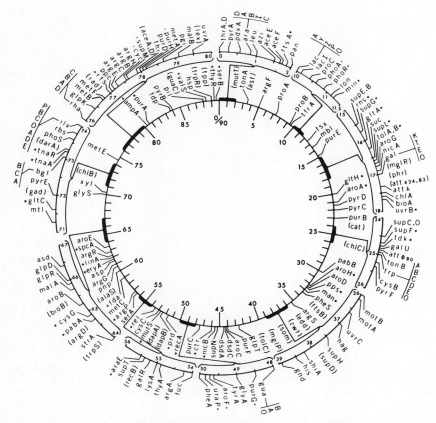

Figure 10-14   A current genetic map of the *Escherichia coli* chromosome. Taylor, A., and Trotter, C. *Bacteriological Reviews*, v. 31, p. 337, 1967. With permission from American Society for Microbiology.

of IS elements in a bacterial chromosome are not constant from strain to strain nor are they maintained within a given strain. A scheme for the movement of IS elements is presented in Figure 10-16 and involves pairing of the complementary ends of an IS element, the detachment of the IS element from the chromosome, and its reinsertion in a new location. There are no demonstrated homologies between IS elements and the random locations in chromosomes where they are inserted. Because the insertion of an IS element into different positions in a chromosome does not involve pairing of homologous regions, it represents *illegitimate* recombination. The insertion of the IS element need not always be between genes but can occur within the boundaries of a gene, thereby destroying its function—the mechanism will become clear by the end of Chapter 13. Finally, IS elements are not grouped but are widely distributed around the chromosome.

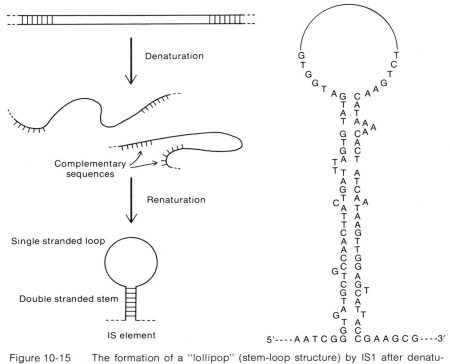

Figure 10-15   The formation of a "lollipop" (stem-loop structure) by IS1 after denaturation and renaturation of DNA. IS Sequence from Ohtsubo and Ohtsubo, *DNA Insertion Elements, Plasmids and Episomes,* ed. Bukhari, Shapiro, and Adhya, Cold Spring Harbor Laboratory, 1977.

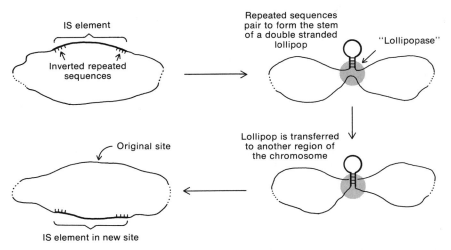

Figure 10-16   A hypothetical scheme for the transposition of an IS element. Although the chromosome is represented as a single strand it is, of course, a double helix.

F contains IS elements and integrates into the bacterial chromosome through homologies with chromosomal IS elements.

IS elements are not limited to the bacterial chromosome. The fertility factor, *F*, contains three different IS sequences, one copy of IS2, and two copies of IS3 as well as one we have not previously mentioned, γ-δ (Figure 10-17). All of these insertion elements are used as sites for recombination with their counterparts in the bacterial chromosomes during integration of *F* and the formation of *Hfr* strains. Because there are multiple IS elements in the *F* factor and multiple IS elements with variable positions in the bacterial chromosome, numerous different *Hfr* strains are possible.

F and λ are examples of episomes.

Similarities between *F* and λ DNA are evident. Both are small, circular, double-helical molecules. Both can replicate independently of the bacterial chromosome, *F* producing a second copy in conjunction with cell division and λ producing multiple copies during the lytic growth cycle. Both *F* and λ insert into the nuclear chromosome by single recombinational exchanges, *F* in multiple sites and λ normally in a single site between the *bio* and *gal* loci. (λ contains IS2 and, in mutants lacking the *att* region, can insert itself via IS2 in other regions of the *E. coli* chromosome.) Both can detach and become independent of the nuclear chromosome. Detachment can be normal or abnormal. In abnormal detachment adjacent regions of the bacterial chromosome can be carried along. In both cases the alternative states are

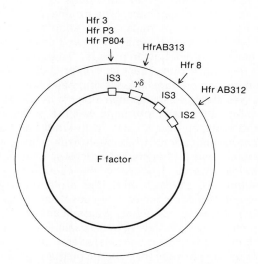

Figure 10-17   The position of IS elements in the *F* plasmid. IS elements used for insertion into various *Hfr* strains are indicated. After Berg in *DNA Insertion Elements, Plasmids and Episomes,* ed. Bukhari, Shapiro and Adhya, Cold Spring Harbor Laboratory, 1977.

mutually exclusive. A cell can be F^+ or *Hfr* but cannot maintain both free and integrated *F* factors. λ also can be integrated or free, with integration providing immunity for lysis by λ. Elements such as *F* and λ that exist alternatively as integrated or autonomously replicating non-integrated factors were named *episomes* by Jacob and Wollman.

Grouping such seemingly diverse items as λ and the *F* factor into the same category calls attention to the possibility that λ and *F* are evolutionarily related. Both may have evolved from similar beginnings as self-replicating elements within a bacterial cell. λ became an infective agent by evolving into what we recognize as a phage, and *F* became infective through the evolution of bacterial conjugation. *F*, of course, is a plasmid, that is, an element capable of autonomous self-replication within a cell. Other plasmids are known, including the *col* plasmids, which contain genes for colicins, a group of specialized bacterial toxins, and resistance, *R* plasmids, which we will now discuss.

R plasmids carry genes for resistance to antibiotics.

Like *F* factors, resistance (*R*) factors or plasmids are small, circular, autonomously replicating elements composed of double-helical DNA, but, unlike *F* factors, *R* factors carry genes of particular importance to human and veterinary medicine. One of the first *R* plasmids identified confers on the bacterial cells that carry it resistance to streptomycin, tetracycline, sulfonamide, and chloramphenicol—all agents commonly used to control bacterial infections. It is not in the least far-fetched to view *R* plasmids as adaptations to the widespread use of antibiotics to control bacterial infections in humans and domestic animals. (The original *R* factor also conferred resistance to ionic mercury, perhaps as an adaptation to environmental pollution.) An *R* factor, like an *F* factor, enables bacteria to conjugate and pass copies of itself to other bacteria. Thus, *R* factors are infectious. Although the drug resistance that *R* factors confer is significant in itself, the fact that the plasmids can be transferred is of greater consequence. The first *R* plasmid, *R222*, was identified in Japan in *Shigella dysenteriae,* a bacterium that causes dysentery. What was intriguing was the possibility that the *Shigella* obtained the *R* factor from *E. coli*, with which it shares a common habitat. Since that identification, it has been experimentally demonstrated that resistance factors can be transferred across species lines: from *E. coli* to *Shigella,* to *Salmonella* (*S. typhosa* is the cause of typhoid fever), to *Klebsiella* (*K. pneumoniae* causes a particular virulent form of pneumonia), and to *Proteus* (some *Proteus* species have been implicated in urinary and intestinal diseases). Recently, gonococci, bacteria causing venereal disease, have been found carrying *R* factors. It was possible and just recently demonstrated that *R* factors could even be transferred from *E. coli* to *Pasteurella pestus* (the bacterium responsible for bubonic plague) because experimental transfer of *F* from *E. coli* to *P. pestus* had been achieved.

The infectivity of *R* plasmids is not merely theoretical. In Japan in 1953, only 0.2 per cent of the *Shigellae* isolated from patients with dysentery were resistant to antibiotics. By 1960, the level had risen to 13 per cent and by

1965, to 58 per cent, with the vast majority of cells having multiple resistance to the four drugs listed above. At the same time 84 per cent of the *E. coli*, 93 per cent of the *Klebsiella*, and 90 per cent of the *Proteus* cultures isolated from humans had multiple drug resistance. Although these bacteria came from hospitalized patients who had received antibiotic therapy, a group that might not be typical of the whole population, there were similar results in the late 1970's in England where the majority of bacteria found in sewer water from both hospitals and private residences had substantial resistance to antibiotics. Thus, it is clear that bacteria are adapting through the infective transfer of autonomously replicating resistance plasmids to the "pollution" in their environment. It is also clear because of the widespread distribution of resistance factors that the efficacy of antibiotic therapy has decreased sharply. The rise of pathogenic bacteria with multiple resistance to antibiotics has become a serious problem in human and veterinary medicine.

The production of plasmids carrying multiple-resistance factors involves the presence of resistance genes in elements called transposons.

We might wonder how a plasmid clearly related to an *F* factor can acquire an array of resistance factors. The secret appears to lie in the way genes for resistance are located in the bacterial genome. Resistance genes are often found in genetic elements that are now called *transposons*. A transposon is constructed so that a gene is flanked on both sides by a common repeated sequence (Figure 10-18). The repeated sequences sometimes have the same orientation but are usually inverted with respect to each other. Thereby, like IS elements (p. 284), they will form "lollipops" upon denaturation and renaturation, the flanking sequences forming the stem and the DNA sequence of the resistant gene forming the head. Thus, the existence and positions of transposons can be identified by the presence of lollipops in renatured DNA (Figure 10-19). In some cases the repeated sequences are IS elements, for example, transposon 9, carrying resistance to chloramphenicol, is flanked by two IS1 sequences that have the same orientation, and transposon 10, carrying tetracycline resistance, is flanked by two IS3 elements that are inverted with respect to each other. The presence of the flanking repeating sequences provides the resistance genes with the opportunity to become translocated to different positions within a bacterial

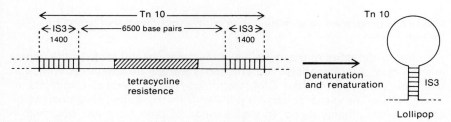

Figure 10-18   The structure of transposon 10 (Tn10) and its conversion into a stem-loop structure.

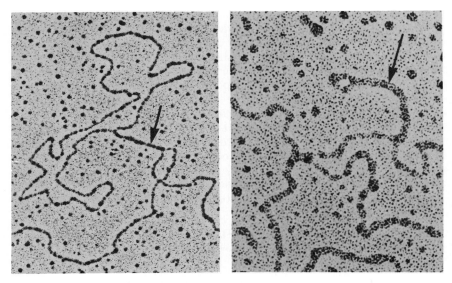

Figure 10-19   Electron micrographs of two stem-loop structures formed by the denaturation and renaturation of transposon DNA. (A) A stem-loop structure containing a large internal sequence between the repeated flanking elements. Photo courtesy of Louise Chow and Tom Broker, Cold Spring Harbor Laboratories, Cold Spring Harbor, New York. From *Insertion Elements, Plasmids and Episomes,* edited by A. Bukhari, J. Shapiro, and S. Adhya, Cold Spring Harbor Laboratory, 1977, with permission. (B) A lollipop-shaped structure formed by a transposon containing a gene for a bacterial toxin. Photo courtesy of Magdalene So, Department of Biochemistry and Biophysics, University of California, San Francisco. From So, Heffron, and McCarthy, *Nature,* v. 277, p. 453, 1979, with permission.

chromosome and to be translocated from the chromosome to a plasmid in a manner similar to the translocation of an IS element (Figure 10-16). Therefore, a mechanism for the insertion of several different resistance genes into a single plasmid is available. In addition to bacteria, there is evidence for the existence of genes in eukaryotes, notably in yeast, maize, and *Drosophila,* that transpose to varying positions in a genome and that may have sequences similar in structure to those found in transposons.

Plasmids can be used in genetic engineering experiments as carriers of foreign DNA.

Other properties of plasmids make them very useful tools to the experimental molecular geneticist. In particular, they can be used in conjunction with so-called recombinant DNA technology as carriers of fragments of foreign DNA. The plasmids can be purified from the *E. coli* chromosome by centrifugation, manipulated so that a piece of DNA from any organism—humans, *Drosophila,* redwood trees—is inserted into the plasmid, and then returned to the bacterial cell. There, because the plasmid replicates autonomously, the inserted DNA fragment is also replicated. The advantage of having a fragment of foreign DNA replicated in a plasmid—or *cloned*, to use the current jargon—is that a particular sequence of DNA that in its normal habitat, say, the human genome, would be present in vanishingly small amounts can now be manufactured in sufficient quantities to be thoroughly characterized.

The manipulation of a plasmid to insert a piece of DNA involves, first, cutting it open so that a foreign piece of DNA can be inserted and, second, assuring that the DNA fragment and the cut ends of the plasmid both have complementary single stranded regions through which to pair. The first step is accomplished using the restriction endonucleases we discussed earlier (pp. 70–72) that have the property of cleaving DNA at specific sites, usually palindromes. As an example, the restriction endonuclease *Hind*III (from *Hemophilus influenzae*) cleaves DNA as follows:

$$\downarrow$$
$$5' \ldots AAGCTT \ldots 3'$$
$$3' \ldots TTCGAA \ldots 5'$$
$$\uparrow$$

so that the following cleavage products are produced:

$$5' \ldots A \qquad AGCTT \ldots 3'$$
$$3' \ldots TTCGA \qquad A \ldots 5'$$

The number of sites in plasmid DNA that have the appropriate base sequence for a restriction endonuclease, in this case the *Hind*III sites, may be few. For example, a small 9,200-base plasmid, pSC101, that is commonly used has only a single *Hind*III site at which the enzyme can operate.

DNA fragments are inserted into cut plasmid DNA through the use of poly A and poly T tails.

One way to insert a fragment of foreign DNA into a plasmid, utilizes the enzyme terminal deoxynucleotidyl transferase. This enzyme can add 50 to 200 deoxyadenosine nucleotides to the 3' ends produced by the cleavage of pSC101 DNA with *Hind*III and, in a separate reaction, add a string of T residues to the 3' ends of DNA fragments from any organism (Figure 10-20). The poly A tails of the plasmid DNA can base pair to the poly T tails of the fragment and reform a circle. Through the use of polymerase I and DNA ligase, the fragment can then be covalently sealed to the plasmid DNA.

Bacteria that pick up plasmids from the medium can be selectively recovered.

In the presence of calcium ions, about one to ten E. coli cells in a million will pick up plasmids carrying foreign DNA from the medium. Identifying and recovering those rare cells that now contain the plasmids is done using selective techniques (Figure 10-21). For example, pSC101 contains a gene conferring resistance to the antibiotic tetracycline. Cells containing the pSC101 plasmid are resistant to the antibiotic and, in its presence, can be

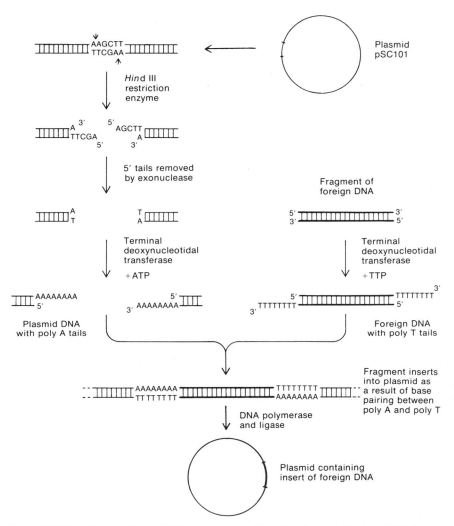

Figure 10-20   Recombinant DNA technology: The insertion of foreign DNA into a plasmid.

identified in a large population of cells that lack resistance. Because the plasmid replicates autonomously, the plasmid and its insert of foreign DNA will be cloned along with the bacterial cells. Thus, a human gene can be replicated in *E. coli*.

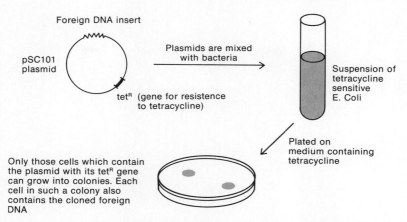

Figure 10-21   The selective recovery of bacteria (transformants) containing plasmids with foreign DNA inserts.

SUMMARY

Genetic exchange between bacteria occurs through transformation, transduction, and conjugation. Transformation and transduction are best suited for mapping over short distances, conjugation for mapping over long distances. Bacteria are transformed by the uptake of DNA. There are two forms of transduction: generalized transduction in which any gene can be transferred and specialized transduction in which only one or two genes can be transferred. Generalized transduction is mediated by temperate phages that accidentally carry a piece of bacterial DNA. Specialized transduction is mediated by phages carrying chromosomes composed of both bacterial and phage DNA, the bacterial DNA being present as a result of an error during prophage detachment. Bacterial conjugation depends on the presence of an *F* plasmid. The *F* plasmid can replicate autonomously or be integrated in the bacterial chromosome. When *F* is integrated in *Hfr* strains, transfer of the bacterial chromosome occurs during conjugation. Integration of *F* depends on homologies between IS elements in chromosomal and plasmid DNAs. In addition to *F*, other plasmids exist. Resistance plasmids are infectious, carrying genes for resistance to antibiotics and conferring the ability to conjugate. Other plasmids, for example pSC101, replicate autonomously but do not allow conjugation. Plasmids like pSC101 are used for the cloning of foreign DNA fragments in bacteria.

III

Gene Function

Genetic Determination of Protein Structure

Studies of gene transmission, particularly recombinational studies, allow us to map chromosomes into discrete regions that we identify as genes. The action of a gene ultimately makes itself known in the appearance of a specific phenotype, such as eye color, hair texture, nose shape. It is, of course, not eye color that is inherited but a particular DNA sequence that controls eye color. It is reasonable to assume, for a variety of reasons, that the DNA sequence composing a gene produces its phenotypic effects indirectly through the manufacture of intermediate products. Identification of these gene products is crucial to an understanding of how DNA exerts its control over phenotypic characteristics.

We have already indicated the nature of one type of gene product in noting that some human genetic diseases are characterized by the absence of a specific enzyme, such as hypoxanthine phosphoribosyl transferase (HPRT) in the Lesch–Nyhan syndrome (p. 162). The absence of specific enzyme activities was also implicated in the inability of bacterial auxotrophs to make all of their required nutrients. Chemically, all enzymes are proteins. Therefore, it is to protein structure that we must look to understand the mechanisms of DNA function. In this chapter we review the studies that led ultimately to the conclusion that structural genes specify the amino acid sequences of proteins. The framework established in this chapter will then be filled out in the following two chapters, in which we consider the step-by-step translation of a DNA nucleotide sequence into the amino acid sequence of a polypeptide.

Garrod noted that errors in enzyme activity are inherited.

In the early 1900's an English physician, Archibald Garrod, first realized that genes control the structure of enzymes. Initially, little importance was attached to Garrod's observations, and his ideas suffered the same fate as Gregor Mendel's. The scientific community was not yet prepared to appreciate the significance of his observations.

A book that Garrod wrote in 1918, *Inborn Errors of Metabolism,* set forth his views. In it, he describes a condition called *alcaptonuria* that, much to the consternation of the affected individual, causes the urine to turn black upon exposure to air. The blackening of the urine results from the oxidation of homogentisic acid, a compound that accumulates in the urine because its breakdown in the liver does not occur.

Garrod noted that several members of one family often had *alcaptonuria* and that the condition was present more often in the offspring of marriages of first cousins than from marriages of unrelated people. Because the condition appeared to be inherited and was present from the time of birth, Garrod concluded that it was an inborn error of metabolism and that it was caused by the absence of a particular enzyme. We now know that *alcaptonuria* is inherited on an autosome, which one is still unknown, and is phenotypically revealed because of the absence of the enzyme, homogentisic acid oxidase. Garrod demonstrated through his studies that the presence or absence of a particular enzyme was an inherited condition, in the same sense as the presence or absence of other phenotypic traits, such as eye color in *Drosophila* or starchy corn kernels in maize, are inherited. Furthermore, if a mutant allele is responsible for the *absence* of enzyme activity, then the normal allele of the gene is responsible for the *presence* of enzyme activity.

A systematic study using nutritional mutants in *Neurospora crassa* provided evidence that enzymes are specified by genes.

In the 1940's, George W. Beadle and Edward L. Tatum set out to gain insight into the nature of the products specified by genes. For their study, they chose to use the pink bread mold, *Neurospora crassa,* which is an excellent organism for genetic research. With a short generation time, two mating types (see Figure 8-2), and ordered meiotic products in a single ascus, *Neurospora* enables the experimenter to do detailed mapping studies. It is also a convenient organism with which to work. Single ascospores can be recovered and grown on a simple medium. Indeed, all that is necessary for the growth of hyphae of normal *N. crassa* is a carbon source, a nitrogen source, inorganic salts, and a single vitamin, biotin. The medium containing these components is called a *minimal medium* because it contains the minimal set of compounds from which *N. crassa* can synthesize all that it needs for growth: amino acids, nucleotides, vitamins, and so forth. The synthesis of these components is not left to chance but is a direct result of the action of biological catalysts, enzymes. Hence, *N. crassa,* like *Escherichia coli* (p. 261), normally comes equipped with all of the necessary enzymes to catalyze the metabolic conversions of the compounds supplied in the minimal medium to the products required for growth. Beadle and Tatum sought to systematically isolate nutritional mutants, strains of *N. crassa* that were incapable of synthesizing compounds necessary for growth from the minimal requirements but that could grow when additional compounds were added to the medium. In other words, they set out to obtain auxotrophs.

Selective techniques were used to recover nutritionally deficient mutants (auxotrophs).

To obtain their nutritional mutants, Beadle and Tatum used the selection scheme illustrated in Figure 11-1. They irradiated wild-type conidia (spores) to produce mutations and, after crossing the irradiated spores with a wild-type strain, they introduced individual ascospores, resulting from meiosis, onto an enriched culture medium, which was the minimal medium plus a yeast extract rich in vitamins. Many of the spores carried lethal mutations and did not germinate, but others did germinate and grew into flourishing mycelia. Growth of the mycelia indicated that the tissue derived from these spores was able either to make all of the components necessary for growth or to obtain them from the enriched medium. Spores from these tubes were then introduced into culture tubes containing the minimal medium lacking the vitamins. Most of the spores grew normally, but some grew not at all or very slowly. The growth of some strains on the vitamin-supplemented medium but not on the minimal medium indicated that they were incapable of synthesizing all of the vitamins necessary for growth. Beadle and Tatum then tested their mutant strains in media supplemented with individual vitamins and isolated three different strains: one requiring pyridoxine (vitamin B-6), one thiamine (vitamin B-1), and one folic acid (a component of another B vitamin). In later work they also isolated mutants requiring various amino acids.

Because each strain required a single specific vitamin, Beadle and Tatum deduced that each had lost the ability to synthesize the particular vitamin it now required because a necessary enzyme was lacking. When they subjected the strains requiring different vitamins to genetic analysis, they found the inheritance of each nutritional requirement involved a different single genetic factor, a gene. It appeared, then, that a particular enzyme activity was associated with a single gene, leading Beadle and Tatum in 1945 to summarize their work with a simple catch phrase, "one gene, one enzyme."

Current procedures greatly facilitate the recovery of auxotrophs.

Beadle and Tatum recovered the nutritional mutants, auxotrophs, at the cost of substantial labor. They analysed about 2,000 ascospores to recover the three vitamin-requiring mutants, all of the transfers being done one at a time. Current technology facilitates the recovery of auxotrophs in fungi, such as *Neurospora* and yeast, and in bacteria. Two techniques, replica plating and enrichment, are especially worth noting.

Replica plating. After studying nutritional mutants in *Neurospora,* Edward Tatum turned his attention to bacteria, in particular to *Escherichia coli.* The first *E. coli* auxotrophs he isolated were recovered in a laborious manner similar to the one originally used for *Neurospora* mutants; individual strains were transferred one at a time to different types of media. Subsequently, in 1952, Tatum's associates, Esther and Joshua Lederberg, introduced the procedure of *replica plating.* This technique is very effective in recovery of

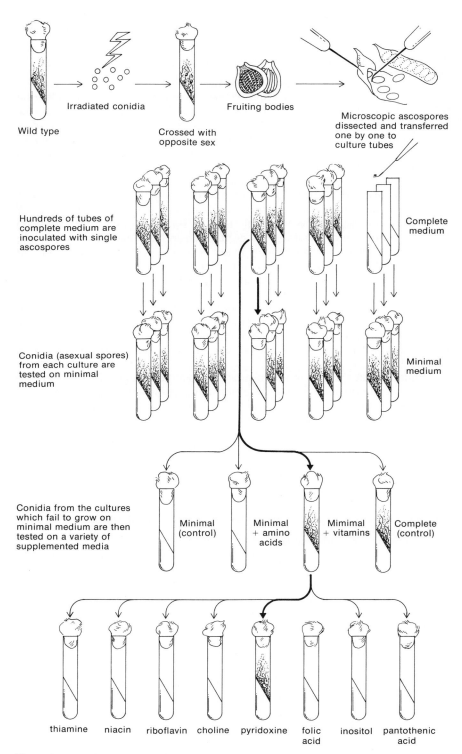

Irradiated conidia

Wild type

Crossed with
opposite sex

Fruiting bodies

Microscopic ascospores
dissected and transferred
one by one to
culture tubes

Hundreds of tubes of
complete medium are
inoculated with single
ascospores

Complete
medium

Conidia (asexual spores)
from each culture are
tested on minimal
medium

Minimal
medium

Conidia from the cultures
which fail to grow on
minimal medium are then
tested on a variety of
supplemented media

Minimal
(control)

Minimal
+ amino
acids

Mimimal
+ vitamins

Complete
(control)

thiamine   niacin   riboflavin   choline   pyridoxine   folic
acid

inositol   pantothenic
acid

Figure 11-1   The recovery of nutritional mutants in *Neurospora*. After Beadle, *American
Scientist,* v. 34, p. 31, 1946.

mutants in organisms, such as bacteria, that produce discrete colonies by division on solid agar medium.

The recovery of mutants using replica plating (Figure 11-2) starts with treating the bacteria with a *mutagen,* that is, an agent such as irradiation that causes mutations. A suspension of bacterial cells is then "plated" on solid enriched medium in a Petri dish so that about 100 cells are randomly distributed across the medium. These cells divide to produce colonies of bacteria in which all of the cells of a colony are genetically identical. The distribution of the colonies on the agar medium is random, but each colony is present in a specific position relative to the others. The Petri dish is then inverted on a circular block, which is covered by a sterile velvet cloth, and small numbers of bacteria from each colony are transferred to the cloth. Several sterile Petri dishes, each containing a different medium, are then pressed against the surface of the cloth. Some bacteria are transferred to the surface of the medium in each dish in a pattern identical to that in the original dish. Thus, the original pattern of colonies is reproduced or "replicated" on the different experimental media.

The different Petri dishes contain media of different compositions: one is minimal; others may be enriched with different vitamins, with amino acids,

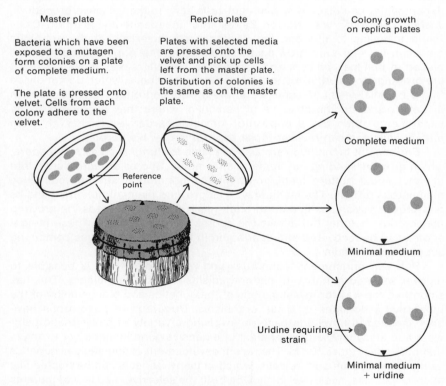

Figure 11-2   The recovery of mutants by replica plating.

or with purines and pyrimidines. After a day or two the dishes are examined and the distribution of colonies recorded. If colonies are found on enriched media that are absent on the minimal medium, new auxotrophs have been identified. One of the auxotrophs might be growing on the medium enriched with amino acids, while another is present on the medium containing purines and pyrimidines. The replica plating procedure allows rapid screening of hundreds of colonies for the possible existence of new mutants. Replica plating is not limited to the recovery of auxotrophs. It may also be used for the selective recovery of mutants that are resistant to antibiotics (different antibiotics are included in the test plates). Similarly, phage-resistant mutants or mutants sensitive to physical agents such as ultraviolet light can be recovered, as can conditional lethals, for example, by growing the bacteria at different temperatures. Indeed, the types of mutants that can be recovered by replica plating are perhaps limited only by the imagination of the experimenter.

Enrichment techniques. An experimenter usually wants to recover only certain types of mutants, those affecting the particular process under study. Despite the virtues of replica plating, finding the desired mutants is often a slow process because of the low frequency at which they are likely to be present even after treatment with mutagens. The search is speeded if the experimenter can discard non-mutant cells early in the selective procedure. Joshua Lederberg with Norton Zinder and, independently, Bernard Davis found a means to this end by utilizing a characteristic of the antibiotic penicillin—it kills only growing bacterial cells. If a population of cells is placed in minimal medium and exposed to penicillin, only those cells that are dividing are killed. Auxotrophs or other mutant cells, which do not divide in the minimal medium, survive and are introduced onto enriched medium in Petri dishes in the absence of penicillin, where they might grow. The experimenter then has a population of colonies derived exclusively from mutant cells and can recover those appropriate for his purpose.

Enrichment techniques are also available for fungi such as *Neurospora,* which grow by sending out numerous filaments. Spores are introduced into minimal liquid culture medium where the nonmutant strains commence filamentous growth. The medium with its growing filamentous forms and its non-growing spores is filtered through glass wool, which traps the growing filamentous material but allows the spores to pass through. These mutant spores can be recovered and distributed in tubes or Petri dishes containing supplemented media.

Between enrichment techniques and replica plating, it is possible to recover selected mutants in microorganisms with high efficiency. This has smoothed the way for investigations of the genetics and biochemistry of the biosynthetic domains of these organisms. Hundreds of auxotrophs have been isolated in fungi and bacteria involving a variety of biosynthetic pathways and allowing the development of extensive genetic maps, for example that of *E. coli* (Figure 10-15). The recent development of somatic cell genetics using human tissue culture cells, which in many respects can be treated like microorganisms, has likewise allowed both the selective recovery of mutants and a dramatic improvement of our understanding of the genetic map of the human genome. By contrast, using whole multicellular organisms, even a genetically well-characterized one like *Drosophila,* diminishes the chances of recovering auxotrophs because the nutritional requirements of multicellular organisms are complex and because the rapid screening of millions of individuals is not yet possible.

Compounds necessary for growth are formed by multi-step biosynthetic pathways.

The hypothesis that a single enzyme is specified by a single gene—the "one gene, one enzyme" hypothesis—proved to be essentially correct although in need of some modification. For instance it was hard to conceive how one of Beadle and Tatum's non-mutant organisms could, in a single enzymatic step, synthesize a vitamin from the few different compounds in the minimal medium. It was already known that some enzymatic conversions, for example, those involved in glycolysis, occur in a series of discrete metabolic steps leading from a precursor to a distant metabolic product. We now know that the synthesis of virtually every compound necessary for growth is the result of a series of linked biosynthetic conversions defining a metabolic pathway. A representative metabolic pathway in Figure 11-3 depicts the biosynthesis of pyrimidine nucleotides. There are six different steps catalyzed by six different enzymes, starting with the formation of carbamoyl phosphate under the direction of carbamoyl phosphate synthase and ending with the formation of uridine monophosphate (UMP). Thymidine monophosphate (TMP) and cytidine monophosphate (CMP) are in turn formed from UMP through the action of two additional enzymes. UMP can also be formed by the phosphorylation of uridine.

The use of auxotrophs facilitates the delineation of biosynthetic pathways.

If we accept for the moment the "one gene, one enzyme" concept, we can suppose that what Beadle and Tatum initially recovered were mutant alleles of genes specifying one of the enzymes in each of the biosynthetic pathways leading to the synthesis of thiamine, pyridoxine, and folic acid. However, selective techniques soon made it possible to obtain mutant alleles of genes for several, and eventually all, of the steps in a particular metabolic pathway. Such alleles, because they eliminate or greatly reduce the amount of a particular enzyme, are called *null alleles.*

As an example we can consider the recovery of mutants in yeast that are pyrimidine auxotrophs. Organisms carrying mutant alleles grow when uridine is added to the culture medium but fail to grow in the absence of uridine. Such mutant strains must contain the enzyme uridine phosphorylase (Figure 11-3), the enzyme necessary to convert uridine to UMP, and must also be able to convert UMP into CMP and TMP. If the synthesis of either CMP or TMP were missing, DNA could not be synthesized and growth could not occur. Not all uridine auxotrophs are defective in the same manner. The synthesis of UMP requires six separate steps, each of which is catalyzed by a particular enzyme and each of which, according to the one gene one

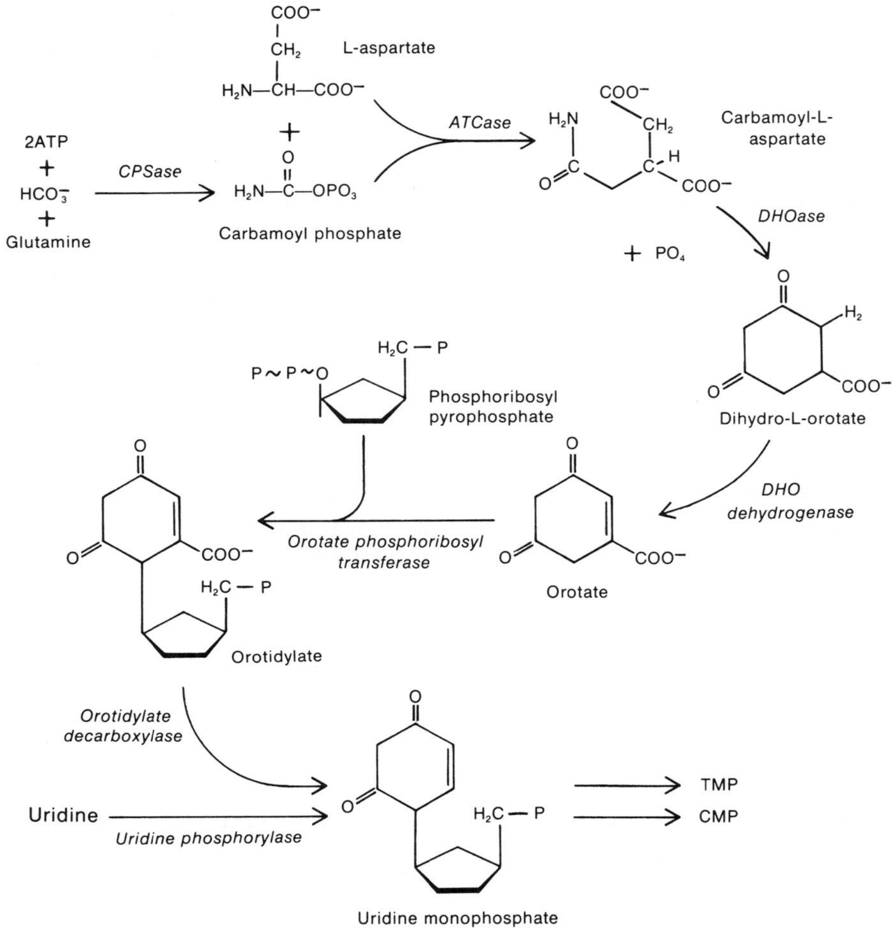

Figure 11-3   The pyrimidine biosynthetic pathway.

enzyme hypothesis, will be specified by its own gene. Thus, for the six steps in the synthesis of uridine monophosphate there should be six different genes, each of which can mutate to cause pyrimidine auxotrophy. We can formulate a generalization: the number of different types of mutants recovered that are auxotrophic for a particular compound will approximate the number of independent enzymatic steps in a biosynthetic pathway. Some enzymes are aggregates of more than one protein. Hence, there may be more genes than enzymatic steps. Some proteins have more than one enzymatic function, and single mutations may destroy all functions (pp. 412–414). Hence, there might be more enzymes than genes.

The nature of the specific enzyme encoded by a particular gene can be determined.

The value of auxotrophs is not limited to telling us the approximate number of steps in a biosynthetic pathway. They may also facilitate the identification of the intermediary metabolites and the enzymes that constitute the pathway as well as the determination of which gene specifies a particular enzyme. We can see how such an identification is made by considering a specific hypothetical mutational loss of one of the enzymes in the pyrimidine biosynthetic pathway—dihydroorotase. We can take three different approaches: analyses for accumulation of an intermediary metabolite, supplementation with intermediary metabolites, and assays of enzyme activity.

Accumulation of an intermediary metabolite. Normally intermediary compounds formed during the synthesis of a required metabolite are present in low amounts because they are immediately converted to the next compound in a pathway. For example, in normal cells, only small quantities of carbamoyl aspartate are present. In a mutant lacking dihydroorotase, carbamoyl aspartate is no longer converted to dihydroorotic acid and will accumulate and can be identified in extracts of mutant cells. The accumulation of an intermediary metabolite indicates that the enzyme that normally converts the accumulated compound to another one is defective.

Supplementation studies. Uridine is not the only compound that can supplement a mutant lacking dihydroorotase. Both dihydroorotic acid and orotic acid (Figure 11-3) can be converted to UMP in a dihydroorotase-deficient mutant and can support growth, the enzymes following dihydroorotase in the pathway being normal. However, the addition of carbamoyl aspartate to the medium will not support growth because the lack of dihydroorotase prevents the conversion of this compound to UMP. The observation that dihydroorotic acid supports growth but carbamoyl aspartate does not indicates that dihydroorotase is the defective enzyme in the mutant cells.

Enzyme assays. The ultimate identification of the nature of a defective enzyme in an auxotroph depends on an assay of enzymatic activity. In the example we are considering, the results from studies of the accumulation of intermediary metabolites and supplementation point to the loss of dihydroorotase in this mutant. A direct assay demonstrating the absence of dihydroorotase enzyme activity then confirms that the enzyme is lacking and allows us to conclude that the gene under study specifies dihydroorotase.

We generally expect null mutations, leading to the loss of enzyme activity, to be recessive.

Enzymes are biological catalysts, greatly accelerating the rates at which chemical reactions occur within cells. In general we expect that most mutants affecting enzyme activities will be inherited as recessives. For example, Tay-Sachs disease, in which hexosaminidase A is lacking, is inherited as a recessive trait carried on chromosome 15. We expect in a heterozygote for a null allele that enzyme activity will be reduced by half because only one gene produces a functioning enzyme. Even with only half of the enzyme activity present, in most cases the rate at which metabolic products are produced is not greatly affected. We have already noted that many intermediary metabolites are present in low concentrations, because they are rapidly converted to the next metabolite in the chain. In such cases the enzyme may be present in excess and not working at full speed. An enzyme might be active only 35 per cent of the time, spending most of its time waiting for metabolic intermediates to be pushed its way. Reducing the number of enzyme molecules by half may require the remaining enzyme to be twice as active—that is, 70 per cent of the time—but it will still be capable of supplying the products required by the organism. From an evolutionary or physiological perspective such a situation seems sensible. There is an adaptive advantage for organisms that produce more enzyme than they normally require, because they are thereby equipped to cope with deleterious mutations or with periods of physiological stress when increased metabolic activities are required.

Proper diet controls the deleterious effects of some human hereditary diseases that result from enzyme deficiencies.

Both deleterious and non-deleterious genetic traits in humans result from specific enzyme deficiencies (Table 11-1). We have already noted examples such as the Lesch-Nyhan syndrome, Tay-Sachs disease, and alcaptonuria. Many others exist, for example, orotic aciduria and phenylketonuria. Orotic aciduria is a disorder in pyrimidine biosynthesis that prevents normal growth and the formation of red blood cells, in which orotic acid is excreted in the urine. The effects are alleviated by feeding uridine or cytidine to the affected individual much as one would to a bacterial or fungal auxotroph. Phenylketonuria is a disease characterized by mental retardation and early death and is caused by the absence of tyrosine synthesis, resulting from a block in the conversion of the amino acid phenylalanine to tyrosine and the accumulation of toxic ketones that are excreted in the urine. The disease is controlled by placing affected children on a diet containing tyrosine but limited amounts of phenylalanine.

Some human genetic diseases resulting from enzyme deficiencies can be detected before birth.

The gene for HPRT is carried on the X chromosome so that hemizygous males are preferentially affected with the Lesch-Nyhan syndrome. Affected males inherit the X chromosome carrying the disease allele from their mothers. We know that heterozygotes who are carriers of diseases involving loss of enzyme activity can often be identified because they show reduction of enzyme activity to half the normal level. Carrier females for the Lesch-Nyhan syndrome can be identified because they exhibit reduced levels of HPRT. Amniocentesis (p. 149) makes it possible to determine whether a carrier female will give birth to a son with the Lesch-Nyhan syndrome because we

Table 11-1   Selected Human Genetic Disorders with Demonstrated Enzyme Deficiencies

Condition	Enzyme Deficiency	Condition	Enzyme Deficiency
Acid phosphatase deficiency	Acid Phosphatase	Leigh's necrotizing encephalomelopathy	Pyruvate carboxylase
Albinism	Tyrosinase	Lesch-Nyhan Syndrome	Hypoxanthine phosphoribosyl transferase
Alcaptonuria	Homogentisic acid oxidase	Lysine intolerance	Lysine:NAD-oxidoreductase
Ataxia, intermittent	Pyruvate decarboxylase	Male pseudohermaphroditism	Testicular 17,20-desmolase
Disaccharide intolerance	Invertase	Maple sugar urine disease	Keto acid decarboxylase
Fructose intolerance	Fructose-1-phosphate aldolase	Orotic acid uria	Orotidylic decarboxylase
Fructosuria	Liver fructokinase	Phenylketonuria	Phenylalanine hydroxylase
G6PD deficiency (favism)	Glucose-6-phosphate dehydrogenase	Porphyria, acute	Uroporphyrinogen I synthetase
Glycogen storage disease	Glucose-6-phosphatase	Porphyria, congenital erythropoietic	Uroporphyrinogen III cosynthetase
Gout, primary	Hypoxanthine phosphoribosyl transferase	Pulmonary emphysema	α-1-antitrypsin
Hemolytic anemia	Glutathione peroxidase	Pyridoxine-dependent infantile convulsions	Glutamic acid decarboxylase
Hemolytic anemia	Hexokinase	Pyridoxine-responsive anemia	γ-Aminolevulinic synthetase
Hemolytic anemia	Pyruvate kinase	Kidney tubular acidosis with deafness	Carbonic anhydrase B
Hypoglycemia and acidosis	Fructose-1,6-diphosphatase	Ricketts, vitamin D-dependent	25-Hydroxycholecalciferol 1-hydroxylase
Immunodeficiency	Adenosine deaminase	Tay-Sachs disease	Hexosaminidase A
Immunodeficiency	Purine nucleoside phosphorylase	Thyroid hormone synthesis, defect in	Iodide peroxidase
Immunodeficiency	Uridine monophosphate kinase	Thyroid hormone synthesis, defect in	Deiodinase
Intestinal lactase deficiency (adult)	Lactase	Tyrosinemia	para-Hydroxyphenyl-pyruvate oxidase
Ketoacidosis	Succinyl CoA:3-ketoacid CoA-transferase	Xeroderma pigmentosum	DNA-specific endonuclease

Note that some similar conditions result from various enzyme deficiencies and single enzyme deficiencies can produce multiple defects.
Modified from Table 1-4, *The Metabolic Basis of Inherited Disease*, J. B. Standbury, J. W. Wyngaarden, and D. S. Fredrickson, Eds. McGraw-Hill Book Company, 4th Edition, 1978.

can determine whether recovered fetal cells have an HPRT deficiency. The karyotype of the fetal cells can first be examined to determine if they are male or female, that is, whether they have a Barr body. If the cells come from a male fetus, the cells can be tested for growth in the presence of aminopterin (which we recall from the HAT technique (pp. 251–252) prevents the synthesis of purines and pyrimidines) because HPRT-deficient cells cannot grow in the presence of this inhibitor. Thus, we can determine whether a fetus has an HPRT deficiency and will after birth be afflicted with the Lesch-Nyhan syndrome. It is also possible, by assaying fetal cells for Hexosaminidase A, to determine whether a fetus after birth will develop the Tay-Sachs disease. At present, no procedures exist to change the disastrous course of either disease, and, like trisomy 21 (p. 143), both will, failing abortion of the fetus, doom affected children to a short and miserable life.

GENES ENCODE THE AMINO ACID SEQUENCES OF POLYPEPTIDES

We will turn now to consider some ways in which mutational changes restrict the enzyme activity of a protein. One obvious way in which this can occur is if the gene—that is, the precise nucleotide sequence—specifying the enzyme is totally absent from the chromosome on which it is normally carried. Other ways in which the activity of an enzyme or the function of a protein can be disrupted do not involve the loss of an entire gene or a portion of a gene but involve, instead, changes in the nucleotide sequence of a gene. The information in the remainder of this chapter and in the next two chapters will allow some appreciation of the means by which such changes can alter the activity of the enzyme specified by a gene. We will see that the genetic language or code is written so that each amino acid in a polypeptide chain is individually encoded. One important change in protein structure that can lead to the loss of normal function is the substitution in a polypeptide of one amino acid for another as a result of a mutation.

There are different levels of protein structure: primary, secondary, tertiary, and quaternary.

The four levels of organization of protein structure are shown in Figure 11-4:

Primary Structure: The amino acid sequence, the order of amino acids in the polypeptide chain or chains of which a protein is composed, constitutes the primary structure of a protein.

Secondary Structure: Polypeptide chains can exist as helices formed through hydrogen bonding or they can be non-helical. The type of configuration along the polypeptide chain constitutes the secondary structure of a polypeptide.

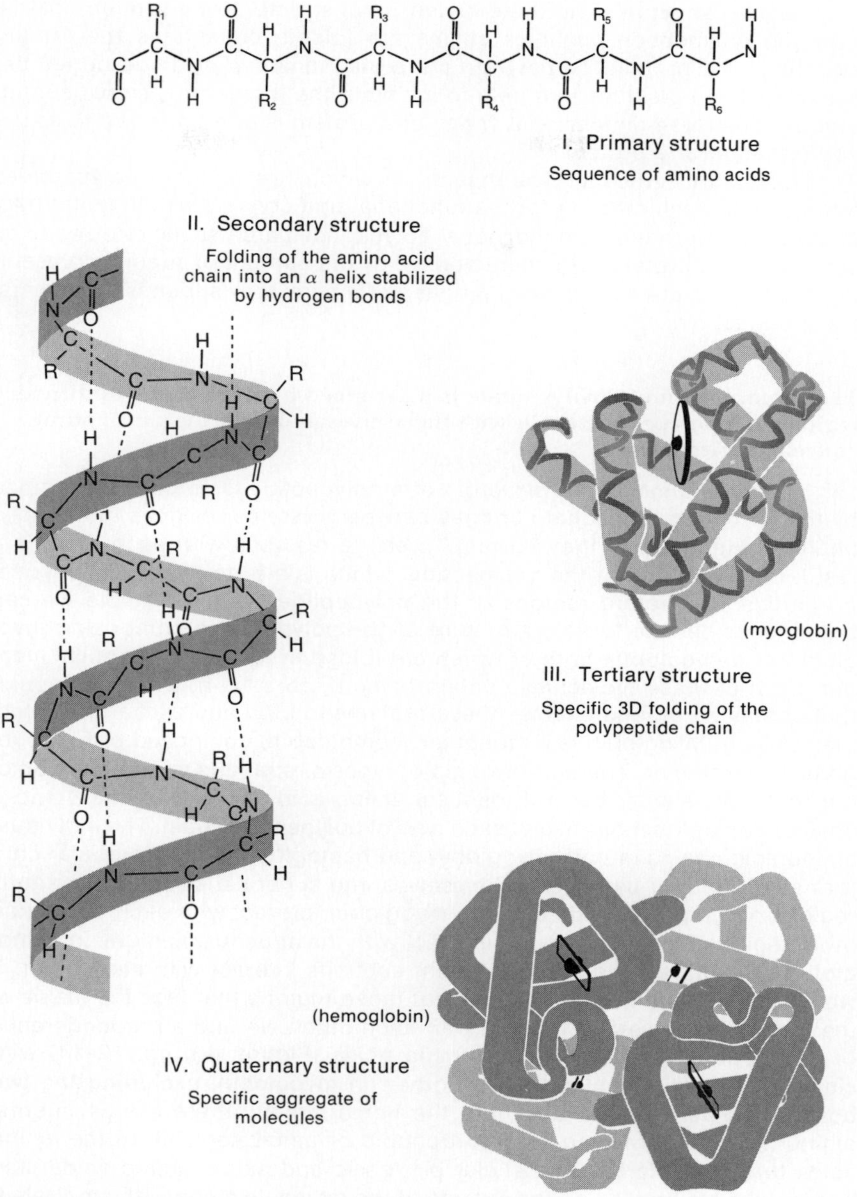

I. Primary structure

Sequence of amino acids

II. Secondary structure

Folding of the amino acid
chain into an α helix stabilized
by hydrogen bonds

(myoglobin)

III. Tertiary structure

Specific 3D folding of the
polypeptide chain

(hemoglobin)

IV. Quaternary structure

Specific aggregate of
molecules

Figure 11-4   The four levels of protein structure. (2) After Corey and Pauling, *Proceedings of the International Wool Texture Resources Conference of Australia*, 1955B, 1956. (3) and (4) After Dickerson and Geis, *The Structure and Action of Proteins*, New York. Harper & Row, 1969.

Tertiary Structure: The three-dimensional structure of a protein, that is, how the polypeptide chain or chains are folded, constitutes the tertiary structure of a protein. Tertiary structure also includes disulfide bonds between cysteine residues that help to maintain the three-dimensional configuration. The three-dimensional shape of a protein is often referred to as the *conformation* of a protein.

Quaternary Structure: The manner in which two or more polypeptides interact with each other to form a functional unit constitutes the quaternary structure of a protein. The individual polypeptides are usually referred to as sub-units, and proteins are characterized by the number of subunits present. For example, dimers have two subunits; tetramers, four subunits; hexamers, six subunits, etc.

The conformation of a polypeptide is determined by interactions of different regions of a polypeptide chain with themselves and with the surrounding aqueous environment.

The three-dimensional structure of a polypeptide is, in part, maintained by the formation of disulfide bridges between cysteine residues in the polypeptide. Alignment of the "correct" cysteine residues with each other is a result of the folding of the polypeptide, which is a function of the physical properties of different regions of the polypeptide. As an example we can briefly consider the tertiary structure of the polypeptides comprising myoglobin or hemoglobin, both of which are folded in an almost identical manner. Both of these molecules contain heme (Figure 11-4), a flat compound that contains iron and enables these proteins to bind and exchange molecular oxygen. Myoglobin is a monomer; hemoglobin, composed of four subunits, is a tetramer. There are two major types of subunits in adult hemoglobin that have similar but not identical amino acid sequences, the α and β chains. Hemoglobin has two of each type of polypeptide chain. The individual amino acid chains in both myoglobin and hemoglobin have a globular conformation, folding back upon themselves into spherical structures. Various regions of the polypeptide chain in myoglobin interact with each other, and other regions face outward and interact with the aqueous medium. In hemoglobin as well, regions in the different subunits interact with each other. If we look at the amino acid sequence of those regions that face the inside of the globin molecule and interact with each other, we find a preponderance of hydrophobic and ambivalent amino acids (Figure 1-3, pp. 12–14) with side chains that can bind to each other. In myoglobin, excluding the two basic histidine residues that hold the heme moiety, there are 34 internal amino acids, all of which are hydrophobic or ambivalent. All of the amino acids that are polarized at cellular pH (basic and acidic amino acids) and hydrophilic are located on the outside of the molecule along with ambivalent amino acids such as glycine and alanine. The conformation of a polypeptide is a direct result of the placement of hydrophobic amino acids, which bind with themselves, on the inside of the molecule and hydrophilic ones facing outward. In hemoglobin, a tetramer, those regions of polypeptides that bind with each other to form the tetrameric structure are similarly rich in hydrophobic and ambivalent amino acids so that they are capable of interacting.

It is only necessary for the primary structure of a protein to be encoded in a gene.

If we were to take one end of a polypeptide chain and stretch it out in a straight line, it would tend to re-form when we let go of it. Those residues that are hydrophobic would be attracted to each other. They would, consequently, form the inside of the polypeptide by "dissolving" in themselves. The hydrophilic side chains would interact with water and remain on the outside. The spatial distributions of particular regions of a polypeptide seem to be inevitable results of the solubility properties of the side chains of their amino acids and their positions in a polypeptide.

The considerations in the preceding paragraphs lead us to expect that the conformation of a protein is a function of its primary structure. In the late 1950's Christian Anfinsen and his colleagues confirmed this expectation in experiments demonstrating that several enzymes, when unfolded, could spontaneously refold to conformations that were enzymatically active (Figure 11-5). One such enzyme was ribonuclease, which, like many enzymes, is a globular protein composed of a single polypeptide chain. Reducing the disulfide bonds between four pairs of cysteine residues altered the tertiary structure of the enzyme. Exposing the enzyme in solution to urea, which eliminated hydrogen bonds, disrupted the secondary structure as well. Only the normal primary structure remained. An unfolded polypeptide with no ribonuclease activity was produced. Dialysis removed the urea and allowed hydrogen bonds to return, and bubbling oxygen through the solution allowed

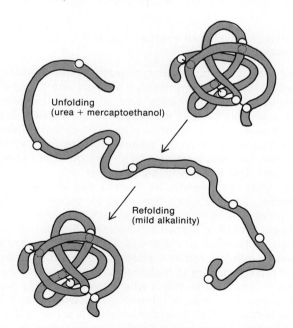

Figure 11-5   Denatured ribonuclease spontaneously refolds. After Epstein, Goldberger, and Anfinsen, *Cold Spring Harbor Symposium,* v. 28, p. 439, 1963.

the disulfide bonds to reform. Ribonuclease activity was restored, and using several physical criteria, Anfinsen and his colleagues demonstrated that the refolded enzyme was indistinguishable from the native one. The normal conformation of the protein spontaneously formed. Thus, we have a direct experimental demonstration that the primary structure of a protein is sufficient to direct the protein to fold into its enzymatically active configuration. If proteins spontaneously fold into their active configurations as a function of their primary structures, it follows that only the primary structure, that is, the amino acid sequence of a protein, need be encoded in the genome.

The genetics of a non-enzymatic protein, hemoglobin in humans, provides insight into gene structure and function.

The one gene, one enzyme hypothesis indicates that in some manner the genes of an organism code for proteins with particular enzymatic activities. There are, however, proteins with no enzymatic activity that are important constituents of cells. Because all proteins are structurally similar, differing only in amino acid sequence and conformation, there is good reason to believe that these proteins, like enzymes, are specified by genes. There is an advantage in studying the genetics of non-enzymatic proteins: They often occur in large quantities, in contrast to enzymes, which, befitting their function as catalysts, typically occur in small amounts. It is much easier to study the normal and mutant forms of a protein if it is present in large quantities. One such protein is human hemoglobin, the study of which has helped to elucidate how a gene specifies the amino acid sequence of a protein.

An inherited disease, sickle cell anemia, results from the presence of an abnormal hemoglobin molecule.

Sickle cell anemia is a deleterious, recessive genetic disease in humans, first described in 1910 by James Herrick. Herrick noted that the red blood cells of individuals with a particular type of anemia underwent dramatic change, when a sample of cells was deprived of oxygen, from the normal puckered shape to a sickle-like shape (Figure 11-6) from which the anemia takes its name. For sufferers of the disease, this means that during physiological stress, for example, running, the red blood cells become sickled in massive numbers. The sickle-shaped cells not only tend to break down sooner than normal ones, hence the anemia, but also are rigid and block capillary networks, causing many tissues to be deprived of oxygen. In addition to the primary effect on red blood cells, the anemia leads to many secondary effects, including kidney and heart malfunction (Figure 11-7). The disease is a good example of *pleiotropy* in which a mutant gene, acting through a primary defect, produces a series of secondary changes.

Sickle cell anemia occurs in high frequencies in humans whose ancestors have lived for many generations where malaria was or continues to be present, such as parts of Africa, southern Europe, and India. (We will discuss the connection in Chapter 21.) Sickle cell anemia is found in Black Americans whose ancestors lived in areas where malaria was present. In technologically

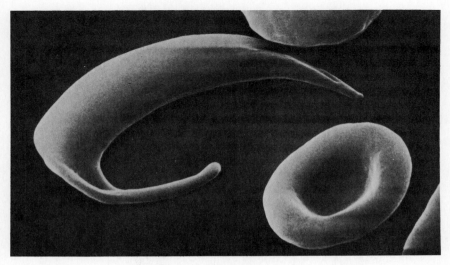

Figure 11-6   Scanning electron micrograph of normal and sickle red blood cells. Photo courtesy of John A. Long, University of California, San Francisco.

underdeveloped regions of the world, sickle cell anemia leads to early death. Only recently, in technologically advanced societies, have the lives of people with sickle cell anemia been substantially prolonged. Even so, the average life expectancy is only about forty years, and a woman with sickle cell anemia is unable to give birth to children because she cannot supply a developing fetus with sufficient oxygen. Sickle cell anemia is inherited as a recessive trait, with heterozygous carriers having the sickle cell trait, that is, their red blood cells undergo some sickling under severe oxygen deprivation.

In the late 1940's, Linus Pauling and his associates Harvey Itano, Seymour Singer, and Ibert Wells, stimulated by Beadle and Tatum's one gene, one enzyme hypothesis, investigated the possibility that sickle cell anemia resulted from an abnormality in hemoglobin, the major protein in red blood cells. To test for abnormal hemoglobin they used the technique of electrophoresis. As we have described electrophoresis (pp. 68–70), charged molecules, in this case proteins, migrate when placed between electrically charged negative and positive poles. In the absence of any retardation owing to size, the rate at which a protein migrates is a function of the net charge on the protein. At pH 7, a protein with excess acidic amino acids is negatively charged and moves towards the positive pole, while a positively charged protein with excess basic amino acids moves towards the negative pole. During electrophoresis, proteins with different net charges will move at different speeds toward the appropriate pole, thereby being separated into individual "bands." Consider now that a particular protein was changed as a result of a gene mutation that left it with an altered number of charged amino acids. The normal and abnormal forms of the protein may migrate at different speeds during electrophoresis.

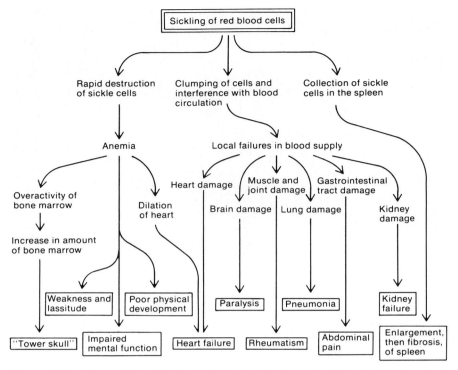

Figure 11-7   The multiple (pleiotropic) effects of sickling of red blood cells. After Neel and Schull, *Human Heredity,* University of Chicago Press, Chicago, 1954.

When Pauling and his co-workers subjected hemoglobin obtained from persons afflicted with sickle cell anemia to electrophoresis they discovered that it did not migrate as rapidly towards the positive pole as hemoglobin obtained from persons without the disease (Figure 11-8). They also discovered that people who are heterozygous for the sickle cell allele and have the sickle cell trait have equal amounts of two types of hemoglobin, the normal one, specified by the normal allele (abbreviated Hb-A) and the slow-moving one, specified by the mutant allele (abbreviated Hb-S). Pauling and his co-workers concluded in a 1949 publication that the cause of sickle cell anemia rested in the abnormal hemoglobin and that sickle cell anemia represented a *bona fide* example of a "molecular disease."

Sickle cell hemoglobin differs from normal hemoglobin by one amino acid, a valine replacing a glutamic acid.

Several years after Pauling and his colleagues discovered that persons with sickle cell anemia had an abnormal form of hemoglobin, Vernon Ingram was able to demonstrate the basis for the altered electrophoretic mobility of

Hb-S molecules. Ingram demonstrated in 1957 that Hb-S and Hb-A differed only in a single amino acid in each of the β chains in the tetrameric molecule. In Hb-A a glutamic acid residue present at a specific although undetermined position in the β chain was replaced with a valine residue in Hb-S. The substitution of valine, a neutral amino acid, for glutamic acid, an acidic one with a negative charge, reduced the negative charge of the protein and accounted for the lower mobility of Hb-S towards the positive pole during electrophoresis.

The two kinds of polypeptide chains in hemoglobin, α and β, are specified by two genes.

Since the time Ingram first studied normal and sickle cell hemoglobins, the techniques for elucidating the primary structures of proteins have improved substantially. Hemoglobin, after the enzyme ribonuclease and some polypeptide hormones such as insulin, was one of the first proteins to have its amino acid sequence completely determined. We now know that the amino acid change in Hb-S, a *substitution,* is at position 6 of the β chain. (Amino acid sequences are numbered starting at the N-terminal end.) There are also inherited variants with amino acid substitutions in the α chain. One of the variants with an α chain substitution is called Hopkins 2 (Hb-HO2). A family pedigree involving individuals having both Hb-S and Hb-HO2 is shown in Figure 11-9. In the third generation, three types of individuals are identified, one with only Hb-S, a second with only HB-HO2, and a third with both HB-S and Hb-HO2. If the two traits were allelic, we would expect all of the offspring to be either Hb-S or Hb-HO2, because the alleles would segregate. Presence in the pedigree of offspring carrying both traits, one or the other, or neither indicates independent assortment of two pairs of alleles. We can conclude that there are two genes, one for the α chain and one for the β chain. The original parent was heterozygous for two pairs of alleles. The genes for the α and β polypeptides have now been located on specific chromosomes. The gene for the β chain is carried on chromosome 11 and that for the α chain on chromosome 16. The point is that proteins composed of multiple polypeptides can be specified by more than one gene, each

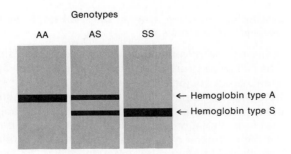

Figure 11-8   Electrophoresis of normal and sickle hemoglobin.

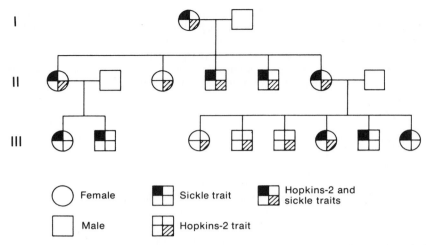

Figure 11-9   Pedigree demonstrating independent assortment of sickle cell (a β-chain defect) and Hopkins-2 (an α-chain defect) traits. Modified from Smith and Torbert, *Bulletin of the Johns Hopkins Hospital,* v. 102, p. 38, 1958.

polypeptide by its own gene. We need to revise the "one gene, one enzyme" hypothesis to read "one gene, one polypeptide."

We emphasize at this point that "one gene, one polypeptide" is not only a statement of relationship between genes and polypeptides but also constitutes one definition of a gene, namely, the region of the DNA that encodes an entire polypeptide. This is a definition, however, that we will need to qualify later in the book (pp. 323–324; 348–349). Also many geneticists would refer to a gene defined in this way as a "structural gene," that is, a gene that specifies the structure of another macromolecule, for example, a polypeptide, using the term "gene" again in other contexts. We have chosen through the rest of the book to use the term "gene," as a synonym for "structural gene," although on occasion, for emphasis, we will use the term "structural gene."

Amino acid substitutions affecting many of the residues in the two hemoglobin polypeptides are known.

We have stressed the advantage of using microorganisms as genetic tools because they allow us to select particular mutants, such as auxotrophs. Surprisingly, humans, as subjects for genetic studies, offer similar advantages. The simple fact that a person will tell a physician when she or he is unwell has led, for example, to the identification of further abnormalities in hemoglobin structure. Variations in hemoglobin structure have also been identified by random screening of human populations for electrophoretic variants of hemoglobins. It should be understood that the two procedures are very different. Random screening of populations will identify variants that are not deleterious or only slightly so as well as recessive deleterious ones present in a heterozygous condition. Obtaining samples from hospital patients tends to identify deleterious genetic variants. As of 1976, using both procedures, more than 200 structural variants of hemoglobin had been identified. Among them, about one third involve amino acid substitutions in the α chain and two thirds, substitutions in the β chain. These substitutions occur at many different positions in the two polypeptide chains. Most amino acid substitutions in hemoglobin are found in very low frequencies, only about one in eight hundred people having a substitution resulting in a change in electrophoretic migration (Hb-S is a major exception). Although substitutions at all possible positions in the polypeptide chains have not been identified, we infer that substitution could in fact occur at any position, typically one amino acid at a time. Because each substitution is inherited, the nature of the information encoded in the DNA must be such that each amino acid is individually specified.

An enormous number of different possible alleles of a gene can exist.

Given that there are twenty different amino acids commonly found in polypeptides and that amino acid substitutions can occur at any position in a polypeptide chain, it follows that an enormous number of alleles can exist for any gene. Consider a gene encoding a polypeptide only 100 amino acids long. Even if we limit ourselves to single amino acid substitutions, 1,900 different forms of the polypeptide are possible—any of 19 amino acids can substitute at any of 100 positions. We can theoretically have at least 1,900 different forms, that is, alleles, of the gene specifying the polypeptide. If we entertain the possibility of multiple amino acid substitutions, for instance, up to 3 positions in any one variant, the number of possible variant polypeptides and the corresponding number of alleles is astronomical (*circa*, 3^{20}, about 3.5×10^9).

Substitution of an amino acid by one with a different affinity for water, particularly at internal positions in a protein, is likely to be deleterious.

Whether an amino acid substitution will be deleterious is, to a degree, predictable. The activity of a polypeptide is a function of its conformation. We expect an amino acid substitution to alter the activity of a polypeptide if it alters the three-dimensional configuration. Such an alteration in conformation could well result if a hydrophilic residue were substituted for a hydrophobic one in the interior of the polypeptide. The hydrophilic side chain of the amino acid would "attempt" to become free of the internal hydrophobic environment by trying to orient itself towards the exterior of the protein. If the hydrophilic side chain is "successful," the reorientation of the affected region in the polypeptide chain will cause a distortion of the normal three-dimensional configuration and result in loss of function. In a review published in 1972, George Stamatoyannopoulos noted that about 50 of 124 amino acid substitutions in both hemoglobin chains were deleterious to some degree. Six instances that involved the replacement of internal, non-polar residues (hydrophobic or ambivalent) with hydrophilic ones invariably produced a deleterious effect on the properties of hemoglobin.

Perhaps even more important than the nature of the substitution is its location. For example, changes of any sort in the interior of a protein, where polypeptide chains fit together in a precise configuration, are more likely to be deleterious than changes on the outside of the molecule. In hemoglobin, changes in regions of the polypeptide chains that contact either the heme moiety or other polypeptide chains in the tetramer are usually deleterious. A summary of the positions of amino acid substitutions in hemoglobin and their effects is provided in Table 11-2. The table confirms that external amino acid substitutions are rarely deleterious, in contrast to internal substitutions or those involving contacts between the different polypeptides in the tetramer or between the heme moiety and the polypeptides.

Sickle cell anemia is an example of the effect of replacing a hydrophilic amino acid with a hydrophobic one.

The N-terminal region of the β chain of hemoglobin, including the amino acid at position 6, glutamic acid, is located on the outside of the globin molecule. Glutamic acid has a negatively charged side chain at cellular pH and is hydrophilic. There are two known substitutions at residue 6; valine is substituted in Hb-S and lysine, a positively charged and hydrophilic basic amino acid, is substituted for glutamic acid in Hb-C. Hb-S, of course, is associated with sickle cell anemia. Hb-C, in homozygotes, causes a mild anemia, much less severe than sickle cell anemia. Thus, when a hydrophilic amino acid on the outside of a protein is replaced by another hydrophilic one, even one with a different charge, there is comparatively little effect. However, replacement with valine, an amino acid with a hydrophobic side chain, causes a dramatic effect. Hemoglobin-S tetramers, under oxygen deprivation, polymerize to form long tubules (Figure 11-10). These long tubules align themselves side by side and produce the sickle shape of the red blood cells. The formation of tubules is thought to result from hydrophobic interactions between tetrameric hemoglobin molecules possibly in-

Table 11-2   Relationship Between Location of Amino Acid Substitutions and Abnormality in Functions in Human Hemoglobins

Location of Amino Acid Substitution in Hemoglobin	Total Number of Mutants	Mutants with Abnormal Function
External	66	8 (12%)
Internal		
Contact with heme	20	19 (95%)
Contacts between α- and β-chains	19	15 (79%)
Other	7	7 (100%)
Total Internal	46	41 (89%)

After G. Stamatoyannopoulis, "Molecular basis of hemoglobin disease," *Annual Review of Genetics*, v. 6, 47–70, 1972.

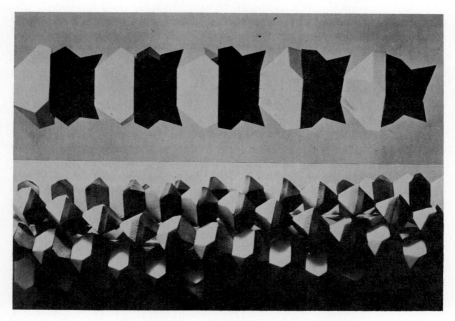

Figure 11-10   A model for the polymerization of sickle cell hemoglobin. Photo courtesy of Makio Murayama, National Institutes of Health, Bethesda, Maryland. From Murayama, "Molecular Mechanism of Human Red Cell Sickling" in *Molecular Aspects of Sickle Cell Hemoglobin: Clinical Applications,* R. Nalbandian, Ed., 1971, courtesy of Charles C Thomas, Publisher.

volving the valine at position 6 in the β chain of Hb-S. If so, then the side chain of the valine residue has satisfied its "need" to be in a hydrophobic environment by binding to another hemoglobin molecule.

Agents that alter protein conformation block the sickling of red blood cells.

Sickle cell anemia is of enormous personal, medical, and social significance. In the United States alone, we estimate that there are 100,000 to 200,000 persons afflicted with the disease. Worldwide, it is probably responsible for more than 100,000 deaths yearly. There has been a recent, belated effort to moderate or eliminate the symptoms of the disease. The primary symptoms characteristic of sickle cell anemia—breakdown of red blood cells and blockage of capillary circulation—result from the sickle shape of the red blood cells. Thus, prevention of sickling would alleviate the symptoms. Inasmuch as sickling results from the polymerization of hemoglobin molecules in the absence of oxygen, if the polymerization of the molecules could be prevented, sickling itself could be prevented. The polymerization, depicted in Figure 11-10, appears to involve specific interactions between different regions of hemoglobin molecules and is dependent on a particular conformation of the molecules. (It is known that hemoglobin has different conformations when it binds oxygen and when it does not.) Agents that could alter the conformation of hemoglobin might interfere with or even reverse the polymerization. Guided by such deductive steps, scientists, in laboratory tests, found that many agents known to affect protein conformation, were effective in preventing sickling of red blood cells. These agents include such compounds as urea, which interferes with hydrogen bonds; cyanate, which binds to free primary amino groups; and a derivative of aspirin, acetyl-3,5-dibromosalicylic acid, which acetylates amino groups. Shigeo Kubota and Jen Tsi Yang took an alternative approach in 1978. They reasoned that polymerization involves the binding of one tetramer to a specific limited polypeptide region of another tetramer. They have, consequently, attempted to prevent the interaction by adding polypeptides to a solution of Hb-S, with the hope that the added polypeptide will "coat" the surface of a hemoglobin-S tetramer and prevent it from binding with another tetramer. In particular, they have investigated peptides with structures similar to the regions of Hb-S that are suspected of binding during polymerization and have found that these peptides interfere with Hb-S polymerization. None of the agents has yet proved in clinical tests to prevent sickling in people, but some seem to be helpful in alleviating sickling.

SUMMARY

An understanding of the nature of the primary products encoded in the genome is sought not in the complex physical characteristics of organisms but in their molecular inventories. The general nature of one primary gene product was anticipated by Beadle and Tatum who demonstrated that deficiencies in biosynthetic activities were inherited as single genes and summarized their findings in the "one gene, one enzyme" theory. The nature of the information in the DNA encoding a protein, however, was mainly appreciated through the analysis of specific proteins. Anfinsen and his co-workers demonstrated that polypeptides spontaneously re-form into normal conformations, indicating that only the amino acid sequence of a polypeptide need be encoded in the genome. Studies on genetically determined variant proteins, such as those of hemoglobin, demonstrated that each polypeptide is encoded by a separate gene and provided one definition for a gene, namely that region of the DNA that encodes a single polypeptide. The information for an amino acid sequence of a polypeptide is encoded in the DNA so that each amino acid is specified separately, allowing any amino acid in a polypeptide chain to be replaced by another amino acid through mutational change. Amino acid substitutions, particularly those in which an amino acid with a certain affinity for water is replaced by an amino acid with a differing affinity, for example, a hydrophobic one by a hydrophilic one, are most likely to change the function and physical properties of a polypeptide. Such a change is dramatically demonstrated in sickle cell anemia, the substitution of a glutamic acid residue by a valine residue causing hemoglobin in the absence of oxygen to polymerize into long filaments that produce the sickle shape of red blood cells.

12

Transcription and Information Transfer

We have now determined that the amino acid sequence of a polypeptide is specified by the information in a gene, that is, by a base sequence in the DNA. Both types of molecules, DNA and polypeptides, are linear polymers and contain complementary information although the information is encoded in two different languages: the Chinese of DNA nucleotide sequences and the English of polypeptide amino acid sequences. The problem at hand is how to convert one language into the other. Like spoken languages, the two languages in the cell are spoken in geographically separate locations; Chinese in the nucleus, English in the cytoplasm. The cell, therefore, makes a transcript of the information encoded in the DNA. The transcript is transported to the cytoplasm where it is translated into a polypeptide. We now know that the translation is done in much the same way that spoken languages are translated, namely, by making use of a translator.

In this chapter we examine the first stages of information transfer, leaving the actual translation process for the next chapter. Here we shall meet some of the characters involved in information transfer and translation. Specifically, we are concerned with the nature of the *ribosome,* the site of polypeptide synthesis, that is, *translation,* and with *transcription,* the process in which information is copied from the DNA into RNA, where it is then ready to be transferred to the ribosome.

RIBOSOMES AND RIBOSOMAL RNA

Synthesis of polypeptides occurs away from the DNA on small cytoplasmic particles called ribosomes.

Before discussing transcription we will digress briefly to the language of the cytoplasm. The synthesis of proteins from amino acids is monitored experimentally by use of radiolabelled amino acids. Using radiolabelled amino acids John Bishop, John Leahey, and Richard Schweet in 1960 and Howard Dintzis, separately, in 1961, demonstrated that the synthesis of polypeptides occurs sequentially from the N-terminal end to the C-terminal end. Experiments in which radiolabelled amino acids were incorporated into *nascent* polypeptides—those in the process of being synthesized—demonstrated that protein synthesis is geographically separate from the DNA, that is, separate from the genes that specify the amino acid sequences of the proteins. In one such experiment, first performed by George Palade and Philip Siekovitz in 1956, liver cells were exposed to radioactive amino acids for a short time; the cells were then broken open and their cellular constituents were separated by centrifugation. Most of these constituents, including the nuclei with their DNA contents, contained few or no radiolabelled proteins. The microsomes, however, had a substantial amount of radioactive protein associated with them. Microsomes, appearing as small spherical vesicles, are composed of membranes derived from the rough endoplasmic reticulum (p. 37) to which small particles, the ribosomes, are attached. The radiolabelled proteins remain associated with the ribosomes when the ribosomes are removed from the membranes with detergent (Figure 12-1). Because the period of protein synthesis in the presence of radiolabelled amino acids was short, it was reasonable to suppose that the radiolabelled polypeptides associated with the ribosomes were being synthesized. This interpretation was lent further support when cells synthesized proteins first from radiolabelled amino acids and then continued protein synthesis in the absence of radiolabelled amino acids. Synthesis of radiolabelled proteins on the ribosomes was completed, and the radioactive proteins were released to other parts of the cell. The ribosomes were found to be free of newly synthesized, radioactive proteins. Palade and Siekovitz's initial observations were subsequently confirmed by John Bishop and his co-workers and by Howard Dintzis, among others. By using cell-free systems in their studies of

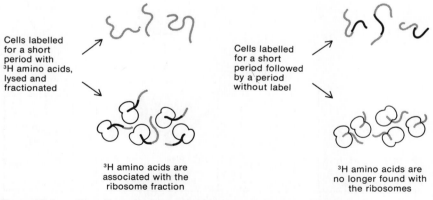

Cells labelled for a short period with ³H amino acids, lysed and fractionated

Cells labelled for a short period followed by a period without label

³H amino acids are associated with the ribosome fraction

³H amino acids are no longer found with the ribosomes

Figure 12-1   Experiment implicating ribosomes as sites of protein synthesis.

hemoglobin synthesis, they demonstrated that synthesis of the protein only occurred when ribosomes were present.

Ribosomes are complex particles containing three different RNA molecules and over fifty different proteins.

The structure of ribosomes has been extensively analyzed, in part by determining the sedimentation coefficients of various ribosomal fractions during centrifugation. Sedimentation coefficients (measured in Svedbergs, abbreviated S) are measures of the speed of sedimentation of a particle or molecule during centrifugation, with larger S values generally but not always reflecting larger molecular or particle weights. The general structure of ribosomes from prokaryotes and eukaryotes is summarized in Figure 12-2. Although ribosomes from prokaryotes such as bacteria are smaller (70S) than those from eukaryotes (80S), all ribosomes share the same basic structure. They are composed of two major subunits of different size and with different sedimentation coefficients. Ribosomes can be completely dissociated into their component parts in solutions of urea or detergents. Each subunit contains at least one RNA molecule and numerous proteins. Naturally enough, the RNA's, which are structural components of the ribosomes, are referred to as *ribosomal RNA's* or *rRNA's.* In both prokaryotes and eu-

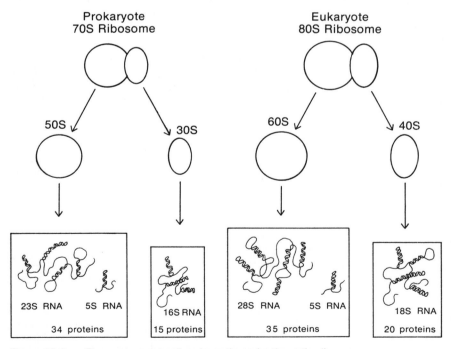

Figure 12-2   The components of prokaryotic and eukaryotic ribosomes.

karyotes the larger ribosomal subunit contains two rRNA molecules, a large one (23S in prokaryotes, 28S in eukaryotes) and a small 5S one. The 23S and the 28S molecules contain about 3,000 and 5,000 nucleotides, respectively. The 5S molecule contains about 85 nucleotides. The sequences of 5S rRNA's from several genera have been determined and found to differ. The small ribosomal subunit contains a single rRNA molecule (16S in prokaryotes, 18S in eukaryotes, composed of 1,500 and 2,000 nucleotides respectively). The two subunits also contain different numbers of proteins.

Ribosomal subunits assemble from their component parts in a test tube: an important example of self-assembly.

Not only can ribosomes be taken apart, they can be put back together. Masayasu Nomura and his associates, working with the ribosomes of *E. coli*, were able to purify the proteins and the rRNA molecules from each subunit. Then, adding the proteins back with the rRNA molecules, they were able to reassemble the ribosomes. The rRNA serves a structural function and is absolutely required for reassembly. The proteins by themselves will not assemble. The reconstituted ribosomes have sedimentation properties identical to native ones isolated from bacteria and, furthermore, are capable of participating in protein synthesis in cell-free systems. The reassembly of ribosomes from their component parts is a spectacular example of self-assembly. We have seen in the previous chapter that unfolded enzymes, for example, ribonuclease, will spontaneously fold back into active forms and that the α and β chains of hemoglobin will spontaneously assemble into a tetramer. In ribosomes, we have a subcellular particle containing not one or four but over 60 components. Nevertheless, we come to a similar conclusion: The architecture of the ribosome is an inherent result of the physical properties of its component rRNA's and proteins. These physical properties are directly attributable to the sequences of nucleotides and amino acids. We have already noted that the amino acid sequence of polypeptides is determined by the DNA. We shall see presently that the sequence of nucleotides in rRNA is also determined by the DNA. Thus, the structure of ribosomes is a necessary consequence of the nucleotide sequences in DNA that encode ribosomal proteins and rRNA's.

Ribosomal RNA's are encoded by specific genes.

The 5S and 16S rRNA's of *E. coli* have specific sequences of 84 and 1520 nucleotides, respectively. It is unreasonable to believe that these molecules are made under the direction of a host of different enzymes, each one responsible for the addition of a single nucleotide at a particular place in the polynucleotide chain. The only reasonable assumption is that rRNA's are synthesized using a DNA template. We have already discussed a technique that can be used to determine whether this explanation is correct, the renaturation or hybridization of one polynucleotide with its complement (pp. 77–78). We incubate radiolabelled *E. coli* rRNA with single stranded *E. coli* DNA. The rRNA forms double helical molecules with the DNA, which on the basis of their respective melting temperatures, (p. 75) makes a good match. Thus, sequences in *E. coli* DNA are complementary to rRNA. Such perfect matching of complementary sequences is conclusive evidence that the ribosomal RNA's are synthesized using DNA as a template, or *transcribed.* There are genes for ribosomal RNA's and we must broaden our definition of a gene to include the genes for ribosomal RNA's. Thus, genes not only encode the amino acid sequences of polypeptides but can also specify the nucleotide sequences of structural RNA molecules such as rRNA.

Multiple copies of rRNA genes are present in the genomes of cellular organisms.

The exact number of DNA sequences complementary to the different rRNAs can be determined by adding to solutions of single-stranded DNA increasing amounts of radiolabelled rRNA until RNA/DNA hybrids are formed with all of the complementary sites on the DNA (Figure 12-3). The number of sites complementary to the rRNA is then determined. In actual experiments, more hybrids are formed than would be expected if there were only a single gene for a particular rRNA in a haploid genome. For example, in *E. coli* the amount of hybrid formed indicates that there are about four genes respectively, for the 23S and 16S rRNA's, in *D. melanogaster* about 120 each for the 28S and 18S rRNA's, and in *Xenopus laevis,* the African horned toad, about 500 of each. We can also see that in *Drosophila* the rRNA genes are located on the X and Y chromosomes, because the number of genes found is a function of the number of X or Y chromosomes present (Figure 12-3).

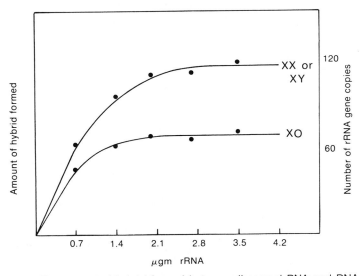

Figure 12-3   The amount of hybrid formed between ribosomal RNA and DNA of *Drosophila* as a function of increasing amounts of ribosomal RNA. After Ritossa and Spiegelman, *Proceedings of the National Academy of Science, U.S.,* v. 53, p. 737, 1965.

The genes for 5S rRNA are also present in multiple copies, *E. coli* having about 6, *D. melanogaster* about 120, *X. laevis* about 20,000, and humans about 2,000. Phages and viruses, however, do not have genes for rRNA's.

In eukaryotes the copies of rRNA genes are clustered at specific chromosomal loci.

The rRNA genes in eukaryotes are clustered. In contrast, the rRNA genes in *E. coli* are scattered at three or four different loci, indicating that, at most, two copies of an rRNA gene are present at a single locus. (We shall see that the genes for the 23S, 16S, and 5S rRNA's in *E. coli* are located side by side at these loci.) In eukaryotes, through the use of *in situ* hybridization (pp. 85–87), it has been possible to map the chromosomal locations of the rRNA genes. Typically, there are a small number of sites at which copies of rRNA genes are found. In *Drosophila melanogaster,* genes for the 28 and 18S rRNA molecules are restricted to the X and Y chromosomes near the centromere. The *Drosophila* 5S rRNA genes are found together at a single locus on the second chromosome, region 56F on the salivary chromosome map (Figure 12-4). In humans, the genes for the 28 and 18S rRNA's are located on the telomeric tips of five chromosomes (13–15, 21, and 22), and the 5S genes are found on chromosome 1. In *Xenopus* the 28 and 18S rRNA genes are all located on chromosome 12; the 5S rRNA genes are located at the telomeric tips of several chromosomes.

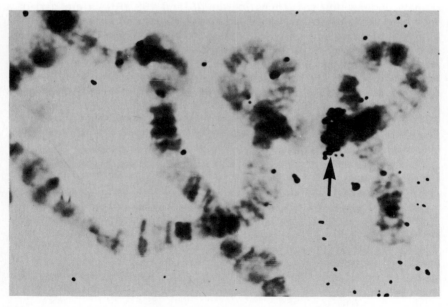

Figure 12-4   The *in situ* hybridization of *Drosophila* 5S rRNA to salivary region 56F. Photo courtesy of Dale M. Steffensen, Department of Botany, University of Illinois, Urbana.

rRNA's are synthesized as large precursors, which are then converted by nucleases into the molecules present in ribosomes.

In both prokaryotes and eukaryotes the initial rRNA molecules synthesized are larger than those found in the ribosomes. These pre-rRNA's are then processed into the mature rRNA's present in ribosomes.

Prokaryotes: In *E. coli* the initial transcript from the DNA is a large 30S molecule that contains regions corresponding to the 23, 16, and 5S rRNA's (Figure 12-5). This 30S molecule is cleaved by nucleases in a series of steps to produce the final separate rRNA's. In contrast to eukaryotes, regions in the *E. coli* chromosome coding for 5S rRNA are adjacent to regions coding for the larger rRNA's.

Eukaryotes: In eukaryotes, a large precursor (45S in mammals, 40S in amphibia, 38S in insects) containing the 28 and 18S rRNA sequences is the initial transcript produced. The precursor is then processed (Figure 12-5) into 28 and 18S molecules by a series of steps involving cleavages by nucleases. The 5S rRNA's are also produced as large precursors, which must be trimmed to produce the molecules found in ribosomes.

The rRNA genes are present in moderately repeated tandem sequences.

We noted in Chapter 3 that, although most of the moderately repeated sequences in eukaryotes are interspersed with single-copy DNA, some sequences existed as tandem repeats. We have already seen that the copies of the rRNA genes in eukaryotes are clustered. They may be present in tandem repeats. Indeed, the sequences for the 28 and 18S rRNA's are part of a system of tandem repeats, which is illustrated for *Xenopus* in Figure 12–6. The regions complementary to the large rRNA precursors are interspersed with a "spacer" region, which is not used as a template for the synthesis of an RNA. The spacer plus the region complementary to the precursor is repeated over and over in a series of tandem repeats, about 60 times on each sex chromosome in *Drosophila* and about 10,000 times in each chromosome 12 of *Xenopus*.

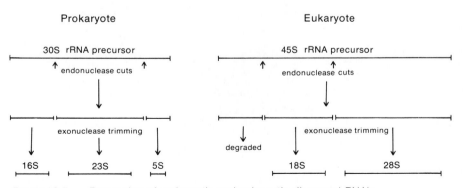

Figure 12-5   Processing of prokaryotic and eukaryotic ribosomal RNA's.

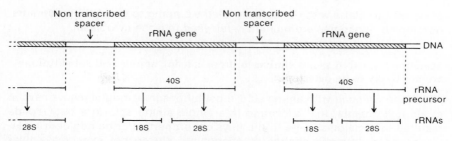

Figure 12-6   The chromosomal arrangement of ribosomal RNA genes in *Xenopus*.

Ribosome assembly in eukaryotes occurs in nucleoli.

In all eukaryotes, ribosomes are assembled in the nucleoli (p. 000). The nucleoli form around the chromosomal regions that contain the sequences for the 28 and 18S ribosomal RNA's. In *Drosophila* John Sedat's cytological observations demonstrate that several chromosomal regions, presumably including the region containing the 5S rRNA genes, also are associated with the nucleolus (Figure 12–7). The transcripts from the rRNA genes are formed directly in the nucleoli. The ribosomal proteins are manufactured on cytoplasmic ribosomes, as are all other proteins. The ribosomal proteins then travel to the nucleoli, which are factories for assembling ribosomes. Even before the rRNA precursors are cleaved into their final form, the assembly

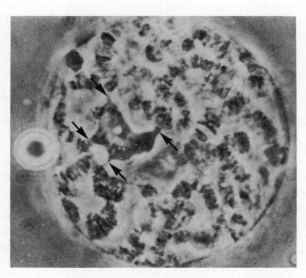

Figure 12-7   The nucleolus, the site of ribosome assembly, rests in a chromosomal basket in a polytene nucleus of *Drosophila*. Arrows show points of attachment of nucleolus to specific chromosomal sites. Photo courtesy of John Sedat, Department of Biochemistry and Biophysics, University of California, San Francisco.

of ribosomes starts with ribosomal proteins binding to the rRNA precursors. Assembled ribosomes are ultimately released into the cytoplasm.

Mutations in rRNA genes similar to those causing amino acid substitutions are not likely to be detected.

When we dealt with amino acid substitutions in hemoglobins we saw, in sickle cell anemia, that a change in a single amino acid in a polypeptide resulted in a major change in the physical properties of hemoglobin molecules and a dramatic change in phenotype. We do not expect a similar situation in the rRNA genes because of their multiplicity. Let us assume, using *Drosophila* as an example, that one of the 120 rRNA genes mutates to a new sequence. If the mutant complementary rRNA sequence is detrimental to the function of the ribosome in which it is contained, then we expect that about 1 ribosome in 120 will be defective, perhaps no longer able to participate in protein synthesis. It is unlikely, however, that the loss of 1/120th of protein-synthesizing capacity will result in a mutant phenotype.

We might note, as an aside, that the genes for rRNA pose an interesting problem. rRNA forms well-matched and very stable hybrids with any rRNA gene-copy in the DNA. Because all these DNA/RNA hybrids are well matched, it follows that all the sequences in the DNA complementary to rRNA are virtually identical to each other. Yet we would expect that during the course of evolution sequences in the rRNA gene copies would have diverged (become different) as a result of mutation. Since this has not happened, we infer that some sort of correction mechanism—which is not yet understood— keeps the different rRNA sequences identical to each other. The mechanism for maintaining identical sequences may depend on the tandem array of the rRNA genes, inasmuch as in *E. coli*, where the rRNA genes are physically separate, some differences exist in nucleotide sequence among the different rRNA genes.

Loss of large numbers of copies of rRNA genes can occur through mutation.

Although mutations altering the nucleotide sequence of an individual rRNA gene may not have a phenotypic effect, mutations involving rRNA genes have been identified in *Xenopus* and in *Drosophila*. In *Xenopus* the anucleolate mutant, *O-nu* involves the rRNA genes. Diploid *Xenopus* cells normally have two nucleoli, but in the *O-nu/+* heterozygote only one nucleolus is present, and in the *O-nu/O-nu* homozygote no nucleoli are found. In the last condition, DNA complementary to the 28 and 18S rRNA's is completely missing. The absence of the nucleoli is a direct result of the absence of the rRNA genes. The *O-nu/O*-nu condition is lethal, death occurring during embryogenesis. A similar, but less severe, mutation, called *bobbed* (*bb*), exists in *D. melanogaster*. The *bobbed* phenotype (slow development, small bristles) results from the loss of rRNA genes, the phenotype being expressed when less than 40 per cent of the normal number of genes are present. Because the Y chromosome contains approximately the same number of rRNA genes as the X, X^{bb}/Y^+ males are normal. If the reduction in the number of rRNA genes is sufficiently great (exceeding 90 per cent), there is lethality. Both *Xenopus* and *Drosophila* develop normally when they have as little as 50 per cent of the normal number of rRNA genes. This normal development confirms our suspicion that a defect in a single rRNA molecule will not be expressed phenotypically. It also demonstrates, as with enzyme activity, that the organism is buffered with excess synthetic capacity.

rRNA cannot be the informational molecule that directs polypeptide synthesis.

We have noted that proteins are synthesized on ribosomes, separate from the DNA. Yet we know that the DNA encodes the primary structure of proteins. Consequently, there must be a means for transferring the information in the DNA to the ribosomes. The function of rRNA was at one time thought to be far different from the structural role in which we have just portrayed it. Originally, rRNA was believed to be the informational molecule that carried the instructions from the DNA to the ribosomes for the synthesis of proteins. According to this notion, different rRNA's existed, each an accurate transcript of different regions of the DNA and each directing the synthesis of a different polypeptide.

From a variety of evidence it is now well established that rRNA cannot play such a role. First, all ribosomes contain the same rRNA molecules, with the same nucleotide sequences. Second, the genes for the rRNA's are limited to specific regions of the genome, and the genes specifying the amino acid sequences of different proteins are distributed throughout the genomes of the organisms in which they are carried. Finally, rRNA has a long life span in cells, which, as we shall discuss, is a serious drawback for the notion that rRNA encodes information for polypeptide synthesis.

MESSENGER RNA AND TRANSCRIPTION

The discovery of messenger RNA (mRNA).

The identification of a more promising candidate for the role of informational molecule for directing protein synthesis resulted from attempts to understand the mechanisms of phage infection. Very shortly after infection by a virulent phage, the host bacterium stops synthesizing its own proteins and switches to the production of phage proteins. The means by which this conversion occurs is dependent on the manner in which ribosomes receive information for the synthesis of specific proteins. According to one model the information for the synthesis of a polypeptide is embodied in the ribosome (for example, in the rRNA) at the time of its formation. To produce a new set of polypeptides after phage infection, it is necessary to make new ribosomes and shut off the synthesis in the old ribosomes (Figure 12-8). In

MODEL I

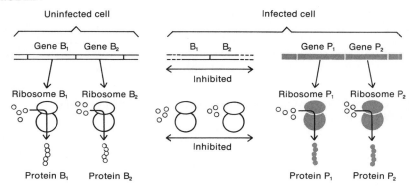

MODEL II

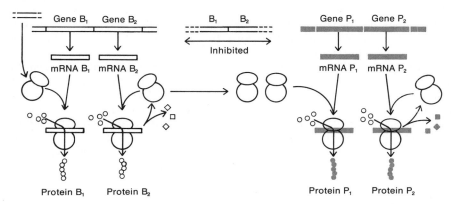

Figure 12-8   Alternative models for the onset of synthesis of phage proteins following phage infection of a bacterial cell. Phage synthesized products are shaded. After Meselson, M., F. Jacob, and S. Brenner, *Nature*, v. 190, p. 576, 1961.

an alternative model, information for the synthesis of polypeptides is carried by a short-lived RNA molecule that attaches transiently to ribosomes and directs the synthesis of proteins. Following phage infection, old bacterial ribosomes are called upon to synthesize phage proteins, using newly produced RNA transcripts of the phage DNA. The bacterial genome is turned off and the existing short-lived bacterial RNA breaks down to be replaced on the ribosomes by new phage RNA.

A set of experiments supporting the second model were reported in 1961 by Sydney Brenner, Francois Jacob, and Mathew Meselson. They utilized the density gradient techniques developed by Meselson and Stahl (pp. 94–95) to separate components with different buoyant densities. When *E. coli* is grown in medium containing ^{15}N and ^{13}C—stable, heavy forms of nitrogen (^{14}N) and carbon (^{12}N)—the ribosomes synthesized by the bacteria have a heavier buoyant density than those made by bacteria growing in control medium containing ^{14}N and ^{12}C. The "heavy" and "light" ribosomes can be separated from each other by equilibrium centrifugation in a cesium chloride gradient.

In the actual experiment, *E. coli* cells were grown in ^{15}N, ^{13}C medium in order to produce "heavy" ribosomes. The cells were then washed free of heavy isotopes and incubated in a medium with light isotopes. At the same time, the bacteria were infected with T4 phage. Seven minutes later, when the bacteria had switched to the synthesis of phage proteins, the ribosomes were recovered from the cells and subjected to equilibrium centrifugation. The recovered ribosomes were still "heavy," indicating that no new ribosomes had been produced as a result of phage infection. In another experiment, shortly before the cells were broken open, they were exposed to ^{35}S to label methionine and cysteine in newly synthesized proteins. The "heavy" ribosomes were found to be labelled, indicating that they still were active in the synthesis of proteins. The results from these experiments demonstrate that at the time phage proteins are being made, only ribosomes made before phage infection are engaged in the synthesis of proteins. Therefore, the synthesis of phage proteins is carried out on the existing bacterial ribosomes.

As a final aspect of their experiments, Brenner, Jacob, and Meselson exposed the phage-infected bacteria to ^{14}C-uracil for two minutes to label newly synthesized RNA. They found that the labelled uracil was incorporated into a very short-lived RNA, having a half-life of only a few minutes, which was associated with the existing ribosomes. They concluded that the newly synthesized, short-lived RNA was directing the synthesis of the phage proteins on the stable bacterial ribosomes. They christened the newly discovered molecule *messenger RNA* or *mRNA,* based on their belief that it transmitted information from the DNA to the ribosomes. At the same time Brenner, Jacob, and Meselson were doing their work, researchers in the laboratory of James Watson demonstrated that an RNA with properties similar to the hypothetical mRNA existed in a variety of sizes different from those of ribosomal RNAs, befitting molecules directing the synthesis of different-sized proteins.

RNA synthesized after phage infection is complementary to phage DNA.

Two important questions were left unanswered by the Brenner, Jacob, and Meselson experiments: Are mRNA molecules transcripts of different genes? Does mRNA indeed direct the synthesis of phage proteins? We will deal with the first of the questions now and the second in the next chapter. The first question was answered in 1961 by Benjamin Hall and Sol Spiegelman, who investigated whether RNA synthesized after phage infection was made using phage DNA as a template. If the RNA synthesized after phage infection is a messenger directing the synthesis of phage proteins, then it must be synthesized using the phage DNA as a template. Hall and Spiegelman incubated phage-infected *E. coli* in the presence of $^{32}PO_4^{--}$. The radiolabelled RNA synthesized by the infected cells was purified and incubated with denatured phage and bacterial DNA. The radiolabelled RNA formed a hybrid with the phage DNA but not with the bacterial DNA. The base sequences of the RNA were complementary to those of the phage DNA, which could only occur if the RNA sequences were made using the phage DNA as template. Thus, following infection of *E. coli* the phage chromosome is actively directing the synthesis of new RNA. The association of this phage RNA with the existing bacterial ribosomes strongly suggests that it functions as our postulated messenger, carrying the information for the synthesis of phage proteins.

The synthesis of large RNA molecules with sequences complementary to DNA sequences requires a DNA-dependent RNA polymerase.

Recall that the semi-conservative replication of DNA requires an enzyme system that polymerizes deoxynucleotides on a template of DNA. In a similar way, the synthesis of rRNA and mRNA requires an enzyme system capable of making ribonucleotide polymers on a DNA template; in other words, a DNA-dependent RNA polymerase. If the analogy to DNA polymerase holds, we would also expect that the enzyme would use nucleotide triphosphates for the synthesis of RNA parallel to the way in which DNA polymerase uses deoxynucleotide triphosphates for the synthesis of DNA. The overall reaction for the synthesis of RNA would then be:

$$n \begin{bmatrix} UTP \\ GTP \\ ATP \\ CTP \end{bmatrix} + DNA \xrightarrow{\text{RNA Polymerase}} RNA + 4n(P-P).$$

in which UTP is uridine triphosphate, GTP is guanosine triphosphate, ATP is adenosine triphosphate, and CTP is cytidine triphosphate.

DNA-dependent RNA polymerases have been isolated from prokaryotes and eukaryotes.

DNA-dependent RNA polymerases with the above properties were first identified in both animal and bacterial cells in the laboratory of Samuel Weiss. Using partially purified bacterial RNA polymerase in a DNA/RNA hybridization experiment, E. Peter Geiduschek, Tokumasa Nakamoto, and Weiss demonstrated in 1961 that RNA synthesized in a cell-free system was complementary to the DNA added to the reaction mixture. They concluded that the RNA was synthesized using the DNA as a template. The synthesis of RNA, like that of DNA, occurs in the 5' to 3' direction with nucleotides being added one after another to a free hydroxyl at the 3' end of the polymer. Also as is true of DNA, the orientation of the template is antiparallel to the polynucleotide being formed. In order for the DNA to act as a template, it is unwound by the polymerase over a short region, so that a single strand of DNA is available as a template for the synthesis of RNA (Figure 12-9). Unlike the synthesis of DNA, however, a primer is not needed for the synthesis of RNA. The reaction can start with the formation of a phosphodiester bond

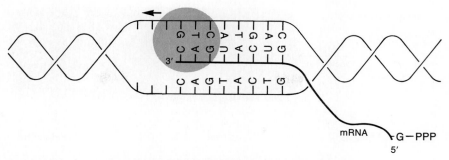

Figure 12-9   The synthesis of RNA by DNA-dependent RNA polymerase.

between the first two nucleotides complementary to the DNA template. The 5′ terminus is a triphosphate (Figure 12-9). In bacteria, the terminus is always a purine, either PPPA or PPPG.

For a given gene we expect that only one strand of the DNA will be used as a template.

Any segment of double helical DNA can theoretically make two distinct transcripts, depending on which of the two strands is used as a template. The two RNA molecules made would have different but complementary base sequences and, presumably, could code for the synthesis of two different polypeptides. Consider the likelihood that both of these polypeptides could be functional. The situation is somewhat comparable to expecting a sentence to make sense when read both forward and backward. Although this can happen—we have discussed the palindrome, "Able was I ere I saw Elba"—the possibilities of this happening are obviously exceedingly limited. By analogy it is likely that only one strand of the DNA, the *sense strand*, serves as a template for the synthesis of a mRNA.

Studies by Musaki Hyashi and Sol Spiegelman in 1961 using *E. coli* cells infected with phage φX174 demonstrated that RNA is synthesized using only one strand of the DNA as a template. We recall that φX174 has a small circular chromosome composed of a single strand of DNA, the (+) strand (Figure 5-6). Following infection, the complement of the (+) strand, the (−) strand, is synthesized, and together they form the double-stranded replicating form, RF (Figure 5-6). Radiolabelled RNA obtained from *E. coli* cells following infection with φX174 was found to form DNA/RNA hybrids with RF DNA but not with the single-stranded DNA found in mature φX174 phages (Figure 12-10). We can conclude that the RNA only hybridizes with the (−) strand of the DNA, which is made after entry of the φX174 chromosome into the bacterial cell. Therefore, only the (−) strand serves as a template for the synthesis of RNA. This result is not unique to φX174 and has been found with other small phages with single-stranded chromosomes.

Although in φX174 only one strand, the (−) strand, is used as a template for RNA synthesis, this is not the case in large phages, bacteria, and eukar-

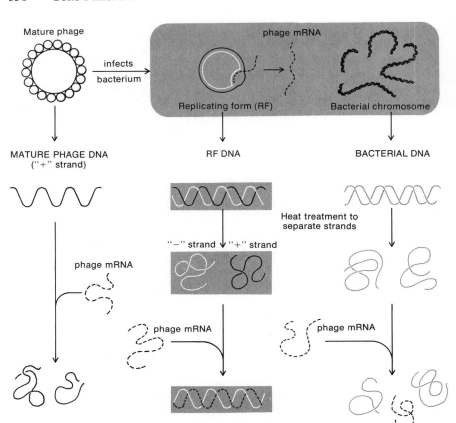

Figure 12-10   Hybridization experiment demonstrating that ØX174 mRNA is transcribed from only one strand of ØX174 DNA.

yotes. In these higher organisms the strand that is read varies in different regions of a chromosome. Nevertheless, for a given gene, only one strand of the DNA has been shown to be transcribed.

DNA-dependent RNA polymerases from bacteria are large complex enzymes composed of several subunits.

We might well wonder how RNA polymerases can find the particular strand and region that they are supposed to transcribe. We do not know yet how this is accomplished, but we gain some insight by studying the structure of RNA polymerases. RNA polymerases from bacteria have been completely purified and extensively characterized. In *E. coli* the entire enzyme, the *holoenzyme,* has a molecular weight of about 490,000 D. and is composed of several different-sized polypeptide subunits. The holoenzyme contains one β, β', ω, and σ subunit and two α subunits (Figure 12-11). The σ subunit separates from the holoenzyme during the action of the enzyme or during purification. The remainder of the enzyme, less the σ subunit, is the *core enzyme.*

The σ subunit is responsible for the specificity of transcription.

Following infection, the ØX174 phage chromosome is transcribed by *E. coli* RNA polymerase. Because only one strand of the double helical RF is used as a transcription template, *E. coli* RNA polymerase must have some way to recognize the strand it is supposed to transcribe, the sense strand. The RNA polymerase, in fact, not only recognizes the sense strand but also specific sites adjacent to structural genes. Starting from these recognition sites, entire structural genes are transcribed. The specificity of recognition, as first suggested by Richard Burgess, Andrew Travers, John Dunn, and Ekkehard Bautz, depends on the presence of the σ factor in the holoenzyme. Using either bacterial or phage DNA as a template, core RNA polymerase, lacking the σ factor, will evidently transcribe either strand in a given region—the sense strand or its complement—and will start at random positions along the DNA. With the σ factor present, the efficiency of transcription is greatly increased and the enzyme transcribes only the sense strand.

Masahiro Sugiura and his colleagues, using RF DNA as a template, demonstrated that if σ factor was present only the (−) strand—the one formed after the single-stranded DNA enters the bacterium—is transcribed. Furthermore, synthesis starts at only three sites on the (−) strand, and the completed RNA hybridizes only with the (−) strand, but not the (+) strand of the mature phage. In the absence of σ factor, both strands are transcribed and transcription begins randomly on the chromosome. The σ factor assures that the correct region and strand of the DNA are transcribed.

As we mentioned, the σ subunit separates from the RNA polymerase during transcription. Although σ is used to find the correct starting place, when transcription starts, it is released and can then bind to another polymerase core, permitting that molecule to initiate transcription (Figure 12-12). Thus, because σ can cycle between several core polymerases and permit

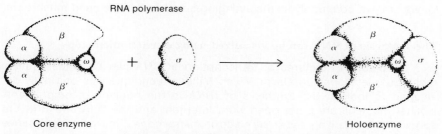

Figure 12-11   The structure of *Escherichia coli* DNA-dependent RNA polymerase. The shape and the arrangement of the subunits are hypothetical.

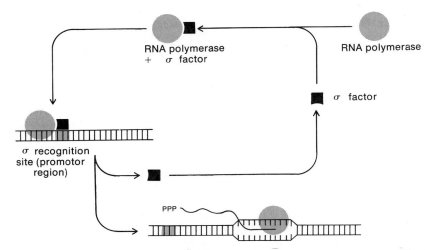

Figure 12-12   The σ factor directs RNA polymerase to specific sites (promoters) for the initiation of transcription.

them to correctly initiate transcription, there can be fewer σ subunits than core polymerase molecules.

Three RNA polymerases have been isolated from eukaryotes.

Eukaryotic DNA-dependent RNA polymerases, like their bacterial counterparts, are large proteins composed of several polypeptide chains. In extracts of animal cells three different RNA polymerases, designated I, II, and III in the United States and a, b, and c in Europe, have been identified and purified. RNA polymerase I, which is found in nucleoli, is the enzyme that transcribes the rRNA genes. RNA polymerase II is responsible for the transcription of genes coding for polypeptides. RNA polymerase III transcribes the genes for 5s rRNA and the genes for another group of RNA's, *transfer* RNA's (tRNA's) (pp. 353–358), that are involved in protein synthesis. RNA polymerase II is sensitive to and is inhibited by the drug alpha-amanitin, a compound extracted from the highly poisonous mushroom, *Amanitis phalloides,* known commonly as the avenging angel or death's head mushroom.

The synthesis of RNA can be visualized using electron microscopy.

"Stop-action" pictures of transcription of DNA can be made using a technique developed by Oscar Miller. Miller's technique was originally developed to look at the synthesis of rRNA in the nucleoli of amphibians. Isolated nucleoli are burst open with detergent and the DNA and RNA are spread out for viewing under the electron microscope (Figure 12-13). A series of "Christmas trees" are seen. Each "Christmas tree" in the picture represents a region of the DNA that is being actively transcribed. DNA forms the

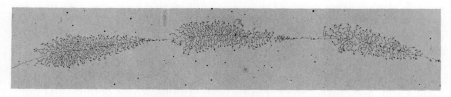

Figure 12-13   Electron micrograph showing the "Christmas tree" pattern of transcription of tandemly repeated rRNA units in *Xenopus*. Photo courtesy of Oscar L. Miller, Jr., Biology Department, University of Virginia, Charlottesville.

trunk of the tree and the branches are rRNA molecules in the process of being transcribed. The direction of transcription is from the top of the Christmas tree toward the bottom (Figure 12-14). Bases complementary to the DNA are added to the 3' end of a growing RNA molecule as the polymerase moves along the DNA. The completed 5' end of the RNA dissociates from the DNA to form a branch. In the stop-action picture, the Christmas tree effect results from a number of different polymerase molecules simultaneously transcribing one segment, a *transcriptional unit*, of the DNA. Different polymerase molecules have moved different distances from the apex of the tree to the bottom and are associated with RNA molecules of different lengths, the length of a particular RNA molecule being a function of the number of nucleotides it contains. We expect the number of nucleotides in each RNA transcript to equal the number of nucleotides in the DNA over which the polymerase has passed.

Electron micrographs of rRNA synthesis reveal specific regions for starting and stopping transcription.

We have already noted that the σ factor directs the RNA polymerase to a specific point in the DNA from which to begin transcription. In Figure 12-13 we see that transcription does indeed have defined starting and stopping

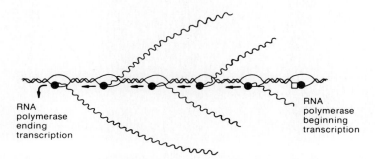

Figure 12-14   Interpretation of the "Christmas tree" pattern of transcription seen by electron microscopy.

points. Transcription starts at the apex of each tree and ends just past the bottom branch. All of the transcriptional units are the same length as are the spaces between the transcriptional units. The photo shows a tandem array of rRNA transcriptional units interspersed with nontranscribed regions. The constant size of the transcriptional unit, as seen by electron microscopy, demonstrates pictorially that there are defined places where transcription begins and, also, where transcription ends. Therefore, we deduce that there are sequences in the DNA that act as signals to start and stop transcription. The signals for starting and stopping transcription are not unique to regions containing rRNA genes but are characteristic of all transcriptional units.

In prokaryotes the nucleotide sequences of regions, *promoters,* where RNA polymerase binds for transcription, have been determined.

It has been possible in bacteria to identify the sequences in the DNA, *promoters,* to which DNA-dependent RNA polymerase binds in preparation for transcription. One method to identify promoters involves the incubation of DNA fragments containing one or a limited number of genes with RNA polymerase holoenzyme in the absence of nucleotide triphosphates. The polymerase binds to the promoters, but because the synthesis of RNA in the absence of the triphosphates is not possible, the polymerase remains in place. Electron microscopy can then demonstrate binding of polymerase molecules to specific regions (Figure 12-15). The nature of the sequences to which the polymerase is bound can be determined by subjecting the DNA to digestion with DNase. The promoter DNA is protected from digestion by the bound polymerase. After digestion the only DNA remaining is the DNA protected by the polymerase, that is, the promoter regions. The regions pro-

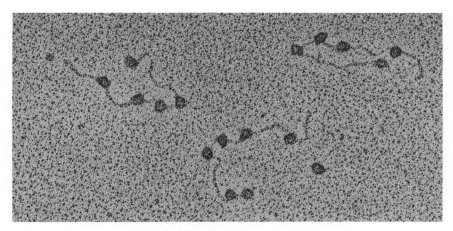

Figure 12-15   An electron micrograph of RNA polymerase molecules bound to promoters on a fragment of DNA produced by restriction endonuclease cleavage. Photo courtesy of Robley Williams, Virus Laboratory, University of California, Berkeley.

tected are short, about 50 nucleotides long. They can be recovered and their sequences determined.

All promoters in prokaryotes contain a related seven-nucleotide sequence.

David Pribnow first pointed out that different promoters in prokaryotes all contain a seven nucleotide sequence, a *Pribnow sequence,* adhering, with some variation, on the sense strand to the formula, 3' A-T-A-pyrimidine-T-A-pyrimidine 5'. Nucleotides 2 and 6 in this sequence have proved to be invariant in all promoters that have been sequenced (Table 12-1). Positions 4 and 7 are occupied by pyrimidines. The other positions are variable. The seven-nucleotide sequence is believed to be particularly important for the binding of the RNA polymerase holoenzyme to the correct position on the DNA. Promoters in general are rich in A-T base pairs, which are believed to facilitate the separation of the two strands of double-helical DNA by the polymerase in preparation for transcription. (Recall that A-T rich DNA is less stable than G-C rich DNA.) Transcription starts 6 to 7 base pairs in the 5' direction on the sense strand from the Pribnow sequence. Promoter sequences have also apparently been identified in eukaryotes. Sequences similar to the Pribnow one, with the basic formula 3' A-T-A-A-A-T-A 5', have been found about 25 bases upstream from where transcription starts for several different genes in eukaryotes (histone genes in sea urchins and *Drosophila,* globin genes in mammals, ovalbumin in chickens, cytochrome C in yeast). The actual binding of RNA polymerase to the regions containing this sequence has also been shown.

Promoters, initiating and terminating regions in prokaryotes, often contain palindromic DNA.

The use of DNA/RNA hybridization and sequencing studies of DNA and RNA transcripts has identified regions in the DNA where transcription is

Table 12-1   Pribnow Sequences in Promoters of Phage and Bacterial Genes

	Pribnow Sequence	First Base Transcribed
phage fd	3'A-C-G-A-A-G-A-C-T-G-*A*-*T*-*A*-*T*-*A*-*T*-C-T-G-T-C-C-*C*-A-T-T-T-C-T-G5'	
T7 A2	3'T-C-A-T-T-G-T-A-C-G-T-C-*A*-*T*-*T*-*C*-*T*-*A*-*T*-G-T-T-T-A-*G*-C-*G*-A-T-C-C-A-T5'	
T7 A3	3'C-A-T-T-T-G-T-G-C-C-*A*-*T*-*G*-*C*-*T*-*A*-*C*-*A*-T-G-G-T-G-*T*-A-C-T-T-T-G-C5'	
λ	3'T-G-G-A-G-A-C-C-G-C-C-*A*-*C*-*T*-*A*-*T*-*T*-A-C-C-A-A-C-G-*T*-A-C-A-T-G-A-T5'	
E. coli lac UV5	3'G-G-C-C-G-A-G-C-*A*-*T*-*A*-*T*-*T*-A-C-A-C-A-C-C-*T*-T-A-A-C-A-C-T5'	
E. coli tRNA gene	3'G-T-A-A-A-C-T-*A*-*T*-*A*-C-T-A-C-G-C-G-G-G-G-*C*-G-A-A-C-G-G-C5'	

Pribnow sequences in promoters from several phage and bacterial genes are depicted. The DNA strand that is used as the template is shown with transcription occurring from left to right starting with the indicated base. A2 and A3 are promoters from transcription of genes in phage T7. UV5 is a mutant form of a promoter at the lactose locus in *E. coli* to which RNA polymerase binds with increased affinity. Modified from W. Gilbert, in *RNA Polymerase,* R. Losick and M. Chamberlin, Eds., Cold Spring Harbor Laboratory, 1976.

initiated and terminated. Nucleotide sequences of promoter, initiating, and terminating regions of DNA are palindromic. We recall that in palindromes in DNA the sequence on one strand is the same as that on the other strand read backwards. For example, the following sequence is a palindrome:

5′ A-T-A-C-G-T-A-T 3′
3′ T-A-T-G-C-A-T-A 5′

An interesting property of palindromes is that, in theory, they permit pairing between bases on the same strand, which results in the formation of possible loops of DNA extending outward from the axis of the chromosome, as shown below:

Examination of nucleotide sequences in promoter regions, in initiating regions, and in terminating regions of transcriptional units all show that substantial extrahelical loops—*Gierer structures,* after Alfred Gierer—can be formed. An example of a Gierer structure, from a terminating region in the λ chromosome, is depicted in Figure 12-16. Although palindromes are common in promoter, initiating, and termination regions for transcription, their significance is not clear. An extra helical loop might extend from the DNA and facilitate the binding of the RNA polymerase or prevent it from traveling farther along the DNA. Alternatively, the loops might form in the presence of the polymerase, for example, facilitating the separation of strands necessary for the initiation of polymerization.

While on the subject, we might mention that there are two types of termination of transcription in *E. Coli.* One type depends on the presence of a protein terminating factor called ρ, which is required for the release of an RNA transcript from the DNA. A second type of termination does not require ρ, termination occurring in the absence of the protein. In the few examples studied, the terminating sequence in the DNA for termination that requires ρ, involves a simple palindrome and possible Gierer structure; in termination not requiring ρ, the palindrome and the possible Gierer structure are extensive.

Prokaryotic mRNA's contain leader sequences, coding sequences, and trailer sequences that are colinear with the DNA.

We will assume for now that mRNA's do, in fact, carry information for the synthesis of polypeptides. As such, all mRNA's have *coding regions* that contain the structural information for the synthesis of a polypeptide (Figure 12-17). All mRNA's also have non-coding regions, *leader sequences,* up-

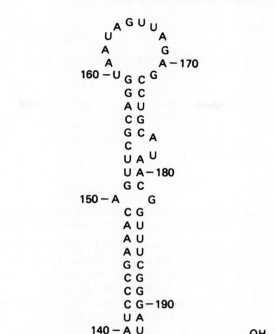

Figure 12-16   A possible extrahelical loop—A Gierer loop—formed by the pairing of inverted repeats similar to those found in IS elements. The structure shown, as found in the RNA transcript, is from a terminating region of a phage gene.

stream of the coding regions, that is, toward the 5' end of the RNA, the end synthesized first. Consequently, the synthesis of an mRNA begins with a non-coding region of the molecule. As we shall see in the following chapter, the leader sequence plays an important role in the initiation of polypeptide synthesis. Downstream from the coding region, that is, towards the 3' end of the molecule, there is another region of non-coding bases, aptly called the *trailer sequence.* We have already noted that in both prokaryotes and eukaryotes the synthesis of rRNA involves the synthesis of an rRNA precursor molecule that is then processed to form the rRNA's found in ribosomes. In prokaryotes, messenger RNA's are not processed after transcription; the molecule that is initially synthesized is the one that is used for protein synthesis. Thus, there is a perfect match between the base sequence in the DNA used for transcription and the base sequence in the RNA transcript, the two sequences being *colinear.* We shall see presently that many eukaryotic mRNA's are processed after being synthesized.

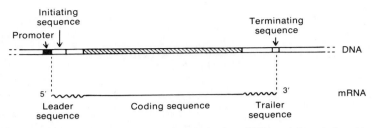

Figure 12-17   The general structure of prokaryotic mRNA and its relationship to the DNA from which it is transcribed.

Leader sequences of eukaryotic mRNA's are often modified through methylation and the addition of a 3' G 5' PPP terminus called a "cap."

Like mRNA's of prokaryotes, mRNA's of eukaryotes also have leader, coding, and trailer sequences. Usually the leader sequences of eukaryotic mRNA's are modified after transcription by the addition of a guanidine residue, a *cap*, to the 5' end of the mRNA. The cap is apparently added to the 5' end of the RNA even before transcription of the 3' is completed. The 5' carbon of the guanidine nucleotide bonds through a triphosphate group in such a way that it faces the 5' carbon of the first nucleotide in the mRNA (Figure 12-18). Besides the addition of the guanidine triphosphate, the two terminal nucleotides are methylated (Figure 12-18). The significance of capping eukaryotic mRNA's remains to be solved. Capping may be necessary in many cases for the mRNA to direct polypeptide synthesis.

Trailer sequences of eukaryotic mRNA's are often modified after transcription with the addition at the 3' end of a tail of poly A.

The trailer sequence at the 3' end of eukaryotic mRNA's is itself often modified by the addition of a polymer composed only of adenosine nucleotides, that is, a poly A tail. The length of a poly A tail is variable, ranging from about 25 to 50 nucleotides. The poly A tail is added enzymatically without the benefit of a template. There is some evidence that mRNA's containing poly A have a longer life in cells than those that lack poly A. However, other mRNA's, such as those for histones, lack poly A tails and function perfectly well. It also appears that some mRNA's exist in a cell with or without poly A tails.

The poly A tail provides a convenient short cut to the separation of mRNA's from other RNA's present in eukaryotic cells. Microscopic beads that have poly U sequences extending from them can be made. When solutions of RNA are mixed with the beads, conditions can be controlled to allow poly A sequences to form hybrids with the poly U sequences attached to the beads. Poly A mRNA's become attached to the beads. After other RNA's are washed away, the hybrids can be dissociated, releasing the poly A containing mRNA's and allowing their recovery.

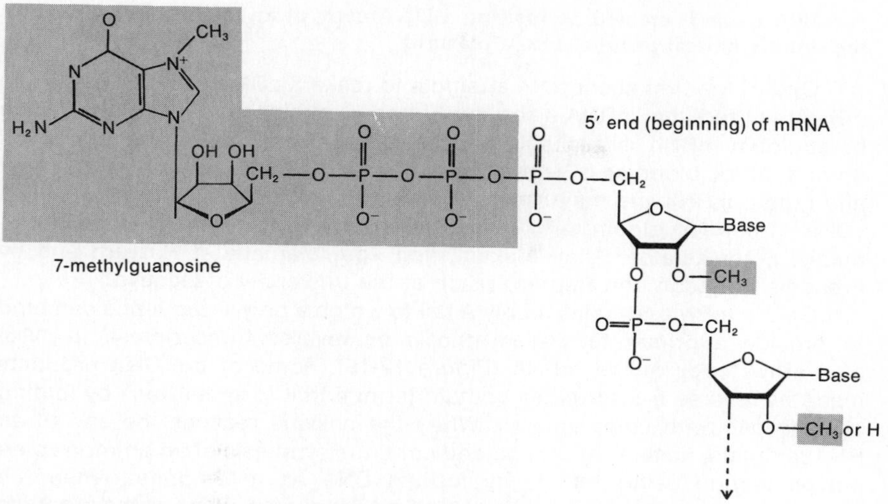

Figure 12-18   The structure of the cap region of a eukaryotic mRNA.

In higher eukaryotes, most mRNA's are synthesized as high molecular weight precursors, which are then processed to smaller mRNA's.

The sizes of RNA molecules in the nuclei of cells of higher eukaryotes, for example, vertebrates, are highly variable and, on average, larger than those of the mRNA's found in association with the ribosomes in the cytoplasm. This variable nuclear RNA fraction is referred to as heterogeneous nuclear RNA (hnRNA). Furthermore, we find that there are more kinds of RNA sequences represented in the nuclear RNA than in the mRNA. In contrast to higher animals, the nuclear RNA of fungi (*Aspergillus, Saccharomyces* and the water mold *Aclya*) is about the same size as the cytoplasmic mRNA, and most of the same kinds of sequences are present in both locations. The function of hnRNA in animals has been debated since its discovery in the mid-1960's. Recently, evidence has indicated that many, perhaps all, hnRNA molecules are precursors to the mRNA's found in the cytoplasm, the hnRNA being processed, like pre-rRNA, to form the mRNA's. The identification of high molecular weight nuclear precursors to mRNA's and knowledge of the steps involved in processing of hnRNA has depended on the isolation of mRNA's for specific proteins and on the use of recombinant DNA technology, without which many of the recent advances could not have been made. Specifically, it has been possible to clone (p. 289) DNA copies of mRNA's (*complementary DNA or cDNA clones*) and fragments of chromosomal DNA (*genomic clones*) containing specific genes and neighboring sequences. These purified mRNAs and cloned DNA's have been used to investigate the nature of transcription and gene structure.

A cDNA clone is created by making a DNA copy of an mRNA and inserting the double helical product into a plasmid.

One of the first successful attempts to make a cDNA clone of a specific mRNA utilized the mRNA's for the α and β chains of hemoglobin. Pure hemoglobin mRNA is available because at some stages during the development of red blood cells hemoglobin is by far the major protein synthesized and globin mRNA the major message present in the cells. A double helical cDNA of purified globin mRNA was first produced in 1976 by Agiris Efstradiadis, Fotis Kafatos, Allan Maxam, and Tom Maniatis at Harvard and by Francois Rougeon and Bernard Mach at the University of Geneva.

Globin mRNA contains a poly A tail to which a poly T sequence can bind to provide a primer for the synthesis by reverse transcriptase (p. 116) of a cDNA copy of the mRNA (Figure 12-19). Some of the DNA products made by reverse transcriptase contain termini that form hairpins by folding back to pair with complements. When the enzyme reaches the end of an RNA template, it can turn around and continue synthesis of an uninterrupted polynucleotide using the newly formed DNA as a template. When the DNA/RNA product is denatured and the DNA recovered, some of the products contain hairpins that, using *E. coli* DNA polymerase, can serve as primers for the synthesis of double helical DNA. The two strands of the double-helical hairpin-shaped product are joined at one end through a short single-

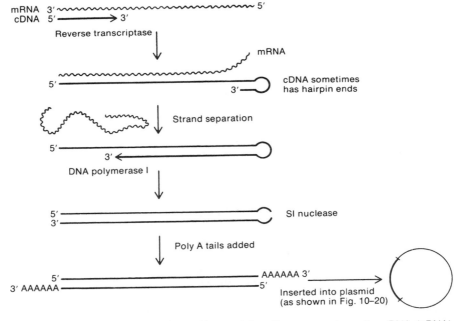

Figure 12-19   A procedure for making and inserting a complementary DNA (cDNA) copy of mRNA into a bacterial plasmid.

stranded DNA loop. This joint can be cut with S1 nuclease, an endonuclease that is specific for single-stranded DNA, to produce a double-helical molecule with two pairs of free ends (Figure 12-19). The double-helical cDNA can then have tails added to it with deoxynucleotidyl transferase and be inserted into a plasmid, as we have previously described (Figure 10-21). Thus, a cDNA copy of globin mRNA can be cloned in *E. coli*.

A modified λ chromosome, λ WES, is particularly useful for making genomic clones.

The insertion of foreign DNA into plasmids offers one means for cloning eukaryotic DNA's. However, the size of the piece that can be inserted while maintaining the plasmid in a clone of bacterial cells is limited to around 10,000 base pairs. Larger pieces of DNA, about 20,000 base pairs, can be inserted into a modified form of the chromosome of phage λ and replicated along with the λ chromosome (Figure 12-20). The special phage λ chromosome used is one in which the attachment region (p. 271) and the genes necessary for integration as a prophage have been removed. The chromosome is also modified so that it now contains only 2 *Eco*RI restriction endonuclease sites (pp. 71–73). *Eco*RI cleaves a short palindrome to produce products with short tails as follows:

$$5'\ G\text{-}A\text{-}A\text{-}T\text{-}T\text{-}C\ 3' \qquad\longrightarrow\qquad 5'\ G\text{-} \qquad \text{-}A\text{-}A\text{-}T\text{-}T\text{-}C\ 3'$$
$$C\text{-}T\text{-}T\text{-}A\text{-}A\text{-}G \qquad\qquad\qquad C\text{-}T\text{-}T\text{-}A\text{-}A\text{-} \qquad \text{-}G$$

To insert a fragment of foreign DNA, the λ DNA is digested with *Eco*RI. Eukaryotic DNA is partially digested with *Eco*RI and large fragments, about 20,000 base pairs long, are preferentially recovered. The λ fragments and the eukaryotic fragments are then mixed together. Because both the λ DNA and the eukaryotic DNA were cleaved with *Eco*RI, both populations of molecules contain the same short complementary single-stranded tails. Although these complementary tails are short, they can pair with each other so that a fragment of eukaryotic DNA becomes inserted between the two arms of the λ chromosome. The fragments can be covalently sealed together using DNA ligase. The sealed DNA pieces are then mixed with λ head proteins so that phages are reassembled much as tobacco mosaic viruses were assembled (Figure 1-14). The reconstituted phages are then allowed to infect *E. coli*. Only those phages that contain both arms of the λ chromosome with a large fragment of foreign DNA in between will produce plaques. Those with only foreign DNA do not have the genes necessary for phage reproduction. Those that lack an insert of foreign DNA have a chromosome so small that it is not properly packaged into phage heads and infective phages are not produced. Instead of being cloned in a bacterium, the foreign DNA is cloned in a phage, so that each plaque produced on a bacterial lawn contains copies of a single sequence of foreign DNA. Furthermore, because each cloned fragment represents a sizeable piece of DNA derived directly from the chromosome of another organism, by cloning enough fragments an entire genome can be cloned.

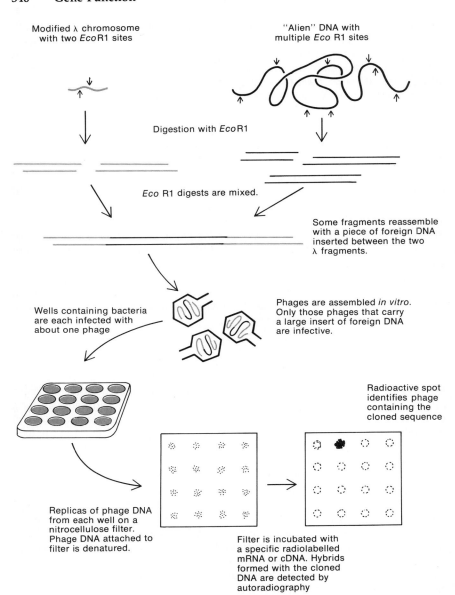

Modified λ chromosome
with two EcoR1 sites

"Alien" DNA with
multiple Eco R1 sites

Digestion with EcoR1

Eco R1 digests are mixed.

Some fragments reassemble
with a piece of foreign DNA
inserted between the two
λ fragments.

Wells containing bacteria
are each infected with
about one phage

Phages are assembled in vitro.
Only those phages that carry
a large insert of foreign DNA
are infective.

Radioactive spot
identifies phage
containing the
cloned sequence

Replicas of phage DNA
from each well on a
nitrocellulose filter.
Phage DNA attached to
filter is denatured.

Filter is incubated with
a specific radiolabelled
mRNA or cDNA. Hybrids
formed with the cloned
DNA are detected by
autoradiography

Figure 12-20   Use of a modified phage λ chromosome to produce genomic clones of
foreign DNA and the method of identifying a genomic clone with a specific fragment of
DNA.

Specific genomic clones are identified by hybridization with a specific mRNA or cDNA.

The genomic clones of foreign DNA carried in the λ chromosome produced by the procedures we have just described constitute a collection of fragments of chromosomal DNA of a particular genome. By recovering large numbers of clones, a *library* (in Switzerland, a *bank*) of cloned DNA fragments of a particular genome is established. The problem of identifying a specific clone carrying a specific gene or chromosomal region remains. A specific radiolabelled mRNA, for example, that of the β chain of hemoglobin, or a cDNA of the mRNA, can serve to make the identification. A large number of λ phages, each carrying a fragment of foreign DNA in its chromosome, are grown in *E. Coli* in tiny wells in a microtiter plate (Figure 12-20). Once the phages have grown, some of the phages from each well are transferred to a membrane filter, providing a replica of the distribution of the phage in the wells in the microtiter plate. The phage DNA is covalently linked to the filter and denatured. The filter is incubated in a solution containing radiolabelled mRNA or denatured cDNA. The radiolabelled mRNA will form a hybrid only with the phage DNA with which it is complementary. The hybrid is identified by autoradiography. The well contains phages carrying the gene with a DNA sequence complementary to a particular mRNA is thereby identified. The phages are recovered from the well and can be grown to produce a clone of genomic DNA for a particular gene.

Primary pre-mRNA transcripts in eukaryotes are colinear with the DNA.

The cDNA and genomic clones provide large quantities of specific nucleotide sequences identical to those found in the cells themselves. These clones—or in cases where an mRNA is plentiful, such as hemoglobin mRNA, the mRNA itself—can be used to probe the structure of genes and the nature of their transcripts. By making hybrids between hnRNA and a DNA copy of globin mRNA, it was possible in 1976 for workers in several laboratories to identify and purify a large 15S precursor, containing about 1,500 nucleotides, to the smaller 10S cytoplasmic globin β-chain mRNA, containing about 700 nucleotides. The 15S precursor has been used by workers in Philip Leder's laboratory for "R loop" experiments (Figure 12-21). In R loop experiments, first described by Marjorie Thomas, Raymond White, and Ronald Davis, DNA's are incubated at temperatures somewhat below the melting temperature (p. 74) in the presence of complementary RNA. Under these conditions RNA/DNA duplexes are more stable than DNA/DNA ones. The RNA binds to its complement in the DNA, displacing a loop of DNA, but otherwise leaving the remainder of the DNA double-helical molecule intact. Under electron microscopy, a loop, representing the position in the DNA to which the RNA hybridized, can be seen. In 1978, when Shirley Tilghman and her colleagues did an R loop experiment using the 15S pre-mRNA for the β chain of hemoglobin and a genomic clone carrying the gene for the β chain, they obtained the result shown in Figures 12-21 and 12-22. A single, continuous R loop was formed, the length of which is equal to the length of the 15S molecule.

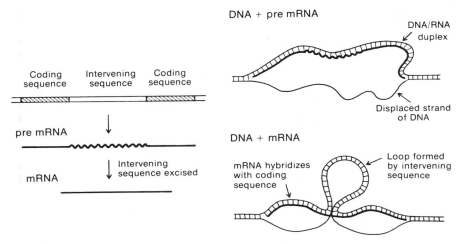

Figure 12-21   The detection of intervening sequences using the R loop technique to make RNA/DNA hybrids.

The perfect match between the DNA and RNA indicates that, over the region where the hybrid is formed, all of the sequences in the DNA are also present in the 15S pre-mRNA. The pre-mRNA is colinear with the DNA from which it is transcribed.

Internal, non-coding intervening sequences that are transcribed but are removed during processing of the pre-mRNA transcript interrupt many genes in higher eukaryotes.

The colinearity of the 15S precursor to globin mRNA and DNA does not seem surprising. However, in earlier R loop experiments with globin mRNA, Leder and his co-workers had been greatly surprised. For when mRNA for the β chain was used in an R loop experiment the results, also depicted in Figures 12-21 and 12-22, were obtained. Instead of a single large loop equal in length to the mRNA, three small loops were found. The interpretation of this result is that there are two sequences in the DNA that are complementary to either end of the mRNA. Between these two sequences in the DNA, however, there is a sequence that is not complementary with the mRNA. A sequence is missing in the middle of the mRNA, actually in the coding region of the mRNA, that is present in the DNA. Such a sequence which interrupts the coding region of a gene is called an *intervening sequence,* or *intron* (expressed coding sequences are called *exons*). The sequences in the mRNA are not colinear with those in the DNA.

Recalling that the 15S pre-mRNA is colinear with the DNA, we realize that the missing sequence has been removed after the RNA was transcribed. A means exists by which a region in the middle of an RNA molecule can be excised with the flanking sequences being joined together (Figure 12-23).

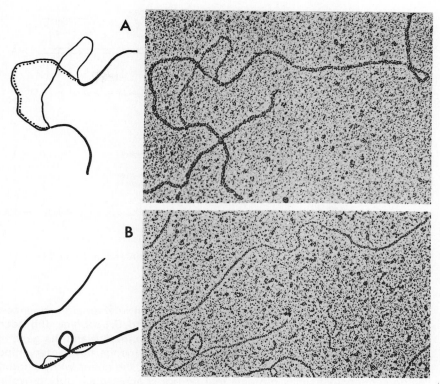

Figure 12-22   Electron micrographs of R loops formed with globin pre-mRNA and mRNA demonstrating that the premRNA is co-linear with the DNA but that an intervening sequence has been removed from the mRNA. Photo courtesy of Philip Leder, National Institutes of Health, Bethesda, Maryland. From Tilghman, Curtis, Tiemeier, Leder, and Weissman, *Proceedings of the National Academy of Sciences,* v. 75, p. 1309, 1978.

The nature of the enzyme, call it an RNA ligase, and the mechanisms by which it works are generally unknown.

The function of intervening sequences is not known.

The function of intervening sequences is not known. Intervening sequences have been found in the β globin genes of humans, rabbits, and mice in virtually identical positions. The ovalbumin gene in chickens contains 7 intervening sequences, the premRNA being about 9 times larger than the mRNA. Intervening sequences have also been found in the ovomucoid gene in chickens, the gene for adrenal cortical tropic hormone (a small peptide) in mammals, and in the genes of many animal viruses. In these examples, each intervening sequence is unique, that is, is present once per genome. The presence of intervening sequences in initial premRNA transcripts may entirely account for the larger size of hnRNA than of mRNA.

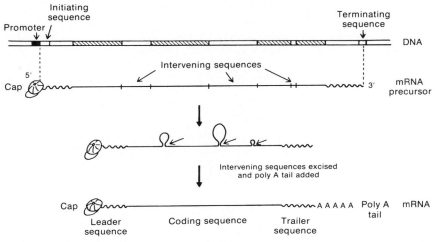

Figure 12-23   The general structure of eukaryotic premRNA and mRNA and their relationships to the DNA from which they are transcribed.

There is also a list of genes that do not contain intervening sequences. Included are: histone genes from sea urchins and *Drosophila,* the actin genes in the cellular slime mold *Dictyostelium discoideum,* and the gene of cytochrome c in yeast. It is not clear why some genes have intervening sequences and others do not. In the few examples, the genes without intervening sequences specify proteins that do not change much during evolution. However, some of the examples are from lower eukaryotes wherein, because of the similarity of nuclear and cytoplasmic RNA's, intervening sequences are, at best, small. One possible explanation for the presence of intervening sequences is that they provide, on an evolutionary time scale, a mechanism by which coding regions of different genes might be recombined to produce new genes and new polypeptides, including polypeptides with multiple functions. Such a suggestion, for now, is speculation.

Our definition of a gene, that region of the DNA containing the sequences encoding the amino acid sequence of a polypeptide or specifying the sequence of an RNA molecule such as rRNA, still holds. But, at least in higher eukaryotes, we must keep in mind that the coding regions of a gene may be interrupted by internal non-coding intervening sequences.

SUMMARY

RNA products are made using the DNA as a template in the first step in the utilization of the information in the genome for the production of the phenotype. We discussed two types of RNA transcripts: ribosomal RNA (rRNA) and messenger RNA (mRNA). Both of these RNA's are made using DNA-dependent RNA polymerases. Ribosomal RNA is transcribed using specific sequences, the rRNA genes, as templates and is a structural element along with proteins in ribosomes. rRNA is made in both prokaryotes and eukaryotes first as a precursor that is then processed into the final molecule found in the ribosomes. Messenger RNA is transcribed using the genes that encode the amino acid sequences of the polypeptides found in cells. The nucleotide sequences of mRNA molecules are subsequently translated into amino acid sequences of polypeptides in association with the ribosomes (to be discussed in Chapter 13). In prokaryotes, mRNA's are used as they are transcribed for the synthesis of polypeptides. In eukaryotes, mRNA's are often synthesized as precursors, which are then processed into mature mRNA molecules. The processing of premRNA's can involve removal of intervening, non-coding sequences and modification of the ends of mRNA molecules: the 5' end with a "cap" and the 3' end with a poly A sequence. Molecules with a poly A tail can be used as templates for reverse transcriptase as the first step in a process leading to the "cloning" of DNA segments complementary to the mRNA's, cDNA clones, in bacteria. These cDNA clones can then be used to recover specific genomic clones, for example, phage carrying a DNA fragment derived from a specific region of the genome of a eukaryote.

13

Polypeptide Synthesis and the Genetic Code

In the previous chapter we discussed some properties of one RNA transcript, mRNA, which apparently serves as the go-between from the gene to the ribosome in the control of protein synthesis. We also discussed the transcription process in some detail, looking at how DNA-dependent RNA polymerases synthesize RNA transcripts. We assumed as a matter of faith, however, that these mRNA molecules contain coding sequences for the formation of polypeptides and that in association with ribosomes, they do indeed direct the synthesis of specific polypeptides. In this chapter we lessen our reliance on faith by presenting a full outline of the mechanics of polypeptide synthesis and the details by which the nucleotide code is translated into an amino acid sequence. With that knowledge, we shall then consider the nature of genes in terms of nucleotide sequences.

A useful model for polypeptide synthesis: the ribosome as a tape reader.

It is useful at this point to draw an analogy between polypeptide synthesis and the production of sound by a standard tape recorder. The tape head or reader of the recorder is the element that converts the electromagnetic data stored on recording tape into meaningful sounds. The nature of information stored on the tape determines the nature of the sounds produced whether an opera by Wagner or a rock concert. As the tape moves across the reader, electrical impulses are instantaneously produced that result in a sequence of sounds. The working of the major elements in polypeptide synthesis is analogous. The mRNA (tape) moves across the ribosome (tape reader), and a polypeptide (music) is produced as a linear sequence of amino acids connected by peptide bonds. In the tape recorder, it is the electronic components of the recorder that translate the information encoded on the tape into music. If the analogy holds, there should be comparable molecular components that translate the nucleotide sequence in the mRNA into an amino acid sequence in the polypeptide.

tRNA'S AND ELONGATION

Adapter molecules, which carry amino acids and bind to the mRNA, are required for polypeptide synthesis.

The amino acids must be physically close to one another and be associated with an mRNA in order for peptide bonds to form. There is no chemical means, however, by which amino acids can be aligned in proper order directly on the surface of an mRNA molecule. There must be a mediator between the amino acids and the mRNA. It would appear, then, that we have some basis in fact for hypothesizing special chemical components for achieving polypeptide synthesis. Francis Crick in 1957 correctly guessed the nature of one of the required components. He proposed the existence of an adapter molecule that could bind to the mRNA and to which an amino acid could also be linked. Hypothetically, each amino acid would have its own adapter molecule that would bind to correspondingly different base sequences in the mRNA. The natural candidate for an adapter molecule was an RNA molecule to which an amino acid could be covalently linked, for example at one of the termini, and which could bind to the mRNA by base pairing. Presumably, as the mRNA moved across the ribosome and as different mRNA sequences became available, the adapter RNA molecule would bind to each sequence by base pairing and deliver its amino acid to the ribosome to be added to the growing polypeptide chain. Only the RNA adapter with the correct antiparallel, complementary sequence would be bound to a particular code sequence in the mRNA.

Transfer RNA (tRNA) is the adapter molecule for polypeptide synthesis.

About the same time that Crick put forward his proposal, Mahlon Hoagland and his associates discovered that soluble RNA's were required for protein synthesis. These soluble RNA's were required over and above the components, such as, ribosomes, mRNA's, and amino acids, which we already know to be necessary for the test tube-synthesis of polypeptides. These soluble RNA's, now called *transfer RNA's* (*RNA*), have two functions: they read the mRNA and they transfer appropriate amino acids to growing polypeptide chains.

tRNA's have specific sequences about 75 nucleotides in length with internal base pairing that produces "clover-leaf" shapes.

The primary structure and three-dimensional configurations of tRNA molecules have been the subject of intense research. As we shall discuss presently, every organism has at least twenty different tRNA's, to which the twenty different amino acids commonly found in polypeptides can be attached. The first sequence of a tRNA was elucidated by Robert Holley and his co-workers. They reported in 1965 the nucleotide sequence of a yeast alanyl-tRNA, a tRNA to which alanine is bound. Since that time, sequences of tRNA's from many different organisms have been elucidated. A tRNA molecule typically is composed of about 75 nucleotides, the sequence of which differs according to the specificity of each molecule. All tRNA's, however, have common regions of internal complementarity so that the molecules fold into a characteristic shape (Figure 13-1), which is represented in two dimensions as a "clover leaf." In three-dimensional representations we see that one leaf and the stem of the clover leaf fold over to produce an L-shaped structure.

The base sequences of different tRNA's are specified by specific genes. There are repeated copies of each tRNA gene.

The specific nucleotide sequence of an RNA molecule 75 nucleotides long must arise from a complementary sequence in the DNA that acts as a template for its synthesis. DNA/RNA hybridization experiments establish that there are sequences in DNA complementary to the nucleotide sequences in tRNA's. We conclude from this evidence that there are specific genes for tRNA's.

In eukaryotes in particular, tRNA genes are present in multiple copies. In *Drosophila melanogaster,* there are on average about 10 copies of each tRNA gene per genome, in *Xenopus,* 180 copies. Furthermore, as shown by *in situ* hybridization experiments with polytene chromosomes, the different types of tRNA genes in *Drosophila* are clustered in one to four locations, presumably as tandem repeats. For example, a valine tRNA hybridizes only to a short region (band 64D) on the left arm of the third chromosome. In both eukaryotes and in bacteria, tRNA's are first synthesized as high molecular weight precursors, which are then processed to produce the tRNA's found in the cells.

All tRNA's have a 3' terminus of 5' C-C-A 3' and contain modified bases.

As one of their common structural features, the 3' terminus of each tRNA has the same nucleotide sequence, 5' C-C-A 3'. The C-C-A terminus, however, is not encoded in the DNA. It is added enzymatically one base at a time without the benefit of a template after transcription is completed. All tRNA's are characterized by the presence of unusual bases, which are derivatives of bases commonly found in polynucleotides. The structures of some of these are presented in Figure 13-2. Like the 3' terminus addition, the modification of bases occurs after transcription of the tRNA molecules. As one function, the unusual bases in tRNA's may protect against nucleases.

Specific regions of the tRNA molecules have specific functions.

To perform their functions, tRNA's must carry amino acids and must bind to specific regions of an mRNA. The amino acid is always attached to the 3' terminus of a tRNA. In a two-dimensional representation, the tRNA binds to the mRNA through a loop of bases, the *anticodon* loop, opposite the place

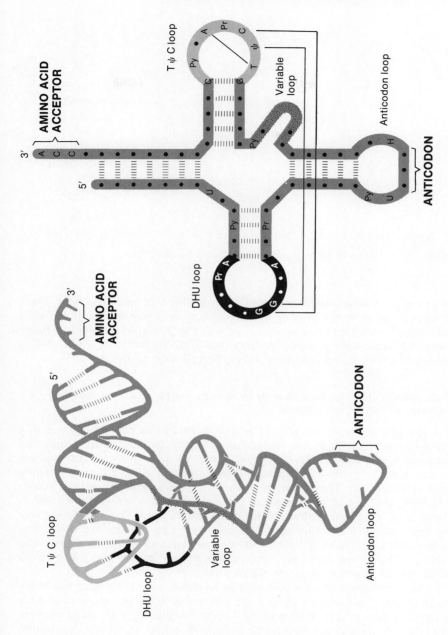

Figure 13-1   The structure of transfer RNA. A three-dimensional representation (left) and a two-dimensional representation (right). After Rich and RajBhandary, *Annual Review of Biochemistry*, v. 45, p. 805, 1976.

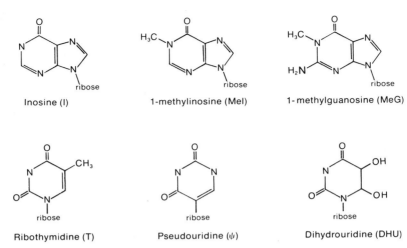

Figure 13-2   Some of the modified bases found in transfer RNA.

of amino acid attachment (Figure 13-1). In addition to these two sites, there is also a region in the DHU or dihydrouridine loop that is thought to bind with the enzyme that adds an amino acid to the 3' end of the tRNA. Finally, there is a region on yet another loop, the TΨC loop (Figure 13-1), that is believed to bind to the ribosome during polypeptide synthesis.

Each transfer RNA is charged with its specific amino acid by a specific enzyme, amino acyl synthetase.

The specificity of insertion of an amino acid in its correct position in a polypeptide chain is a two-step process. First, the amino acid must be attached to the correct tRNA. Second, the tRNA, by virtue of a complementary sequence to the mRNA, must insert the amino acid into a growing polypeptide chain. The first step depends for its specificity on a group of enzymes, *amino acyl synthetases,* that add amino acids to their tRNA's. There is at least one synthetase for every amino acid. The synthetase recognizes its specific amino acid and attaches it to the correct tRNA. The valine synthetase picks up valine and attaches it to a valine tRNA (abbreviated $tRNA_{val}$). The valine amino acyl synthetase will not attach alanine to valine tRNA nor will it attach valine to alanyl tRNA.

The attachment of an amino acid to its tRNA, called *charging* the tRNA, involves two steps (Figure 13-3). First, the amino acyl synthetase covalently links the amino acid to AMP with the concurrent cleavage of pyrophosphate from ATP. As a result the amino acid becomes "activated," that is, it is bound to AMP through a high-energy bond. It is now in a condition in which it can be added to a tRNA in an energetically favored reaction. Without the amino acid-AMP complex dissociating from the enzyme, the amino acid is transferred to the adenosine at the 3' terminus of its specific tRNA. The attach-

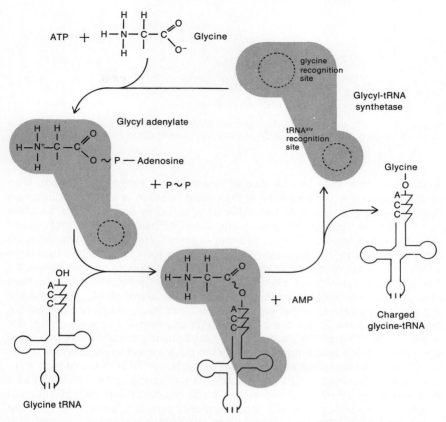

Figure 13-3   The process of charging a transfer RNA with its specific amino acid.

ment is through an ester bond between the carbonyl carbon of the amino acid and the 3' hydroxyl of the terminal ribose of the tRNA.

The correct insertion of an amino acid into a polypeptide depends on the base sequence of the tRNA.

According to the model we are considering, once the tRNA is charged, the specificity of insertion of the amino acid into a growing polypeptide chain resides in a base sequence in the tRNA that is complementary to a specific sequence in the mRNA. The validity of the model can be tested if a tRNA is charged with the wrong amino acid. For example, if a phenylalanine tRNA is charged with leucine, then we expect to find leucine inserted into a growing polypeptide chain in the position normally occupied by phenyl-alanine. In contrast, if the specificity of insertion into a polypeptide chain resides in the amino acid and tRNA is simply a device to bring amino acids

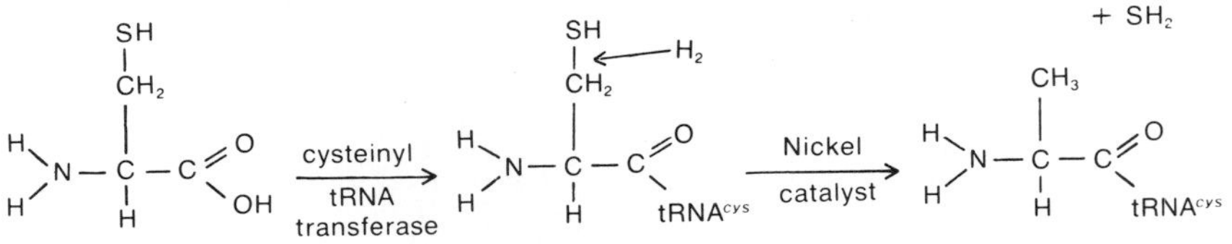

Figure 13-4   Cysteine, attached to cysteinyl tRNA, is reduced to alanine. The cysteinyl tRNA then inserts alanine in place of cysteine into polypeptides.

together so that they can be zipped together in a polypeptide chain, then amino acids will never be added at the wrong position in a polypeptide chain regardless of the nature of the tRNA to which they are attached. Gunther von Ehrenstein, Bernard Weisblum, and Seymour Benzer convincingly tested the alternative models. They used the cysteine amino acyl synthetase to attach cysteine to the cysteinyl-tRNA. They then took the charged cysteinyl-tRNA and, in the presence of molecular hydrogen and a catalyst, converted the attached cysteine to alanine (Figure 13-4). Cysteinyl-tRNA was then charged with alanine, the wrong amino acid. The alanyl-tRNA$_{cys}$ was used in the test tube synthesis of hemoglobin. The α and the β chains of hemoglobin each contain a single residue of cysteine (at position 112 in the β chain, 104 in the α chain). When the products synthesized using the alanyl-tRNA$_{cys}$ were analyzed, alanine was found to occupy the position in the polypeptide chain where cysteine was normally located. The cysteinyl-tRNA charged with alanine inserted alanine into the polypeptide as a function of the tRNA, not as a function of the attached amino acid. We can conclude that the specificity of insertion of an amino acid into a polypeptide chain is a function of the base sequence of the tRNA, presumably a function of the ability of the tRNA to base pair with an appropriate sequence in the mRNA.

To reiterate, the two major steps responsible for the specificity of insertion of amino acids into polypeptides are:

1. Attachment of amino acids to tRNAs, with specificity of attachment depending on an amino acyl synthetase recognizing and matching its appropriate tRNA with its appropriate amino acid.

2. Insertion of the amino acid into a growing polypeptide chain, with specificity of insertion dependent on the structure of the tRNA and its presumed correct binding to a complementary sequence in the mRNA.

Peptide bond formation is a cyclic process involving a tRNA carrying a growing polypeptide and a tRNA carrying an amino acid.

We have already noted that polypeptides are synthesized by the addition of one amino acid at a time starting with the N-terminal end of the polypeptide. As with the synthesis of any linear polymer, the synthesis of polypeptides can be divided into initiation, elongation, and termination. We will describe elongation here and return to the processes of initiation and termination after we have described the elements of the genetic code.

The processes by which a polypeptide chain is elongated, one amino acid after another, have been worked out in detail. In addition to the components we have already discussed (ribosomes, tRNA's, mRNA's, and amino acids), the process involves different proteins, enzymes, and nucleotide triphosphates as energy sources. The process is depicted as follows.

1. There are two tRNA sites present in the ribosome/mRNA complex: a peptide site for a tRNA carrying the growing polypeptide and an amino acid site for the tRNA carrying the next amino acid to be added to the growing polypeptide. The peptide is bound to its tRNA through the C-terminal amino acid. Both the peptide and the amino acid are attached to the 3' ends of their tRNA's as we have previously shown (Figure 13-1). The tRNA carrying the peptide is hydrogen bonded through antiparallel base pairing to the mRNA.

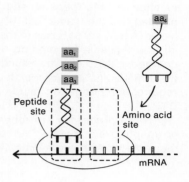

2. The insertion of a specific charged tRNA into the amino acid site requires GTP and a specific protein, an *elongation factor.* The tRNA is paired with the mRNA through specific base pairs.

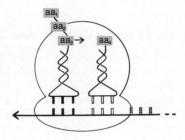

3. A peptide bond is formed in the presence of an enzyme called peptidyl transferase, which is one of the proteins of the 50S ribosomal subunit (p. 322). The peptide is transferred from the tRNA in the peptide site to the tRNA in the amino acid site by the formation of a peptide bond involving the carbonyl group of the C-terminal amino acid of the polypeptide and the amino group of the amino acid. The tRNA that occupied the peptide site is released from the ribosome. It is now available to be recharged with an amino acid and reused in the synthesis of polypeptides. Thus, tRNA's are recycled.

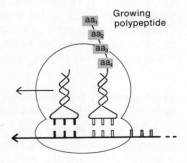

4. The tRNA in the amino acid site, now carrying the growing polypeptide, is translocated from the amino acid site to the peptide site. The transfer is accomplished with the concurrent movement of the mRNA across the ribosome, so that the tRNA in its new location in the peptide site is still bonded to the same set of mRNA bases. A new set of bases in the mRNA is now available for binding the next charged tRNA. The translocation of the mRNA and the associated tRNA requires an elongation factor, G, and GTP.

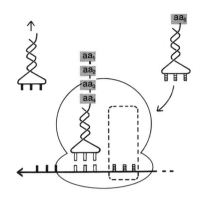

These processes are repeated over and over until the synthesis of the polypeptide is complete. The rate of peptide elongation is rapid, being about the same in most organisms, two amino acids per second.

THE GENETIC LANGUAGE

mRNA code words for amino acids must contain at least three nucleotides.

We have now reviewed the basic components and mechanisms involved in the synthesis of polypeptides. Central to the process is the binding of tRNA molecules to mRNA via complementary bases. We have not yet discussed how many bases in the mRNA are involved in pairing to tRNA, in other words, how many bases are involved in each amino acid code word or *codon* in the mRNA. Because there are twenty amino acids commonly found in polypeptides, there must be at least that many different code words. If we assume that a codon is composed of a single base, it is only possible to code four amino acids, A for one, C for a second, G for a third, U for a fourth, but no more. With code words of two bases, only sixteen amino acids can be encoded ($4 \times 4 = 16$) which, alas, is four less than the needed twenty. Therefore, assuming codons of constant size, there must be at least three bases in each to encode twenty amino acids. Clearly, codons of three bases provide an excess of combinations ($4 \times 4 \times 4 = 64$).

Specific mRNA code words for amino acids were determined by synthesizing polypeptides using synthetic RNA's with simple sequences.

The first test-tube syntheses of proteins, for example, those in which hemoglobin was synthesized, used mRNA's derived from cells. These mRNA's, including the mRNA for hemoglobin, had at the time unknown nucleotide sequences. It was discovered that synthetic polynucleotides, made in a test tube by a chemist, could serve in place of cellular mRNA in a protein-synthesizing system isolated from bacterial cells. Synthesis of polypeptides using synthetic polynucleotides paved the way for identifying specific code words for specific amino acids. Indeed, the first code word, reported by Marshall Nirenberg and J. Heinrich Matthaei in 1961, was identified on the basis of experiments demonstrating that polyuridine caused the synthesis of a specific polypeptide containing only phenylalanine. Assuming a three-letter code, at least one code word for phenylalanine is UUU. Other homogeneous synthetic mRNA's direct the synthesis of polypeptides containing only one amino acid; polyadenosine produces a polypeptide composed only of lysine (AAA = lysine) and polycytosine one composed only of proline (CCC = proline). Polyguanosine does not function as an mRNA because it forms multistranded aggregates that prevent its translation into a polypeptide.

mRNA is read in a 5′ to 3′ direction during polypeptide synthesis.

Synthetic polynucleotides were also used to determine the direction in which the mRNA is read, that is, its orientation as it moves across the ribosome. The direction of movement was determined by using a modified poly A polymer. The modification consisted of adding a single C at the 5′ end and following it with five more A residues (5′ A-A-A-A-A-C-A-A.....A-A 3′). If the molecule were read in the 5′ to 3′ direction, we would expect production of a polypeptide that had a single different amino acid at or near its N terminal end—remember polypeptides are synthesized from the N-terminal to the C-terminal amino acid—followed by a long chain of lysines. However, if the mRNA were translated in the 3′ to 5′ direction we would expect a polypeptide to be produced containing a long chain of lysines at the N-terminal end, then a single different amino acid at or near the C-terminal end. The results of this experiment (Figure 13-5), reported by Albert Wahba, Margarita Salas, and Wendell Stanley, Jr. in 1966 support the first hypothesis. These workers found that a polypeptide was produced with asparagine at the N-terminal end followed by a long series of lysine residues. When they used a poly A that had a single C at the 3′ end of the molecule, a polypeptide was produced with a series of lysines at the N-terminal end and a single

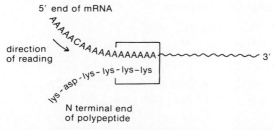

Figure 13-5   An experiment demonstrating that mRNA is translated from the 5′ end towards the 3′ end.

asparagine residue at the C terminal end. We can conclude that, in synthesizing polypeptides, mRNA is read in a 5' to 3' direction.

mRNA is read three bases at a time from a fixed starting point.

Further experiments providing direct information about the genetic code used other synthetic polynucleotides. One type of polynucleotide consisted of two alternating nucleotides that were extensively repeated (...UCUCU-CUCUCUC...). A second type comprised extensive repeats of three bases (...AGCAGCAGCAGCAGC...). When these polynucleotides were used as mRNA's, the first type, the U-C polymer, produced a polypeptide consisting of two alternating amino acids, serine and leucine. The second type, the AGC polymer, produced three different polypeptides, one composed only of serine, a second only of alanine, and a third only of glutamine.

These results are understandable if we assume that the synthetic polynucleotides are "read" three bases at a time from a *fixed starting point.* For the two-base polymer, alternating sets of three bases are separated by imaginary partitions, *reading frames,* as follows:

5'... UCU CUC UCU CUC UCU CUC UCU CUC ...3'

or if reading started with a C:

5'... CUC UCU CUC UCU CUC UCU CUC UCU ...3'.

In either case we end up with two alternating three-letter code words. If each alternating triplet stands for a different amino acid, a polypeptide consisting of the same two alternating amino acids would result. When H. Gobind Khorana and his associates used this polynucleotide in a protein-synthesizing system, they isolated a polypeptide composed of alternating serine and leucine residues (5' UCU 3' is a code word for serine, and 5' CUC 3' is a code word for leucine). These workers went on to determine that poly AG produced a co-polypeptide of arginine and glutamic acid; poly UG, a co-polypeptide of cysteine and valine; and poly AC, a co-polypeptide of threonine and histidine. In each case only one polypeptide, containing two alternating amino acids, was produced.

The results using the three-base synthetic polymer are also consistent with the hypothesis that the mRNA is read three bases at a time from a fixed starting point. Starting with A we set up the following reading frames and produce one possible polypeptide:

5'... AGC AGC AGC AGC AGC AGC ...3'
N ... ser   ser   ser   ser   ser   ser ...C,

starting with G:

5'... GCA GCA GCA GCA GCA GCA ...3'
N ... ala   ala   ala   ala   ala   ala ...C

and, finally, starting with C:

5'... CAG CAG CAG CAG CAG CAG ...3'
N ... glun glun glun glun glun glun ...C.

Three different polypeptides are produced because the message can be read from three starting points. If we assume instead that code words are longer than three bases, the data summarized above become largely uninterpretable. For example, a code composed of four letters would not fit the results obtained with the two-base polymers. If we read four bases at a time, then two sets of reading frames would be produced:

GAGA GAGA GAGA GAGA GAGA

and

AGAG AGAG AGAG AGAG AGAG

There would be no alternating code words, and we would expect to isolate two polypeptides each composed of a different amino acid and not one composed of alternating amino acids. All possible lengths of code words other than three-letter code words are not, however, automatically eliminated. Some multiples of three, nine being the smallest, are consistent with the results. A nine-letter code would alternate two sets of bases using the two base polymers and give rise to three separate readings using three base polymers. A nine-letter code, on the other hand, is inconsistent with physical data on the size of mRNA's and the polypeptides they produce. For example, the mRNA for the α chain of hemoglobin is composed of about 650 nucleotides and codes for a polypeptide containing 141 amino acids. With a three-letter code, 141 amino acids would require 423 nucleotides, leaving 227 of the 650 for the leader and trailer sequences of the mRNA. If each codon were 9 bases long, there would have to be 1,269 nucleotides in the single-stranded mRNA that codes for the α chain. Because this figure greatly exceeds the known number, we can eliminate code words that are multiples of three. The code consists of three-letter words read as such from a fixed starting point.

Many mRNA code words were assigned by binding specific, charged tRNA's to ribosomal complexes in the presence of trinucleotides.

The use of synthetic polymers as mRNA's directing polypeptide synthesis eventually specified many codons although a large number of others remained ambiguous. For example, how do we know that UCU in the above example is a code word for serine and not leucine? Although some ambiguities were resolved through the use of different polynucleotides, a very different procedure was followed to eliminate the ambiguity from other codon assignments. The new procedure, initially reported by Akai Kaji and Hideko Kaji in 1963 and later extended by Marshall Nirenberg and Philip Leder in 1964, does not require the use of polynucleotide-directed synthesis of polypeptides. Instead of long polynucleotides, short trinucleotides were used (Figure 13-6). If a trinucleotide—initially UUU was used by Kaji and Kaji—is added with ribosomes and the correct charged tRNA, all three components will form a complex. The trinucleotide will become associated with a ribosome, and the correct tRNA will then base pair with the trinucleotide. In the case of UUU, charged phenylalanyl-tRNA will form a complex with it in association with ribosomes but other tRNA's will not do so. Because the ribosomes are large they cannot pass through membrane filters. If the

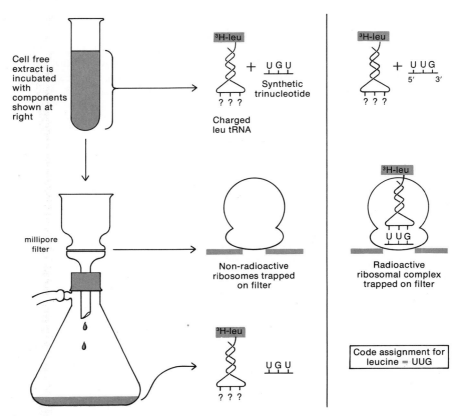

Figure 13-6   The tRNA binding technique for codon assignment.

charged tRNA is bound to the ribosome in association with an appropriate trinucleotide, it, too, is retained on the filter. By using radiolabelled tRNA's, then measuring the radioactivity retained on the filter one can determine if the ribosome-tRNA-trinucleotide complex is formed. Using this technique, Nirenberg and Leder were able to assign 31 code words to amino acids.

The genetic code is degenerate; many amino acids have multiple code words.

The two techniques described have together fully elucidated the code. Before discussing some properties of the code, we should emphasize that the codon assignments refer to the base sequence in the mRNA and are read from the 5' end to the 3' end. Of the 64 possible three-letter code words, 61 are known to specify particular amino acids. Three remaining code words, UGA, UAA, and UAG, do not code for amino acids, but partici- pate in the termination of polypeptide synthesis (pp. 368–369). (Note that code words are always written from left to right, 5' to 3', the same direction as the mRNA is read.)

An inspection of the coding assignments in Table 13-1 reveals the *degeneracy* of the code, the existence of multiple code words for many amino acids. For example, leucine, arginine, and serine each have six codons; five other amino acids have four codons each; isoleucine has three; and most of the others have two code words. Only two amino acids, methionine and tryptophan, have single code words. Most of the degeneracy involves the third letter in the codon. For example, proline is coded by CC followed by any other nucleotide. As we shall see, this is probably of great importance for the reading of the code.

tRNA's have sequences, *anti-codons*, that pair with the codon in the mRNA. One amino acid may have more than one tRNA.

We have noted that one region in the tRNA molecule base pairs with mRNA. If we examine the bottom loops (Figure 13-1) of different tRNA's, we find trinucleotide sequences that are complementary in an antiparallel arrangement to at least one appropriate mRNA code word. The three-letter sequences in the tRNA that are complementary to the codons are called *anticodons.*

The study of tRNA primary structure also reveals one basis for the redundancy of the genetic code. We find in a yeast leucyl-tRNA the anticodon sequence 3' GAA 5', which can base pair to the leucine mRNA code word 5' CUU 3'. However, there are six codons for leucine and, not surprisingly, a second yeast tRNA for leucine has been isolated with a different

Table 13-1   The Genetic Code: mRNA Codons

First Ribonucleotide (5')		Second Ribonucleotide				Third Ribonucleotide (3')
		U	C	A	G	
U	U	PHE	SER	TYR	CYS	U
		PHE	SER	TYR	CYS	C
		LEU	SER	*term.*	*term.*	A
		LEU	SER	*term.*	TRYP	G
C	C	LEU	PRO	HIS	ARG	U
		LEU	PRO	HIS	ARG	C
		LEU	PRO	GLUN	ARG	A
		LEU	PRO	GLUN	ARG	G
A	A	ILEU	THR	ASPN	SER	U
		ILEU	THR	ASPN	SER	C
		ILEU	THR	LYS	ARG	A
		MET	THR	LYS	ARG	G
G	G	VAL	ALA	ASP	GLY	U
		VAL	ALA	ASP	GLY	C
		VAL	ALA	GLU	GLY	A
		VAL	ALA	GLU	GLY	G

term. = terminating codon. MET = initiating codon.

anticodon sequence, 3′ AAC 5′, appropriate for the leucine code word, 5′ UUG 3′. One way in which to read the degenerate code is evidently by using tRNA's with different anticodons but with the same amino acid specificity.

Many tRNA's read several codons: "wobble" or two out of three reading.

We were careful to say in the last paragraph that multiple tRNA's provide *one* way of understanding the translation of a degenerate code. If this were the only way, we should expect to find 61 different tRNA's, corresponding to the 61 different code words for amino acids. This is not the case. In any organism, the number of different tRNA's, and consequently different anti-codons, is closer to 30 than to 61. There are far fewer anticodons than there are codons. The explanation for the discrepancy between the number of codons and the number of tRNA's available to read the codons lies in the observation that single tRNA's are able to read several different codons.

We noted that most of the degeneracy in the code words differs in the third position of the mRNA codon. Perfect matches apparently exist between the first two bases in each codon and their counterparts in the tRNA's. However, the tRNA's can read mRNA codons for which there is mismatching in the third base. Francis Crick proposed in 1966 an explanation for the mismatching in which different bases in the tRNA could pair through unusual base-pairing arrangements with the third base in the codon. Such unusual pairing configurations, "wobble," explained how a single anticodon could pair with different codons. For example, according to the wobble hypothesis, a U in the third position in the anticodon may pair with A or G in the mRNA, or inosine (a modified base, Figure 13-2) may pair with U, C, or A in the third position of an mRNA codon. In no case, according to the "wobble" hypothesis, does a base in the anticodon pair with more than three bases in the codon.

An alternative explanation was suggested in 1978 by Ulf Lagerkvist and his co-workers. They discovered in a cell-free protein synthesizing system that a single valyl-tRNA, necessarily with a single anticodon, could read all four of the valine mRNA codons, a result not predicted from any of the hypothetical "wobble" pairing relationships. These workers came to the unexpected conclusion that although three bases in the mRNA are used at a time during translation to determine which amino acid is inserted in the growing polypeptide, at least in some cases, only the first two are used to pair with the tRNA anticodon. For those eight amino acids for which the codon assignments differ only in the third position, such a model provides no difficulty. But for those amino acids for which there are only two codons and in which two amino acids share in common the first two bases of the codon, for example, aspartic acid and glutamic acid (Table 13-1), such a model creates a difficulty. It appears that it is not only the sequence in the anticodon but the sequence in other parts of a tRNA that will determine which codon is read. For instance, a change in the base sequence outside of the anticodon of a tryptophan tRNA will change the codon that that tRNA reads. Therefore, two tRNA's with identical anticodons can as a function of the remainder of their sequences pair with different codons. We presume that differences in the structures of aspartyl tRNA and glutamyl tRNA allow them to pair with their respective mRNA codons even though base pairing involves only two out of three bases. Because only two out of three bases in the anticodon are involved in pairing, it follows that far fewer than 61 tRNA's are needed to read all of the codons. Although the validity of two out of three pairing remains to be confirmed and may not be operative in all interactions between tRNA's and mRNA's, for now it appears to be the best explanation for the observation that a single tRNA can read more than one codon.

Peptides are initiated with methionine and the AUG codon.

We have reviewed evidence demonstrating that translation of mRNA starts at a fixed point and proceeds by moving the mRNA across the ribosome three bases at a time. In the test-tube experiments involving an artificial mRNA with a repeating sequence of three different bases, the results demonstrated that the polynucleotide could be translated starting with any one of three bases (pp. 362–363). For the successful synthesis of polypeptides in a cell, such permissiveness in the reading of mRNA's would be disastrous. If reading could start anywhere, only an occasional polypeptide would be produced with the correct and complete amino acid sequence. A mechanism for initiating the reading of the mRNA at the correct position is an absolute requirement.

The initiating mechanism operates by locating a specific AUG codon in the mRNA, from which the reading will progress. Inasmuch as AUG is a methionine code word, all polypeptides undergoing synthesis start with methionine at their N-terminal ends. More accurately, this generalization applies to eukaryotic proteins. Prokaryotic proteins begin with N-formylmethionine (Figure 13-6), created by adding a formyl group to a methionine that is already bound to its tRNA. Following synthesis, the N-terminal methionine is often cleaved from the polypeptide. Therefore, all polypeptides found in cells do not have N-terminal methionine residues.

The initiating mechanism makes use of a special methionyl-tRNA, labelled $tRNA_{meth}^{init.}$, differing from the methionyl-tRNA ($tRNA_{meth.}$) that inserts methionine in the middle of growing polypeptide chains (Figure 13-7). The anticodons of the two tRNA's are the same; differences occurring in the dihydrouridine and tΨC loops and the stem of the anticodon loop. The initiating methionyl-tRNA differs from other tRNA's in that it can occupy the peptide site of the mRNA-ribosome complex and from this position can participate in the formation of the first peptide bond by donating its methionine to the charged tRNA occupying the amino acid site. The insertion of the initiating methionyl-tRNA into the peptide site depends on the presence of proteins called *initiating factors.*

Initiation may depend on the basepairing of a region in the leader sequence of the mRNA to 16S rRNA (in eukaryotes 18S rRNA).

The primary structure of most proteins includes several internal methionine residues. Because methionine has a single code word, AUG, a means must exist to single out the one methionine codon used for initiation. The

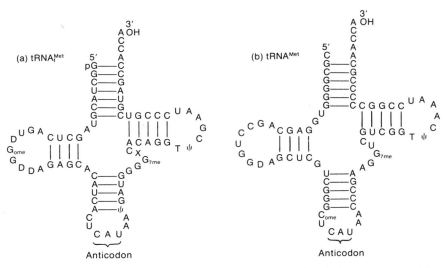

Figure 13-7   The primary structures of methionyl (standard) and N-formylmethionyl (initiating) tRNA's of *Escherichia coli*.

leader sequences of many mRNA's have recently been found to contain sequences located about six bases upstream from the initiating codon that are complementary to a sequence of bases at the 3' end of the 16S rRNA. It is likely, therefore, that the correct positioning of the initiating methionine codon in preparation for polypeptide synthesis results from the binding of the leader sequence to the 16S rRNA molecule in prokaryotes. This is schematically demonstrated in Figure 13-8. Also eukaryotic mRNA's contain sequences in the leader regions that are complementary to the 18S rRNA and that may be involved in initiation. Perhaps of more importance, however, is the cap on the 5' end of eukaryotic mRNA's (p. 342). Aron Shatkin demonstrated in 1978 that there is a protein in the small subunit of eukaryotic ribosomes that binds the cap region and that, therefore, may be important to initiation.

Termination of polypeptide synthesis is accomplished using terminating codons that do not code for any amino acid.

Synthesis of a polypeptide must not only be initiated properly but must also be correctly terminated. During elongation of a growing polypeptide, charged tRNA's bind to the amino-acid binding position on the ribosome and bring the next amino acid into position for its addition to the growing polymer. As long as a code word in the mRNA is recognized by a tRNA, elongation of the polypeptide continues. Termination of synthesis occurs when an mRNA triplet is placed in the amino acid site for which there is no tRNA with an appropriate anticodon (Figure 13-9). Hence, when the noncoding sequences UAA, UAG, or UGA, all called *terminating* codons (formerly

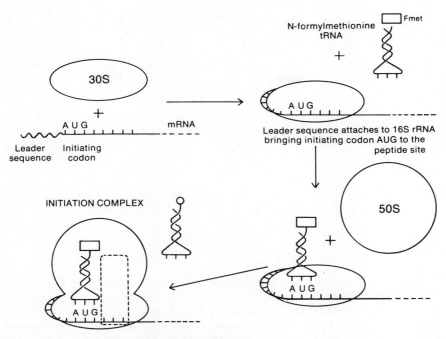

Figure 13-8   Initiation of polypeptide synthesis.

called *nonsense* codons), occupy the amino acid site, no additional charged tRNA binds because none has the correct anticodon. The result is that elongation ceases. The completed polypeptide, attached to the last tRNA, is cleaved from the tRNA and released from the ribosome-mRNA complex. Termination is accomplished through the intervention of special proteins, the *termination factors.* The terminating codons for several polypeptides are known. For example, the terminating codon for the α chain of human hemoglobin is UAA. Some phage proteins, such as the coat protein of a small phage called MS2, are terminated using two terminating codons in a row, UAA-UAG.

The genetic code is universal.

We can easily persuade ourselves that the genetic code must be the same for all species that inhabit our planet, that it is universal (or at least global). Consider what would happen if the code were altered by, say, changing a critical base in the anticodon of a tRNA. The altered tRNA would read a different code word and insert its amino acid—assuming that it could be charged—into new positions in polypeptides. The result would be a series of amino acid substitutions in most of the proteins in that organism. Many of these substitutions, as in the case of altered hemoglobins, would be deleterious. Perhaps even more important, the codon that was formerly read

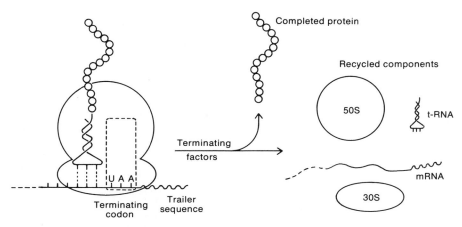

Figure 13-9   Termination of polypeptide synthesis.

by the mutant tRNA could no longer be read. In other words, a new terminating codon would be created, capable of prematurely aborting the synthesis of many polypeptides. Because of the evident deleterious nature of changes in tRNA anticodons, we expect that the code will be conserved during evolution and that all species will use the same code.

The evidence at hand supports this conclusion. John Gurdon and his colleagues among others have been able to inject the large eggs of *Xenopus*, the African clawed toad, with a variety of cellular components, such as nuclei from other cells. When mRNA's from different sources are injected into newly fertilized eggs, which synthesize proteins at a higher rate than unfertilized ones, these mRNA's are translated into their specific proteins. The mRNA's for chick hemoglobin and sea urchin histones have been successfully translated in *Xenopus* eggs, demonstrating that all three of these distantly related organisms use the same genetic code. In other work, the codons in sequenced eukaryotic mRNA's, such as hemoglobin mRNA, have been found to correspond exactly to those codons used by *E. coli*.

Polypeptide synthesis involves the cyclic use of ribosomes.

The same frugal *E. coli* cell that recycles its tRNA's also recycles its ribosomes. Before polypeptide synthesis commences, the 70S ribosome is dissociated into its 50S and 30S subunits. At the initiation of synthesis (Figure 13-8) mRNA is first bound, in association with the initiation factors, to the smaller subunit. This binding involves, in part, the base pairing of the region in the leader sequence to its complement in the 18S rRNA of the 30S subunit. The initiating N-formylmethionyl-tRNA is added with the assistance of another protein factor using energy provided by GTP. Finally, after the binding of still another protein factor, the large 50S ribosomal subunit is added to complete the translational complex. A tRNA bearing an amino acid then binds to the amino acid site of the complex, and the first peptide bond is formed. When a terminating codon appears in the amino acid site, the polypeptide is released, and the ribosome dissociates into two subunits, each of which is available to be used again in the synthesis of another polypeptide. Ribosome cycles similar to the one described here for bacteria also are operative during protein synthesis in eukaryotic cells.

During polypeptide synthesis several ribosomes are attached to a single mRNA and constitute a polysome.

In Oscar Miller's study of the active transcription of DNA, we saw that several transcripts are made from the same transcriptional unit at one time (Figure 12-13). As an RNA polymerase moves away from the promoter region, it leaves the promoter available for the binding of another polymerase that can start transcribing the region long before the first polymerase molecule has completed its travels. Clearly, numerous polymerases can start before the first one finishes. Miller's pictures reveal several polymerase molecules in a row each actively transcribing the same transcriptional unit.

Polypeptide synthesis is similarly efficient. As the mRNA travels across the ribosome, its 5' end becomes progressively extruded from the ribosomal complex and becomes long enough to make itself available for the attachment of another ribosome and the synthesis of another copy of the encoded polypeptide. Such progressive addition of ribosomes to the free 5' end of the mRNA occurs over and over with the result that several ribosomes are bound to the same mRNA in a complex called a *polysome* (Figures 13-10 and 13-11). The lengths of the growing polypeptides on the ribosomes vary as a function of how far an mRNA has traversed a particular ribosome. The number of ribosomes attached to an mRNA is a function of the length of the coding sequence of the mRNA, larger polysomes resulting from larger mRNAs.

In bacteria, synthesis of a polypeptide using a particular mRNA commences before synthesis of the mRNA is completed.

Bacteria further increase the efficiency of translation by starting to read mRNA molecules even before they are completely transcribed. Because bacteria lack nuclear membranes, the ribosomes are in close association with the DNA, and the 5' end of a growing mRNA transcript can become available for attachment to ribosomes. Because mRNA is translated starting at the 5' end, polypeptide synthesis can commence even before transcription is completed. (Is it an accident that transcription and reading of the mRNA both occur in 5' to 3' direction, or is it evolutionarily adaptive?) Pictures taken by Barbara Hamkalo, Oscar Miller and Charles Thomas, Jr., show several bacterial mRNA molecules being simultaneously transcribed and translated (Figure 13-12). We can see that in this picture the branches of the rather sparse Christmas tree are decorated with ribosomes. The overall efficiency of the processes leading to the formation of polypeptides is impressive. One transcriptional region is simultaneously transcribed with the production of multiple transcripts, each of which is simultaneously translated on several ribosomes.

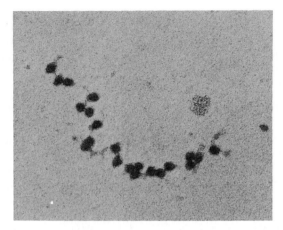

Figure 13-10   An electron micrograph of a polysome. Photo courtesy of Yean Choii, Department of Biology, University of Indiana, Bloomington.

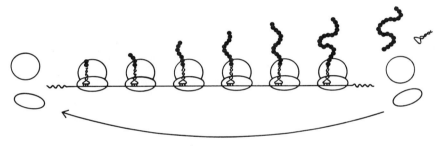

Figure 13-11   An interpretation of protein synthesis along a polysome.

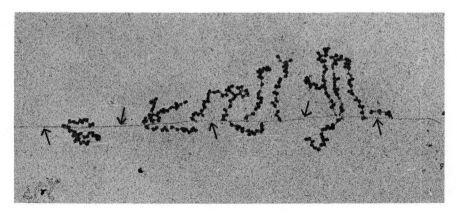

Figure 13-12   Association of polysomes to the DNA (arrows) from which the mRNA is being transcribed in a bacterial cell. Photo courtesy of Oscar L. Miller, Jr., Biology Department, University of Virginia, Charlottesville. From Miller, Hamkalo and Thomas, *Science,* v. 169, p. 392, 1970, with permission.

SUMMARY

The translation of the information encoded in the DNA into the amino acid sequences of polypeptides occurs through the cooperative activity of messenger RNA (mRNA), transfer RNA (tRNA), ribosomes, and various proteins and enzymes. The mRNA carries the information from the DNA to the ribosomes, the sites of polypeptide synthesis. The tRNA has a dual function in protein synthesis, carrying amino acids to the ribosomes for assembly into polypeptides and binding to the mRNA code words to assure that the correct amino acid is inserted properly into the growing polypeptide. There is at least one tRNA with a unique nucleotide sequence for each amino acid. The attachment of an amino acid to its specific tRNA is accomplished by a specific amino acyl synthetase.

The reading of the mRNA occurs three bases at a time in a 5' to 3' direction from a fixed starting point, an AUG initiating codon specifying methionine. As the mRNA progresses over the ribosomal surface, amino acids are linked together one after another through peptide bonds starting from the amino terminus of the polypeptide. Synthesis of a polypeptide ceases when one of the three terminating codons is encountered (UAA, UAG, or UGA). Polypeptide synthesis is efficient. Several ribosomes work on the same mRNA molecule at one time. In prokaryotes, translation starts at the 5' end of an mRNA molecule before transcription of the message is completed.

14

Of Mutations and Mutagens

We have stressed the need for accurate replication and transmission of genes. Yet we are aware of errors in replication and of the production, as recorded in ancient literature, of occasional aberrant types. For example, we find in the Bible (2 *Sam.* 21:20) that, in a war between Israel and Gath, "there appeared a giant with six fingers on each hand and six toes on each foot." The giant of Gath, as a result of a mutation, evidently exhibited polydactyly, a trait inherited as an autosomal dominant.

We now recognize that mutations, errors in the fidelity of replication, are continuously occurring. Traditionally, a mutation was defined as a newly arisen heritable change in phenotype. For our purposes, we will consider a mutation to be any heritable change in the base sequence of an organism's DNA. Not all mutations result in phenotypically detectable changes, but, by and large, those mutations that do are deleterious, causing the organisms carrying them to be less fit than organisms in which the traits are not phenotypically manifested. It is not surprising that mutations are generally deleterious. An enzyme, like a watch, is designed for a particular activity. It is unlikely, though not impossible, that a random change in the amino acid sequence of an enzyme will make it more capable of carrying out its catalytic activity, any more than randomly substituting one part of a watch for another will improve its ability to keep time. Yet mutation provides the variability without which selection and evolution would not be possible. Even so, with too much mutation, no population of organisms could survive. Consequently, mutation rates must be delicately balanced.

SPONTANEOUS MUTATION RATES AND INCIDENCE

New mutations are continually arising at very low rates.

We can determine the rate at which new mutations arise. During cell division, mutation rates are expressed in generations of cells, for example, mutations per gene per generation. During sexual reproduction, mutation rates are expressed in numbers of gametes produced, for example, mutations per gene per gamete. In addition to procedures for determining the rate at which new mutations arise, there are also procedures for determining the incidence at which mutant genes exist in populations of organisms, which is, for example, expressed as mutations per gene or per genome.

One of the earliest procedures for estimation of mutation rates during sexual reproduction in diploid organisms, the *ClB* method, was devised in the 1920's by Hermann J. Muller. This technique measures the rate at which new lethal mutations arise on the X chromosome of *Drosophila melanogaster*. *ClB* is a special X chromosome that contains a large inverted segment to suppress the recovery of recombinant progeny that is called a cross-over suppressor (*C*) (pp. 226–229), a recessive lethal (*l*), and a dominant trait *Bar* (*B*), which causes the round eyes of the fly to be reduced to a narrow bar.

Muller crossed males carrying a normal X chromosome to females heterozygous for the *ClB* chromosome. Males cannot carry the *ClB* chromosome because, in the hemizygous condition, it would cause the death of the fly. Similarly, females cannot be homozygous. Muller then singled out thousands of those females from the cross that displayed the *Bar* condition (Figure 14-1). These females have a *ClB* chromosome from their mothers and an X chromosome from their fathers. Each female was mated—again, to a normal male—placed in a separate vial and left to produce eggs. Because the recovery of recombinant X chromosomes in the progeny is eliminated by the cross-over suppressor, any male offspring must be +/Y because *ClB*/Y is lethal. Most vials yielded the expected standard + males. Some vials, however, contained no male offspring at all. We interpret this to mean that an X chromosome derived from the original male now carries a *new* recessive lethal. This mutation had to arise before or during gametogenesis in that male, because only normal males, with normal X chromosomes, were used. In some of Muller's early experiments about 0.2 per cent of the vials lacked male progeny. Thus, about 0.2 per cent of the X chromosomes produced during spermatogenesis acquired new lethal traits, a mutation rate of 2×10^{-3} mutations per X chromosome per gamete.

The hemizygous condition of the X chromosome in the male has also been exploited to measure spontaneous mutation rates for single genes in *D. melanogaster*. In nonlethal mutations that cause visible changes in phenotype, the estimation of mutation rates per gene is comparatively straightforward. Wild-type males with X chromosomes initially free of visible mutations are mated to females homozygous for one or more recessive traits. By examining the female offspring for one or more of these traits, new mutant alleles that arose during gametogenesis in the male parent can be identified. Traits that are readily picked out—*yellow* body color, *white* eyes, or *singed* bristles—are particularly useful for determining mutation rates, inasmuch as an occasional mutant can be readily identified in the midst of as many as 100,000 normal offspring. In *Drosophila*, the spontaneous mutation rates for single genes have been found to be about 10^{-4} to 10^{-5} per gene per gamete (Table 14-1). Spontaneous mutation rates have also been determined for mice to be about 10^{-5} per gene per gamete and for various genes in man

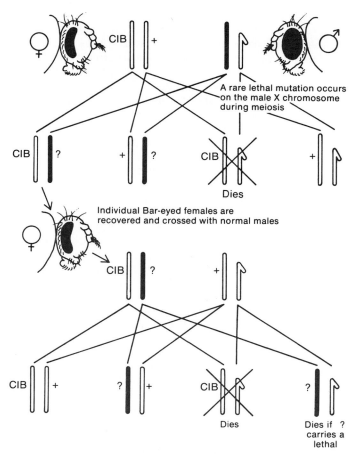

Figure 14-1   The *CIB* method for estimating the rate at which spontaneous lethal mutations arise on the X chromosome of *Drosophila*.

where the frequencies range from 4×10^{-6} to 10^{-4} (Table 14-1). In corn the spontaneous mutation rates seem to be somewhat lower, around 10^{-6} and in *Neurospora,* as low as 10^{-8}.

Estimation of mutation rates in logarithmically growing organisms presents special problems.

The two examples in the preceding section for determining mutation rates in *Drosophila* involved the X chromosome for good reason. Detection of mutants in, or starting with, hemizygous males where X chromosomes are essentially in the haploid state is easy compared to identifying mutants in autosomes, which are present in two copies and where recessive mutations

Table 14-1   Spontaneous Mutation Rates in Haploid and Diploid
Organisms.

Organism	Mutant Character	Mutation Rate*
BACTERIA		4×10^{-10} to 2×10^{-6}
E. coli	to streptomycin resistance	4×10^{-10}
	to T_1 resistance	3×10^{-8}
	to arabinose dependence	2×10^{-6}
S. typhimurium	to threonine resistance	4×10^{-6}
	to histidine dependence	2×10^{-6}
D. pneumoniae	to penicillin resistance	1×10^{-7}
EUKARYOTES		
N. crassa		8×10^{-9} to 1×10^{-7}
	to adenine dependence	4×10^{-7}
	to adenine independence	$8 \times 10^{-9} - 3 \times 10^{-7}$
	to inositol independence	$1 \times 10^{-8} - 1 \times 10^{-7}$
D. melanogaster		6×10^{-5} to 1.2×10^{-4}
	to yellow	1.2×10^{-4}
	to brown	3×10^{-5}
	to ebony	2×10^{-5}
	to eyeless	6×10^{-5}
Maize (corn)		$< 7 \times 10^{-7}$ to 1×10^{-4}
	to waxy	$< 7 \times 10^{-7}$
	to shrunken	1.2×10^{-6}
	to colorless	2.3×10^{-6}
	to purple	1×10^{-5}
	to *i*	1×10^{-4}
Mouse		4×10^{-6} to 3×10^{-5}
	to nonagouti	3×10^{-5}
	to brown	4×10^{-6}
	to albino	1×10^{-5}
	to leaden	8×10^{-6}
Human		4×10^{-6} to 2×10^{-4}
	to epiloia	4×10^{-6}
	to retinoblastoma	2×10^{-5}
	to chondrodystrophy	9×10^{-5}
	to Pelger's anomaly	2×10^{-5}
	to neurofibromatosis	2×10^{-4}
	to Huntington's chorea	5×10^{-6}

* Mutation rate expressed in mutations per cell division or per gamete.
  Modified from Table 23-1, Monroe W. Strickberger, *Genetics,* 2nd Edition, Macmillan
Publishing Company, Inc., 1976.

can go undetected. We might expect that detection of mutations in haploid
organisms might be facilitated because we have to contend with only one
copy of a chromosome or particular gene. Indeed, in *Neurospora* mutation
rates can be determined by counting the mutations in a population of haploid
spores resulting from meiosis.

However, in bacteria and other organisms or cells that grow logarithmically, each cell dividing to produce two, estimating mutation rates per
gene per cell division (or DNA replication) presents some problems. Large
numbers of cells must be grown, and a mutation that occurs early in the
growth of a population of cells would be passed on to the progeny during

subsequent cell divisions. For example, a mutation during the first division could be present in one of two daughter cells and in half of all other cells derived from the first one. Furthermore, as additional mutations occur, the number of mutant cells in a population will increase. For logarithmically growing cells the mutation rate (m) per gene per generation equals the number of mutant cells (n) divided by the number of generations (g) times the total number of cells (N): $m = n/gN$ (Table 14-2). Note that n/N (the fraction of mutant cells) equals mg, the mutation rate times the number of generations; with each generation of growth, the fraction of mutant cells increases. These procedures have given us estimates of mutation rates per gene per generation in bacteria, the rates ranging from about 10^{-10} to 10^{-6} (Table 14-1).

Mutation rates may be lower in haploid than in diploid organisms.

Comparisons between mutation rates in haploid and in multicellular diploid organisms are difficult to make. In higher organisms there are numerous mitotic divisions from the formation of the zygote to the production of germ cells so that mutations in gametes may result from a mutation that occurred long before the meiotic divisions. However, in some cases it is possible to distinguish between mutations in mitotic cells that give rise to gonial cells and mutations that occur only during meiosis. The data in Table 14-1, at least for *Drosophila,* exclude mitotic mutations. The mutation rates

Table 14-2   Determination of Mutation Rate for Logarithmic Growth.

DEFINITIONS:

g = Age of the colony in generations.
N = Total number of cells present (Generation g).
n = Number of *mutant* cells present (Generation g).
m = Mutation rate per cell per generation.
n_i^* = Number of mutant cells present in generation g that *first arose* in generation i.

DERIVATION:

1. With logarithmic growth the number of cells in
   generation i is: 2^i
   In particular, $N = 2^g$

2. The number of new mutants that arise in generation i
   is (approximately): $m2^i$

3. For *any one cell* that was present in generation i, the
   number of its descendants present in generation g
   is: 2^{g-i}

4. From 2. and 3. we get: $n_i^* = (m2^i)(2^{g-i})$
                                                                $= m2^g$
                                                                $= mN$

5. The total number of mutant cells in generation g is: $n = \sum\limits_{i=1}^{g} n_i^*$
                                                                $= gmN$

6. By rearranging 5: $m = \dfrac{n}{gN}$

Thus, if N, n, and g are known, we can determine the mutation rate, m.

in *Drosophila* (per gene per gamete) seem higher than those in bacteria (per gene per generation). Because the units in which the mutation rates are expressed differ, we cannot draw conclusions from these data about the relative mutation rates per replication of DNA. We can, however, draw conclusions regarding the likelihood in each instance of a new mutation resulting in a new mutant organism. In a bacterium each new mutation will result immediately in a mutant organism, about once in every 10^7 cells. In a diploid organism, the likelihood of a newly mutant organism being produced, assuming the trait is recessive, is the product of two independent events, for example, the rate of production of mutant sperm times that for ova. For a mutation rate of 10^{-4}, a new mutant would result about one time in 10^8 zygotes. Thus, the rate of production of mutant organisms, as a result of new mutations, may be similar in bacteria and in multicellular organisms. The ability to tolerate such high mutation rates (per gene per gamete) in higher eukaryotes depends on the diploid nature of the organisms. The ability to tolerate higher mutation rates presumably affords diploid organisms greater genetic heterogeneity and, therefore, greater evolutionary adaptability. We must consider the possibility that each species may be able to regulate, at least to some degree, the level at which mutations occur and that mutation rate, like other characteristics, is in part a phenotypic expression of a genetic endowment.

Mutations occur randomly.

Some early evolutionary theories suggested that mutational changes were responses to particular environments; for example, drought-resistant corn is produced when lack of water causes genetic changes that allow growth even when water is limited. This view contrasts with the one presented in Chapter 1 where we suggested that adaptation involves existing genetic heterogeneity, that is, forms of genes that happen, by accident, to be suitable for a new environment. As we pointed out in Chapter 10, mutants, particularly in prokaryotes, are recovered with selective techniques. For example, mutants resistant to streptomycin are recovered by growing cells in a medium containing streptomycin, and leucine auxotrophs are revealed by their inability to grow on a medium lacking leucine. An early concern of some geneticists was that the selective techniques used for recovering bacterial mutants might be responsible in some manner for the induction of new mutations. Was it possible that streptomycin itself induced streptomycin-resistant mutations, or did the mutants already exist, waiting only to be identified by the selective procedure? Joshua and Esther Lederberg provided one simple and elegant answer to the question by examining the origin of *E. coli* mutants resistant to lysis by T1 phage.

They first grew and plated approximately 100,000 cells in the absence of T1 and then replica plated (pp. 297–300) the colonies to plates containing T1, where only cells resistant to the phage would grow into colonies. The positions of the T1-resistant colonies were the same in all of the replicas, indicating that the mutants were already present on the master plate prior to exposure to T1. If the phages themselves had induced T1-resistance, resistant colonies would have been found in different places on the replicas.

Recessive lethal mutations accumulate in diploid species.

Mutant bacterial cells will accumulate during the growth of a culture, but any mutation that prevents the continued division of the bacteria, for example, a mutation in the gene for DNA polymerase III, would be eliminated. In diploids, however, such a mutation would linger, because a second, wild-type allele would counter its effect. It would only be eliminated when made homozygous or hemizygous. Therefore, we expect that recessive, deleterious mutant genes will accumulate in the genetic constitutions of populations of diploid organisms.

The techniques developed by Muller to determine spontaneous mutation rates in *D. melanogaster* can be modified to determine the frequency of mutations—for example, recessive lethals—present in wild populations of flies. Flies found in a favorite habitat of *D. melanogaster,* such as wineries, are mated to females carrying *balancer chromosomes.* Balancer chromosomes have a structure similar to the *CIB* chromosome; they contain an inversion to prevent recombination, a dominant marker, and a recessive lethal. In the case of an autosome, for example, the second chromosome, a fly can have two different balancer chromosomes, each carrying a different lethal gene, inversion, and dominant marker (Figure 14-2). Because each chromosome is lethal when homozygous, only heterozygotes carrying each of the two balancer chromosomes are viable. Wild males and balancer females are used in a mating scheme designed to produce flies homozygous for second chromosomes originally carried by the wild flies (Figure 14-2). If, starting with a particular wild-type chromosome, no progeny homozygous for the chromosome are produced, we know that the chromosome originally carried a recessive lethal. Results from this kind of experiment (Table 14-3) for several *Drosophila* species reveal that from 9.5 per cent to 61 per cent of the second and third chromosomes from wild populations carry lethals or

Table 14-3   Frequencies of Lethal and Semilethal Chromosomes for Several Species and Geographic Populations of *Drosophila.*

Species	Chromosome	Population	Per cent lethals and semilethals
D. prosaltans	II	Brazil	32.6
	III	Brazil	9.5
D. willistoni	II	Brazil	41.2
	III	Brazil	32.1
	II	Cuba	36.0
	III	Cuba	25.6
D. melanogaster	II	Caucasus	15.5
	II	Ukraine	24.3
	II	Cannonsburg, Pa.	28.2
	II	Amherst, Mass.	36.3
	II	Wooster, Ohio	43.1
	II	Winter Park, Fla.	61.3
D. pseudoobscura	III	Yosemite, Calif.	33.0
	III	San Jacinto, Calif.	21.3
D. persimilis	III	Yosemite, Calif.	25.5

Modified from Th. Dobzhansky and B. Spassky, *Genetics,* v. 39, p. 472, 1954.

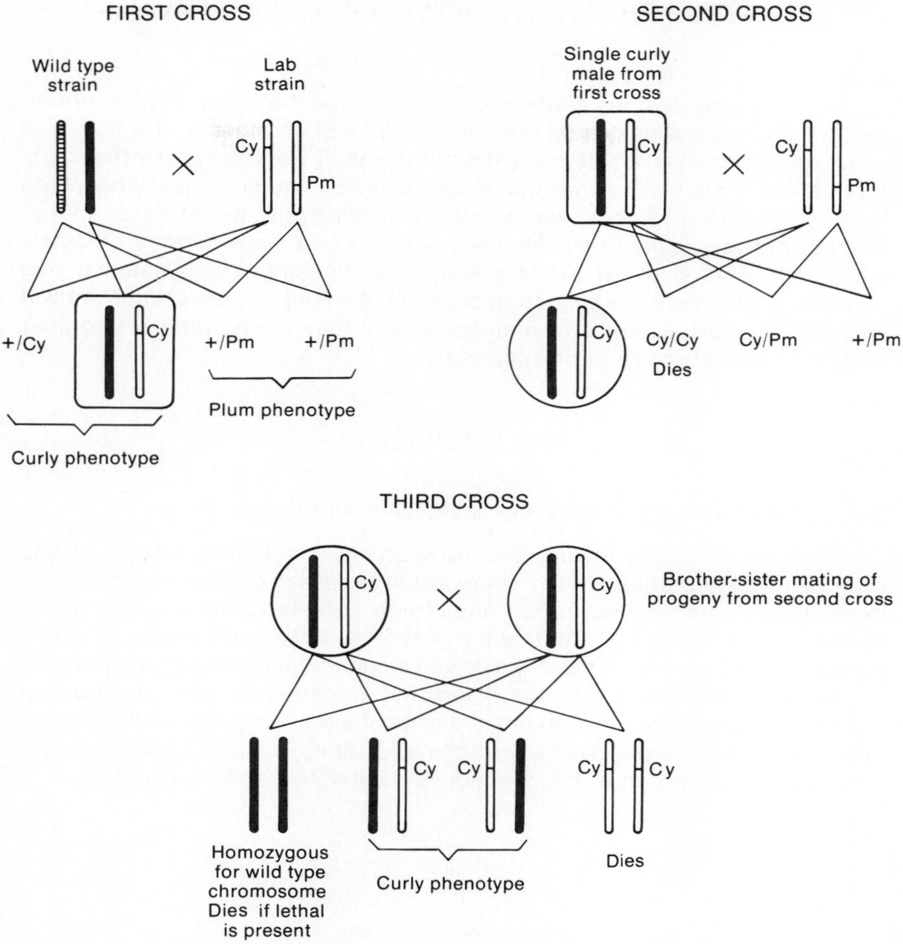

Figure 14-2   The balanced lethal chromosome method for determining whether a second chromosome of *Drosophila* carries a lethal mutation.

semi-lethals (Table 14-3), 10 to 100 times more than those found carrying lethals when only newly arisen mutations are considered. We conclude that most lethal mutations were already present on the chromosomes of the flies from wild populations. Thus, there can be a sizable number of lethals present in populations of diploid organisms, reflecting the genetic heterogeneity of these organisms. The level of lethal mutations varies among different populations. From these data we can estimate that the average fly carries one to two lethal recessive mutations. Similar calculations have been made for human populations, and it is estimated that the average human carries between three and five recessive lethals in his or her diploid genome.

THE STRUCTURAL BASIS OF MUTATION

Our knowledge of the genetic code and the mechanisms of transcription and translation enables us to predict the types of changes in the DNA that will manifest themselves as mutant phenotypes. If we simply take the structure of DNA, possible changes that might result in a mutant condition include base substitutions; single base additions or deletions; deletions of a large number of bases called deficiencies; additions of a large number of bases, which, if identical to an existing sequence, constitute duplications; and inversions, in which a sequence of bases is reversed. In this section we will consider the various kinds of mutations, how they are produced, and their effects on the structure of polypeptides.

Base Substitutions

Base substitutions are of two types: transitions and transversions.

Base substitutions involve the replacement in a DNA molecule of one base pair by another base pair. There are two classes of base substitutions: *transitions,* in which a purine on one strand is replaced by a purine and a pyrimidine on the other strand by a pyrimidine, and *transversions,* in which on each strand purines replace pyrimidines and pyrimidines replace purines.

General mechanisms of transition and transversion are presented in Figure 14-3. During replication one member of a correctly matched base pair mispairs. Following the next round of replication, in which pairing occurs normally, the transition or transversion is established and propagated.

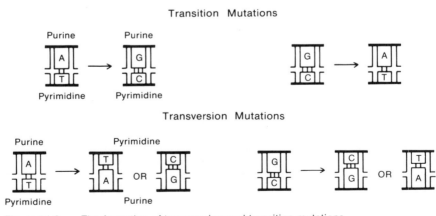

Figure 14-3   The formation of transversion and transition mutations.

Both chemical and physical agents produce base substitutions.

Several mutagens that cause base substitutions have been identified. Two such agents, nitrous acid and ethylmethane sulfonate, act by chemically altering bases in the DNA. Bromodeoxyuridine is a thymidine analog that can replace thymidine in the DNA and cause base substitutions. These three chemical mutagens produce transitions. Ultraviolet light also produces base substitutions.

Nitrous Acid. The mutagenic activity of nitrous acid involves the oxidation of primary amines of cytosine or adenine to keto groups (Figure 14-4), converting cytosine to uracil and adenine to hypoxanthine. During the

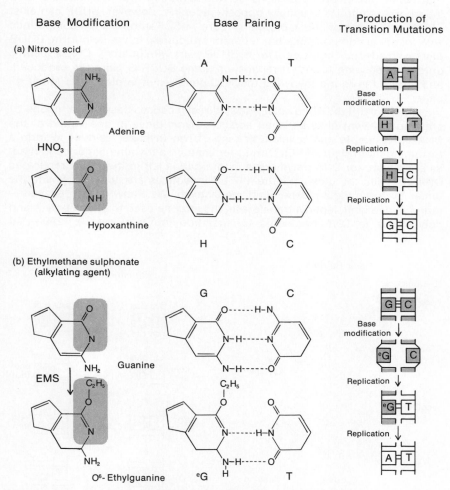

Figure 14-4   Production of transition mutations by chemical modification of bases in DNA.

next round of replication, uracil pairs with adenine, and in all subsequent rounds adenine pairs with thymine. Hypoxanthine pairs with cytosine, and in all subsequent rounds cytosine pairs with guanosine.

Ethylmethane sulfonate. Ethylmethane sulfonate (EMS) is an alkylating agent that, through ether linkages, places an alkyl, in this case ethane, on guanine and thymine (Figure 14-4), producing O^6-ethylguanine or O^4-ethylthymine, respectively. Ethylguanine pairs with thymine, resulting, after two rounds of replication, in the substitution of an AT pair for a GC pair. Ethylthymine pairs with guanine, resulting after two rounds of replication in the substitution of a GC pair for an AT pair.

Bromodeoxyuridine. Bromodeoxyuridine (BUDR) is a pyrimidine analog that, when taken up by cells, is converted into a triphosphate and then incorporated into DNA, usually opposite adenine. However, BUDR can exist as either the keto or enol *tautomer* (Figure 14-5). Tautomers are compounds with identical compositions but different structures. In the keto form, BUDR pairs with adenine. In the enol form, BUDR pairs with guanine. Consequently, during an ensuing round of replication some BUDR:G pairs are produced, and during the round following that, GC pairs result in place of original AT pairs.

Ultraviolet light. Ultraviolet light indirectly causes base substitutions by creating covalent links between adjacent pyrimidines in one strand of the DNA, for example, the covalent linkage of two thymine residues to form a thymine dimer (Figure 14-6). During replication, a thymine dimer cannot act as a template for DNA polymerase, and a gap is left in the newly replicated DNA opposite the dimer (Figure 14-7). Nucleotides are added opposite the dimer by a template-independent mechanism. We know of one enzyme—terminal deoxynucleotidyl transferase, the enzyme used to attach poly-A and poly-T tails to DNA fragments for insertion into plasmids, p. 290—that

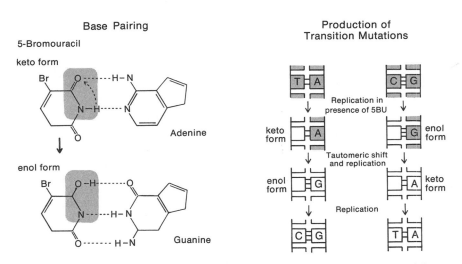

Figure 14-5   Production of a transition mutation with a base analog, 5-bromouridine.

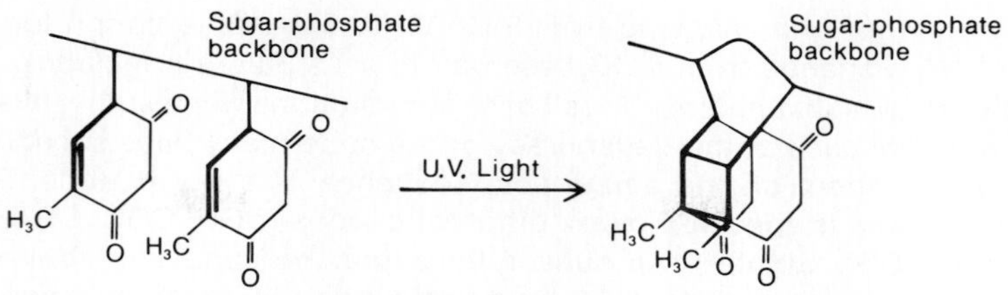

Figure 14-6   Production of a thymine dimer by ultraviolet radiation.

might carry out the addition of bases opposite a thymine dimer. Because terminal deoxynucleotidyl transferase does not use a template, any bases might be inserted opposite the dimer. Consequently, base substitutions, leading to transitions or transversions, may result.

Base substitutions can lead to changes in codons and produce amino acid substitutions among other effects.

Our understanding of the genetic code allows us to predict the various possible effects of base substitutions. They fall into the following categories:

Missense mutations. The proper insertion of an amino acid in a polypeptide depends on the base sequence of an mRNA codon. A base substitution in the DNA can result in a base change in a codon in an mRNA and may lead to the insertion of the wrong amino acid in a polypeptide chain. A base substitution that results in an amino acid substitution is called a *missense mutation,* because it changes the "sense" of the codon for one amino acid to another amino acid. Thus, the insertion of valine for glutamic acid in the β-chain of hemoglobin stems from a missense mutation with an mRNA codon being changed by a single base from 5'GAA3' to 5'GUA3' (in the DNA a TA pair has been substituted by an AT pair, a transversion). Hemoglobin C, containing lysine in place of glutamic acid, probably stems from a change

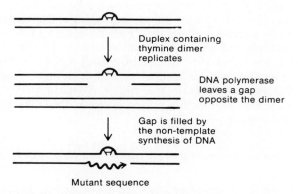

Figure 14-7   Mutagenic effect of a thymine dimer.

in the codon for glutamic acid from 5'GAA3' to 5'AAA3', a codon for lysine (in the DNA a change from a CG base pair to a TA pair, a transition).

Silent base substitutions. Not all base substitutions will result in missense mutations. Because of the degeneracy of the code many base substitutions will have no effect on the amino acid sequence of a polypeptide. For instance, arginine is specified by six different codons (CGU, CGC, CGA, CGG, AGA, and ACG). Single base substitutions that produce interconversions among any of these codons will have no effect on the amino acid sequence of a polypeptide. Base substitutions that do not alter the amino acid sequences of polypeptides are called *silent* mutations. Because of the redundancy of the code, 21 per cent of possible base substitutions will leave amino acid sequences unchanged, thereby providing some insurance against mutations producing deleterious effects.

There is a second group of silent mutations that result in the replacement of an amino acid by another with like properties. Recall, for example, that the substitution of the hydrophilic amino acid lysine for the hydrophilic amino acid glutamic acid at position 6 in the β-chain has mild physiological effects compared to the substitution of the hydrophobic amino acid valine, which produces sickle cell anemia. We can consider amino acid substitutions of the first type, which have little or no effect on the functioning of a polypeptide, as "quiet" mutations.

Chain-terminating substitutions. Three codons (UAA, UAG, and UGA) cause termination of polypeptide synthesis. Base substitutions that convert a codon for an amino acid into a terminating codon, for example, the tyrosine codon UAC to the terminating codon UAG, will result in premature termination of a polypeptide. Any such termination, particularly when near the N-terminal end of a polypeptide (the region synthesized first), can destroy the function of the polypeptide. The amino acid sequences responsible for enzymatic or other functions simply are not present.

Chain-elongating substitutions. The terminating codons responsible for the normal termination of a polypeptide can themselves undergo base substitutions with the result that the polypeptide is no longer terminated in its appropriate position. If the mRNA contains a trailer sequence, an elongated polypeptide results. An excellent example of a base substitution affecting a terminating codon is the mutant human hemoglobin Constant Spring (Hb-CS), which has an α-chain containing 172 amino acids instead of the normal 141. The first extra amino acid is a glutamine residue resulting from a single base change in the terminating codon from UAA to CAA, a codon for glutamine. The elongated α-chain results from the continued reading of what is normally a trailer sequence in the mRNA, 31 more amino acids being added before another terminating codon is encountered by chance in the trailer sequence (Figure 14-8).

The mRNA codon assignments protect against deleterious base substitutions.

It is fascinating to consider the "logic" of codon assignments. The genetic code, itself a product of evolution, seems to conspire to reduce the

	...	137	138	139	140	141	142	143	144	...	170	171	172	173
Normal α-chain		thr	ser	lys	tyr	arg	**term**							
α-chain mRNA		ACX	UCU	AAA	UAC	CGU	UAA CAA	GCU	GGA	...	GUC	UUU	GAA	UAA
Hb Constant Spring		thr	ser	lys	tyr	arg	gln	ala	gly	...	val	phe	glu	**term**

Figure 14-8   Hemoglobin Constant Spring: The effects of an amino acid substitution in the terminating codon of the gene for the α-chain of hemoglobin.

likelihood of base substitutions producing detrimental amino acid replacements. In Figure 14-9 the genetic code is presented in a manner that identifies the hydrophobicity of the amino acid specified by each codon. The hydrophilic amino acids (arginine, lysine, histidine, glutamic acid, and aspartic acid) all have similar codons, often differing from each other by a single base. The hydrophobic amino acids (phenylalanine, valine, leucine, and isoleucine) also have highly similar codons. To change a hydrophilic amino acid to a hydrophobic one would often involve two or more base substitutions, for example, phenylalanine, UUU, or UUA, to lysine, AAA, or AAG. Because it is unlikely that two base substitutions could occur simultaneously in the same codon, even in the presence of a mutagen, replacement of phenylalanine by lysine would be extremely rare. Thus, the structure of the code itself protects against detrimental effects from base substitutions.

Reversion restores a mutant to a normal phenotype.

Up to now we have been concerned only with *forward mutations,* changes from a normal base sequence that convert an inherited phenotype to an abnormal one. But changes also occur in the opposite direction—from mutant to normal phenotype—as a result of *back* or *reverse mutations.* We should emphasize that *reversion* is the result of a second mutational change, often occurring several generations after the initial mutation that caused the abnormal condition, and, like forward mutation, is a rare event. Thus, in a population of organisms with a common mutant defect, we expect only one or two out of thousands to revert spontaneously. Not all kinds of mutational changes can revert. Revertants, by definition, are cells or organisms that have changed from the mutant condition back to the normal or nearly normal one. Revertants can be broadly separated into two classes: true revertants and *suppressors.* True revertants are those in which the genotype of the normal state is exactly restored, for example, a base substitution mutation can undergo true reversion by reconverting a GC to AT base substitution back to the original GC. Suppressors likewise eliminate or reduce mutant phenotypes, but they do so by changes other than those that restore the original base sequence. In contrast to true revertants, the phenotypes of suppressors are sometimes subnormal. For example, an auxotroph may revert to prototrophy but still grow more slowly than the original normal

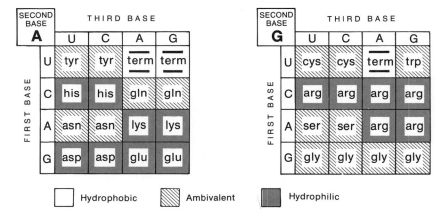

Figure 14-9   The genetic code is displayed showing the properties of amino acid side chains. After Dickerson and Geis, *The Structure and Actions of Proteins,* New York: Harper & Row, 1969.

strain. There are two types of suppressor mutations, *intragenic* (within a gene) and *intergenic* (between genes).

In intragenic suppression of base substitutions one mutation suppresses the effects of another mutation in the same gene.

We envision that base substitution mutations produce their negative effects by creating an abnormally folded enzyme with an attendant loss of catalytic activity. Any additional amino acid substitution in an enzyme that permits the molecule to assume its original conformation could restore activity and return the organism to a normal or near-normal phenotype. This type of reversion is called *intragenic suppression* because both the initial and correcting mutations reside in the same gene. For example, Charles Yanofsky and his co-workers used the *E. coli* enzyme tryptophan synthetase, which catalyzes the conversion of indole glycerol phosphate to tryptophan. The enzyme is composed of two different polypeptides, A and B. In the normal tryptophan A polypeptide a region from amino acid residue 173 through 177 seems to be paired with a region from residue 209 through 213, both of which contain hydrophobic and ambivalent amino acids (Figure 14-10). Mutations at residue 210 that substitute hydrophilic arginine (AGA) or glutamic acid (GAA) for ambivalent glycine (CGA) render the enzyme inactive and cause tryptophan auxotrophy. Any substitution of this type can be expected to interfere with the folding and function of a polypeptide. But a second substitution at position 174 from tyrosine (UAU) to cysteine (UGU), an amino acid more hydrophilic and with a less bulky side chain than tyrosine suppresses the effects of the amino acid substitution at residue 210, presumably because the regions are now able to pair with each other again.

A second type of intragenic suppression of base substitution mutations involves the substitution of a third, different amino acid at the position of the original mutation. For example, several revertants involving position 210 in the A polypeptide of tryptophan synthetase have been recovered. In addition to the true revertant in which glycine is restored, other revertants,

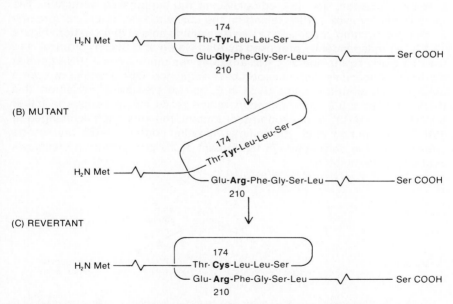

Figure 14-10   Possible relationships between two regions of the polypeptide chain in tryptophan synthetase A. (A) Wild Type. (B) Amino acid substitution at position 210. (C) A second site revertant with a compensating amino acid substitution at position 174. After Yanofsky, Horn, and Thorpe, *Science*, v. 146, p. 1593, 1964.

derived from the arginine mutant, have threonine, serine, or isoleucine, and some, derived from the glutamic acid mutant, have valine or alanine at position 210 (Figure 14-11). The three revertants from the arginine mutant have base substitutions at codon positions different from the original base substitution. The two glutamic acid revertants, however, have base substitutions at the same position in the codon but with bases different from the original. Serine, threonine, and alanine are ambivalent amino acids and isoleucine and valine are hydrophobic. Consequently, the region around residue 210 is again able to pair with the region around residue 174 and restore the normal conformation of the polypeptide.

Intergenic suppression of base substitutions can involve base substitutions in genes for tRNA's.

A change in a second gene can alleviate a mutant condition in several ways and thereby cause *intergenic suppression*. For example, in molecular aggregates, such as hemoglobin, a tetramer, an amino acid substitution in the α-chain, might suppress the deleterious effects of a substitution in the β-chain by allowing normal interaction of the subunits in a manner similar to intragenic suppression.

In one important type of intergenic suppression, the effects of a substitution mutation creating a terminating codon are reversed. Polypeptides, which in the mutant condition were prematurely terminated because of a terminating codon, are now completed in the suppressed condition. The mechanism involves a tRNA taking on the ability to read what was formerly a chain-terminating codon and to insert an amino acid at this location (Figure 14-12). As independently proposed by Alan Garen and Sydney Brenner and their co-workers, the altered tRNA arises from a mutation in a tRNA gene at a point that specifies the anticodon base sequence. One example of such a change is an altered tyrosine tRNA in *E. coli*. In the normal condition, this tRNA contains a 3'AUG5' anticodon, which reads mRNA codons of either 5'UAU3' or 5'UAC3'. In the suppressor mutant, the anticodon is changed to 3'AUC5', which now reads the chain terminating codon 5'UAG3' and inserts tyrosine into the polypeptide in the mutant at the point where it previously would have terminated.

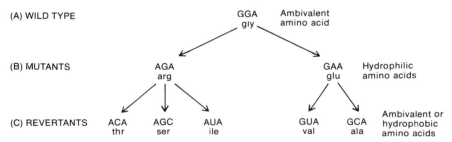

Figure 14-11   Amino acid substitutions at position 210 of tryptophan synthetase A resulting in loss of enzyme activity and suppression of the mutant effect.

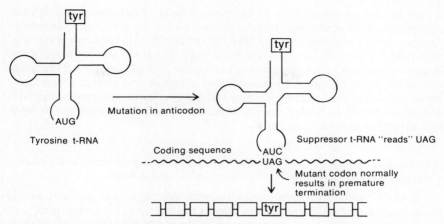

Figure 14-12   A change in the anticodon of tyrosine tRNA causes intergenic suppression of a terminating base substitution.

A remarkable aspect of suppression of terminating codons by altered tRNA's is that defects in several different genes can be cured simultaneously. In yeast, a multiply-auxotrophic strain, requiring histidine, arginine, lysine and tryptophan for growth, can be cured by a single tRNA suppressor. The defect in all of these genes is the same misplaced chain-terminating codon, 5'UAA3'. It is corrected by a tRNA suppressor inserting tyrosine into the polypeptides. The appropriate tRNA anticodon is created in a manner similar to that described for the *E. coli* tyrosine tRNA. Suppression by abnormal tRNA's does not, however, have to involve the anticodon, inasmuch as base changes in tRNA sequences outside of the anticodon have also resulted in the ability of tRNA's to read terminating codons.

Changes in the anticodons of tRNA's do not eliminate normal tRNA's from a cell because, even in *E. coli*, tRNA's are specified by multiple genes. Normal codons, such as those for tyrosine in the preceding examples, are still read correctly. Variant tRNA's, however, can cause trouble for cells. For, although the new tRNA allows the completion of a prematurely terminated polypeptide, it also can allow extension of other polypeptides past the normal terminating position. Possibly for this reason, bacterial and yeast cells containing tRNA suppressors often grow more slowly than normal cells.

Single Base Deletions and Additions

Base additions or deletions arise as a result of mispairing near gaps in one strand of double helical DNA.

George Streissinger and his colleagues in 1966 proposed a model that accounts for the spontaneous origin of base additions or deletions. According to the model, base additions or deletions occur because of mispairing in regions where the stability of the double helix is reduced, such as near single-strand breaks or near termini (Figure 14-13). Single-strand breaks are present during replication (pp. 108–109, Figure 4-14), recombination (pp. 182–183, Figure 7-1), and, as we shall see, in connection with repair of structurally altered DNA. Mispairing can occur in these regions (Figure 14-13), resulting in an extrahelical loop and a gap in one strand of the DNA. When the gap is filled in, the loop may still remain. During the next replication one of the newly replicated double helices may contain an addition or deletion of one or several base pairs.

Intercalating mutagens insert themselves between stacked bases and cause base additions or deletions by stabilizing extrahelical loops.

Certain compounds, *intercalating agents,* are capable of inducing single base additions or deletions. These agents are planar compounds with multiple rings and can insert themselves between stacked bases of double-

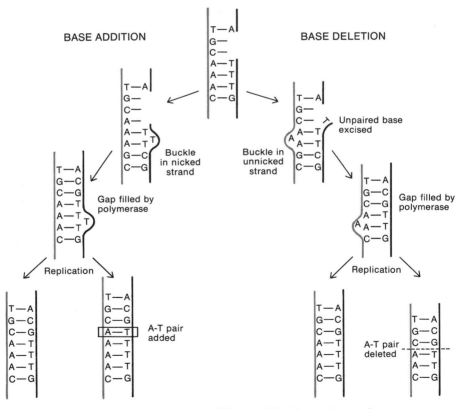

Figure 14-13   A mechanism for base addition and deletion mutagenesis.

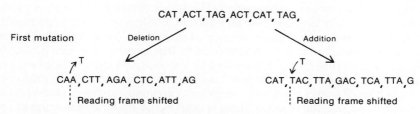

H₂N [structure] NH₂   Figure 14-14   Proflavin, an intercalating mutagen.

helical DNA. Acridines, such as proflavin and acridine orange are examples of such compounds (Figure 14-14). The mutagenic action of intercalating agents is not fully understood, but recent evidence supports the following proposal: Intercalating agents bind to and stabilize extrahelical loops of DNA that occur near single-stranded gaps. In this way they increase both the duration of these loops and the likelihood that DNA replication will occur while the loop still exists. Thus, they increase the likelihood that newly replicated DNA will contain an addition or deletion.

Single base deletions or additions can be frame-shift mutations and cause extensive misreading of an mRNA.

The proper synthesis of a polypeptide results from the correct reading of a series of codons in an mRNA. In a *frame-shift mutation* an mRNA is translated using the wrong series of three-letter code words, that is, using the wrong reading frames (p. 362). Single base deletions or additions can cause frame-shift mutations and scramble the amino acid sequence of a polypeptide. The mutational effect stems from the manner in which the genetic code is read—three bases at a time from a fixed starting point. If a single base is added or deleted, the reading frame of the message is shifted by one base so that the entire message downstream from the addition or deletion is read in wrong reading frames (Figure 14-15). An addition or deletion of a single base near the start of the coding region, usually produces a shortened polypeptide because a terminating codon is encountered accidently ahead of the position of the normal terminator. An addition or deletion of a single base near the end of the coding region produces an abnormally long polypeptide because the normal terminating codon is not read. Hb Wayne is a mutant form of human hemoglobin in which the amino acid sequence of the α-chain is extended because of a single base deletion near the end of the α-chain gene (Figure 14-16). In Hb Wayne one of four bases is missing in the mRNA—either the third base in the codon specifying amino acid 138 (serine) or any one of the bases is missing in the

CAT‚ACT‚TAG‚ACT‚CAT‚TAG‚

First mutation   Deletion   Addition

                    ↗T   ↙T
CAA‚CTT‚AGA‚CTC‚ATT‚AG   CAT‚TAC‚TTA‚GAC‚TCA‚TTA‚G
┊ Reading frame shifted   ┊ Reading frame shifted

Figure 14-15   A base deletion (or addition) acts as a frame-shift mutation.

	137	138	139	140	141	142	143	144	145	146	147	148
Normal α-chain	*thr*	*ser*	*lys*	*tyr*	*arg*	**term**						
	ACX	UC[**U**]	AAA	UAC	CGU	UAA	GCU	GGA	GCC	UCG	GUA	GCA
Hb Wayne	ACX	UCA	AAU	AAC	GUU	AAG	CUG	GAG	CCU	CGG	UAG	CA-
	thr	*ser*	*asn*	*thr*	*val*	*lys*	*leu*	*glu*	*pro*	*arg*	**term**	

Figure 14-16   Hemoglobin Wayne: The effect of a frame-shift mutation near the end of a gene.

codon that specifies amino acid 139 (lysine) in the normal polypeptide. All of the codons upstream (towards the 5' end of the mRNA) of the single-base deletion are read correctly in their normal reading frames. The loss of the base in one codon results in the tRNA reading what would normally be the first base of the next codon. The codons downstream (towards the 3' end of the mRNA) are then read using incorrect reading frames. Additional amino acids are added because the normal terminating codon is no longer read and the polypeptide is extended five amino acids before another terminating codon is accidentally encountered using the altered reading frame.

In general, the addition or deletion of any small number of bases not divisible by three will result in a reading-frame shift. If a contiguous group of three bases (or a multiple of three) is deleted, the effect will be to remove one or more amino acids from the encoded polypeptide chain. Human hemoglobin again provides several examples; in one, hemoglobin-Leslie, a glutamic acid is missing from position 131 of the β-chain and in another, hemoglobin-Tochigi, the tetrapeptide glycine-asparagine-proline-lysine is deleted from positions 56 to 59 in the β-chain. These amino acid deletions result from the deletion of their codons in the mRNA.

Frame-shift mutations revert, usually by intragenic suppression.

Just as base substitutions can revert, so can base additions or deletions. True revertants involve the deletion of an added base pair or the addition of a deleted base pair. Most revertants of single base additions or deletions involve intragenic suppression. For example, a single base deletion can be suppressed by a single base addition, either upstream or downstream but near the site of the original deletion. The effect of neighboring single base additions and deletions is shown in Figure 14-17. Upstream of the site closest to the 5' end of the mRNA, the message is read in the correct reading frame. Between the deletion and the addition the mRNA is read in the wrong reading frame, but, because of the degeneracy of the code and the low frequency of terminating codons, it is likely that the intervening codons will be read with the insertion of amino acids into the growing polypeptide chain. Downstream of the second site, the reading frame is restored so that the remainder of the amino acid sequence is normal. As long as the altered polypeptide sequence

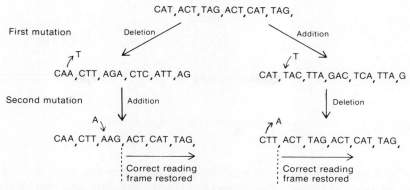

Figure 14-17   The mechanism of suppression of a base deletion or addition by a compensating addition or deletion.

between the addition and the deletion does not eliminate the activity of the polypeptide, suppression of the mutant effect occurs. A single base deletion can be suppressed by any change that restores the original reading frame of the mRNA. For example, two additional, neighboring single base deletions or a single double-base deletion would result in a deletion of three bases and restore the reading frame downstream from the position of the third deletion. Intragenic suppression of a single base addition has also been reported, in which a mutant tRNA reads a four-base codon and thereby restores the reading frame.

Deficiencies

Deletions of large numbers of bases, deficiencies, do not readily revert.

Deletion mutations are not restricted to the loss of single bases. Changes occur in which hundreds, thousands, even tens of thousands of base pairs are lost. Deletions of this magnitude are called *deficiencies.* Deficiencies are induced by ionizing radiation (X-rays, gamma rays) that breaks the DNA backbone. The ends created can subsequently heal together but with the loss of certain segments (Figure 14-18). Such breakages are also the basis for inversions and translocations. If the deficiencies are sufficiently large, then several genes, a single gene, or a substantial part of a single gene can be eliminated from the chromosome, the loss being propagated during subsequent rounds of DNA replication.

Because entire genes may be lost, there is usually no easy means by which reversion can take place as it does in single-base substitutions, additions, or deletions. The absence of reversion back to wild type or pseudo-wild type provides a diagnostic criterion for a deficiency. In many cases the presence of the deficiency can be confirmed by light or electron microscopy. In eukaryotes, particularly those with polytene chromosomes, deficiencies

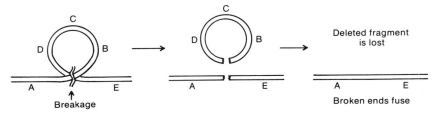

Figure 14-18   Production of chromosomal deficiencies by irradiation.

can be detected as regions of mispairing in synapsed chromosomes (p. 229), and we have seen that heteroduplexes between deficient and normal DNA complements can reveal the presence of a deficiency in λ (pp. 202–203).

Deficiencies can occur as a result of unequal crossing over between chromosomes with tandem repeats.

One type of deficiency that occurs spontaneously at high frequency is exemplified by *bobbed* (*bb*) mutations on the X chromosome of *Drosophila melanogaster,* which appear at a rate of 10^{-3} per gamete, 10 to 100 times more frequently than other mutations arise in *Drosophila.* The *bb* condition results from a reduction in the number of rRNA gene copies (pp. 328–329) as a result of *unequal crossing over.* Because rRNA genes are tandemly repeated, mispairing during synapsis is likely (Figure 14-19). If recombination occurs two chromosomes with different numbers of rRNA genes are produced, one chromosome being deficient and the second replete with extra copies of rRNA genes. The ease of mispairing is responsible for the high frequency of mutation to the *bb* condition. Deficiencies in tandem repeats can revert—in contrast to other deficiencies—because, as we just saw, unequal crossing over can also produce chromosomes with increased numbers of rRNA genes. Unequal crossing over is not limited to rRNA genes but can occur wherever there are tandem repeats including 5S rRNA genes, tRNA genes, and repeated genes encoding polypeptides. In Chapter 17 we will consider an example involving hemoglobin genes.

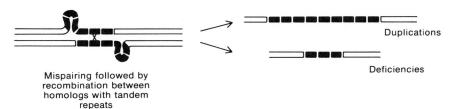

Figure 14-19   Production of chromosomal deficiencies and duplications by unequal crossing over.

DETERMINANTS OF THE RATE OF MUTAGENESIS

The number of mutants induced by X-radiation increases linearly with dose.

We measure radiation (such as X-rays) that causes the formation of ions in *roentgens* (*r*). One *r* produces one ionization in 1 cm³ of air. In tissue or water one *r* produces about two ionizations per cubic micrometer. Studies have been conducted in which the percentage of lethals produced on the X chromosome of *Drosophila* is expressed as a function of radiation dosage. In the 1930's several investigators demonstrated, for levels from about 500 to 6,000 *r*, a linear relationship between dose and the induction of lethal mutations (Figure 14-20). Extrapolation of the data suggested that an increase in mutation frequency is linear with any dose of radiation from zero up to 6,000 *r*. Bentley Glass investigated whether the data could legitimately be extrapolated to low levels of radiation in the early 1960's. Over a period of several years, Glass examined the induction of dominant mutants in *Drosophila* by using 5 *r* X-ray doses. Glass's results indicated that the extrapolation of the linear relationship between mutation induction and dose of X-radiation was valid down to at least 5 *r*. An important fact emerging from studies of X-ray-induced mutants is a determination of the relationship between dose of X-rays and the number of mutations induced. From such data we can calculate that, for example, in mice the mutation rate per gene per roentgen is 22×10^{-8} and that the dose of radiation required to double the spontaneous mutation rate is about 45 *r*. In humans the dose of X-rays

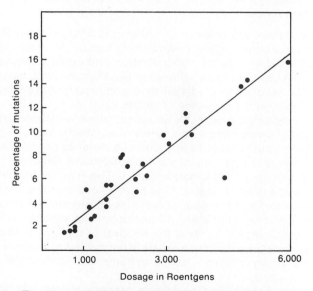

Figure 14-20   The linear relationship between X-ray dose and the induction of lethal mutations. After Schultz in *Biological Effects of Radiation*, B. M. Duggar (ed.), New York: McGraw-Hill, 1936.

that doubles the mutation rate is unknown but presumably is similar to that of our rodent relatives. As we discuss later, the time over which radiation is given and time that elapses between a mutagenic treatment and the next round of DNA replication can influence the number of new mutants recovered.

Enzymes repair alterations in DNA structure and protect against mutation.

The DNA in normal prokaryotic and eukaryotic cells is under constant surveillance. Some enzymes continually search the DNA for abnormalities in pairing, such as might be caused by thymine dimers or mismatched base pairs, and for chemically modified bases, such as would be caused by alkylating agents. When these enzymes encounter abnormalities, they initiate a variety of processes to repair the DNA, typically by excising the offending bases from one strand of the double helix and using the other strand as a template for insertion of an appropriate base sequence.

Repair is not reversion. Repair involves the programmed correction of a potential mutational lesion before it is propagated as an inherited mutation. True reversion represents a mutation of a mutation, that is, the correction of a mutant sequence through another random mutational event. DNA repair can be separated into two distinct categories along lines proposed by A. John Clark, *extrareplication* repair, occurring independently of DNA replication, and *intrareplication* repair, occurring during DNA replication.

Photoreactivation and excision of thymine dimers are examples of extrareplication repair.

Thymine dimers are induced by ultraviolet light and can lead to base substitution mutations (Figure 14-6). Dimers can be removed from DNA by two different mechanisms: *photoreactivation* and *excision* repair. In photoreactivation the covalent linkage between two thymine residues of the dimer is broken. The reaction is mediated by a photoreactivating enzyme, which can function only in the presence of visible light.

A second repair process, excision repair, which is independent of visible light, removes thymine dimers by a series of enzymatic steps (Figure 14-21). In the first step a cut is made to one side of a dimer to produce a nick in the DNA. The dimer and some attached nucleotides are removed using the 3′ exonuclease activity of DNA polymerase I. The gap is then filled by the polymerase and sealed by ligase. One form of excision repair mediated by polymerase I, *short-patch* repair, constitutes 95 per cent of excision repair and involves the removal of about 10 nucleotides. A second excision mechanism, *long-patch* repair, involves the removal of about 100 nucleotides by a different exonuclease. Excision repair is not limited to thymine dimers but also operates, using different enzyme systems, on mismatched and chemically modified bases.

Mutants exist in which the excision process is defective, for instance, mutations that affect the ultraviolet resistant (*uvr*) genes A, B, and C of *E. coli*. These genes encode the enzymes involved in the first steps of excision

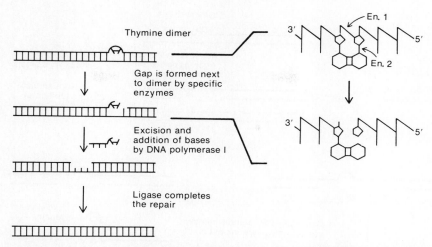

Figure 14-21   Excision and repair of thymine dimers. According to Hazeltine and Grossman, En. 2 (a pyrimidine dimer DNA-glycosylase, encoded by the *uvrB* gene) acts first to cleave the bond between the sugar and base. En. 1, an endonuclease, apparently encoded by *uvrA* then cleaves the phosphodiester bond.

(Figure 14-21), and cells carrying these mutations accumulate dimers in large numbers upon exposure to ultraviolet light. Because polymerases are unable to replicate bases opposite these dimers and because the excision process is defective, there is increased cell death in response to treatment with UV light. Any cells with mutations in the *uvr* genes that survive exposure to UV light have large numbers of new base substitution mutations because of the non-template synthesis of DNA opposite thymine dimers. Other mutants affecting the gene for polymerase I, the *pol A* gene, are defective in short-patch repair.

Intrareplication repair involves recombination or the error-prone, template independent synthesis of DNA opposite thymine dimers.

Repair can occur not only independently of replication but also during the replication process itself. Among the repair mechanisms that occur during replication, we should include the error-prone, non-template addition of bases opposite thymine dimers (Figure 14-22). If such gaps were not filled, the DNA would fragment and maintenance of the genome would be impossible. There is, however, a mechanism of repair, *recombinational repair,* that is error free (Figure 14-22). Unreplicated gaps opposite thymine dimers and regions in the process of being excised create nicks or gaps in the DNA not unlike those produced at the start of recombination (Figure 7-1, p. 182). Such gaps can serve as points for the initiation of recombination. Recombination initiated by the presence of a mutagen-caused gap results in the repair of the DNA and the prevention of mutations. The ability of a cell to carry out recombinational repair depends, of course, on the cell having the

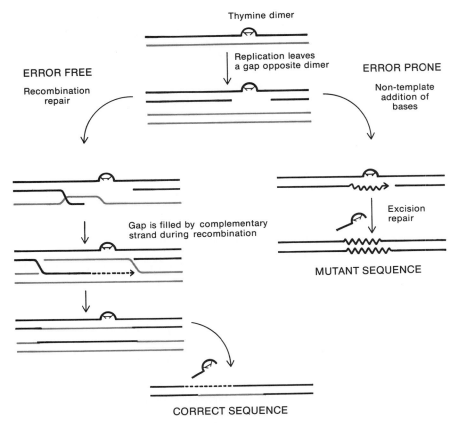

Figure 14-22   Error-free and error-prone intrareplication repair of thymine dimers.

enzymes and the proteins necessary for recombination. Again, mutants have been recovered that are incapable of recombination and, so, are incapable of preventing the propagation of mutational damage by recombinational repair.

As a result of repair mechanisms, the number of mutations induced can depend on the duration of radiation.

The repair mechanisms correct mutational lesions before they are inherited and become part of the established genetic inventory of an organism. Given that enzymes mediate repair processes and that the number of enzyme molecules in a cell is limited, more lesions in the DNA may accumulate from a severe exposure to radiation than the repair enzymes can handle before the next round of replication is completed. We expect, then, that exposure to a large dose of radiation over a short time can induce more mutations than several exposures, equal in total dose to the large one, over an extended period of time. Thus, intermittent exposure to low levels of mutagens is less likely to produce a mutation than a single exposure to a high level of mutagen. The apparent effects of repair were revealed in experiments with mice in 1963 in which William Russell exposed mice to the same total dose of irradiation but delivered it at very different rates. Russell found that in spermatogonia or oocytes but not in spermatocytes, short, high-level exposures to radiation produced large numbers of mutations, but long, low levels of exposure did not. In spermatocytes, the number of mutations produced was independent of the rate of exposure. Apparently repair occurs in spermatogonia and oocytes but not in spermatocytes.

Inability to repair thymine dimers causes the human genetic disease *Xeroderma pigmentosum.*

Cells in animals are divided into two distinct populations: germ line cells, which eventually give rise to gametes (eggs and sperm), and somatic cells, which give rise to different body tissues. Dividing somatic cells, like germ line cells, are susceptible to mutation. Somatic mutations, of course, will not be passed to subsequent generations, but they can have dramatic effects on the phenotypes of the cells themselves. One effect is the conversion of a normal cell into a cancerous one. In humans the bombardment of skin by ultraviolet light from the sun causes the continuous production of thymine dimers and the production of mutations. A particularly serious result of UV exposure is the production, apparently through mutagenesis, of pigmented malignant skin cancers called *melanomas.* The incidence of melanomas is particularly high among individuals who indulge in the tanning effects of the sun. The distribution of melanomas even reflects certain social habits, for example, in Australia where backless sun dresses were once popular, there is a high incidence of melanomas on women's backs. In the United States alone it is estimated that there are 25,000 deaths per year attributable to melanomas induced by sun bathing.

Humans have evolved for long periods of time in the presence of ultraviolet light from the sun, and mechanisms do exist to limit the mutagenicity of sunlight. In tropical areas the indigenous population has darkly pigmented skin, an evolutionary adaptation preventing penetration of UV light into the live cells underlying the surface. Suntan itself is a mechanism to limit penetration of ultraviolet light. Man-made devices, such as suntan lotions, that encourage tanning and filter out UV rays also protect against melanomas . . . clothes, of course, work much better! At another level of protection, somatic cells, like germ cells, have repair enzymes to eliminate abnormalities in DNA structure such as thymine dimers. However, some individuals are homozygous for one of many recessive mutations, each of which eliminates some part of a fully functioning repair system. These persons suffer from different forms of a collection of genetic diseases called *xeroderma pigmentosum* that cause them to have an usually high incidence of UV-induced melanomas. The disease is ultimately lethal, the only therapy so far developed being to keep affected individuals out of sunlight and to remove each new melanoma as it appears.

Most mutagens are also cancer-causing agents called carcinogens.

The induction of melanomas by UV light is not a unique example of mutagen causing cancer. It appears that most, if not all, mutagens are carcinogens, and most carcinogens, but not all, are also mutagens. In a study of 175 known cancer-causing agents, Joyce McCann and Bruce Ames demonstrated that 157 (about 90 per cent) of the agents, or their metabolic products, also caused mutation in the bacterium *Salmonella typhimurium.* Agents that were not known to be carcinogenic were not found to be mutagenic. Although there are other causes of cancer besides mutation—for example, large doses of steroid sex hormones can be carcinogenic but are not mutagenic—mutation is clearly one, perhaps the major, means by which cancers are produced.

Many known and undoubtedly many unknown mutagens and carcinogens are present in our environment.

Many mechanisms have evolved by which organisms deal with the hazards of their environments, DNA repair mechanisms and pigments that screen out UV light among them. However, advances made by modern technological societies have introduced over a short period of time—too short for evolution to respond—a great variety of additional physical and chemical agents that are mutagenic. Many are no doubt beneficial. Few question the careful use of X-rays by medical practitioners. Tuberculosis can be diagnosed, fractures detected, and unsuspected dental cavities located. It is commendable that the dose used in many diagnostic procedures has been markedly reduced in the past two decades. For example, a chest X-ray in 1960 required exposure to about 0.1 roentgens; in the 1970's one tenth of this dose was used (both levels well below the doubling dose). Moreover, many unwarranted and excessive uses have been discontinued. No longer can young children see how their new shoes fit by using the shoe store's fluoroscope (an X-ray machine in which the rays bombard a fluorescent screen, something like TV). Nor is X-radiation used to cure acne (500 to 1,000 r used to be given in a course of treatments) or as a cure for throat infections, and X-ray examination of pregnant women has been reduced (20 to 65 r used to be accepted). But, although public awareness and protest have reduced environmental exposure to physical mutagens and created deserved suspicions about nuclear power generators, the possibility of exposure to chemical mutagens has increased. It is estimated that 30,000 new chemicals are produced in industrialized countries each year. Many of these are produced in enormous quantities, for example, as insecticides. Although these chemicals were and still are tested for physiological toxicity or, at least, for acute toxicity, they were formerly not tested for mutagenicity or carcinogenicity.

Rapid tests to monitor the mutagenicity of chemicals have been developed.

Two genetic dangers can be attributed to chemicals, namely, that they cause mutations in germ cells, which are then inherited, and that they may produce mutations in somatic cells that result in cancer. It is difficult and expensive to determine whether a particular compound is carcinogenic, because tests with animals—often rats and sometimes lower primates—are expensive and time-consuming. It is unacceptable, however, to wait until a chemical shows its effects in humans, particularly because there is a long time-lag—on the order of 25 years—between exposure to some mutagens and the development of cancer. Consequently, the effects of continuing exposure to mutagens may be revealed only when epidemiological studies, a polite term for the counting of corpses, demonstrate the mutagenicity and carcinogenicity of a compound. The mutagenicity of carcinogens, however, affords the opportunity to identify those agents that owe their carcinogenicity to their mutagenicity. Several rapid screens for detecting mutagenic compounds have recently been developed, one using bacteria to detect mutagenicity and another depending on the detection of increased levels of sister-strand exchanges in cultured cells. We have already demonstrated that a carcinogen increases sister-strand exchanges (Fig. 4-4, p. 98). At this juncture we will discuss the use of bacteria to test for carcinogens.

Bacterial test for mutagenicity. Bruce Ames and his collaborators have developed a simple but elegant test for mutagens. The test utilizes a strain of *Salmonella typhimurium* that is defective in excision repair and carries a plasmid that enhances the action of some mutagenic agents, so that the majority of primary genetic lesions are passed to subsequent generations. The strain is also genetically endowed with a cell membrane that is highly permeable to substances added to the medium. In consequence the strain is highly sensitive to mutagens. Mutagen-sensitive strains have been constructed carrying either frame-shift or base-substitution mutations that confer histidine auxotrophy. A concentrated suspension of these bacteria is spread on solid medium in a Petri dish in the absence of histidine so that the cells will not grow. A sample of the substance to be tested is introduced into the center of the solid medium and diffuses outward. If the substance is mutagenic, it will kill most of the bacteria near the center of the dish through the induction of large numbers of mutations. Cells that are progressively farther from the center of the dish receive progressively lower doses of mutagen. Some of these cells revert in response to the mutagen and produce histidine prototrophs, which can now grow in the absence of histidine and produce colonies. Therefore, the appearance of a ring of colonies on the plate demonstrates the presence of a mutagen.

Using this bacterial test Ames and his colleagues, among others, have identified the mutagenicity of chemicals found in manufactured goods. For example, a compound used as a fire retardent in children's clothing (tris) was found to be mutagenic and its use has been discontinued. Compounds produced by the burning of cigarettes are mutagenic and the urine of cigarette smokers contains mutagenic compounds. Many hair dyes, which contain synthetic planar compounds that are putative intercalating agents, have been found to be frame-shift mutagens. As in the case of cigarette smokers, urine samples from users of these hair dyes contain mutagenic compounds. Some insecticides, even DDT, are mutagenic. Saccharin, the artificial sweetener, has also been shown to be mutagenic although its mutagenicity is low.

SUMMARY

Spontaneous mutations continually occur at low rates (10^{-10} to 10^{-4} per gene per generation) and provide the genetic variability on which natural selection acts. Mutation rates can be dramatically increased, over 10,000-fold, by both physical and chemical agents. Some agents, for example, nitrous acid and ethylmethane sulfonate, cause base substitutions, one base pair replacing another in the DNA. Base substitutions, in turn, produce amino acid substitutions in polypeptides and can affect termination of polypeptide synthesis. The mutational effects of amino acid substitutions can be suppressed by a second amino acid substitution in the same polypeptide. Other agents, for example, acridines, cause deletions or additions of single base pairs. Single base pair changes can be frame-shift mutations causing mRNAs to be translated in incorrect reading frames. A single base pair addition (deletion) can be suppressed by a nearby single base pair deletion (addition) that restores the reading frame. Deletions of multiple base pairs, deficiencies, are produced by irradiation and some chemicals. There is a linear relationship over much of the dose range between the induction of mutation and the dose of X-radiation. To protect against the inheritance of newly caused changes in DNA structure, organisms possess repair systems that place the DNA under surveillance and correct abnormalities, often by excising the offending alteration and resynthesizing a short segment of DNA. Agents that cause mutations (mutagens) are also agents that cause cancer (carcinogens). Sensitive tests, such as that developed by Ames, for mutagenicity, provide the opportunity of rapidly identifying compounds that may be carcinogenic.

Genetic Dissection of the Gene

We have described, from the perspective of molecular genetics, the nature of genes and the mechanisms by which the nucleotide sequence in DNA is converted into the amino acid sequence of a polypeptide. mRNA, which is made as an initial transcript containing but not limited to the base sequence necessary for the synthesis of a polypeptide, plays a central role in the formation of a polypeptide. In a normal diploid cell RNA polymerases separately transcribe the two copies of a particular gene carried on homologous chromosomes. The transcripts derived from both copies of the gene participate in protein synthesis. In a cell containing two allelic forms of a gene, a heterozygote, both forms serve as templates for the synthesis of two types of mRNA, which, in turn, direct the synthesis of two slightly different forms of a polypeptide. Alleles of human hemoglobin genes are a typical instance. An individual carrying the Hb-S allele on one chromosome and an Hb-A allele on its homolog produces two types of mRNAs and two types of β-chains; two different globin genes produce two correspondingly different types of globin polypeptides.

The allelism of mutants affecting the primary structure of the hemoglobin β-chain is determined by the identification and inheritance of variant β-chain polypeptides. Because a gene is defined through its specification of the amino acid sequence of a polypeptide, all mutations that alter the amino acid sequence of the β-chain are *alleles,* variant forms of a single gene. Mutants of structural genes are often initially recovered without knowledge of the specific polypeptide encoded by the gene. For example, in phage numerous conditional mutants prevent the lysis of a bacterial cell under restrictive conditions. All these mutants have a similar phenotype, the absence of lysis. Do all of these mutants derive from mutations in a single gene, or are products encoded by several different phage genes required to produce lysis of a bacterial cell? Many animal mutants prevent the production of offspring by causing embryonic death. In *Drosophila* such mutants would all be characterized by failure of the larva to hatch from the egg. Again, all these mutants have a common phenotype. Does it seem reasonable that mutations in only a single gene are always responsible for embryonic death?

COMPLEMENTATION

The complementation test suggests whether two mutants are allelic.

In the instances just described more than one gene is probably capable of mutating to produce a single mutant phenotype. Geneticists use the *complementation* test to estimate the number of genes involved in a process such as lysis of bacteria by phages and to determine whether mutants that cannot carry out the process are allelic. Two mutations of independent origin—that is, resulting from separate mutational changes but producing a similar phenotype—are introduced by mating into a single cell or organism in a diploid or pseudodiploid condition. The phenotype of the cell or organism carrying the mutations is evaluated. If the phenotype is normal, the mutations are said to *complement* each other and we conclude that they involve different genes and are nonallelic. If the phenotype is mutant, the two mutations are non-complementing, and we conclude that the same gene is involved and, therefore, the mutations are allelic. The bases for these conclusions are as follows.

Complementing mutants. Figure 15-1 (top) shows two mutations resulting in similar phenotypes but involving different structural genes, (gene A and gene B) that are introduced in the same cell in a diploid condition. Because the mutations involve two different genes, the diploid cell also contains the nonmutant forms of genes A and B. A double heterozygote results and, provided the mutant conditions are recessive, the phenotype of the cell or organism is normal. The two mutations complement each other. Although neither mutant gene can direct the formation of a normal polypeptide, there is complementation because the transcripts of the normal forms of genes A and B direct the synthesis of normal polypeptides, which results in a normal phenotype. Deductively, then, complementation indicates that the mutations involve two different genes.

Non-complementing mutations. Figure 15-1 (bottom) depicts two independently isolated mutations producing like phenotypes, each the result of a change in the base sequences at two separate positions, *mutation sites,* within gene A. Because both mutations involve gene A, only mutant forms of gene A are present when a diploid is formed. Transcription using either allele of gene A as a template must lead to the formation of an abnormal and defective polypeptide. With normal polypeptides absent we expect the mutant phenotype to be expressed. Although we will see on pp. 410–414 that our expectations can be wrong, we can generalize that in the absence of complementation the mutations involve the same gene, that is, they are allelic.

Complementation tests help to determine the number of genes producing a phenotype and, in conjunction with mapping studies, to define gene boundaries.

Complementation tests are necessarily done by the union of pairs of mutant genomes to produce a diploid condition. The tests often involve the

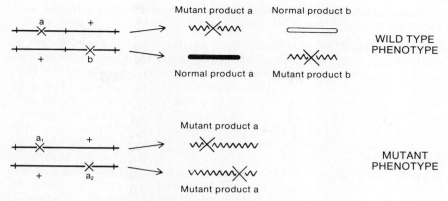

Figure 15-1.   Expected results of complementation tests.

testing by pairs of numerous different mutants sharing a highly similar phenotype. By determining which of the mutants complement each other, it is possible to determine, in turn, how many separate genes are involved in the production of a particular phenotype. Often such studies are done in conjunction with detailed mapping studies. An excellent example is provided by the work of Seymour Benzer, in the late 1950's and early 1960's, on *rII* mutants in T4 phage.

The *rII* phenotype. The *r* or *rapid lysis* phenotype of T4 phage results in the production of plaques on bacterial lawns that are larger than normal. Three separate loci have been identified on the T4 chromosome, *rI*, *rII* and *rIII*, each of which in mutant form produces the rapid lysis phenotype on a lawn of *E. coli* B cells (Table 15-1). *rII* mutants differ from *rI* and *rIII* mutants in that an *rII* mutant produces no plaques on *E. coli* strain K (carrying λ) and produces normal plaques on *E. coli* strain S. Mutants with the *rII* phenotype all map to a single locus on the T4 chromosome, which is physically separate from the *rI* and *rIII* loci. Benzer recovered approximately 3,000 independent *rII* mutants and tested many of them for complementation.

Table 15-1   Phenotype Plaque Morphology of *r* Mutants in Various *E. coli* Strains.

Phage Strain	Bacterial Strain		
	B	S	K(λ)
Normal	normal	normal	normal
rI	r	r	r
rII	r	normal	no plaques
rIII	r	normal	normal

After Benzer, "The elementary units of heredity" in *The Chemical Basis of Heredity*, W. D. McElroy and B. Glass, Eds., Johns Hopkins Press, Baltimore, MD, 1957, p. 70.

Complementation tests with phage are made by simultaneous infection of bacteria with two different mutant phages.

To perform complementation tests, Benzer infected *E. coli* K cells simultaneously with two *rII* mutant phages of independent origin. Because the *rII* mutants were recovered in separate mutagenesis experiments, it is likely that the positions of the mutation sites in the DNA, even if within one gene, are different. Simultaneous infection of a bacterium with two phage chromosomes creates a diploid condition in which the two phage chromosomes exist together in the same cell. If two different *rII* strains are used to infect the same cell, then a test for complementation can be conducted. Inasmuch as *rII* mutants cannot produce plaques on *E. coli* strain K, the absence of plaques following simultaneous infection of the bacteria with DNA from two *rII* mutants indicates that both mutants derive from changes in a single gene. But if there is more than one gene at the *rII* locus, some of the pair tests should involve nonallelic mutants, and plaque formation should occur. Benzer found that most of his mutants fell into two distinct classes or *complementation groups*. Complementation group 1 is a set of *rII* mutants that do not complement each other but do complement any mutant in the other set of mutants, those forming complementation group 2. Group 2 mutants do not complement each other but do complement group 1. These findings indicate that the two groups of mutants originate from separate modifications in two different genes at the *rII* locus. We arbitrarily call these gene A (for group 1) and gene B (for group 2).

Non-complementing *rII* alleles map contiguously.

We expect the mutation sites of gene A alleles to occupy one region of the DNA at the *rII* locus and the mutation sites of the gene B alleles to occupy a neighboring region at the locus. In a genetic map, then, gene A sites should map together and gene B sites should map together with no intermixing. The actual genetic map of the different alleles (Figure 15-2) shows that the mutation sites for all gene A alleles map on one side of the locus and all mutation sites of gene B alleles map on the other side. There is no intermixing in the map positions of the mutation sites of the two groups. The distribution of the mutation sites into two separate but contiguous complementation groups is consistent with a hypothesis that two genes exist side by side at the *rII* locus. Furthermore, these studies pinpoint the boundary between the A and B genes somewhere between the rightmost mutation site of the A gene and the leftmost one of the B gene (Figure 15-2). We can also assume, because of the large number of mutation sites mapped, that the two most widely separated sites are close to the outside boundaries of the A and B genes.

A genetic synonym for gene is *cistron*.

A structural gene is a sequence of nucleotides encoding a polypeptide or an RNA molecule (rRNA or tRNA). So defined, a gene has an unambiguous

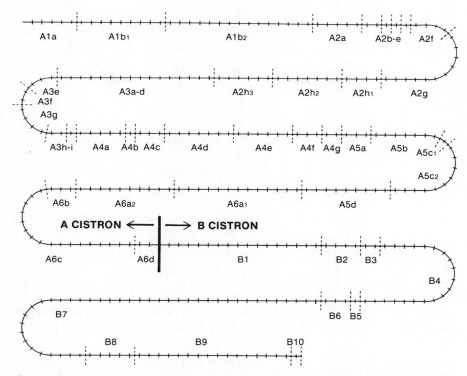

A CISTRON ← | → B CISTRON

Figure 15-2.   Map of the *rII* region of the T4 chromosome. After Benzer, *Proceedings of the National Academy of Sciences, U.S.,* v. 47, p. 403, 1961.

physical reality with determinable boundaries—although intervening sequences within eukaryotic genes leaves some internal ambiguity. In prokaryotes, at least, complementation and mapping studies identify a gene as a contiguous set of non-complementing mutants. Such a set of contiguously distributed, non-complementing mutants is said to belong to the same *cistron,* a term coined by Benzer as a genetic synonym for gene (hence the expression, one cistron, one polypeptide). Benzer observed that, in a diploid condition, two alleles involving different mutation sites in a single gene produce a mutant phenotype when the sites are in the "trans" configuration on opposite chromosomes (Figure 15-3). The term "trans" is borrowed from organic chemistry where it denotes two radicals occupying opposite sides across a double bond. When the two mutation sites are carried in the diploid condition on a single chromosome (Figure 15-3) in the "cis" configuration, the phenotype of the cell is normal. Similarly borrowed from organic chemistry, the term "cis" denotes two radicals occupying the same side across a double bond. In the trans configuration neither copy of the gene can direct the synthesis of a normal polypeptide (Figure 15-3). When both mutation sites are on one chromosome in the cis configuration, the other copy of the

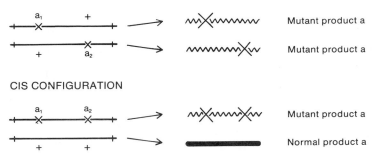

Figure 15-3.   A cis-trans test.

gene has a normal base sequence and can direct the formation of a normal polypeptide (Figure 15-3). The word "cistron" is a *ménage à trois* of "cis" and "trans" and the now-familiar "on" ending (recall codon and replicon). The term characterizes a group of mutation sites on a genetic map belonging to a common complementation group, for example, the A cistron on the *rII* locus. But, as we shall see presently, a cistron does not always correspond to a single structural gene, nor are alleles always non-complementing.

Complementation tests do not always meet expectations (cistrons are not always equivalent to genes).

We have so far considered only situations in which results accord with expectations. Alleles do not complement each other; nonalleles complement each other. Results from complementation tests are not, however, always in accord with expectations. Let us look now at three types of exceptions that, if only complementation data were available, would be difficult to interpret.

A. Intragenic or allelic complementation: genes coding for polypeptides with single enzymatic functions. Alleles should not complement because, in a diploid condition, both gene copies are mutant and neither can direct the formation of a normal polypeptide. For example, in yeast, all mutants at the *histidine-1* locus lack phosphoribosyl transferase enzyme activity, an inability to form phosphoribosyl-ATP from 5-phosphoribosyl-1-pyrophosphate and ATP, the first step in the synthesis of histidine. *Histidine-1* mutants are therefore histidine auxotrophs. Although a diploid yeast cell containing two different *his*×1 alleles should be incapable of growth in the absence of histidine and contain no phosphoribosyl transferase activity, such is not always the case. Some diploid cells carrying two different alleles are able to grow in the absence of histidine and contain substantial enzyme activity. Because these alleles demonstrably affect the activity of the same enzyme, presumably of the same polypeptide, *intragenic* or *allelic* complementation between alleles was at first a mystery. Several investigators suggested that complementation between two alleles occurs because of interactions between the two types of polypeptides specified by the two forms of a gene. Indeed, enzymes for which allelic complementation has been found are usually present in cells as aggregates, called *multimers,* consisting usually of groups of two (*dimers*) or four (*tetramers*) identical polypeptide chains. In normal cells the individual polypeptide chains within the multimer are folded to yield a conformation that has enzyme activity (Figure 15-4). Loss of enzyme activity can result from amino acid substitutions that change the folding of the polypeptides or the shape and stability of the aggregate itself. We know that amino acid substitutions in hemoglobin in the regions of contact between the different polypeptides in the tetramer are often deleterious (p. 316). We can imagine, then, recovering many mutations that cause amino acid substitutions in regions of contact between polypeptide chains in a multimeric enzyme. For example, one allele (a_1) may cause one amino acid substitution and a second allele (a_2) another, each leading to inactivity of the enzyme (Figure 15-4). In a diploid cell heterozygous for both alleles, multimers are composed of two kinds of polypeptides (a_1 and a_2) specified by the two different alleles. We noted before (p. 389) that two different amino acid substitutions at different positions in a single polypeptide chain can restore enzyme activity by compensating for each other. In the case of multimers, the amino acid substitutions on separate polypeptide chains can sometimes compensate for each other by allowing the chains to interact and form a multimer with a normal or near normal conformation (Figure 15-4). The conjecture that intragenic complementation results from interactions between different polypeptide chains is strongly supported by demonstrations that inactive polypeptides extracted from two different allelic strains assemble in a test tube to become enzymatically active.

Among those genes in which allelic complementation occurs, not all alleles are complementing. The sites of complementing alleles, which specify regions of polypeptide chains that are in contact with other polypeptides, tend to be clustered inasmuch as limited regions of a gene code for limited regions of a polypeptide product. Interspersed between these alleles are others that do not complement, presumably because they encode regions of the polypeptides that do not contact other polypeptide regions. Thus, in contrast to intergenic complementation, as at the *rII* locus where complementing mutations are divided in two physically separate complementation groups, intragenic complementation is characterized by the intermixing of

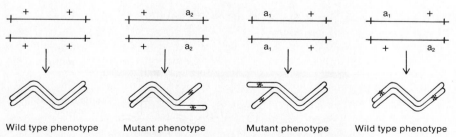

Figure 15-4.   Intragenic (allelic) complementation involving a dimeric protein.

complementing and noncomplementing mutation sites. Other alleles may involve frame-shift mutations, which scramble the amino acid sequence of a polypeptide. These alleles will not complement; an amino acid substitution in another chain in the multimer cannot restore normal folding.

Intragenic or allelic complementation: genes coding for polypeptides with multiple function. In standard *Salmonella typhimurium* the genes for amino acid transferase (C) and histidinol dehydrogenase (D), two of the enzymes involved in the biosynthesis of histidine, are located side by side. These genes, along with other neighboring histidine genes, are transcribed in the same direction using a single mRNA (Figure 15-5). We will see later (pp. 437–439) that single mRNA transcripts for several adjacent structural genes are common in bacteria. Normally the two enzymes are synthesized as separate polypeptides. It is possible, however, to join the two enzymes together and to synthesize a single, bifunctional polypeptide. Joseph Yourano, Tadahiko Kohno, and John Roth constructed a bacterial strain in which a single base deletion existed near the 3′ terminus in the mRNA of the first gene transcribed (D). The second gene transcribed (C) had a single base addition near its 5′ terminus. The first deletion causes a frame-shift mutation in which the terminating codon is no longer recognized, extending the synthesis of the polypeptide into the second gene and scrambling the amino acid sequence near the N-terminal end of the second polypeptide. The base addition in the second gene suppresses the effect of the base deletion and restores the

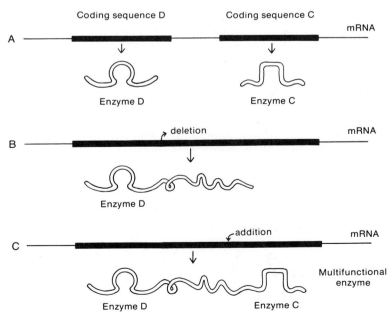

Figure 15-5.   A multifunctional enzyme produced by fusion of two genes at the *histidine* locus.

reading frame so that most of the amino acid sequence of the second polypeptide is normal and amino acid transferase activity is present. The result is two active enzymes joined together by a polypeptide bridge. Although there are two genes in the standard strain, the mutant strain has but one gene responsible for the synthesis of a single polypeptide with two enzymatic activities.

Single polypeptides carrying more than one enzymatic activity are not simply the product of manipulations by inventive geneticists, they also result from natural selection. In eukaryotes especially, some, or all, of the enzymes in a particular biosynthetic pathway can be physically joined in a multienzyme aggregate. Such aggregates, although not common, are far from rare. A major selective advantage of multi-enzyme aggregates is that intermediate compounds in the pathway can be "channelled" so that the product of one reaction is immediately available to the next enzyme in the pathway. Channelling of metabolites through a synthetic pathway can increase the rate of formation of necessary products, reduce the amount of enzyme needed, and be more evolutionarily advantageous than a situation in which the enzymes are physically separate.

There is growing evidence that many, but not all, such aggregates are or were originally synthesized as single polypeptides with multiple functions. For example, the eight enzymes involved in the synthesis of fatty acids in yeast and in animals are contained in only two different polypeptides. The first three enzymes in the synthesis of pyrimidines (Figure 11-3) appear to be synthesized as a single polypeptide in animals although only the first two enzymes reside in a single polypeptide in fungi. An excellent example of a multifunctional polypeptide for which detailed genetic analyses, both complementation and mapping studies, are available is the synthesis of aromatic amino acids in *Neurospora crassa.* In *Neurospora* five enzymes involved in the synthesis of aromatic amino acids are encoded at a single locus, the *arom* locus, and all five are contained in one polypeptide. Thus, we have one gene, one polypeptide, five enzymes.

The complementation relationships at a locus encoding a multifunctional polypeptide are complex. Certain base substitution mutations can eliminate the function of one enzyme but have little or no effect on the function of the other enzymes in the complex. A multifunctional protein consists of a set of globular regions, each responsible for a particular enzymatic activity, joined together by polypeptide bridges. Loss of function of one of these regions could occur without loss of function of other regions, and mutations involving two different regions—two different enzymatic activities—would complement each other (Figure 15-6). Note, too, that frameshift mutations near the beginning of the structural gene will scramble the entire amino acid sequence of the polypeptide and eliminate all enzymatic activities. Mutations eliminating all of the enzymatic activity would not complement any other mutation (Figure 15-6). The complementation and genetic maps of the *arom* locus reveal the existence of both types of mutations; complementing and non-complementing. Because only a single polypeptide is involved, all of these mutants, including the complementing ones, are, by definition, alleles. Thus, we have a second exception to the hypothesis that alleles will not complement.

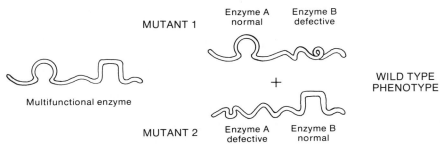

Figure 15-6.   Intragenic (allelic) complementation involving a multifunctional polypeptide.

C. Failure of complementation between mutations in cis-acting regions and adjacent structural genes. In our discussion of transcription we noted a region called the promoter where DNA-dependent RNA polymerase binds upstream from where transcription is initiated. We can imagine a mutational change, for example, a small deletion that eliminates the promoter, eliminates polymerase binding, and results in the absence of both transcription of the adjacent gene and synthesis of its encoded polypeptide. Such a mutation in the promoter region (or promoter gene) prevents the transcription of only the adjacent gene residing on the same chromosome. In a diploid cell a promoter deficiency would prevent transcription of a gene on one chromosome but would have no effect on transcription of the gene on the other chromosome (Figure 15-7) and is said to be *cis-acting.* Imagine a diploid cell with a promoter mutation on one chromosome and a mutation of the structural gene on the other chromosome (Figure 15-7). The gene in the chromosome with the promoter mutation would not be transcribed and no polypeptide would be produced. The other chromosome would be transcribed but a non-functional product would be produced. Consequently, no functioning polypeptide would be present and there would be no complementation even though the mutation sites do not reside in a single gene.

FINE-STRUCTURE MAPPING

Recombination occurs between mutation sites within a gene.

In our discussion of complementation at the *rII* locus in T4 phage we presented a genetic map of the region (Figure 15-2) that placed the different mutation sites in a linear array. The map of the *rII* sites is, in essence, no different from other genetic maps that we have looked at, except that the numerous sites reside within two genes instead of each being in a different gene. In theory, recombination between mutation sites within a gene is possible. The mechanisms of recombination (Figure 7-1) involve the forma-

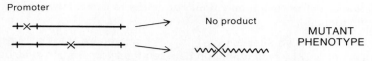

Figure 15-7.   Absence of complementation between promoter and structural gene mutations.

tion of heteroduplexes and polynucleotide bridges between paired homologous regions of chromosomes. Recombination occurs when—after isomerization of the DNA molecules—the polynucleotide bridges are cleaved. Because the structure of DNA is the same everywhere, there is no reason recombinational exchanges cannot occur within the boundaries of a gene as well as between genes. Therefore, by using the same basic techniques employed in mapping the chromosomal positions of different genes, it is possible to map mutation sites within genes.

A major problem confronts the geneticist who undertakes the *fine-structure mapping* of mutation sites within a single locus: the detection of rare recombinants among large numbers of progeny. The entire recombinational length of the *rII* locus, that is, the recombination frequency between the two most distant mutation sites, is about 6 per cent but recombinational frequencies between neighboring mutation sites are as low as 0.02 per cent (2 recombinants in 10,000 progeny). In *Drosophila melanogaster* recombinational lengths of loci can be very small, being as low as 0.01 per cent (1 recombinant in 10,000 progeny) with recombinational frequencies between neighboring sites being as low as 1 in 10^5 progeny. Our ability to resolve neighboring mutation sites in *Drosophila* is less than that in phage, and detailed fine-structure maps using standard genetic techniques would be impossible in mammals.

Typically only + + intragenic recombinants can be identified in a cross between two different alleles.

A cross involving two different mutation sites within the same gene, expressed as (− +) and (+ −), produces two possible types of intragenic recombinants. One recombinant is the (+ +) wild-type one with the two sites that we consider to be normal. The other recombinant, the (− −) one, has two mutation sites within the same gene. The (− −) recombinant usually has the same phenotype as either of the two original alleles and cannot be identified as a recombinant. Only the (+ +) recombinant can be identified. We assume, however, that the two reciprocal recombination classes, (+ +) and (− −), are produced in equal frequency just as in regular mapping. Under this assumption we take the actual number of recombinants to be twice the number of (+ +) recombinants. Therefore, to determine a recombinational frequency the number of (+ +) recombinants is doubled and then divided by the total number of progeny.

The frequency of reversion may limit our ability to detect intragenic recombination.

In a cross of two alleles, $(+ -) \times (- +)$, we realize that $(+ +)$ progeny can arise not only by recombination between the two mutation sites but also by reversion of either mutant site back to the $+ +$ condition. Because frequencies of recombination are very low, in fine-structure mapping it is necessary to distinguish between wild-type progeny resulting from recombination and those resulting from reversion. The frequency at which $(+ +)$ progeny are produced from a cross of $(+ -) \times (- +)$, where $(+ +)$ can result from recombination or reversion, is compared with the frequency with which $(+ +)$ progeny are produced in crosses of $(- +) \times (- +)$ and $(+ -) \times (+ -)$, where $(+ +)$ can only be produced by reversion. The frequency of recombination is then determined by subtracting the frequency of reversion from the frequency at which $(+ +)$ is produced in the $(+ -) \times (- +)$ cross. Recombination can be detected statistically only when the frequency of recombination is significantly greater than the frequency of reversion.

The *rII* locus in T4 phage has been particularly exploited for fine-structure mapping.

The requirements for fine-structure mapping are satisfied among eukaryotes in those organisms such as yeast that can be handled as microorganisms. Among higher animals, fine-structure mapping has been done only in *Drosophila*. In plants, pollen-grain phenotypes have made it possible to study intragenic recombination for some genes in maize.

We find the most detailed fine-structure maps in prokaryotes, especially in crosses between phages and in crosses between bacteria using transduction. The phenotypic characteristics of *rII* mutants in T4 coupled with the life cycle of the phage provide an excellent opportunity for fine-structure mapping, first exploited by Benzer in the 1950's. Recall that *rII* mutants grow normally and produce plaques on *E. coli* S but fail to produce plaques on *E. coli* K. To determine intragenic recombination frequencies, *E. coli* S is infected simultaneously with two different *rII* mutants (Figure 15-8). The phage DNA replicates normally in *E. coli* S and because of the dual infection recombination can take place between the two *rII* mutation sites. Following lysis and the release of phage progeny, one sample of the phages in the lysate is plated on a lawn of *E. coli* S. All of the phages, recombinant or not, produce plaques, and the total number of progeny phages, of which there can be tens or hundreds of thousands, is determined. A second sample of phages from the same lysate is plated on a lawn of *E. coli* K on which *rII* mutants do not grow. The only plaques produced carry chromosomes that are $(+ +)$ at the *rII* locus. Consequently, it is possible to determine the total number of progeny and to detect readily the $(+ +)$ recombinants. Because recombinants with two *rII* mutation sites, $(- -)$, will not produce plaques on *E. coli* K, only half of the recombinant progeny are detected. The frequency of recombination between any *rII* alleles is determined by multiplying by two the number of plaques formed on *E. coli* K and dividing by the number of plaques found on *E. coli* S.

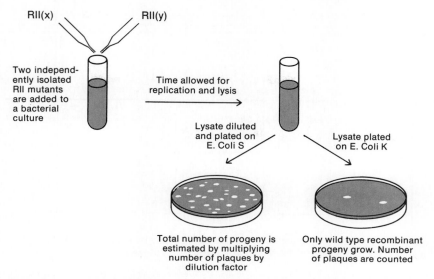

Figure 15-8.   Determination of recombination frequency between two mutation sites at the *rII* locus in T4 phage.

The order of mutation sites within a gene, determined using only two-factor crosses, can be ambiguous.

Fine-structure maps, like the genetic maps of different genes, can be produced by making a series of two-factor crosses. We can take as an example a fine-structure map for 12 different *rII*A alleles that was constructed by Benzer using data obtained from two-factor crosses (Figure 15-9). The 12 different sites have been placed in a linear order. We see that the frequencies are approximately additive, although frequencies of recombination over long distances are smaller than the summation of frequencies over the shorter intervening distances. For example, the frequency of recombination between r240 and r106 is 0.81 per cent, but the summed intervening frequencies equal 1.32 per cent. Also, although the order of genes as presented is unambiguous, we realize that the positions of some of the sites are in doubt. For example, r131 is placed to the left of r973. Because recombinational frequencies are estimates of real frequencies and subject to statistical error, allowing r131 to be placed to the left of r973 may not be significant. Therefore, the order of these two mutation sites is unknown.

The difficulties in constructing fine-structure maps are further demonstrated by recombination data for a group of alleles of the gene for alcohol dehydrogenase (ADH) in maize. The ADH gene is expressed in the haploid pollen grains, and recombination during meiosis between alleles lacking ADH can lead to the production of rare pollen grains in which ADH activity is present (Figure 15-10). Recombination frequencies obtained by Michael Freeling for different ADH allele pairs are shown in Table 15-2. It is impossible to construct a map of the region on the basis of these data because the

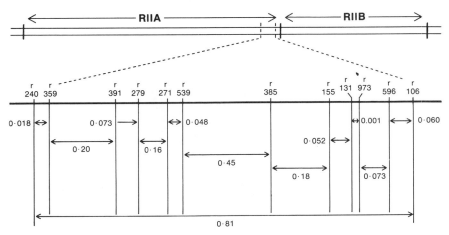

Figure 15-9.   Fine-structure map of part of the *rII* locus made using two-factor crosses. After Benzer, *Proceedings of the National Academy of Sciences, U.S.,* v. 41, p. 344, 1955.

recombination frequencies are not additive. The difficulties arise because the alleles themselves, that is, the base sequences associated with particular alleles, can greatly influence recombinational frequencies. For example, in Table 15-2, we see in the mating between alleles 908 and 1015 that the frequency of recombination is unaccountably high. Thus, two-factor crosses may be inadequate for the construction of fine-structure maps. Unambiguous maps can be constructed, however, by using two additional tools, deletion mapping and three-factor crosses.

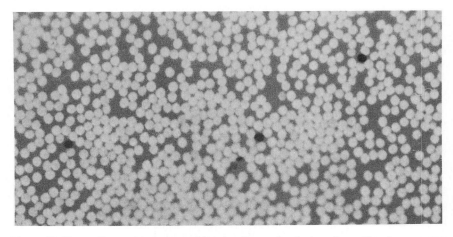

Figure 15-10.   Photo of three recombinant pollen grains (colored) resulting from a cross between two alcohol dehydrogenase null alleles in Maize. Courtesy of Michael Freeling, Department of Genetics, University of California, Berkeley.

Table 15-2   Recombination Frequencies of ADH Alleles in Maize

	5,657	664	296	908	W	667	1,015	719
5,657	0.2*							
664	7.5	0.7*						
296	11.7	19.6	0.2*					
908	10.0	4.9	3.6	0.7*				
W	26.2	35.5	N.D.	33.9	N.D.			
667	9.8	54.1	1.9	4.6	19.8	0.4*		
1,015	5.6	18.2	16.2	66.3	N.D.	13.1	0.1*	
719	16.0	8.2	13.3	3.3	14.1	7.1	8.7	0.2*

From Michael Freeling, *Genetics,* v. *89,* pp. 211–224, 1978.

Results are expressed in terms of number of ADH-positive pollen grains recovered per 100,000. N.D. = Not determined.

* reversion rate.

Deletions within a locus can be genetically mapped in an unambiguous order.

A technique called *deletion mapping* greatly simplifies the task of constructing a fine-structure map. Deletion mapping allows us to determine the order of a group of deletions at a locus. The deletions used for mapping are properly called deficiencies inasmuch as a loss of multiple bases is involved. Such deletions are identified by two properties: They do not revert and overlapping deletions do not produce recombinant wild-type progeny. As we have noted (p. 395), no mechanism exists to allow the reversion of a mutation in which multiple base pairs are lost. The absence of reversion identifies a possible deficiency. Moreover, two deficiencies that have lost a common DNA segment cannot recombine to produce wild-type progeny. There is no way in which recombination can result in the production of a normal chromosome. In contrast, if no common region is missing in both deletions, the deletions do not "overlap," and recombination can produce a completely normal DNA sequence (Figure 15-11). If we have a set of deletions at the *rII* locus, for example, we can order the deletions in an unambiguous linear array by a series of two-factor crosses. Figure 15-12 demonstrates that, among four deletions, deletions 1 and 3 can recombine to produce wild-type progeny. Therefore, deletions 1 and 3 do not overlap. We are not interested in how often they recombine, only if they do recombine. Deletion 2 recombines with neither deletion 1 nor deletion 3 to produce wild-type progeny; therefore, deletion 2 must overlap with both deletions 1 and 3. Because deletion 2 lacks DNA sequences found in both 1 and 3, we conclude that it also lacks the region between 1 and 3 and must lie between 1 and 3. Thus, we have established an order for the three deletions. Similarly, we can demonstrate that deletion 3 must lie between deletions 2 and 4 and that the order of the four deletions must be as shown in Figure 15-12. If four deletions can be placed in order, it follows that a larger number of deletions can also be ordered, a feat that Benzer accomplished for the *rII* locus (Figure 15-13).

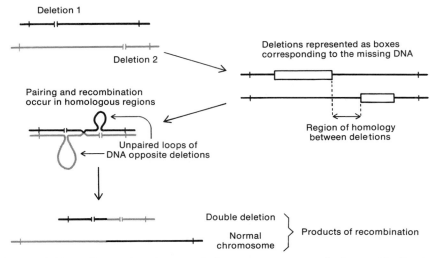

Figure 15-11.   Restoration of a normal chromosome as a result of recombination between non-overlapping deficiencies.

Deletions can be used to determine the location of mutation sites of different alleles.

Once a set of deletions covering a particular region has been recovered, the deletions can be used to locate the positions in a genetic map of mutation sites for any number of alleles. If a mutation site of a particular allele resides outside of the region of the DNA missing in a deletion, then recombination between the deletion and the mutation site can occur with the production of wild-type progeny (Figure 15-14). If the mutation site resides within the region of the DNA missing in the deletion, then recombination cannot produce wild-type progeny; both chromosomes lack the normal base composition at the mutation site (Figure 15-14). If a mating between strains with a deletion and a particular *rII point mutation*—that is, a mutation involving one or few base pairs—fails to produce wild-type phage progeny at a level

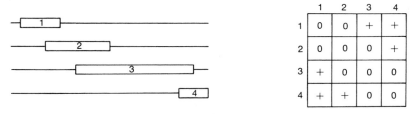

	1	2	3	4
1	0	0	+	+
2	0	0	0	+
3	+	0	0	0
4	+	+	0	0

Figure 15-12.   Deletion mapping. The production of wild-type progeny (+) as a result of recombination in a cross between two strains indicates that two deficiencies do not overlap. The absence of wild-type progeny (−) indicates that the deficiencies overlap.

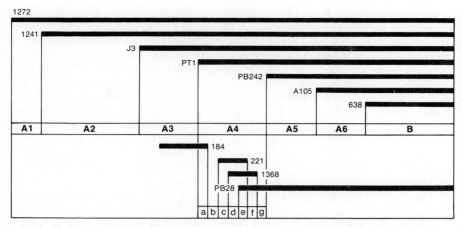

Figure 15-13.   A deletion map of the *rII* locus in T4 phage. After Benzer, *Proceedings of the National Academy of Sciences, U.S.,* v. 47, p. 403, 1961.

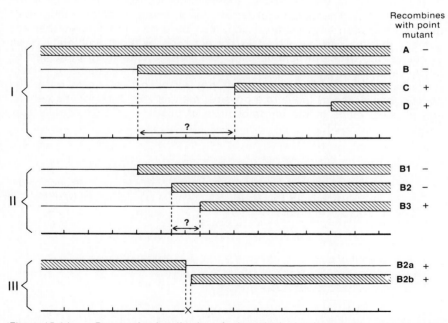

Figure 15-14.   Progressive localization of a hypothetical point mutation using a series of known deficiencies.

above the reversion rate, the mutation site resides within the deletion. In practice, a particular mutation site is placed in a map by testing for recombination between the site and a series of ordered deletions. Again, we are not concerned with the frequency of recombination (as long as it is above the reversion rate); we wish only to know whether recombination produces wild-type progeny. By identifying the deletions that allow the production of wild-type recombinant progeny and those that do not, we find the approximate location of a mutation site. It is then necessary, in order to complete the map, to determine the order of the mutation sites at a particular location.

The unambiguous ordering of mutation sites requires the use of three-point crosses.

We have already discussed the problems of ordering mutation sites using only pair-wise crosses between organisms differing at two pairs of alleles. These difficulties are eliminated if the cross includes a genetic difference at a third site. With only two mutation sites we cannot determine whether one site is to the left or to the right of the other. If we introduce a difference at a third site, say, near to but to the right and outside of the gene being mapped, it is then possible to determine the order of the two sites within the gene (Figure 15-15). If mutation site a_2 resides to the left of site a_1 the wild-type recombinants will carry allele B^- of the *flanking marker*, that is, the nearby site outside of the gene being studied. In contrast, if site a_2 is to the right of a_1, the wild-type recombinants will carry allele B^+ of the flanking marker. Thereby, the order of sites a_1 and a_2 is determined. Given that two mutation sites within a gene can be ordered, it follows that any number of sites can be placed in correct order without ambiguity in an extensive fine-structure genetic map. The fine-structure map produced by Benzer and his co-workers (Figure 15-2) is an excellent example of cartography within a gene. Ironically, it has not yet been possible to relate the genetic map of the *rII* locus to a physical map as was done for the genetic and physical maps of the chromosomes of phage λ and *Drosophila* (pp. 202–203 and 229–232).

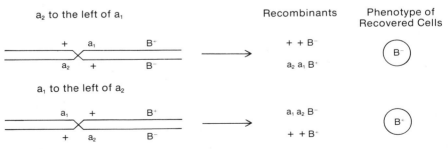

Figure 15-15.   Unambiguous ordering of two mutation sites within a gene using a three-factor cross.

A fine-structure map can be compared to a physical one; the order of mutation sites in the tryp A gene is colinear with the order of amino acid substitutions in the tryp A polypeptide.

An indirect comparison between a fine-structure map and a physical one was accomplished in the laboratory of Charles Yanofsky by utilizing one of the two genes, gene A, that encode the two polypeptides of tryptophan synthetase in *E. coli*. Tryptophan synthetase, we know to be the last enzyme leading to the synthesis of tryptophan. Mutants in the A gene are deficient in tryptophan-synthetase activity and are unable to grow without the amino acid in the culture medium. Consequently, rare (+ +) recombinants involving two different alleles of the A gene can be identified as tryptophan proto-trophs. Yanofsky's recombinational study utilized a generalized transducing phage (Plkc) that carried tryptophan DNA from a bacterial strain with one gene A allele to a second strain with a different allele. Recombination between the mutation sites can result in the production of a tryptophan proto-troph capable of growing in the absence of tryptophan. Through deletion mapping and three-factor crosses, Yanofsky and his colleagues were able to produce a fine-structure map of the tryp A gene in which the mutation sites were placed in an unambiguous order (Figure 15-16).

Yanofsky and his colleagues were working with a gene with a known polypeptide product. They were able to determine the amino acid sequence of the tryp A polypeptide and to recover a group of base substitution mutations that both eliminated tryptophan synthetase activity and caused amino acid substitutions in the tryp A polypeptide. These mutants are placed in an unambiguous order in the genetic map in Figure 15-16. Yanofsky also determined the positions of the amino acid substitutions in the tryp A polypeptide corresponding to each of the mutations, permitting a comparison of the order of the mutation sites in the genetic map with the order of the amino acid substitutions in the polypeptide (Figure 15-16). We notice two condi-

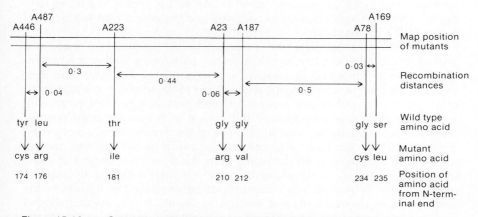

Figure 15-16.   Colinearity of the fine structure genetic map and the amino acid sequence of the A protein of tryptophan synthetase. After Yanofsky et al., *Proceedings of the National Academy of Sciences, U.S.*, v. 57, p. 296, 1967.

tions: 1) the linear order of the mutation sites is the same as the linear order of the amino acid substitutions; 2) the relative distances between the amino acid substitutions and the mutation sites are similar. The genetic map is colinear with the structure of the polypeptide. We know already from our considerations of transcription and translation in prokaryotes that the order of the amino acids in a polypeptide results from the nucleotide sequence in the mRNA, which is a faithful transcript, base by base, of the sequence of the DNA (no pesky intervening sequences being present to complicate matters). Changes in base compositions in the DNA, and hence in the RNA, should be reflected by corresponding changes in the amino acid sequence of the tryp A polypeptide. Thus, the colinearity of the fine-structure map with the positions of the amino acid substitutions in the polypeptide confirms the validity of the fine-structure mapping procedures. At the same time, the colinearity also provides a genetic demonstration that a unique sequence of bases in the DNA corresponds to a unique amino acid sequence.

Fine-structure maps have been constructed for some genes in eukaryotes.

The fine-structure maps considered so far involve prokaryotic systems in which large numbers of progeny can be routinely produced and rare recombinants detected by selective techniques. Fine-structure mapping has also been accomplished in eukaryotes, particularly in yeast. Fred Sherman and his associates have produced a fine-structure map for the gene encoding iso-cytochrome c, which is colinear with the polypeptide. This is a particularly interesting example, because the entire base sequence of this gene and some of its surrounding regions have recently been elucidated by Michael Smith and his colleagues. Thus, we can compare the base sequence, the genetic map, and the structure of the polypeptide and the positions of the amino acid substitutions. The iso-cytochrome c gene does not contain any intervening sequences so that the order of bases in the DNA and mRNA are colinear. The order of mutation sites as placed by genetic mapping is in accord with the base sequence of the gene and the amino acid sequence of the encoded polypeptide.

In multicellular animals, the construction of fine-structure maps has proved possible in Drosophila melanogaster, particularly where rare recombinants can be identified by selective techniques. One extensive map is of the X-linked rudimentary wing (r) locus. The r locus encodes the first three enzymes in the synthesis of pyrimidines and, hence, r flies are pyrimidine auxotrophs. Embryos formed from the mating of homozygous r/r females and r males are unable to develop because of the internal absence of pyrimidines. But if the female is heterozygous for two different r alleles, recombination can produce a chromosome with the + + composition at the r locus. The resulting r/+ female embryos or +/Y male embryos are capable of developing. Therefore, only recombinants between the two mutation sites are capable of development and emerge as adults, thus providing a rapid and sensitive screen for the presence of recombinant progeny in thousands of non-recombinant embryos. The resulting fine-structure map produced by Peter Carlson is pictured in Figure 15-17 along with the positions of the

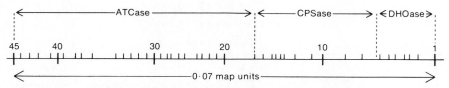

Figure 15-17.   A fine structure map of the *rudimentary wing* locus in *Drosophila* encoding a multifunctional polypeptide. ATCase, acetyltranscarbamoylase; CPSase, carbamoylphosphate synthase; DHOase, dihydroorotase. After Carlson, *Genetical Research,* v. 17, p. 53, 1971, and Rawls and Fristrom, *Nature,* v. 255, p. 739, 1975.

regions that code for the different enzymatic activities. It is likely though not yet proved that the *r* locus encodes a trifunctional polypeptide containing the three different enzyme activities.

POSTMEIOTIC SEGREGATION AND GENE CONVERSION

Mutant sites included in heteroduplexes formed during recombination produce mismatches in the pairing of the two strands.

The model of recombination introduced earlier (Figure 7-1) involves the formation of a region of heteroduplex DNA in which a strand of DNA from one chromosome pairs with its complement on the other chromosome. When mutation sites are distant, the heteroduplex region will probably fall between the two mutation sites. But when the mutation sites are physically close (within a few hundred nucleotide pairs) and recombination occurs between them, it is likely that one site or the other or both will be included in the heteroduplex. Because the base sequence at the site on one strand is different from the other, a mismatch in base pairing must occur (Figure 15-18). The sections that follow describe two ways in which the presence of the mismatch in the heteroduplex can be detected genetically.

Mismatches in heteroduplexes can be detected genetically as postmeiotic segregants.

In discussing transformation (pp. 263–264) we stated that mismatched base pairs are present in the heteroduplex that forms between the strand of transforming DNA and the complement in the host chromosome with which it pairs. These mismatches are eliminated after the next round of replication when each strand serves as a template for the synthesis of its true complement. A mismatch in a heteroduplex region can also be eliminated during the first round of replication following recombination. In haploid organisms, such as *Neurospora* and *Chlamydomonas,* and during the haploid phase of the yeast life cycle, where four haploid spores are produced by meiosis, the presence of mismatched heteroduplexes can be detected. Recombination

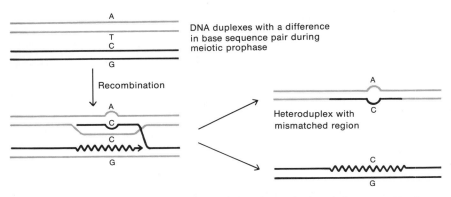

Figure 15-18. The formation of a mismatch in a heteroduplex during recombination.

occurs during a stage of meiosis when four chromatids are present, following which the four haploid spores are produced without any further DNA replication. If a mismatched heteroduplex from recombination is present, then the replication of DNA for the first mitotic division will create two different chromosomes, each with a different DNA sequence where the heteroduplex was present (Figure 15-18). During the first mitotic division the two chromosomes will separate to different daughter cells. Cells descended from one of these daughter cells will be of one genotype, those from the other, of a different genotype. This process, *postrecombinational* or *postmeiotic segregation,* can be detected in *N. crassa* where four spore pairs are present in an ascus as a result of two meiotic divisions and a single subsequent mitotic division. We can see whether two spores in a spore pair differ genetically. For example, in crosses involving spore-color alleles one member of a spore pair might be colored and the other uncolored (Figure 15-19). Postrecom-

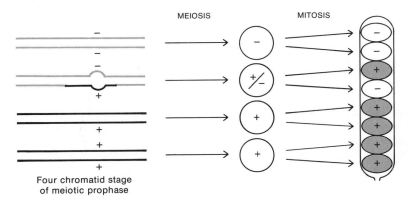

Figure 15-19. Post-meiotic segregation. A mismatch in a heteroduplex is eliminated by replication and segregation during the first mitotic division following meiosis.

binational or postmeiotic segregation can be detected in yeast because half of one colony derived from a spore will be of one genotype and half of another genotype. Thus, again we can see alleles segregating during a mitotic division immediately after meiosis. From this evidence we would infer that recombination during meiosis created a mismatched heteroduplex in the region of the gene being studied. In a phage cross, however, postrecombinational segregation cannot be recognized. For example, in a cross of different *rII* alleles, only + + recombinants can be detected. Such + + phages can result from postrecombinational segregation, but we have no way of knowing because we cannot recover the products of any *particular* recombination.

Mismatches in heteroduplexes can be repaired prior to replication and cause gene conversion.

In the preceding example a mismatch in the heteroduplex is eliminated during the next round of replication. From our discussion on mutation (pp. 398–400) we recall that mismatches in DNA can be repaired by local excision of one of the strands of the double helix, followed by synthesis of a new polynucleotide segment using the remaining strand as a template. Mismatched regions created during recombination can also be repaired. For example, in a cross involving two alleles with different mutation sites (Figure 15-20), a heteroduplex including one mutation site is formed with a mismatched base pair (AC). The strand containing the C is derived from one chromosome and that with the A from the other chromosome. The strand containing the A is excised, resulting in the presence of a GC base pair in this chromosome as well as in its homolog. The DNA composition in the two homologs at this site is now identical although originally it was different. *Gene conversion* changed one site into the other. In association with repair

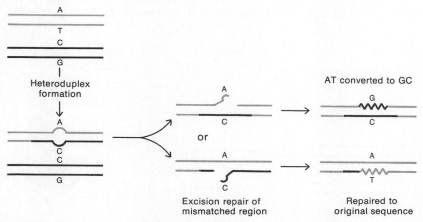

Figure 15-20.   Gene conversion. A mismatched heteroduplex is repaired prior to meiosis.

of the mismatch, the recombinational process continues, resulting either in recombination or in its absence (Figure 7-1), recombination being detected by using flanking markers. Thus, gene conversion can occur with or without detectable recombination of outside markers.

Gene conversion can be detected genetically because of abnormal segregation ratios in tetrads.

The inclusion of a mutation site in a heteroduplex and the repair of the mismatch can be detected genetically. In organisms, such as yeast and *Neurospora,* from which the products from a single meiosis can be recovered and analysed, gene conversion leads to abnormal numbers of meiotic segregants as demonstrated by the work of Mary Mitchell in 1954. Mitchell crossed two strains carrying different alleles, *pdx* and *pdxp*, at the pyridoxine locus in *N. crassa*. In this case the double mutant (pdx pdxp) can be recognized. By examining spore pairs within single asci produced from the cross *pdx +× + pdxp*, she identified many asci in which the *pdx* and *pdxp* mutants and their respective wild-type counterparts were recovered in the expected 2:2 ratio. However, exceptional asci were recovered in which mutant and wild-type alleles were present in 3:1 or 1:3 ratios (Figure 15-21). It appeared as though one mutation site had been converted into its wild-type counterpart. Indeed, these aberrant ratios are the result of gene conversion. In a tetrad, if only two of the chromatids are involved in recombination, the genetic constitutions of the genes on the chromatids not involved in recombination will remain unchanged, one mutant and one wild type. In chromatids undergoing recombination in which a heteroduplex has been formed gene conversion can occur to produce two chromosomes, each carrying a wild type of mutation site. This will result in three chromosomes (or three spore pairs) with one allele and one chromosome with another allele.

Gene conversion is a common result any time a mismatch occurs in a heteroduplex formed during recombination. It is detected genetically only where the altered base sequences result in a detectable phenotypic differ-

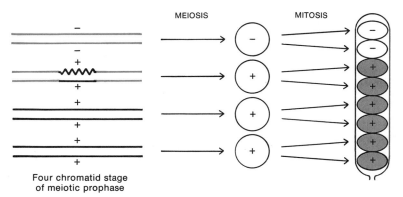

Four chromatid stage
of meiotic prophase

Figure 15-21.   Gene conversion. The repair of a mismatch in a heteroduplex is revealed by 3:1 or 1:3 ratios in meiotic segregants.

ence. Although gene conversion is readily recognized only where the recovery of four products from a single meiosis is possible, it undoubtedly exists in all organisms that undergo recombination and are capable of repair.

The detection of gene conversion demonstrates that during meiosis the mutation site of an allele under study was included in a heteroduplex. Analyses of gene conversion and postmeiotic segregation have allowed geneticists to study the mechanisms of recombination and have been instrumental in the formation of models of recombination, including the one presented in Figure 7-1. A description of how the analysis by Seymour Fogel of postmeiotic segregation has contributed to the development of the model for recombination follows.

The absence of postmeiotic segregation in more than one spore in a yeast tetrad requires the presence of an asymmetrical heteroduplex.

Postmeiotic segregation results from the production of two genetically different chromatids by replication of a DNA double helix containing a mismatch in a heteroduplex. During recombination heteroduplexes are formed asymmetrically on only one chromatid (Figure 15-18). If a heteroduplex is formed on only one chromatid, an asymmetric heteroduplex, then we expect to find tetrads in which only one spore undergoes postmeiotic segregation (Figure 15-19). In contrast, if heteroduplex formation were symmetric, then when a mismatch occurred in one heteroduplex, there would also be a mismatch on the other heteroduplex. Hence, two mismatches would be produced at one time, and we would find a significant number of tetrads in which postmeiotic segregation occurs in two spores. In yeast, tetrads in which postmeiotic segregation occurs in two spores are virtually non-existent. Therefore, we can conclude that heteroduplex formation in yeast is highly asymmetric. In contrast, tetrads containing two spores that undergo postmeiotic segregation occur in another fungus, *Ascobolus.* In *Ascobolus,* we deduce that two heteroduplexes are formed, but one is shorter than the other. Tetrads in which two spores are involved in postmeiotic segregation are found but at much lower frequencies than expected if heteroduplex formation were symmetric. Thus, from the frequency of postmeiotic segregation we can deduce aspects of the mechanism of recombination. In this particular instance, the physical proof is yet to come. Although there is physical proof for heteroduplex formation, there is no physical demonstration of whether the heteroduplexes are formed symmetrically or asymmetrically.

SUMMARY

Genetic analysis can be extended to dissect the structure of genes and to provide insight into events occurring at the level of the DNA. Complementation tests in which two mutations are introduced into a single cell or organism indicate whether the mutation sites reside within the same gene.

By detecting the presence of allelic complementation we can tell whether encoded polypeptides exist in cells as multimers or monomers. We can even determine if a gene encodes a multifunctional polypeptide. Fine-structure mapping, including deletion mapping makes it possible to determine the order of mutation sites within a gene and in conjunction with complementation tests to identify the genetic boundaries of a gene. Finally, by analyzing postmeiotic segregation it is possible from genetic perspectives alone to deduce that recombination in yeast involves the formation of asymmetric heteroduplexes.

Structure and Function of Genomes: Prokaryotes

Prior to 1950, organisms were traditionally separated into two major kingdoms, the plant kingdom and the animal kingdom, with bacteria included among the plants. We now realize that in regard to structure and function organisms are more correctly divided into prokaryotes and eukaryotes. Eukaryotic plants are functionally more closely related to animals than either are related to prokaryotes, such as bacteria or blue-green algae. When we consider the functional and structural organization of genomes, we find that various prokaryotes share characteristics that differ in some aspects from those of eukaryotes. For example, all prokaryotic chromosomes lack histones, in contrast to all eukaryotic chromosomes. Furthermore, the viruses that parasitize the cells of prokaryotes and eukaryotes and utilize the synthetic machinery of the cells of their hosts have the same functional organization as their hosts. Therefore, the genomes of phages function like those of prokaryotic cells, and those of eukaryotic viruses function like eukaryotic genomes.

THE GENOME OF ØX174

The entire nucleotide sequence of the chromosome of ØX174 has been determined.

The bacterial phage ØX174 has been extensively investigated because its small genome size, embodied in a small chromosome of single-stranded DNA, makes it particularly accessible for both genetic and molecular studies. We have, for example, already used it to demonstrate that only one strand of a particular genetic region is transcribed (pp. 333–334). We argued that it would be evolutionarily unadaptive for one region to encode two different proteins because mutational changes that might be advantageous for one protein would probably be disadvantageous for a second protein encoded by the same region. We shall see now, to our embarrassment, the limitations of this argument.

We know ØX174 has nine proteins, encoded by nine genes, A–H and J. The small size of the genome of ØX174 made it an attractive subject for techniques that determine base sequences in DNA. In 1977, Fred Sanger and his colleagues reported the entire base sequence—all 5,375 nucleotides—of the ØX174 chromosome. The results of their study, coupled with the known amino acid sequences of some of the polypeptide products, allow detailed insight into the organization of a prokaryotic genome.

ØX174 has overlapping genes.

The determination of the base sequence of the ØX174 chromosome confirmed that only one strand, the one made after infection of host bacteria, is used as a template for the synthesis of mRNA. Therefore, the direction of transcription must be the same for every gene. There is one outstanding surprise, however, in the functional organization of the ØX174 genome. Even before the sequencing studies were completed, geneticists realized that a genome containing about 5,400 nucleotides encoded polypeptides containing approximately 1,900 amino acids and requiring 5,700 nucleotides for coding. It appeared more nucleotides were required for coding than were available. The sequence data, as well as some genetic data, provide an explanation for the dilemma. There are two pairs of overlapping genes in the ØX174 genome (Figure 16-1). Gene B is entirely included within gene A, and gene E resides within gene D. Also, the initiating methionine codon for gene J uses the final A in the terminating codon of gene D. These overlaps are possible because the different sequences are translated by use of different reading frames. For example, as depicted in Figure 16-2, gene A starts

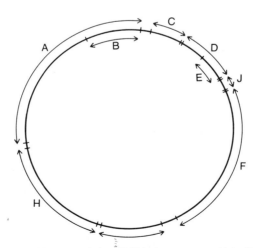

Figure 16-1.   The genetic map of the φX174 chromosome. Note that gene B is within gene A; E within D. After Sanger et al., *Nature*, v. 265, p. 687, 1977.

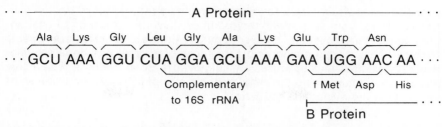

Figure 16-2.   The initial nucleotide sequence of gene B within gene A in a single ϕX174 mRNA. After Sanger et al., *Nature*, v. 265, p. 687, 1977.

with nucleotide number 1, and gene B with nucleotide number 972. Nucleotide number 972 is the third nucleotide in a GAA (glu) codon for residue number 324 of the A protein *and* is the first base in the initiating AUG (meth) codon of the B gene. Hence, the two polypeptides are formed with different reading frames. Upstream from the initiating codons for both the B and A genes, we can also detect the sequences complementary to the 3' end of the 16S rRNA, which are necessary for initiation of polypeptide synthesis:

```
Gene B:      5' A G G A G C T A A A G A A T G
Gene A:      5'   G G A G G C T T T T T A T G
16S rRNA:  A U U C C U C C A C U A G . . . 5'
```

The regions of complementarity between the mRNA's and the rRNA are enclosed in the box, and the ATG sequence corresponding to the initiating AUG codon is italicized. The sequences necessary for initiating polypeptide synthesis are demonstrably present for both genes. Keep in mind, however, that the gene B sequence, which interacts with the rRNA, also codes for a segment of the gene A polypeptide. A promoter sequence, the region where RNA polymerase binds in preparation for transcription, has only been identified upstream from the start of the A gene. Thus, only one transcript of the region is made. That transcript contains the appropriate nucleotide sequences for the initiation of synthesis of two different polypeptides. In other words, a single mRNA transcript can be used to synthesize two polypeptides.

Overlapping genes restrict the evolution of the shared nucleotide sequences.

In those regions of the ØX174 chromosome where two genes share a common nucleotide sequence, there are restrictions, in addition to those that usually exist for polypeptides, on the types of mutational changes that can be tolerated. For example, the region of the A gene that is necessary for the initiation of translation of the B gene (the AGGAG sequence shown here) must be evolutionarily conserved if translation of the B gene is to be retained. And the ATG sequence, corresponding to the AUG initiating sequence needed for initiation of synthesis of the B polypeptide, cannot be changed. Finally, base substitutions in the overlapping coding regions may result in amino acid substitutions in two different polypeptides (a condition that allowed the genetic detection of the overlaps, that is, single base substitutions changed two polypeptides), thereby compounding the effect of base substitutions. Thus, from an evolutionary point of view, the existence of overlapping genes is unexpected. The overlapping regions are tolerated presumably because the encoded amino acid sequences of at least one gene are relatively unessential and because, for unknown reasons, there is a selective advantage in maintaining the small genome size of ØX174 at the expense of flexibility in the nucleotide sequence.

Prokaryotes generally devote most of their genomes to structural information.

In ØX174 about 95 per cent of the genome is used to encode structural information. This is typical of both prokaryotes and eukaryotic viruses; in both the vast majority of sequences are devoted to encoding structural information. Thus, if we know the average size of a gene, we can estimate the number of genes present from the size of a prokaryotic genome. Often a gene is taken to be about 1,000 base pairs long, sufficient to encode a polypeptide containing 333 amino acids. Using this figure as an average, we estimate that the number of genes present in prokaryotes ranges from less than 10 in small phages up to several thousand in bacteria. For example, we estimate that *E. coli* has about 4,000 genes based on its genome size. Where extensive genetic data are available, we find that the number of genes in a genome estimated from genetic data agrees well with the number estimated on the basis of genome size.

BACTERIAL OPERONS

Some regions of prokaryotic genomes are devoted to regulatory functions.

In ØX174 about 300 nucleotides, about 5 per cent of the total, do not encode structural information. Some of them make up promoter regions, which like all promoters perform two functions: They bring DNA-dependent RNA polymerases into the correct position for the transcription of structural genes, and they determine how frequently a gene is transcribed by determining how frequently RNA polymerase binds. The promoter sequences for genes A and D in ØX174 contain the sequences TTTCATG and TATCTGA, respectively, in reasonable agreement with the Pribnow sequence, TAT-purine-AT-purine, found in promoters in other systems (p. 339). Note that the Pribnow sequence here is given for the opposite strand of DNA to those in Table 12-1. The difference in sequence between the two phage promoters may affect the efficiency with which polymerases bind. One promoter may bind more polymerase molecules over a given period of time than the other with the result that one region is transcribed more than the other. In fact, base substitutions within promoters are known to greatly influence the number of transcripts from a given region. Differences in the efficiency with which polymerases bind to promoters may reflect different physical requirements. For example, more mRNA will be needed for a phage-coat protein than for an enzyme to lyse cells.

Other regulatory regions are present in prokaryotic genomes in addition to promoters. One such region, an *operator*, is involved in the transcriptional changes that bacteria make in response to changes in their nutritional environment. Operators are characteristic of genes that are subject to regulation, that is, are turned "on" and "off," in contrast to *constitutive* genes that encode proteins essential for growth and are always "on."

Bacteria cope with changing environments through physiological adaptation.

A chart showing the metabolic conversions of compounds demonstrates that a variety of different compounds can be used as sources of energy for growth. Bacteria can use both five- and six-carbon sugars, monosaccharides, and disaccharides as well as such other compounds as glycerol as sources of energy. These compounds, *catabolites,* are broken down to provide energy for the bacteria. The bacteria must contain the appropriate enzyme system for the metabolism of the substance to be able to use a particular compound. There are specific enzyme systems for the utilization of the five-carbon sugar, arabinose, others for the utilization of six-carbon sugars, such as glucose and galactose, and still others that initially break down linked sugars into their monosaccharide parts. The enzymes for metabolizing these different substances can be provided in two ways. A bacterium could constantly synthesize all of the enzymes or else could activate enzyme synthesis only as necessary to metabolize whatever particular catabolite happens to be present. In the second case, the cell adapts physiologically to its changing environments.

If a cell continually synthesizes all of its enzymes, it can manufacture comparatively few molecules of each type because its available synthetic capabilities are fragmented among all of the different enzymes. If a cell adapts to the presence of a new nutrient by preferentially synthesizing only those enzymes needed to metabolize that nutrient, it can make a comparatively large amount of the enzyme and, thereby, increase its efficiency in utilizing the nutrient.

From an evolutionary perspective, cells that live in changing nutritional environments and can adapt physiologically to the changes will have an advantage over those that cannot. When a new nutrient becomes available, a cell that synthesizes all its enzymes continually will be able to respond immediately but inefficiently with increased growth. A cell that adapts to nutritional changes will not be able to utilize the nutrient immediately. There will be a *time lag* while it senses the presence of the new component and starts preferentially to synthesize the enzymes needed to metabolize it. But once the cell adapts, it will be able to utilize the catabolite more efficiently than cells that cannot adapt. Eventually, it will outgrow and replace the other cells. If we accept the general argument about the selective value of physiological adaptation—without forgetting the shortcomings of such arguments—we should not be surprised to find that bacteria adapt physiologically. Particularly good examples of adaptation are provided in *E. coli*, the common gut bacterium, which, as a function of its host's eating habits, can go from feast to famine. The classic example of adaptation is the response of the bacterium to the presence of the disaccharide, lactose, in the medium in which it is growing.

In *E. coli* the genes for the three enzymes involved in lactose utilization map together at the *lac* locus.

Lactose is a disaccharide composed of galactose and glucose, which are joined together via a β-linkage between the 1 carbon of galactose and the 4 carbon of glucose (Figure 16-3). In other words, lactose is a β-galactoside. In order for lactose to be utilized it is first broken down into its constituent parts, glucose and galactose, either of which can be further catabolized to produce energy.

E. coli has three proteins that are involved in lactose metabolism (Figure 16-4). One, β-galactoside permease (or just galactoside permease), has a molecular weight of about 30,000 D (252 amino acids), is located in the cell membrane, and transports β-galactosides such as lactose into the cell. A second protein, β-galactosidase, cleaves the disaccharide into glucose and galactose residues. β-galactosidase is a tetramer, composed of four identical polypeptides each with a molecular weight of about 132,000 D (1,173 amino acids). The amino acid sequence of β-galactosidase has been entirely elucidated. The third enzyme, β-galactoside transacetylase, adds an acetyl group to the hydroxyl at the 6 position in the galactose molecule. The transacetylase is a dimer, with two identical polypeptides each having a molecular weight of about 32,000 D (268 amino acids). The physiological significance of the transacetylase is not understood.

Mutants have been isolated that each eliminate one of the three enzymes involved in lactose metabolism but do not affect the other two. Mapping experiments using Hfr and F⁻ strains demonstrate that each of the mutants maps to a single locus at 10 mins on the *E. coli* chromosome (Figure 10-15).

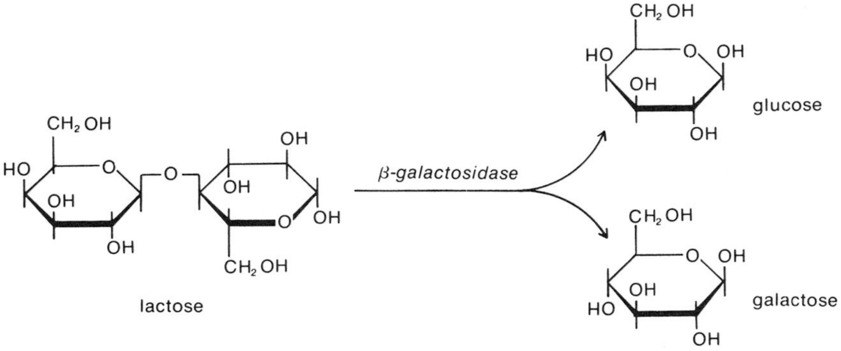

Figure 16-3.   The structure of lactose and its cleavage by β-galactosidase.

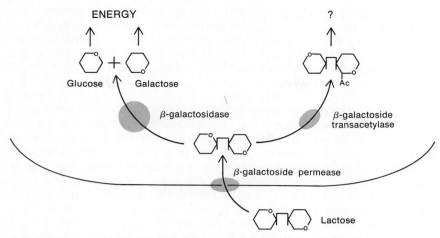

Figure 16-4.   The functions of the enzymes encoded at the *lactose* locus.

Further mapping experiments using transduction revealed that mutants lacking permease activity map between those lacking β-galactosidase and transacetylase activity. Hence, we have a picture of the *lactose* (*lac*) locus with three structural genes, abbreviated *z*, *y*, and *a*, respectively, encoding β-galactosidase, galactoside permease, and galactoside transacetylase (Figure 16-5).

Normal bacteria adapt to lactose in the culture medium by coordinately increasing the synthesis of the three *lac* enzymes, a process called enzyme induction.

When *E. coli* is grown in a medium that does not contain a β-galactoside, only low, *uninduced* levels of the three enzymes are present. When lactose or another β-galactoside is added to the medium, the level of β-galactosidase increases markedly as do the levels of the other two *lac* proteins. This

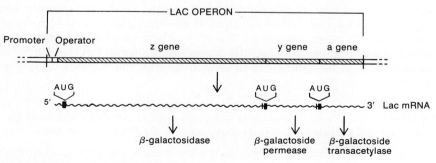

Figure 16-5.   The *lactose* operon.

process, in which levels of enzymes increase markedly in response to a compound in the medium, is called *enzyme induction*. The molecule that causes an increase in enzyme levels, in this case a β-galactoside, is the *inducer*. After the addition of the inducer, there is a lag before the levels of the enzymes start to increase, followed by a period during which the levels of the enzymes increase linearly.

Several explanations are possible for the induced increase in the level of activity of the three *lac* enzymes. One maintains that inactive forms of the enzymes are already present in the cells and are then activated in response to the inducer. An alternative explanation holds that the three enzymes are newly synthesized. Experiments performed in the 1950's demonstrated that the increase in enzyme activity was in fact the result of new synthesis of the three different *lac* proteins. Furthermore, the synthesis of the *lac* proteins is coordinated, the relative amount of synthesis being the same under different conditions. In cells that are weakly induced, making only a comparatively small amount of *lac* enzymes, the molar ratio of β-galactosidase to trans-acetylase, for example, is about 4.5:1. In cells that are strongly induced, with large amounts of the proteins being made, the ratio is the same. A mechanism exists by which synthesis of the three enzymes is coordinated. In strongly induced cells the increase is marked, going from about 10 to 15 molecules of β-galactosidase per cell (there are always a few molecules around) to several thousands.

Jacob and Monod formulated the "operon" model to explain the coordinate induction of the *lac* enzymes.

The basis of the coordinate induction of the *lac* enzymes was intensively studied by Francois Jacob and Jacques Monod and their collaborators at the Pasteur Institute in Paris. They produced and recovered mutants that affected the ability of *E. coli* to utilize lactose as a catabolite. Among the mutants they discovered two interesting types that they could attribute to point mutations (mapping as though only one or a few base pairs were affected). One type of these mutations eliminated the presence of all three enzymes; the other type caused the continual or constitutive synthesis of the three enzymes, that is, the *lac* enzymes were synthesized in the presence or absence of inducer. Based on the analysis of their mutants and physiological data Jacob and Monod proposed in 1960 a model for the organization and regulation of function of the *lac* locus involving a genetic unit of function and regulation called the *operon*. We will first present an up-to-date version of the model and then consider the properties of the mutants that led to the development of the model. Historically, of course, the mutants came first, the model followed.

A single mRNA, a *polycistronic* or *polygenic mRNA*, is transcribed from the *lac* locus and contains the coding sequences for the three *lac* enzymes.

According to the operon model, the *lac* locus is transcribed as a single unit (Figure 16-5). The mRNA produced, a *polycistronic* or *polygenic* mRNA, contains, without overlaps, the coding sequences for the separate synthesis of the three *lac* polypeptides. There is a start codon and a stop codon in the mRNA for each of the polypeptides. The promoter for the synthesis of the mRNA is located next to the *z* (β-galactosidase) gene (Figure 16-5). The *z* gene is transcribed first, the *y* gene second, and the *a* gene last. In the translation of the polycistronic mRNA, the ribosomes can translate any of the three coding sequences. A polycistronic mRNA containing the coding sequences for all three *lac* enzymes has been isolated from *E. coli* cells. From such an mRNA we might expect that the three polypeptides would be produced in a 1:1 molar ratio instead of the 4.5:1 ratio found for β-galactosidase and galactoside permease. We must assume that there is preferential initiation and translation of the β-galactosidase coding sequence or preferential survival of different regions of the mRNA. In other polycistronic mRNA's that have been identified, for example, an mRNA transcribed from the *galactose* locus containing coding sequences for 3 polypeptides and one from the *tryptophan* locus containing coding sequences for 5 polypeptides, the different polypeptides are produced in equal molar amounts.

Regulation at the *lac* locus is effected by a protein, the *lac* repressor that binds to a regulatory region, the *operator*, located between the promoter and the *z* gene.

Between the promoter, the site where polymerase binds, and the *z* gene, a region called the *operator* is located. The operator and promoter regions are sometimes referred to as genes. As noted before, we are reserving the term "gene" for sequences that encode structural information. The operator is a region where a regulatory protein, the *lac repressor protein*, binds (Figure 16-6). In the absence of an inducer, the regulatory protein stays bound to the operator region and interferes with the binding of the polymerase to the promoter, preventing the transcription of the locus. When present the inducer binds to the regulatory protein and causes it to dissociate from the operator, thereby permitting transcription of the three *lac* genes to take place. Originally an operon was defined as an operator region plus the structural genes next to it and under its control. Now the promoter is included within the boundaries of the operon as well. The operon is a unit of genetic function and regulation, and the operator and the promoter determine, respectively, the actual occurrence of transcription and the degree of transcription.

The *lac* repressor, the protein that regulates transcription of the *lac* operon, is encoded by the *i* gene, which maps at the *lac* locus to the left of the operon (Figure 16-6). The *lac* repressor protein is synthesized constitutively at very low levels in normal cells. The *lac* repressor protein can exist in either of two conformations. In one conformation, in the absence of inducer, it binds to the operator and inhibits transcription. In a second conformation, when it binds an inducer, the repressor no longer binds to the operator region. The binding of the inducer to the protein causes the repressor to undergo a change in conformation, an *allosteric transition*, so that it no longer binds to *lac* operator DNA.

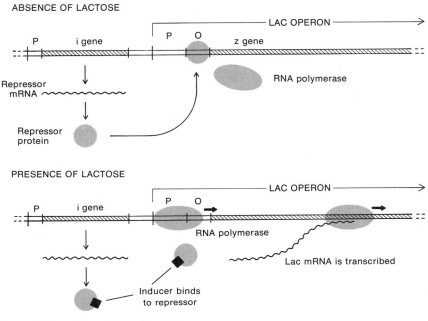

Figure 16-6.   Regulation of the lactose operon.

The properties of the repressor and the *lac* operon effect the induction and coordinate synthesis of the enzymes specified by the *lac* locus. In the absence of a β-galactoside inducer little mRNA for the three enzymes is synthesized. Transcription is said to be *repressed.* When inducer is added and binds to the repressor protein, causing it to dissociate from the operator, the polycistronic mRNA is synthesized. Transcription is *derepressed.* Because the coding sequences for the three polypeptides are all in the same mRNA, there is coordinate induction of synthesis of the three *lac* enzymes. When the inducer is removed from the medium or used up through metabolism, no inducer is available to bind to the repressor, which, again, is free to bind to the operator. Synthesis of the *lac* mRNA ceases and so, subsequently, does the synthesis of the three enzymes (bacterial mRNA's have short life spans).

Properties of *lac* mutants led to the formulation of the operon model of regulation.

   1. *Operator mutations.* We have noted already that at the *lac* locus point mutations were recovered that caused the constitutive synthesis of the three *lac* proteins. It is not difficult to account for this condition if we imagine a mutant in the operator region, a base substitution, or even a small deletion that eliminates the binding of the repressor. Such mutants, *operator consti-*

tutive mutants or O^c mutants, map adjacent to the z gene, in what we have defined as the operator region. These mutants, because the repressor no longer binds, allow the constitutive synthesis of the *lac* mRNA and enzymes. Such O^c mutants are, like the promoter mutations discussed previously (p. 414), cis-acting. In a diploid condition (Figure 16-7), an O^c mutant affects the transcription of the adjacent structural genes on the same chromosome but has no effect on the transcription of the *lac* structural genes on the other chromosome. In a partial O^c/O^+ diploid, created through the use of an F' *lac* plasmid, and in a system lacking inducer, transcription of the *lac* genes will occur using only the DNA carrying the O^c mutation as a template. The O^+ region will bind *lac* repressor so that transcription will not occur.

The cis-acting nature of the O^c mutation was demonstrated by the construction of a partial diploid in which one chromosome carried an O^c mutation along with a z gene that encoded an electrophoretic variant form of β-galactosidase. The other chromosome carried a normal operator and normal z gene. In the absence of an inducer only the electrophoretic variant form of β-galactosidase was synthesized, demonstrating the cis-acting nature of the O^c mutation. The O^c mutation is dominant; in the O^c/O^+ cell there is constitutive synthesis of the *lac* enzymes in the absence of inducer.

2. *Promoter mutations.* Mutations that affect the transcription of the operon can also occur in the promoter region. Promoter mutations can be fully repressed; that is, in the absence of inducer there is little or no transcription of the *lac* genes. There can be mutations that completely eliminate transcription—polymerase no longer binding—or that increase or decrease transcription—polymerase binding more or less efficiently. These mutations, like operator mutations, are cis-acting. Genetically, they define the promoter region.

3. *i gene mutations: i⁻ mutants.* According to the model (Figure 16-6), the i gene encodes a protein that binds to the *lac* operator and prevents transcription in the absence of inducer. The elimination of a functional repressor

ABSENCE OF LACTOSE

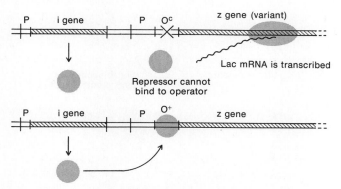

Figure 16-7.   The dominant expression of an O^c mutant in a partial diploid.

through mutation, the i^- mutation, results in an operator that is unoccupied by repressor in the absence or presence of inducer. In a partial diploid cell i^-/i^+, the i^- allele is recessive to the i^+ allele. In the absence of inducer, the repressor protein encoded by the i^+ allele will bind to the operator regions on both chromosomes, and the i^- allele product binds to neither (Figure 16-8). Synthesis of the *lac* mRNA does not occur. It was the recessive nature of i^- alleles in partial diploids, requiring that there be a product of the normal i gene free in the cytoplasm and able to attach to the operator sites on both chromosomes, that led Jacob and Monod to hypothesize the existence of a repressor. The protein nature of the i gene product was demonstrated genetically through the recovery of an i^- allele that was suppressed by a tRNA that reads a terminating codon. Because the action of a suppressor tRNA is limited to protein synthesis, the repressor had to be a protein.

i^s mutants. The binding of an inducer causes the repressor protein to dissociate from operator DNA. Consequently there must be a site or sites on the repressor for binding the inducer and a site or sites for binding to operator DNA. A mutant, i^s, changes the amino acid sequence of the repressor so that it no longer binds inducer but still retains its ability to bind to operator DNA. In such a mutant, the repressor will remain associated with operator DNA whether or not inducer is present. In either case, transcription of the *lac* genes will be blocked.

In a partial diploid, i^s/i^+, in the absence of inducer, both the altered and normal forms of the repressor will bind to operator DNA. When inducer is added to the partially diploid cell, normal repressor will dissociate from operator regions but will be replaced by the altered repressor to which the inducer does not bind (Figure 16-9). Because no induction of the *lac* enzymes occurs, the i^s allele is dominant. The dominance of the i^s allele also provides evidence for the existence of a regulatory substance that is free in the cytoplasm and can bind to operator regions on both chromosomes.

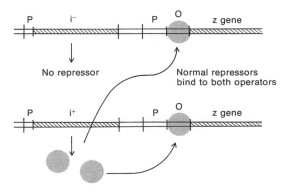

ABSENCE OF LACTOSE

Figure 16-8.   The recessive expression of an i^- mutant in a partial diploid.

PRESENCE OF LACTOSE

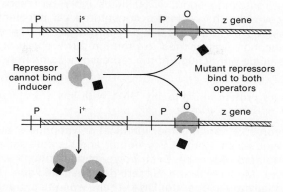

Figure 16-9.   The dominant expression of an i^s mutant in a partial diploid.

In review, we see that the properties of the different mutants, the O^c, i^-, and i^s, are consistent with the model of regulation for the *lac* locus depicted in Figure 16-6. Such consistency is necessary because the model of the *lac* operon in Figure 16-5 was deduced by Jacob and Monod from the properties of these mutants. It is extraordinary that properties the model predicted have all been established as true by subsequent elegant molecular investigations, some of which are described in the following section.

The repressor protein was purified by using mutants with increased amounts of repressor.

The existence of the *lac* repressor was hypothesized because of the properties of the i^- and i^s mutations. One proposed property of the repressor protein was that it be able to bind inducer. The binding of radiolabelled inducer by the repressor was used as an assay for the purification of the receptor. Walter Gilbert and Benno Müller-Hill discovered that the repressor protein constitutes only about 0.002 per cent of the total protein in a normal *E. coli* cell. Therefore, to obtain pure repressor for characterization, a 50,000-fold purification would have been necessary, a significant feat for a protein chemist. To simplify the purification of the receptor, Gilbert and Müller-Hill turned to genetic manipulations. They were able to recover a mutant in the promoter region of the *i* gene, called i^Q (Q standing for quantity), which increases the amount of transcription ten-fold. Subsequently, a second mutation was isolated, called i^{superQ}, which increased the amount of transcription of the *i* gene fifty-fold. The i^{superQ} mutant and its adjacent *i* gene were introduced into a defective λ chromosome that was established as a prophage. The prophage was induced to enter the lytic cycle, resulting in the production of 200 copies per cell of a λ chromosome carrying the i^{superQ}, *i* gene combination. Because the λ chromosome was defective, lysis did not occur. Thus, there was potentially a 10,000-fold increase in the amount of repressor produced (fifty-fold increase in transcription × 200 copies = 10,000). In practice, cells containing about 1,250 times more repressor than normal cells were recovered. The repressor constituted 2.5 per cent of the total protein in these cells (1,250 × 0.002 per cent = 2.5 per cent), and only a forty-fold purification was needed to produce pure repressor. The genetic manipulations made purification of the repressor a comparatively easy task.

The *lac* repressor protein binds to any DNA with low affinity and to operator DNA with high affinity.

Purified repressor contains four identical polypeptide chains, each containing 347 amino acids. Konrad Beyreuther and his colleagues have determined its amino acid sequence. Each polypeptide contains a single site for binding inducer. Müller-Hill and Gilbert demonstrated that the repressor binds preferentially to *lac* operator DNA. These workers extracted DNA from a cell carrying a bacterial chromosome lacking the *lac* region and mixed that DNA with defective λ DNA carrying the *lac* region. The repressor was found to bind to both DNA's but preferentially with the DNA carrying the *lac* operator. If an inducer was added, the repressor did not preferentially associate with the λ DNA. This and other experiments have led Gilbert and Müller-Hill as well as Suzanne Bourgeois and Arthur Riggs and others to the realization that in the absence of inducer, the *lac* repressor has low affinity for any DNA but high affinity for *lac* operator DNA, the affinity being over a thousand-fold greater for *lac* operator DNA. In the presence of inducer, the repressor no longer binds to the operator with high affinity but still binds with low affinity to any DNA. In the cell there is an equilibrium between repressor bound to any DNA and that free in the cytoplasm. In the absence of inducer, the repressor is preferentially associated with *lac* operator DNA.

Other inducible operons have properties similar to those of the lac operon; all are repressed in the presence of glucose.

Other operons besides *lac* are involved in the utilization of catabolites, such as galactose and arabinose. As with *lac*, the genes specifying the enzymes involved in galactose or arabinose metabolism are located together at a single locus. The levels of enzymes are coordinately induced at each locus by galactose and arabinose, respectively. The genetic organizaton and regulation are designed to effect optimal utilization of the different sugars.

In addition to being functionally organized into operons, the *lactose, galactose,* and *arabinose* loci have another property in common, namely, that in the presence of glucose none of the inducers is very effective in eliciting enzyme induction. The basis of the glucose effect, *catabolite repression,* can be understood within the economics of cellular metabolism. Lactose, arabinose, galactose, and glucose, among others, are catabolites used by the cell to produce energy. Yet, of these, glucose is used more efficiently, and mechanisms exist to reduce the expenditure of energy for the synthesis of inducible enzymes if glucose is already present. These mechanisms involve the purine nucleotide cyclic-3′, 5′-monophosphate, cAMP (Figure 16-10).

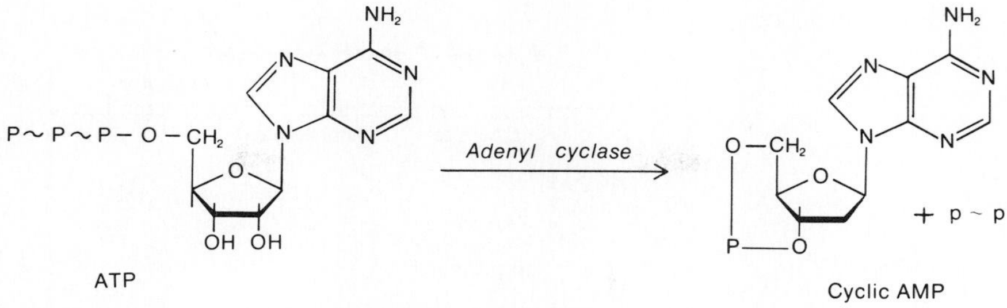

Figure 16-10.   The synthesis of cyclic-AMP (cAMP).

Cyclic AMP is synthesized from ATP by an enzyme called adenyl cyclase. Cyclic AMP is, in turn, degraded by a nuclease to produce AMP. The level of cAMP in a cell is controlled by the relative activities of these two enzymes. In the presence of glucose, the activity of the cyclase decreases, and the concentration of cAMP decreases. When glucose is absent, cyclase activity increases as does the amount of cAMP.

The catabolite activator protein (CAP) binds cAMP and affects the binding of RNA polymerase to the lac promoter.

Now let us consider the effect of cAMP on the *lac* operon. The *lac* promoter has two regions. One region, immediately adjacent to the *lac* operator, is where RNA polymerase binds. A second region between the *i* gene and the polymerase binding site is a site for the binding of the *catabolite activator protein* (CAP) (Figure 16-11), When CAP is bound at this site, the RNA polymerase binds to the polymerase site. When CAP is absent, the polymerase cannot bind. CAP itself cannot bind to the *lac* promoter (or to the *gal* or *arab* promoters) unless cAMP is bound to it. In the presence of glucose, cAMP levels are reduced so that CAP does not bind to the promoter. Because the polymerase cannot bind in the absence of CAP, the *lac* genes are not transcribed. The effect is not limited to the *lac* operon but includes other operons as well, for example, the *galactose* and *arabinose* operons. Thus, superimposed on the individual operons is a regulatory system to assure efficient utilization of nutrients.

There is a basic difference in the mechanisms by which the *lac* repressor and CAP regulate the *lac* operon. The *lac* repressor operates as a *negative* control agent; it "senses" the level of a β-galactoside in the cell and turns off transcription of the *lac* genes by binding to the operator if β-galactosides are absent. A mutant lacking repressor transcribes the *lac* operon constitutively. CAP, in contrast, is required for transcription to occur. In a mutant lacking CAP, RNA polymerase would be unable to bind to the promoter and transcription would not occur. Thus, CAP is required for transcription and is a *positive* regulator; the more CAP molecules bound, the more transcription occurs.

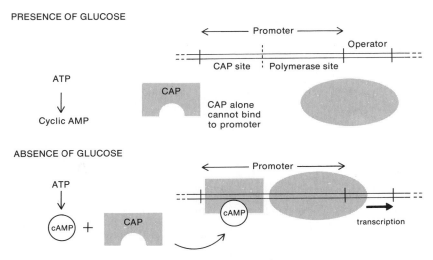

Figure 16-11.   The role of glucose in regulating the *lactose* operon.

The regulated transcription of the *lac* operon has been produced in a test tube.

An organic chemist confirms a proposed chemical structure deduced by analysis by synthesizing the compound, that is, by putting it together. A molecular biologist confirms a proposed functional mechanism by having the mechanism operate when the component parts are reconstituted in a test tube under defined conditions. All of the components necessary for the function of the *lac* operon have been isolated from cells, or obtained otherwise, and reconstituted in a test tube. These include the repressor, CAP, cAMP, an inducer, DNA-dependent RNA polymerase (complete with the σ factor), all four nucleotide triphosphates, and *lac* DNA. The *in vitro* components assembled together are found to synthesize *lac* mRNA in the manner predicted by the operon model. For example, in the absence of repressor, *lac* mRNA is synthesized. When the repressor is added, no *lac* mRNA is made. But when an inducer is then added, *lac* mRNA is made again. Also, if no cAMP or CAP are present, no *lac* mRNA is produced. Only when both cAMP and CAP are present does transcription of the *lac* DNA occur. The control of *lac* mRNA synthesis is exactly as expected.

The complete base sequence of the *lac* promoter and operator regions has been determined. The operator and CAP binding sites contain palindromes.

The base sequence from the end of the *i* gene into the *z* gene at the *lac* locus is depicted in Figure 16-12. The regions in the operator and the promoter that, respectively, bind the repressor and CAP have been identified. In both these regions palindromes (in boxes) have been identified. And so,

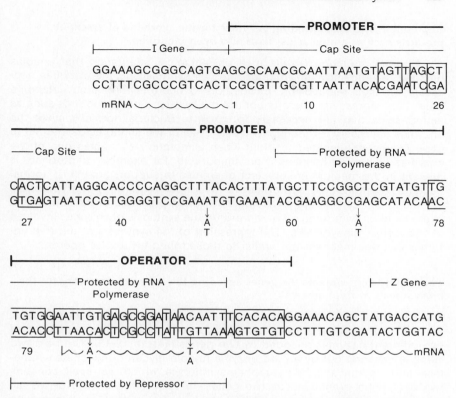

Figure 16-12.   The base sequence of the promoter and the operator regions of the *lactose* operon. After Dickson, et al., *Science*, v. 187, p. 27, 1975, and Johnsrud, L., *Proceedings of the National Academy of Sciences, U.S.*, v. 75, p. 5314, 1978.

again we find that regions in the DNA that interact with specific proteins contain palindromes. Base substitution mutations have been identified in the operator region that result in the operator constitutive phenotype and clearly identify the position of the operator. The position at which transcription starts is indicated, and we can identify the position of the AGGA sequence that pairs with the 16S rRNA during the initiation of translation. It is located 7 bases upstream from the position of the AUG polypeptide initiating codon. Note that the operator lies within the region of the DNA that encodes the leader sequence. Because of the palindromic DNA, the leader region of the mRNA contains base sequences that can base pair to form a hairpin. The entire region between the end of the *i* gene and the start of the *z* gene is composed of only 131 nucleotide pairs, but the genes (*i*, *z*, *y*, and *a*) for the four proteins comprise about 6,300 nucleotide pairs. At least within this limited region of the *E. coli* chromosome, most of the DNA is utilized for structural information, a finding also true of ØX174.

Repressible operons, encoding enzymes for the synthesis of essential products such as amino acids, have also been identified.

Enzyme induction results in a physiological adaptation that enables bacteria to efficiently utilize catabolites by synthesizing the enzymes necessary for their metabolism. In addition to operons regulating catabolite utilization, *repressible operons* control the synthesis of compounds such as amino acids that are necessary for growth. Bacteria normally make the enzymes necessary for the biosynthesis of all of the amino acids and fail to grow in unsupplemented medium when a mutation eliminates one of the enzymes. But what happens if an amino acid, for example, tryptophan, is present in the medium in sufficient quantities to support growth of normal bacteria? Synthesis of tryptophan is no longer required. For tryptophan, normal *E. coli* cells adapt physiologically to the presence of the amino acid in the medium by turning off or *repressing* the synthesis of enzymes involved in tryptophan biosynthesis. The repression of the synthesis of the *tryp* enzymes involves mechanisms similar to those found for the *lac* operon.

The *tryp* operon contains the genes encoding the enzymes necessary for the biosynthesis of tryptophan.

The biosynthesis of tryptophan involves five different enzymatic steps, the first two of which are carried out by an enzyme complex composed of two different enzymes encoded by two genes (Figure 16-13). The next two steps are carried out by a third enzyme (indole glycerol phosphate synthetase) and the final step by tryptophan synthetase, which, we recall, contains two polypeptides encoded by two genes. In total there are five genes, all located together in an operon, placed at 27.4 mins. on the *E. coli* map, with the order of genes paralleling the order of the biosynthetic conversions

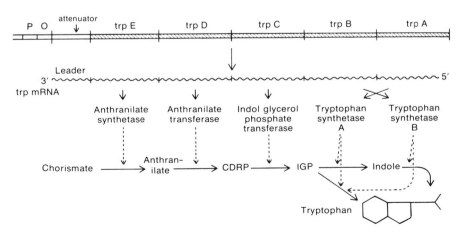

Figure 16-13.   The *tryptophan* operon and its encoded products. CDRP, carboxy phenylamino deoxyribulose phosphate; IGP, indole glycerol phosphate.

(Figure 16-13). (Such coincidence between the map order of genes and the order of steps in a biosynthetic pathway does not always prevail. In the leucine operon, for example, the second enzyme in the pathway is encoded by the third gene in the operon.) A *tryp* repressor, encoded by the *tryp R* gene at 99.8 mins. on the genetic map, is produced; it binds to the *tryp* operator next to the E gene. The repressor does not bind to the operator in the absence of tryptophan. Therefore, in the absence of tryptophan, there is synthesis of the enzymes needed for tryptophan biosynthesis.

The presence of exogenous tryptophan represses the *tryp* operon.

When the concentration of tryptophan increases in the cell, for example, by addition of tryptophan to the medium, the synthesis of the polycistronic mRNA for the *tryp* enzymes is repressed. Repression occurs because tryptophan acts as a *corepressor*. It binds to the repressor protein which, in turn, binds to the *tryp* operator, thereby preventing transcription of the *tryp* operon (Figure 16-14). This mechanism contrasts to the *lac* repressor, which binds to the *lac* operator in the absence of inducer.

When the concentration of tryptophan in a cell is sufficient to occupy the repressor, then further synthesis of the *tryp* mRNA is turned off. When the concentration of tryptophan drops, for example, because of the use of tryptophan in protein synthesis, then synthesis of the *tryp* mRNA will resume. The cell "senses" the amount of tryptophan present and responds by increased or decreased synthesis of *tryp* mRNA and enzymes. Energy is expended to synthesize *tryp* mRNA and proteins only when necessary.

Other operons that encode enzymes necessary for amino acid synthesis have been identified in *E. coli* and *Salmonella typhimurium*. Of particular interest is the histidine operon in which charged histidyl-tRNA is the corepressor for the histidine repressor protein. Synthesis of the mRNA and the enzymes for histidine biosynthesis is only repressed when there is sufficient charged histidyl-tRNA present. For this reason, mutants that affect the production of histidyl-tRNA, for example, the histidyl amino acid acylase, also affect the transcription of the *his* operon. Conditional mutants of the histidyl acylase, under restrictive conditions in which no histidyl-tRNA is formed,

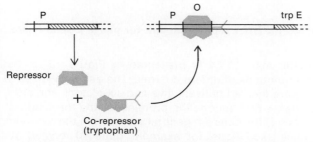

Figure 16-14.   Regulation of the *tryptophan* operon. Transcription is inhibited when the repressor–co-repressor complex binds to the operator.

cause the constitutive transcription of the *his* operon even in the presence of excess exogenous histidine.

Even when repressed, operons are not entirely turned off; some transcription still occurs. Repressible operons have a second region, the *attenuator,* that increases the efficiency of transcription repression. The attenuator region is located beyond the operator between the site where transcription is initiated and the initiating codon of the first encoded protein, that is, in a region corresponding to the leader sequence of the mRNA (Figure 16-13). Attenuators are regulated transcription terminators. In the *tryp* operon, the presence of tryptophan causes termination in the attenuator region of those few *tryp* transcripts that are made. Termination occurs before a coding region is transcribed. When tryptophan is absent, transcription proceeds through the attentuator. The attenuator functions in a manner somewhat analogous to turning off a leaky faucet.

GENE REGULATION IN BACTERIOPHAGES

In bacteria genes encoding enzymes with related functions are grouped in functional clusters that provide the coordinate regulation of their transcription. We find similar clusters of genes with related functions when we look at the genetic maps of phages. In bacteria the operons facilitate the response of the bacterium to fluctuations in available nutrients. In phages, the operons facilitate the passage of the phage through a programmed developmental sequence leading to the production of new phages. The life cycle of phages falls into identifiable phases. For example, the temperate phage λ has a lytic cycle of growth leading to the rapid formation of new phages distinct from a lysogenic cycle of growth. Within the lytic cycle we find two different temporal stages. The first stage, occurring shortly after infection, involves the replication of phage DNA. The second stage involves the synthesis of various components and the assembly and release of new phages. In the life cycle of a virulent phage, such as T4, only the replication and assembly phases are present.

In T4 the genes for DNA replication and for phage assembly are respectively clustered.

The genetic map of T4 is presented in Figure 16-15, indicating the positions and functions of different genes. The genes involved in replication of phage DNA are located between five and six o'clock and eight and eleven o'clock. The genes that encode the proteins for the formation of the mature phages and lysis of the cells are generally located in the remaining positions. There are some exceptions, for example, gene 40 located at ten o'clock specifies a protein that is involved in the formation of the head of the phage.

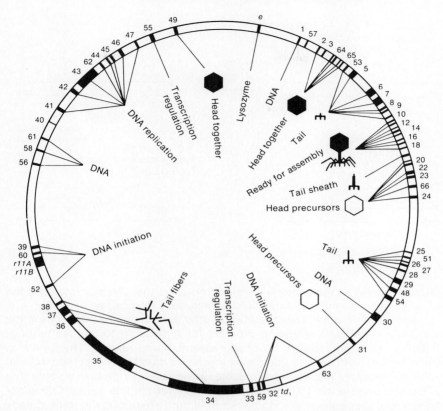

Figure 16-15.   Functional organization of the T4 chromosome. Modified from Edgar and Wood, *Proceedings of the National Academy of Sciences, U.S.,* v. 55, p. 498, 1966.

Transcription of the T4 chromosome is regulated by changing the specificity of the bacterial RNA polymerase for phage promoters.

The two regions of the genome—one specifying replication functions, the other phage assembly functions—are transcribed at different times in the phage life cycle. Those genes that encode proteins necessary for replication are transcribed early, during the first few minutes of infection, and are called *early* genes. Genes that encode proteins for assembly and lysis are transcribed after the early genes and are called *middle* and *late* genes. Transcription of the early genes occurs exclusively in a counterclockwise direction using only one strand of the DNA as a template (Figure 16-16). Four polycistronic mRNA's are synthesized, each containing coding sequences for approximately five polypeptides. The early genes are transcribed by the bacterial DNA-dependent RNA polymerase containing the bacterial σ

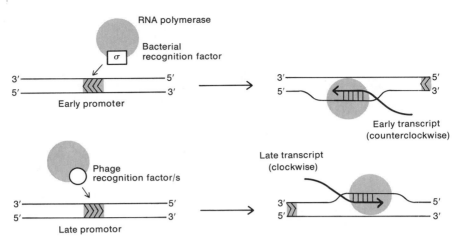

Figure 16-16.   Regulation of transcription in T4.

factor. The use of the bacterial σ factor indicates that the promoters for the early genes are similar to those in the bacterial chromosome.

About ten minutes after the start of infection, there is a switch from the synthesis of mRNA's encoding proteins for replication to the synthesis of mRNA's encoding proteins for phage assembly and release. The switch in transcription from early to late genes involves a modification of the bacterial RNA polymerase. The bacterial σ factor is replaced by phage proteins that cause the polymerase to recognize promoters for the late genes. Transcription of the late genes is in a clockwise direction. Therefore, different strands of DNA are used as templates for early and late genes. This has been demonstrated dramatically by hybridization experiments in which 98 per cent of the early transcripts hybridize to one strand and 80 per cent of the late transcripts hybridize to the other. Two of these phage proteins are encoded by early genes (33 and 55). These proteins specified by early genes modify the RNA polymerase so that it no longer transcribes early genes, but now transcribes late genes. The regulation of transcription of the T4 genome is different from that of the bacterium. Repressor proteins are not involved. Instead changes in transcriptional specificity result from changes in the RNA polymerase.

Functionally related genes are in clusters on the λ chromosome.

The genetic map of λ in Figure 16-17 shows the functional nature of many of the different genes. In the λ life cycle there are alternatives between lytic and lysogenic growth. During the lytic cycle there are stages when replication occurs and when phage assembly occurs. As with T4, in the λ chromosome genes that function at common stages are clustered. Genes involved in the lysogenic cycle of growth, such as those responsible for integration and excision from the bacterial chromosome, are grouped in the

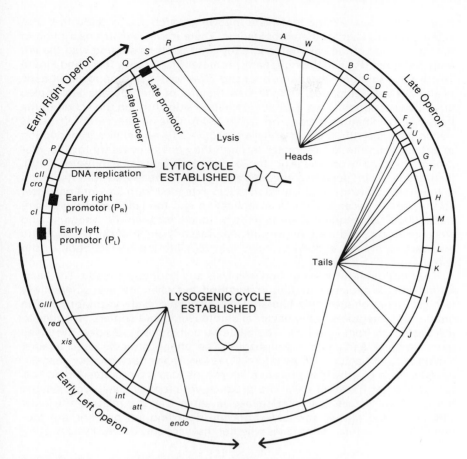

Figure 16-17.   Functional organization of the λ chromosome.

left early operon. The genes involved in DNA replication during the lytic cycle (genes 0 and P) are in the right early operon. The genes encoding the structural elements of mature phages are likewise clustered, with the head genes in one region, the tail genes in another. The transcriptional organization of λ is simpler than that of T4. There are only three major transcriptional units in the λ chromosome, early left and early right operons and an operon for the late genes.

The choice between entering the lysogenic or lytic cycle depends on the balance of transcription between two early operons.

Like T4, transcription in λ is divided into early and late genes. A decision must be made, however, during the initial stages of infection: to lysogenize or not to lysogenize? The first event following infection of a cell with λ DNA is the circularization of the chromosome. There is also early transcription of limited regions of the early left and early right operons starting with the left early promoter (P_L) and the right early promoter (P_R) (Figure 16-17 and Figure 16-18). Transcription occurs in opposite directions using opposite strands of the DNA as templates. Only limited regions of the operons are copied before transcription is blocked at termination sites through the action of the bacterial termination factor, rho (ρ) (p. 340). One of the early left genes transcribed is N, the product of which inactivates ρ and allows transcription to extend from the left promoter through the genes necessary for the integration of λ as a prophage in the bacterial chromosome (Figure 16-18). Included is the *int* gene that encodes a polypeptide necessary for integration at the attachment site. Transcription is also extended from the right promoter so that genes necessary for DNA replication and the lytic growth cycle are transcribed. Consequently, some gene products for both the lysogenic and lytic cycles are produced. As we will see presently some of the genes in the two early operators encode proteins that regulate the transcription of the two operons.

At this point the "choice" between lysis and lysogeny is made. The basis for making the choice is not fully understood, but when the level of infectivity is high (many phages, few bacteria) the lysogenic cycle is preferred. The phage DNA integrates in the bacterial chromosome and confers immunity to the bacterium from lysis by λ. The bacterial cell survives, divides, produces more bacteria and more λ prophages. The phage protects itself against running out of bacteria. When infectivity is low (many bacteria, few phages), the lytic cycle of growth is preferred. More λ phages are produced, going on to infect other bacterial cells and produce yet more phages. We presume that the level of infectivity determines in some way the relative efficiency with which the protein products from the right and left operons are produced, which, in turn, determines whether the lysogenic or lytic cycle occurs.

The lysogenic cycle depends on the regulatory products of the *cI, cII,* and *cIII* genes.

Lysogenic growth depends initially on the products encoded in the genes of the left operon that allow the phage chromosome to integrate into the bacterial chromosome. Maintenance of the lysogenic state at first depends on the products encoded by the *cII* and *cIII* genes. These proteins combine to activate transcription of the *cI* gene and inhibit transcription of the late genes. The cI protein is a repressor, called the λ *repressor,* which prevents transcription from the early promoter sites, P_L and P_R (Figure 16-18). In so doing, the λ repressor maintains the lysogenic state by preventing the autonomous replication of the integrated λ prophage. In addition, the repressor provides the cell with immunity against further infection by other λ phages. If another λ chromosome enters the cell, the λ repressor binds to the operator sites O_R and O_L adjacent to the early promoters on the invading chromosome and prevents synthesis of the products necessary for it to be replicated. The O_L and O_R regions have been characterized by Mark Ptashne

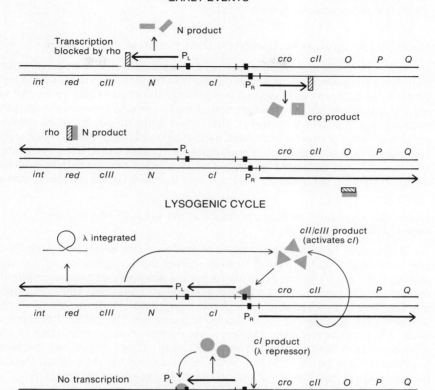

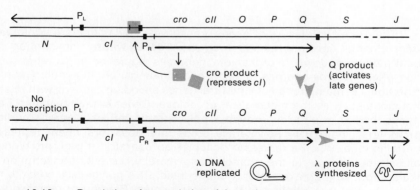

Figure 16-18.   Regulation of transcription of the λ chromosome.

and his associates as having three sites adjacent to the early promoters to which the repressor can bind.

The *cI* protein also acts to regulate the transcription of its own gene.

Not only does the product of the *cI* gene act to prevent transcription from the early promoters, it also regulates the transcription of its own gene. When small amounts of *cI* protein are present, the protein acts as a positive regulator to stimulate the transcription of the *cI* gene so that more *cI* protein is produced. It appears, however, that when large quantities of *cI* protein are present, the *cI* protein represses transcription of the *cI* gene. The protein consequently regulates its own production, a process called *autoregulation,* and assures that sufficient quantities of *cI* protein are present to prevent infection by another invading λ phage and to maintain the lysogenic state.

Two major regulatory proteins, products of the genes *Q* and *cro,* are required for the lytic cycle.

Production of new phages requires the products encoded by genes *Q* and *cro* (Figure 16-13). The Q protein causes the transcription of the late genes, those genes encoding proteins for the assembly of phages and their release into the medium. The late genes are all contained within a single operon. The Q protein acts as a positive regulator to direct the DNA-dependent RNA polymerase to transcribe this operon. The product of the *cro* gene is a negative regulator that prevents transcription from starting at the promoter for *cI* and at the left early promoter. Thus, these two proteins facilitate the synthesis of components necessary for the production of new phages and prevent the transcription of genes necessary for lysogeny.

SUMMARY

There are two classes of genes: constitutive genes that are continually transcribed at constant rates and regulated genes whose transcription can be turned up or down. In prokaryotic genomes regulation of gene transcription often involves operons, units of coordinate genetic function. Operons commonly contain groups of structural genes encoding proteins with related functions such as the metabolism of lactose. Genes within an operon are transcribed as a unit to produce a polycistronic mRNA. Operons contain promoters where RNA polymerase binds and operators where repressor proteins bind. When a repressor occupies an operator, it prevents binding of RNA polymerase to the promoter and thereby prevents transcription of the operon. When the operator is unoccupied, RNA polymerase binds to the promoter and transcription of the operon occurs. Binding of repressors to operators is regulated by metabolites, for example, binding of a β-galactoside to the *lac* repressor causes it to dissociate from the operator. Operons are also found in phages. In λ the binding of regulatory proteins, for example, the λ repressor, to operators is instrumental in determining whether the phage enters the lytic or lysogenic cycle and in maintaining the lysogenic state.

17

Structure and Function of Genomes: Eukaryotes

We have recited the litany that the exactness of DNA replication and mitosis assures that daughter cells are genetically identical. During animal development we expect that all cells derived from an original zygote will be genetically identical, constituting a clone. We shall see in Chapter 18 that this is not, however, always the case. Yet obviously all cells in a higher animal are not physically identical. Our consideration of regulation in prokaryotes provides us with one possible answer to the seeming enigma that cells with common genotypes have different phenotypes. Namely, all genes may not function in all cells or the encoded products may not be translated into functional proteins in all cells: There may be regulation of gene expression. In this chapter and in part of the next we will deal with genome structure and regulation in eukaryotic cells.

Eukaryotes, particularly higher, multicellular ones, have genomes with strikingly different structure and regulation from prokaryotic genomes. We find little or no evidence for polycistronic mRNA's in eukaryotes; each mRNA is apparently responsible for the synthesis of only one polypeptide. We do find that through judicious processing, particular primary transcripts can form mRNA's that direct the synthesis of differing polypeptides. Another striking difference between prokaryotes and eukaryotes is that in higher eukaryotes very little of the DNA actually encodes structural information.

As we did with prokaryotes, we will start by reviewing the structure and regulation of the genome of a virus, in this case, Simian virus 40, SV40. We will see that the structure and regulation of the virus genome reflect, in general, the structure and regulation of the eukaryotic cells that the virus parasitizes.

SV40 lyses monkey cells but can cause oncogenic transformation of other mammalian cells.

SV40 is a small spherical virus (diameter 45 nm) with a protein shell composed of 70 subunits and a circular, double-stranded DNA chromosome containing 5,226 base pairs (Figure 2-9). Nucleosomes are present, containing histones H2A, H2B, H3, and H4—all provided by the host cell—but lacking

H1. The SV40 chromosome is replicated bidirectionally from a single origin approximately at nucleotide$_0$ on the standard representation of the chromosome (Figure 17-1). In about 1 per cent of infected monkey cells the SV40 chromosome is rapidly replicated. About one million viruses, centered in the nucleus, are produced per cell. From 24 to 48 hours after infection, the cells lyse and the viruses are released into the medium. When tissue culture cells of other mammals, such as those of humans or mice, are infected, oncogenic transformation occurs. The viral DNA becomes integrated into the DNA of at least one chromosome (chromosome 7 in human cells) presumably by a recombination process similar to that responsible for integrating the λ chromosome into the chromosome of *E. coli* (Figure 7-4). As in the case of RNA oncogenic viruses, such as mammary tumor virus (pp. 244–246), integration is essential for oncogenic transformation with about one cell in 10^7 becoming transformed. Furthermore, when the virus is injected into newborn mice, it induces tumors in the mice.

The SV40 genome is divided into early and late genes, which encode, respectively, proteins for DNA replication and virus assembly.

The complete nucleotide sequence of the SV40 chromosome was published in 1978 by two groups, one in New Haven (Sherman Weissman and his colleagues) and one in Ghent, Belgium (Walter Fiers and his co-workers). The SV40 chromosome contains only two transcriptional units, one for early genes and one for late genes. The early gene transcript is about 2,650 nucleotides long and is transcribed in a counter-clockwise direction starting approximately at nucleotide pair 5215 and extending to nucleotide pair 2585 (Figure 17-1). The promoter for the early gene transcript resides near the

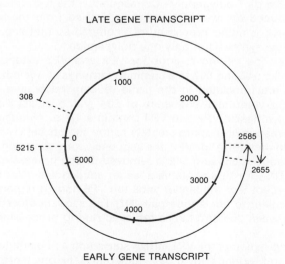

Figure 17-1.   The functional organization of the SV40 chromosome.

origin of replication. The late gene transcript is about 2,350 nucleotides long and is formed in the clockwise direction starting approximately at nucleotide pair 308 and extending to nucleotide pair 2655. The templates for the two transcripts overlap at their ends by 80 base pairs. The early genes, as in phages, encode proteins *t* and *T* (*T* for tumor) that function in replication of viral DNA. *T* is also required for the oncogenic transformation of host cells. The late genes encode viral coat proteins, called viral proteins 1, 2, and 3 (VP1, VP2, and VP3).

The synthesis of each viral protein depends on the correct processing of the primary RNA transcripts.

Although there are five known viral proteins and only two primary RNA transcripts, none of the proteins is synthesized using a polycistronic mRNA. Instead, synthesis of the proteins depends on the processing—that is, the cutting and ligating—of the primary transcripts into differing functional mRNA's.

Early genes. The coding region for the *t* protein extends from nucleotide 5146 to 4622 and for the *T* protein from nucleotide 5146 to 2674 (Figure 17-2). Both proteins are initiated using the same AUG codon and share a common N-terminal amino acid sequence of about 80 amino acids, *t* having a total of 146 amino acids and *T* about 726. The coding sequence of the mRNA that produces *t* is co-linear with the DNA, no non-coding, intervening sequences being removed after transcription. For the synthesis of *T*, however, a sequence of about 350 nucleotides, corresponding approximately to base pairs 4901 to 4556, is removed from the primary RNA transcript. This sequence contains the terminating codon for the *t* protein. Removal of the terminating codon allows polypeptide synthesis to continue with the production of the large *T* polypeptide. Consequently, the two different polypeptides are produced on two mRNA's, each derived from the same primary transcript, which is either processed or unaltered so that each mRNA molecule directs the synthesis of only one polypeptide.

Late genes. The coding regions for VP2 and VP3 extend respectively from nucleotides 542 and 897 to a common terminus at nucleotide 1601. The two peptides are translated in the same reading frame and they share a common carboxyl-terminal sequence of 234 amino acids. The processed mRNA's that synthesize VP2 and VP3 contain a 47- or 48-nucleotide leader sequence corresponding approximately to the region between nucleotides 308 and 354. In the mRNA for VP2, the sequence corresponding to the region between nucleotides 354 and 542 is removed during processing of the transcript. The mRNA for VP3 contains a leader sequence similar to that of the VP2 mRNA, except that the leader joins the VP3 coding region starting at a position equivalent to nucleotide pair 897. The sequence corresponding to the region between 355 and 897 is removed during processing of the transcript.

The coding region of the VP1 mRNA starts with a nucleotide corresponding to 1480 and extends for 1,094 nucleotides to one corresponding to nucleotide 2574. The beginning of the VP1 coding sequence overlaps with

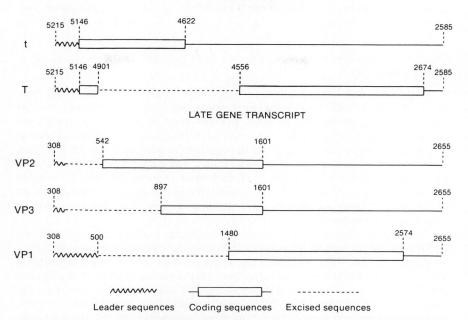

EARLY GENE TRANSCRIPT

LATE GENE TRANSCRIPT

Leader sequences   Coding sequences   Excised sequences

Figure 17-2.   Processing of SV40 transcripts to produce various mRNA's. After Fiers, W., et al., *Nature,* v. 273, p. 113, 1978, and Reddy, V. B., et al., *Science,* v. 200, p. 494, 1978.

the end of the common coding sequence for VP2 and VP3. The VP1 sequence is read in a different reading frame from that of VP2 and VP3, so that polypeptide synthesis extends past the terminating codon for VP2 and VP3. The VP3 mRNA has a leader sequence corresponding to the region between nucleotide pairs 308 and 500 that is longer than the leaders of the VP2 and VP3 mRNA's but shares a common 5′ sequence with them. During the processing that produces the VP1 mRNA, the nucleotide sequence corresponding to the regions between nucleotides 500 to 1480 is removed. As in the case of the early genes, each protein is synthesized on a separate mRNA.

The synthesis of early and late gene products in SV40 is first controlled by regulating the transcription of the early and late genes. The removal of different internal sequences in the transcripts determines which protein is synthesized by a particular mRNA. Thus, the production of the viral proteins involves regulation at two levels, first, at the level of transcription, and, second, at the level of processing. In no case does the final mRNA direct the synthesis of more than one polypeptide product. The mRNA's are not polycistronic.

The SV40 chromosome is slightly smaller than that of ØX174. Nevertheless, only about 77 per cent of the SV40 genome encodes structural information in contrast to ØX174, in which about 95 per cent of the genome is devoted to structural information. A substantial proportion of the SV40 genome is non-coding, foreshadowing a situation common in eukaryotes, particularly higher eukaryotes.

Structural genes appear to occupy only 5 to 10 per cent of the genomes of higher eukaryotes.

We know that the genomes of higher eukaryotes differ strikingly from those of prokaryotes in two ways. First, these genomes are immensely larger than those of prokaryotes. Second, the genomes of higher eukaryotes are composed of highly repeated and moderately repeated sequences as well as single-copy sequences in contrast to prokaryotic genomes, which are composed essentially of single-copy sequences. We might suspect that eukaryotes require much more structural information than do bacteria in order to live and maintain themselves. But that suspicion seems to be wrong. Much of the excess DNA appears to have other functions than the encoding of structural information.

The highly repeated sequences, for example, located principally in the centric heterochromatin of eukaryotic chromosomes, do not encode structural information. First, although some genes map to the heterochromatin, they are comparatively few. Second, the heterochromatic state of the region, like that of heterochromatized X chromosomes in mammalian females, suggests lack of function. Third, DNA/RNA hybridization studies using highly repeated sequences demonstrate that these sequences do not serve as templates for the synthesis of RNA, there being no RNA molecules complementary to the highly repeated sequences. We can now conclude that highly repeated sequences, constituting 20 per cent to 40 per cent of the total DNA in higher eukaryotes, do not encode structural information. The nature of the function of highly repeated sequences remains open for discussion, their localization near centromeres suggesting a role in chromosome movement during mitosis.

We are still left with 60 per cent to 80 per cent of higher eukaryotic genomes available for encoding structural information. Yet among moderately repeated and single-copy sequences, only 5 per cent to 10 per cent of those available appear to encode structural information. We might well ask about the function of the remainder of the DNA. Some of it is immediately adjacent to structural genes and so may be devoted to regulation. Although the idea is attractive, there is little evidence, except in promoters (p. 339), for the presence of regulatory regions next to structural genes in higher eukaryotic DNA. Such regions, however, *must* be present. Is it likely that most of the DNA in eukaryotes is devoted to regulation when we know that bacteria function very well using, at most, only a small percentage of their sequences for regulation? Maybe much of the excess DNA in eukaryotes is useless, a genomic appendix. Perhaps extra sequences readily accumulate during evolutionary history, but genomic appendectomies are rare.

But we are getting ahead of our story. Let us consider some of the evidence indicating a low concentration of structural information in single-copy and moderately repeated sequences of higher eukaryotes. As we do so we will learn that although much has been discovered about eukaryotic genome structure and regulation, much, much more remains a mystery.

Each chromomere in *Drosophila* contains enough DNA for 20 to 30 genes but only one or a few cistrons.

We know (pp. 58–59) that polytene chromosomes are the result of endoreplication, DNA replication occurring without separation of newly replicated chromatids. Because of endoreplication and the precise pairing of the replicated strands, the individual chromomeres, resulting possibly from superhelical coiling, are readily identified (Figure 2-16). In *Drosophila melanogaster,* where about 5,000 bands or chromomeres have been identified, the DNA content of individual chromomeres varies from about 10,000 to 70,000 base pairs per chromosome strand, with the average chromomere containing 20,000 to 30,000 base pairs. Thus, each chromomere has enough DNA to encode 20 to 30 different polypeptides, each with an average length of 333 amino acids.

The polytene chromosomes provide exceptional opportunities for the localization of genes using *in situ* hybridization. By allowing the precise cytological localization of deficiencies, they also permit the precise localization of mutant traits. By determining the position of different deficiencies we were able to limit the *facet* locus to a single band, 3C7, on the X chromosome in *Drosophila* (pp. 230–232). These analyses can be used to ask not where a particular gene is located but how many genes are located in a particular chromosomal region. Several laboratories have been engaged in doing so.

Ben Hochman and his colleagues have been studying the small fourth chromosome of *D. melanogaster,* which contains about 50 chromomeres, in order to learn how many genes, that is, how many *cistrons* or complementation groups, are located on the chromosome and how they are distributed. Other geneticists have studied limited regions of other chromosomes. One of the most detailed studies, conducted by Burke Judd and his co-workers, has been of a 13-chromomere region, 3A2 to 3C2, between the *zeste* and *white* loci on the X chromosome (Figure 17-3). Judd and his colleagues first recovered a large number of recessive lethal mutations that mapped to the region. Through the use of a group of deficiencies, like those we have discussed earlier (pp. 58–59), they assigned particular mutations to limited regions between bands 3A2 and 3C2, in many cases to particular bands. In addition they performed complementation tests between various pairs of lethals and determined the complementation pattern. In the initial study over 100 mutations were put into 13 complementation groups and most complementation groups were assigned to a single band within the region. In no case was there more than one complementation group per band. Because a group of mutants belonging to a single complementation group, that is, a single cistron, may be all alleles, there may be but one gene per band. Since the initial study, however, a few additional mutations that complement all of the originally characterized mutations have been identified. These mutations are not recessive lethals. One causes sterility in females and another affects

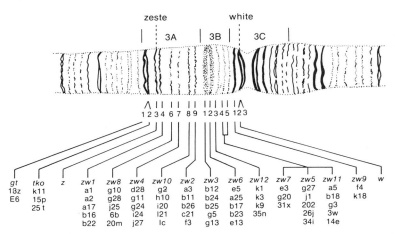

Figure 17-3.   Complementation groups, cistrons, in the *Zeste-white* region of the X chromosome in *Drosophila*. After Judd et al., *Genetics*, v. 71, p. 139, 1972.

the time of day that new adult flies emerge from pupa cases. Nevertheless, based on complementation, only one or a few genes appear to be present in each chromomere, about 15–20 in all, in a region that has enough DNA to contain 250 to 400 genes. We deduce then that only a small proportion of the DNA encodes structural information.

For what, then, is the rest of the DNA used? Judd suggests that much of the DNA is composed of cis-acting regulators. We already know that mutations in cis-acting regulatory regions will fail to complement with ones in adjacent structural genes (p. 414). Therefore, we can entertain the idea that most of the mutational sites in the collection of mutants recovered by Judd and his colleagues map outside of the boundaries of the structural genes. If this were the case, we would have some ideas about the function of the DNA adjoining a structural gene in a eukaryote. Alas, most mutations recovered in *Drosophila* seem to reside within structural genes.

Induced point mutations at the *rosy* locus in *Drosophila* map within the boundaries of the structural gene.

None of the products encoded by the genes that map in the *zeste-white* region is known. There is no molecular proof of whether a particular mutant site resides within a coding sequence or within an adjacent cis-acting regulatory region. We can, however, consider studies of another locus in *Drosophila* for which the encoded product is known. The *rosy* locus, which maps at position 52.0 on the third chromosome, has been located cytologically between polytene bands 37D8-12. The *rosy* gene, so named because its mutant form affects eye color, encodes the enzyme xanthine dehydrogenase (XDH), which converts xanthine to uric acid. Arthur Chovnik and his co-workers have long studied the fine structure of the *rosy* gene and have determined the map positions of a group of null alleles, which lack enzyme activity, and some electrophoretic variants. We deduce that the alleles that cause altered electrophoretic mobility of XDH do so by changing the amino acid sequence of the XDH polypeptide in a manner analogous to the effects of the sickle cell allele on the electrophoretic mobility of hemoglobin. If we deduce correctly, the mutant sites of these alleles must reside within the boundaries of the *rosy* structural gene. The question is whether the null alleles that eliminate XDH activity map at or between the mutant sites that alter electrophoretic mobility and, therefore, are presumably in the coding region of the *rosy* gene. (We still must worry about mutant sites located at intervening sequences). In a study of approximately 20 ethylmethanesulfonate- and X-ray-induced null alleles of *rosy*, all of the mutant sites mapped near or between the structural gene boundaries defined by the most widely separated alleles producing alterations in electrophoretic mobility. Thus, in this set of *rosy* mutations, we can deduce that each is a change in the structural gene, none a defect in a regulatory region. If we can generalize from the results at *rosy*, we must conclude that most of the mutations isolated by Judd and his co-workers in the *zeste-white* region are structural, not regulatory, mutations. Most mutants isolated in *Drosophila melanogaster* and, presumably, in other eukaryotes appear to be defects in structural genes. Thus we are again left wondering where the regulatory regions are and what the rest of the DNA is used for.

Apparent regulatory mutants have been isolated at *rosy*.

Although the evidence discussed indicates that the majority of mutations induced in *Drosophila* are in structural genes, there is evidence for some cis-acting regulatory mutations. Chovnik and his collaborators have identified strains, *overproducers*, initially found in populations of flies living in the "wild" (that often means, these days, in the fruit section of a supermarket or in a winery), in which the number of XDH molecules is greater than normal. The map positions of some of the mutant sites responsible for overproduction of XDH have been found to reside, apparently, next to but just outside of the XDH structural gene. Estimates based on recombination frequency and the known size of the XDH polypeptide (XDH is an enormous polypeptide of about 160,000 D or 1,454 amino acids, requiring a coding region with 4,400 base pairs) suggest that the mutant sites responsible for the increase in XDH protein reside about 1,500 base pairs from the structural gene. If there is indeed a cis-acting regulatory element 1,500 base pairs from the structural boundary of the *rosy* gene, then it differs from the promoter and operator regions found in bacteria, all of which are closely adjacent to structural genes. Despite the mapping data and the elegant genetic evidence, we cannot conclude that these mutant sites are really distant from the structural gene. There may be a non-coding intervening sequence between the apparent regulatory sites and the sites within the coding region. Unequivocal demonstration that these putative regulatory sites are distant from a boundary of the *rosy* structural gene awaits analysis of a genomic clone containing *rosy* DNA.

Hybridization experiments with mRNA's indicate that only a small percentage of eukaryotic genomes encode structural information.

One of the objections to Judd's genetic analysis, which suggested that each chromomere contains only one or a few genes, is that there may be many genes that cannot readily be identified by mutational analysis. For example, the enzyme alcohol dehydrogenase in *Drosophila,* although no doubt of value in many natural environments (*Drosophila* used to be called *Enophilia,* the wine lover, and thrives amidst fermenting fruit), is not required for survival under standard laboratory conditions. A mutation in the alcohol dehydrogenase gene would go unnoticed. In addition, when genes are present in multiple copies, a mutation in one copy would not eliminate the production of normal product by the other gene copies, and the gene might not be detected by standard mutational analysis. Mutations in many genes may go undetected, and many genes may exist in each chromomere or, at least, some chromomeres may contain multiple genes. Direct molecular analysis does provide a conclusion similar to the results of genetic analysis: a limited amount of DNA is devoted to encoding structural information for the synthesis of polypeptides.

We know, however, that mRNA's contain leader and trailer sequences as well as coding sequences. By determining the total percentage of the single-copy DNA to which total mRNA hybridizes, we obtain an estimate of the amount of DNA in an organism that is devoted to encoding structural information. Such experiments have been conducted in several laboratories, notably in Eric Davidson's, using the sea urchin *Strongylocentratus purpuratus,* and in John Bishop's, using *Drosophila melanogaster.* Davidson's results indicate that the amount of DNA forming hybrids with mRNA is equivalent to 20,000 to 30,000 genes and corresponds to less than 10 per cent of the total single-copy DNA. These results are of particular value because they represent the sum of the hybrids formed with the mRNA population isolated from various stages of sea urchin development as well as from various adult tissues. Thus, there is adequate structural information in less than 10 per cent of the genome of the sea urchin for normal development. Bishop's results with *Drosophila* also indicate that only 5 to 10 per cent of the genome of that organism encodes structural information. We can relate Bishop's numbers to Judd's proposal that there is but one or a few genes per chromomere. Each chromomere on average contains 20,000 to 30,000 base pairs. If the average structural gene has a coding region 1,000 base pairs long, then according to Judd's suggestion, less than 5 per cent of the DNA of the average chromomere would encode structural information. Therefore, Bishop's results on the hybridization of mRNA are reasonably consistent with the genetic results of Judd and his co-workers.

Specific coding sequences in vertebrate DNA are located in the midst of non-coding regions.

The non-coding, intervening sequences in the chicken ovalbumin gene are several times longer than the coding sequences (p. 349). We must consider the possibility that all or most of the non-coding DNA in an animal genome is contained within intervening regions. There are few examples on which to base conclusions about the distribution of coding and non-coding regions. In those few, two of which we will consider, the coding sequences are surrounded on both sides by extensive, single-copy, non-coding sequences.

Chicken ovalbumin gene. The ovalbumin gene contains extensive non-coding intervening sequences that interrupt the coding regions (Figure 17-4). Knowledge of the presence of the intervening sequences was obtained by cDNA clones made from ovalbumin mRNA and the recovery of genomic clones (pp. 343–347) containing the ovalbumin gene .Several ovalbumin genomic clones contain not only coding and intervening sequences but also sequences adjacent to the ovalbumin gene. These genomic clones provide an opportunity to study the nature of the adjacent sequences. Bert O'Malley and his colleagues have investigated whether sequences adjacent to the ovalbumin gene are templates for the synthesis of mRNA by determining whether any mRNA found in chick cells form RNA/DNA hybrids with these DNA sequences. They used single-copy sequences up to 10,000 nucleotides long from both sides of the ovalbumin gene. If the sequences on either side coded for the synthesis of polypeptides, we would expect RNA/DNA hybrids to form. O'Malley detected no RNA/DNA hybrids, and we can conclude that the adjacent sequences do not serve as templates for the synthesis of mRNA.

Globin genes. We turn yet again to the genetics of hemoglobins for investigation of the structural organization of higher animal genomes. We have indicated that hemoglobin is composed of two α- and two β-chains, but the situation is more complex. In humans, additional β-like or non-α chains are found; that is, certain chains form hemoglobin tetramers along with two α-chains. In adults, δ-chains, structurally similar but slightly different in amino acid sequence from β-chains, replace about one out of every forty β-chains in hemoglobin. For the six months prior to birth, β-chains are normally absent from hemoglobin; their place is taken by γ-chains of which there are two kinds, G_γ and A_γ. The amino acid sequences of each γ chain differ but both are similar to the sequence of the β-chain. In very early embryos, when hemoglobin is first formed, two ϵ-chains are present instead

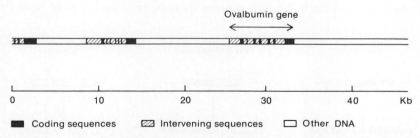

Figure 17-4.   The ovalbumin gene and its neighborhood. Unknown genes with structures similar to the ovalbumin gene are over 10,000 base pairs away. After Royal et al., *Nature,* v. 279, p. 125, 1979.

of the γ-chains. The ε-chains have amino acid sequences similar to β-chains and can form tetramers with α-like chains.

Analysis of genetic variants indicates that the structural genes for three of the β-like genes are closely linked. Of particular interest are two variants called hemoglobin-Lepore (Hb_{Lepore}) and hemoglobin-Kenya (Hb_{Kenya}). In Hb_{Lepore}, the N-terminal end of the polypeptide has a δ-chain amino acid sequence, but the C-terminal end has the β-chain amino acid sequence (Figure 17-7.) Hb_{Lepore} could result from a fusion of two ends of the δ- and β-genes (Figure 17-5) that arose from unequal crossing over. Because no normal δ- or β-chains are found in the Hb_{Lepore} variant, we can propose that, with respect to the direction of transcription, the genes are located 5′ . . . $Hb_δ$ . . . $Hb_β$ . . . 3′. This orientation allows the production through unequal crossing over of variants lacking normal δ- and β-chains but having the Lepore variant (Figure 17-6). This orientation is also suggested by another variant, an anti-Lepore variant, in which the C-terminal end of the polypeptide has the δ-chain sequence and the N-terminal end has the β-chain sequence (Figure 17-6). In the anti-Lepore variant, normal δ- and β-chains are present. In Hb_{Kenya}, the variant globin polypeptide has an $A_γ$ N-terminal sequence and a β-chain C-terminal sequence (Figure 17-5). No δ-chains are present. We can propose that the $A_γ$ gene lies farther upstream from the δ-chain gene and that the unequal crossing over that fused $A_γ$ and β-genes also eliminated the δ-gene (Figure 17-6). An anti-Kenya variant is possible but has not yet been found. Nevertheless, from genetic data we can align the genes 5′ . . . $HbA_γ$ . . . $Hb_δ$ . . . $Hb_β$ . . . 3′ (Figure 17-6). We cannot however, tell how close the genes are to each other.

Researchers in several laboratories, including those of Richard Flavel, Philip Leder, and Tom Maniatis, have obtained genomic clones of the DNA of the region of Chromosome 11 encoding the β-like globin chains. Analysis of the genomic clones in these laboratories has directly determined the order of the β-, δ-, and γ-genes and their positions in the DNA (Figure 17-7) and the ε-genes as well. The order of three of the genes determined with recombinant DNA techniques is the same as that deduced from the genetic data. In addition, the $G_γ$ gene is located to the left of the $A_γ$ gene and the ε-gene lies even further to the left. The direction of transcription, from left to right, is the same for all five genes. In DNA obtained from an individual with the Lepore variant, the sequence between the δ- and β-genes is missing as we would expect if the variant arose by unequal crossing over. Note that

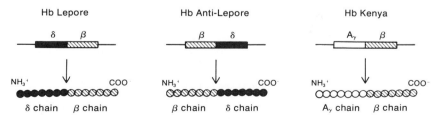

Figure 17-5.   The β-chains of Lepore, Anti-Lepore and Kenya hemoglobins.

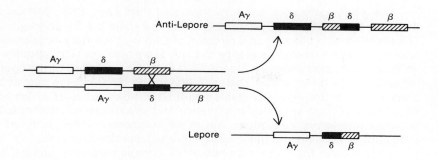

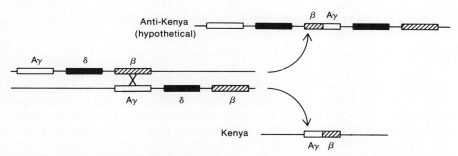

Figure 17-6.   The production of Lepore, anti-Lepore, and Kenya hemoglobins as a result of unequal crossing over. After Weatherall and Clegg, *Cell*, v. 16, p. 467, 1979.

the position of the major intervening sequence is the same in all five genes. It is not known whether the small intervening sequence in the β-gene is present in any of the others. Of particular note is the possibility that the unequal crossing over that produced the Lepore variant may have occurred through or at the junction with the intervening sequences of each gene. If true, we must entertain the possibility that intervening sequences may promote recombination between differing but related genes.

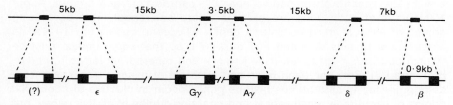

Figure 17-7.   The genetic organization of human β-like globin genes. The ? denotes a β-like gene that apparently does not encode a synthesized polypeptide. After Tom Maniatis.

The globin genes are distinctly separated from each other, particularly those for the δ and the Aγ polypeptides and the ε and the Gγ polypeptides. The presence of a pseudogene, a region of DNA with a sequence characteristic of a β-chain coding sequence but not expressed in the synthesis of a polypeptide, is also shown in Figure 17-7. With the exception of the coding regions for the β-like globin chains, there are no other coding sequences present. The entire region contains about 50,000 base pairs of which only about 3,000 encode structural information and approximately another 6,600 compose intervening sequences. Again, we find genes placed in the midst of regions of DNA that do not encode other polypeptide products.

In summary, from the distribution of complementation groups, the amount of DNA forming hybrids with total mRNA, and the analysis of the nucleotidic neighborhoods in which the ovalbumin and β-like globin genes reside, we conclude that much of the DNA in higher animals does not encode structural information. The function of the non-coding DNA remains obscure.

Some polypeptides are encoded by repeated gene copies.

The examples we have considered all involve genes constructed of single-copy sequences. Even the genes encoding the five β-like globin chains have nucleotide sequences that differ, though we can surmise that the five genes arose during evolution from a common predecessor gene that became duplicated and from which the duplicated copies then diverged (see Chapter 25). Genomes of higher organisms have numerous gene families in which the individual members are no longer identical. In Drosophila, Sarah Tobin and her co-workers in San Francisco and Eric Fyrberg and his collaborators in Pasadena have identified six regions scattered on three chromosomes that encode the contractile protein actin. These genes differ only slightly in nucleotide sequence. In the moth, Antheraea polyphemus, Fotis Kafatos and his collaborators have demonstrated that the egg shell is constructed of a series of related polypeptides that differ from each other by amino acid substitutions in an otherwise common polypeptide sequence. These polypeptides differ in primary structure and are encoded by a series of related genes with different base substitutions. Some genes encoding polypeptides have nucleotide sequences that are very similar if not identical. Perhaps the outstanding examples are the genes that encode histones. The histone gene copies are distributed as tandem repeats. In Drosophila melanogaster the histone gene copies are located between polytene bands 39D-E on the left arm of the second chromosome near the centric heterochromatin. About 100 copies of each gene are present. There are repeat units containing copies of each of the five histone genes. The gene copies are arrayed in each repeat unit as depicted in Figure 17-8 so that some genes must be transcribed from right to left (H2b and H4) and others from left to right (H1, H2a, and H3). Between each gene and each set of histone genes is a noncoding spacer. In sea urchins histone genes occur in clusters of about 300 copies of each gene. Each repeat unit contains copies of all five genes, but their orientation differs from Drosophila, with all of the genes having the same orientation (Figure 17-8). Although in sea urchins each repeat unit may

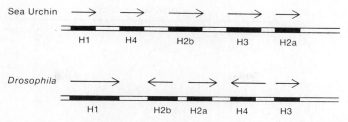

Figure 17-8.   The organization of histone genes in sea urchins and *Drosophila*. After Schaffer et al., *Cell*, v. 14, p. 655, 1978 and Lifton et al., *Cold Spring Harbor Symposium on Quantitative Biology*, v. 42, p. 1047, 1977.

be transcribed as a single unit, in both *Drosophila* and sea urchins each gene copy appears to be transcribed separately. The histone genes differ from the ovalbumin gene and the five β-like chain genes in that the individual genes are comparatively close together.

We do not yet really know whether genes in higher animals will, in general, be isolated amongst non-coding sequences or have structural genes as comparatively near neighbors. With the exception of gene copies that are distributed as tandem repeats, the genetic and molecular data so far suggest that genes will generally be well spaced along the DNA in eukaryotes.

The synthesis of the β-like globin chains is regulated during development.

Hemoglobin is synthesized in different tissues at various stages of mammalian development. In humans, embryonic hemoglobin, consists of two ζ- and two ϵ-chains, the ζ-chains having α-like sequences; the ϵ-chains, β-like sequences. At this stage of development hemoglobin is synthesized in the yolk sac, a structure evolutionarily homologous to the yolk sac in bird eggs, although mammalian embryos are essentially yolkless. After about three months of development the site of hemoglobin synthesis switches to the liver and spleen. ζ- and ϵ-chains are no longer made. Instead α-chains are synthesized along with Aγ and Gγ chains and comprise fetal hemoglobin (Figure 17-9). Synthesis of fetal hemoglobin persists until shortly after birth, when synthesis of γ-chains ceases, and the synthesis of β-chains and some δ-chains commences. The site of hemoglobin synthesis is now the bone marrow. The synthesis of β- and δ-chains also differs, δ-chains are primarily made in young red blood cell precursors, and β-chains in old ones. There is a tantalizing aspect to the genetic organization of the β-like genes. From analysis of genetic data and the genomic clones, we learned that all five genes are oriented in the same direction so that, as pictured, transcription is from left to right (Figure 17-7). The order in which the genes are expressed developmentally is the same as the order of genes in the chromosome with respect to the direction of transcription. The ϵ-gene is first, then the γ-genes and, finally, the δ- and β-genes. Whether the correspondence between the arrangement of these genes on the chromosome and their order of devel-

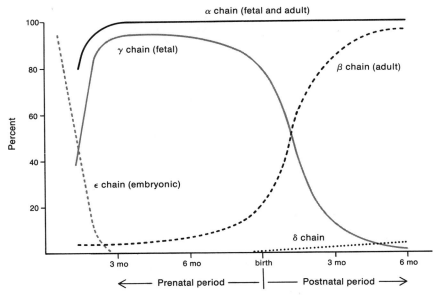

Figure 17-9.   The synthesis of globins during human embryonic development. Embryonic globin contains ζ-chains in addition to ε-chains. After Huehns et al., *Cold Spring Harbor Symposium on Quantitative Biology,* v. 29, p. 327, 1964.

opmental expression is coincidental or reflects some secret of the regulation of eukaryotic gene expression remains to be determined.

The mechanisms by which the sites of synthesis and the nature of the globin polypeptides synthesized are selected is not understood. We presume that all of the globin genes are present in all cells regardless of whether they synthesize globins or not. Mechanisms must exist that determine which globin polypeptides are made in each tissue and when during development they are made. If we were to borrow from phage and virus development, we would immediately suspect that there is regulated transcription of the globin genes. For example, the γ-gene would be transcribed during the latter part of fetal development and the β-gene after birth. We would hypothesize that *transcriptional regulation* (also called *differential gene function*) is responsible for the changes in globin polypeptide synthesis that occur during development.

Even so, we have learned from our discussion of SV40 gene function that the nature of the polypeptide product need not only result from the regulation of transcription but can also result from differential processing of primary RNA transcripts. Whether VP1 or VP2 of SV40 is synthesized is not only a function of the synthesis of the original transcript but also depends on the processing of the transcript (pp. 460–462). With one form of processing, VP1 is made, and with the other, VP2 is produced. We can imagine that processing of a pre-mRNA can determine whether *any* polypeptide is made using a particular transcript. Thus, we recognize in eukaryotes that *post-*

transcriptional regulation may be responsible for determining whether the information encoded in a particular gene sequence is translated into a polypeptide. Post-transcriptional regulation may not only involve the nature of processing of a particular transcript but may determine whether the transcript ever escapes from the nucleus into the cytoplasm to be translated. Both forms of regulation, transcriptional and post-transcriptional, appear to be important for the regulation of gene expression in higher eukaryotes.

The state of chromatin coiling regulates transcription.

In our early discussions of chromosome structure (pp. 52–61), we indicated that eukaryotic chromosomes exist in interphase nuclei in various helical arrangements ranging from fundamental nucleosome structures through solenoids and supersolenoids. The different amounts of coiling can reflect different amounts of transcriptional activity. At one extreme, in inactive chromosomal regions, such as the centric heterochromatin or the highly condensed and heterochromatic X chromosomes (Barr bodies) found in mammalian cells, the chromatin is transcriptionally inactive. The DNA is physically isolated, as it were, from the RNA polymerases necessary for transcription. At the other extreme is very uncoiled chromatin, consisting, in essence, of a string of nucleosomes. Because the DNA is on the outside of each nucleosome, we believe that such DNA is accessible to RNA polymerases and can serve as a template for the synthesis of RNA transcripts. The two outstanding examples of transcriptionally active, uncoiled regions of chromosomes are the loops found in lampbrush chromosomes of amphibian oocytes and insect spermatocytes (Figure 2-14) and the puffs found in various positions of the polytene chromosomes of insects such as *Chironomus* (Figure 2-18). In both these examples, uncoiled, transcriptionally active loops of DNA containing nucleosomes extend from the central axis of the chromosome. In the polytene chromosomes, because over 1,000 copies of each chromatid are present, the transcriptional activity of the puff is readily demonstrated. We incubate polytene tissues with ³H-uridine so that newly synthesized RNA is radioactive. Following a brief incubation of a few minutes, the tissue is fixed; the chromosomes are spread on a slide; and an autoradiograph is made. Places along the chromosome that were being transcribed are readily revealed by silver grains in the photographic emulsion resulting from β-particles given off by the newly synthesized RNA. Large numbers of silver grains are seen over the regions where the puffs are located, indicating sites of particularly intense RNA synthesis. By determining the number and location of puffs, *the puffing pattern,* under various physiological conditions or at various developmental stages, we can discern changes in gene transcription and relate these changes to developmental changes in the insect.

Metamorphosis in insects is under the control of the steroid hormone, ecdysone.

Among the most spectacular developmental sequences are those associated with *metamorphosis,* in which an organism of one shape and physiology changes over a limited time into one of another shape and physiology.

Metamorphosis is brought about by hormones manufactured in endocrine glands and released into the blood from which they act on target tissues to elicit physiological or developmental changes. The tadpole, a filter-feeder swimming about a pond and respiring through gills, changes, in response to the hormone thyroxine, into a frog that moves on land, eats flies and other insects, and breathes through lungs. A caterpillar, in response to steroid hormone 20-hydroxyecdysone (Figure 17-10), which we will call ecdysone for short, forms a cocoon from which it later emerges as a butterfly. In *Drosophila,* the larva responds to ecdysone by forming a hard outer skin from which 4 or 5 days later a fly emerges. We might wonder how these hormones bring about their developmental changes from immature forms into reproductively mature adults. By observing the puffs in polytene chromosomes, investigators in several laboratories have learned that the insect hormone acts by changing the transcriptional activity of different loci along the chromosomes.

Ecdysone produces a specific pattern of gene transcription (puffs) in polytene chromosomes.

In 1959 Hans Becker reported that at the beginning of metamorphosis in *Drosophila melanogaster* an impressive series of new puffs is seen in the polytene chromosomes of larval salivary glands. Later Becker demonstrated that the puffs depend on the presence of the ring gland, an endocrine organ that produces ecdysone, and Ulrich Clever demonstrated, using larvae of the midge, *Chironomus tentans,* that injection of ecdysone also produces a series of puffs in the salivary gland polytene chromosomes. The development of *in vitro* culture systems by Hans Berendes and Michael Ashburner in the late 1960's, in which isolated salivary glands responded to incubation with ecdysone, allowed a detailed investigation of the induction of puffs, or gene transcription, by ecdysone. Ashburner and his colleagues have been particularly successful in elucidating the steps that occur in the induction of puffs by ecdysone in cultured salivary glands of *Drosophila melanogaster.*

Demonstrable changes in puffing in the salivary glands occur within 3 to 5 minutes after the addition of ecdysone to the culture medium. These changes involve, on the one hand, the regression of certain puffs, for ex-

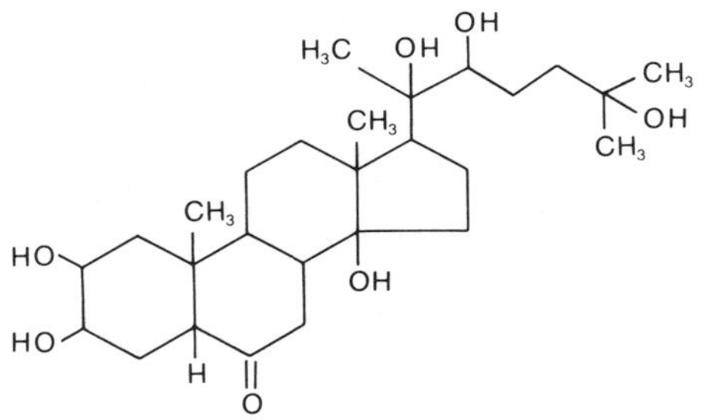

Figure 17-10.   Structure of 20-hydroxyecdysone.

ample, at regions 25A–C and 68C, that are normally active prior to meta-morphosis, and, on the other hand, the appearance of puffs that normally appear at the beginning of metamorphosis. Puffs that appear rapidly in response to the steroid hormone are "early" puffs analogous to the early genes that function at the start of phage infection. The early puffs are not limited to a single region of the *Drosophila* genome but are distributed at several locations on three chromosomes, including salivary bands 2B1–10 on the X chromosome, 23E on the left arm of the second chromosome, and two puffs at 74E–F and 75B on the right arm of the third chromosome. Other puffs appear in response to ecdysone at later times, for example, one at 78D after three hours, and one at 63E after five hours of the addition of the hormone. The changes in transcription induced by ecdysone are not re-stricted to limited regions of the genome.

There is a marked difference between the early and the late puffs. The early ones still appear in response to ecdysone when protein synthesis is inhibited, but the late ones fail to appear. This observation, made originally by Ulrich Clever working with *Chironomus* and also by Ashburner and his co-workers, has been interpreted to suggest that the genes at the site of the early puffs encode regulatory proteins that elicit the late puffs. Such proteins may be analogous to the regulatory proteins in bacterial systems (for ex-ample, CAP, p. 445) that cause transcription by binding to the DNA. There is no direct evidence, however, about the nature of the proteins encoded by the genes at early puff sites. What is clear is that the hormone responsible for metamorphosis in insects acts, at least in part, by changing the tran-scriptional activity of different genes. Furthermore, the changes in gene function constitute a temporal series of changes.

Steroid hormones act by binding to a receptor protein that, in turn, binds to the chromatin.

How does ecdysone elicit the specific transcriptional activity manifested by the puffs in polytene chromosomes? Hormone action appears to emulate the mechanism found in bacteria in which small molecules, such as β-gal-actosides and cyclic-AMP, alter transcription by binding to proteins that, in turn, bind to the DNA. Generally, cells that respond to steroid hormones do so through a specific protein, a *receptor protein,* that binds a specific steroid, such as ecdysone in insects or an estrogen in vertebrates, to form a steroid-receptor complex. The receptor-hormone complex then binds to the chro-matin and, in an unknown way, elicits transcription of specific genes. Little is known about the receptor protein that binds ecdysone; much more is known about the proteins that bind vertebrate steroid hormones. A partic-ularly good example is the estrogen receptor in the chicken oviduct.

The oviduct is the tube through which the egg passes on its way to being laid. *Estrogens,* female sex hormones, have two effects on the oviduct. In immature chickens, estrogen causes the oviduct to develop the charac-teristics of the mature hen in which the oviduct produces egg-white proteins. In mature chickens, the hormone stimulates the production of some of these proteins, particularly ovalbumin, in huge amounts and other egg-white proteins in lesser amounts, for example, ovomucoid. There are about 10,000 molecules per cell of a specific receptor for estrogen in the oviduct. The receptor is a dimer composed of one chain with a molecular weight of about 90,000 D and another chain of about 100,000 D. Electron micrographs reveal that the dimer has a cigar-like shape. In the absence of estrogen the receptor proteins are located primarily in the cytoplasm. Following binding with estrogen, the protein changes in conformation so that the estrogen-receptor complex enters the nucleus and binds to the chromatin, thus increasing transcription. In particular, the transcription of the genes encoding ovalbumin and other egg-white proteins is increased. The effect of estrogen on the transcription of these genes is not immediate, but occurs after a few hours. It is possible, therefore, that the increased transcription of the ovalbumin gene represents a "late" gene and that the effect of the hormone is indirect.

The examples of steroid hormone action in insects and chickens both involve substantial increases in transcription of specific genes in response to specific stimuli. These examples parallel those in prokaryotes, where increased transcription of specific genes is a common response to an environmental stimulus such as the addition of an inducer to the culture medium, or, in phages, to a particular stage in the phage life cycle. We would be ill-advised, however, to extend these observations to all higher eukaryotic development. That is to say, post-transcriptional regulation may also play an important role in the changes in protein synthesis that occur during animal development, a role we will now discuss.

At various developmental stages in sea urchins, the mRNA sequences differ markedly, but the nuclear RNA sequences are similar.

We recall from our discussion of transcription (p. 343), that in higher eukaryotes such as vertebrates, mRNA's are synthesized as high molecular weight precursors. These RNA's, *heterogeneous nuclear RNA's,* are then processed mainly, it seems, through the excision of non-coding intervening sequences to form the mRNA's in polysomes. The frequency of the mRNA's and nuclear RNA's specified by single-copy DNA varies (Table 17-1). Some mRNA's, the *superprevalent* or abundant class, are present in thousands of copies per cell. Globin mRNA in red blood cells is an example. Other mRNA's, the *prevalent* class, are present in from 20 to 60 copies per cell. Still others, the *rare* class, are found in only 1 to 10 copies per cell. The abundant class is commonly found in differentiating tissues, for example, during muscle cell development or in developed tissues that are synthetically active, for example, in *Drosophila* salivary gland cells. In all tissues, however,

Table 17-1   Classes of mRNA's in Cells

Class of mRNA	Number of Molecules per Cell
Superprevalent (abundant)	>1,000
Prevalent	20–60
Rare	<10

the rare class, constitutes 80 per cent or more of the mass of the RNA present. For, although each mRNA is present in only a few copies, there are hundreds of thousands of different mRNA molecules.

In the examples considered, the induction of puffing by ecdysone and the induction of synthesis of ovalbumin mRNA by estrogen, the mRNA sequences belong or probably belong to the superprevalent class of mRNA's. Thus, we looked at a highly selected class of mRNA's. We must now consider the regulation of synthesis of mRNA's in the prevalent and rare classes. Several investigators, particularly those in Tom Humphrey's and Eric Davidson's laboratories, have compared the RNA populations specified by single-copy DNA sequences and found at various stages of development or in different adult tissues. A single example suffices to demonstrate the type of result from such comparisons. mRNA's from early embryos and those from adult intestines in sea urchins have 15 per cent of their mRNA sequences in common and differ in 85 per cent. Consequently, the proteins synthesized in intestinal cells and those made in early embryos differ markedly. Moreover, because rare and prevalent RNA classes constitute the majority of the mRNA's, we know that the two tissues differ in these classes of sequences. Now, if the differences in mRNA populations between these two tissues are a result of transcriptional regulation—the embryonic genes being transcribed in one case, the intestinal genes in the other—we reasonably expect the nuclear RNA's, the precursors of the mRNA's, to differ in a similar way. Surprisingly, as first noted in the laboratory of Tom Humphrey and confirmed in detail later by Davidson's group, the nuclear RNA's in various tissues differ, at most, only moderately. Davidson's laboratory reports that only 20 per cent of the nuclear RNA sequences differ between embryonic and intestinal cells. In other words, 80 per cent of the sequences found in the nuclear RNA of intestinal cells are also present in embryonic cells. Recall, however, that 85 per cent of the mRNA sequences differ between these tissues. We are hard pressed to explain the differences in mRNA's in intestinal and embryonic cells on the basis of transcriptional regulation. Indeed, it appears that most of the RNA's are synthesized in both tissues. The differences in mRNA populations must result from selective release or processing of the nuclear RNA's in the two tissues. In embryos one set of transcripts is processed and released into the cytoplasm. In the intestinal cells another set of transcripts is processed and released. In sea urchins the difference between tissues in mRNA's belonging to the prevalent and rare classes may primarily result from post-transcriptional regulation.

At present no one knows the basis for selective processing of mRNA's. We recall that the primary transcripts in eukaryotes are large, being several thousand nucleotides long, and five to ten times longer than the mRNA's. We recall also that single-copy sequences in sea urchins and in most vertebrates are mainly interspersed with moderately repeated sequences (pp. 82–83). On average, moderately repeated sequences of 200 to 400 nucleotide pairs alternate with single-copy sequences of 1,000 to 3,000 nucleotide pairs. Thus, on the average, we expect the repeated sequences to be represented in all or most of the primary RNA transcripts. Davidson and his colleagues have suggested that the selection of which transcripts are processed may depend in part on the moderately repeated sequence with which a single-copy sequence is associated. For example, single-copy sequences associated with moderately repeated sequences 0–100 may be processed in embryos, and those associated with moderately repeated sequences 85–125 may be processed in intestinal cells. The result is that only a small fraction of mRNA's would be shared in the two tissues. Whether the moderately repeated sequences function to regulate processing of nuclear RNA is for now speculation. If they do, there is one important implication: Those organisms such as *Drosophila* that have long-period interspersion—10 to 30 thousand base pairs of single-copy sequence between moderately repeated sequences—may have comparatively little or no post-transcriptional regulation.

SUMMARY

The state of knowledge about regulation of gene expression in higher eukaryotes lags far behind that of prokaryotes. Our first glimpses of gene regulation in eukaryotes indicate that regulation in eukaryotes is more complicated than in prokaryotes. The successful expression of genetic information in eukaryotes is regulated at both transcriptional and post-transcriptional levels. This is suggested by the mechanics of gene expression in the animal virus SV40 where synthesis of a particular polypeptide requires, first, the transcription of the gene, and, second, the correct processing of the RNA transcript into a functional mRNA. In animal cells themselves we find transcriptional regulation in the induction of transcription of specific chromosomal regions by steroid hormones acting in conjunction with receptor proteins. In addition, as revealed by studies on sea urchin development, post-transcriptional regulation occurs with primary transcripts in nuclei being very similar at two different developmental stages but with the mRNA's in the cytoplasm differing markedly. Also, in higher eukaryotes we find situations in which multiple genes, each encoding a similar, but slightly different polypeptide, are expressed at different developmental stages, for example, the genes for the β-like globin chains. Overriding all of these observations is the enigma of excess DNA in higher eukaryotes, where we find far more DNA, about a ten-fold excess, than is required to encode the number of genes *apparently* present. The function of the non-coding DNA, both within and between coding sequences, remains a mystery. Perhaps it plays a major role in regulation of gene expression; perhaps its role is yet to be defined; or even, most distressingly, perhaps it has no function at all and is evolutionary junk.

Genetic Frontiers

Aldous Huxley described in *Brave New World* a society of increased efficiency and conformity resulting from cloning and developmental and behavioral manipulation. We have described the creation of genetically doctored bacterial cells with increased synthesis of a specific polypeptide, the *lac* repressor. As our understanding of genetic processes in eukaryotes increases, we will undoubtedly achieve the technology necessary for the genetic manipulation of higher animals. Already we have the means to allow parents to choose the sex of their children because amniocentesis allows early identification of the sex of a fetus that could, if unwanted, be aborted. We will review several areas of current genetic research that will ultimately provide the knowledge and technology necessary for genetic manipulations of higher animals. As a consequence, this chapter will be somewhat amorphous, a genetic stew. Its topics will include the rise of phenotypically different organs and tissues during the development of multicellular organisms; the role of genes in behavior; antibody production; and the use of so-called genetic engineering to solve practical problems in medicine and agriculture. The first three topics are developmental ones, dealing with the ontogeny of complex phenotypes (behavior is a phenotype); the last one touches on genetic manipulations using recombinant DNA technology.

THE GENETICS OF DEVELOPMENT

Changes in cellular phenotypes occur during development in both time and space.

Gene regulation in prokaryotes is characterized by changes in gene transcription over time. During phage growth, *early* genes are responsible for the synthesis of products needed for DNA replication and *late* genes, for the synthesis of proteins needed for phage assembly. The *development* of a multicellular organism, that is, the series of changes that convert the zygote into an adult, is characterized by changes over both time and space. Early development commonly involves rapid DNA replication and cell division. Late development involves *differentiation,* the conversion of generalized cells into structurally and functionally diverse entities such as muscles, nerves, kidneys, and skin.

Differentiation is also associated with developmental changes in space as well as time. Eyes, after all, always form in roughly the same position in our heads, not in our torsos, and our hands arise at the ends of our arms, not protruding from our knees. How is it that cells that develop into eyes are always located in their characteristic positions and not elsewhere? How do the cells that form the eyes "learn" that they are to do so? There are no answers yet to these questions, only hypotheses.

Development of cells into different phenotypes results, in theory, from exposure to different local signals.

All of the cells in a multicellular organism are derived by mitoses from the zygote, and we assume that they are all genetically identical. To understand, in theory, how genotypically identical cells develop into cells that are physically, that is, phenotypically, distinct, we need only to consider the lessons learned from prokaryotic gene regulation. The induction of the *lac* enzymes in *E. coli* results from the addition of an inducer to the culture medium. Bacteria grown in the presence of lactose become phenotypically different from bacteria grown in the absence of lactose. The alternative phenotypes result from subjecting the cells to alternative environmental stimuli. We can surmise that, during the development of multicellular organisms, the differing phenotypes that genotypically identical cells achieve result from exposure of separate cells or cell groups to varying local environments. Put simply, cells that develop into muscle are exposed to a "muscle-directing" environment or signal, cells that develop into neurons receive a "neuron-directing" signal, and so on.

We can pose two schemes whereby cells receive signals directing them to develop into particular tissues. We can imagine that the daughter cells formed by cell division during early development in a multicellular organism, though presumably equal in nuclear genome content, may differ subtly in cytoplasmic contents. Furthermore, such differences may arise at every division. The cytoplasmic contents of cells may progressively diverge, and the genetic activity of each cell may also change as a result of exposure of nuclei to diverging microenvironments. Such divergent genetic activity may lead eventually to cells with different phenotypes. The developmental fate of any particular cell may be a direct result of cytoplasmic differences that arise internally as a result of a specific cell lineage. We will borrow from Sydney Brenner and call this developmental scheme the *European Plan*: Each cell knows its place in cellular society as a function of its ancestry.

An alternative scheme, called, of course, the *American Plan,* is one in which the particular place a cell resides, its external environment, is the determining factor. As an extreme example, we can imagine that any cell in a developing organism, when placed into a particular position, develops into the cell type normally occupying that position. For example, tissue removed from a region of an embryo that normally forms skin, if placed in the region normally occupied by an eye, develops into parts of an eye. The position must be changed sufficiently early in development; no change in development occurs if the eye has already formed in its normal position and the transplanted tissue has already taken on the characteristics of adult skin. In the American Plan a cell learns its developmental role while growing up, so to speak, as a result of its immediate environment. In an enriched environment, it may form part of an eye. In a deprived one, being only one of the "masses," it simply forms a bit of skin.

Probably no organism develops entirely on the European or on the American plan. We believe much of human, indeed all vertebrate development uses the American Plan. Human embryos are not, obviously, particularly choice subjects for experimentation. We turn to development in two organisms that in some aspects serve as models for the two developmental schemes. The first, perhaps exemplifying the European Plan, is *Caenorhabditis elegans,* a free-living nematode. The second, with some parts of its development exemplifying the American Plan, is our ubiquitous friend, *Drosophila melanogaster.*

The nature and life cycle of *Caenorhabditis elegans* are highly suitable for genetic and developmental studies.

The nematode, *Caenorhabditis elegans,* unlike its large cousin, the intestinal parasite *Ascaris lumbricoides,* which reaches a length of over 30 cm, is a tiny, 1–1.5 mm long, free-living roundworm that feeds on bacteria in soil. There are two adult forms, self-fertile hermaphrodites that produce both male and female germ cells, and slightly smaller males (Figure 18-1). The diploid adults have five pairs of autosomes. The hermaphrodite is XX and the male, produced as a result of non-disjunction, is XO. Development from fertilization through three larval stages to the adult worm takes about 3.5 days at 20° C (Figure 18-2). The small roundworms are easily grown in Petri dishes on a solid medium where—shades of phages—they feed on a lawn of *Escherichia coli.* A series of mutants have been recovered and characterized largely through the efforts of Sydney Brenner in Cambridge, England, and his students and associates. Two outstanding traits of the organism make it particularly suitable for developmental studies. First, it has a very small number of cells. At hatching the young larva contains only about 550 cells, and the adult, only 880 non-gonadal (somatic) cells along with about 2,500 gonadal cells. In contrast, tiny *Drosophila melanogaster* has in just one wing several thousand cells. The second useful attribute of *C. elegans* is that its skin is transparent so that by using light microscopy all of the cell divisions from fertilization to the adult organism can be observed. The combination of low cell number and transparent skin provides a unique opportunity to study the process by which cells come to differ from each other during development.

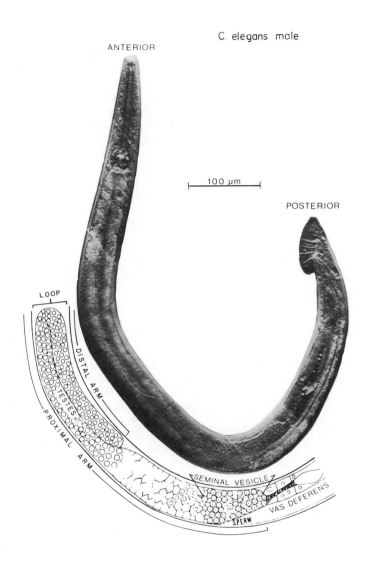

Figure 18-1.   Light micrographs and line drawings of the nematode *Caenorhabditis elegans*. Above, male. Facing, hermaphrodite. Above: Courtesy of David Hirsh, Department of Cell and Molecular Biology, University of Colorado, Boulder. Facing: From Hirsh et al., *Developmental Biology*, v. 49, p. 200, 1976, with permission from Academic Press.

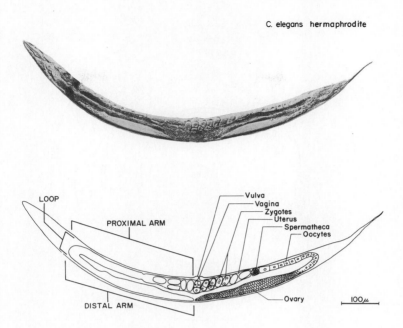

C. elegans hermaphrodite

The developmental fates of embryonic cells in *Caenorhabditis elegans* may be determined by their cell lineage.

Two laboratories have been particularly involved in tracing the lineage of the cells that give rise to specific adult structures during the embryonic and post-embryonic development of *C. elegans,* that of Gunther von Ehrenstein in Germany (recall his early work on tRNA function, p. 358), studying embryonic divisions, and that of John Sulston in England, studying divisions of larval cells. These biologists have determined, by observing cell divisions in living embryos and larvae, that in all specimens, the embryonic or larval cells that give rise to particular structures have essentially invariant cell lineages. For example, researchers in von Ehrenstein's laboratory have demonstrated that the 2,500 gonadal cells found in the adult hermaphrodite are always derived from cell P_4 shown on the cell lineage scheme in Figure 18-3. P_4 itself is, in turn, derived from the zygote by an invariant lineage. So, too, other structures are derived from other cells with invariant lineages (Figure 18-3).

The invariance of the cell lineage patterns during nematode development is not new knowledge; it has been recognized since the late nineteenth century when Theodor Boveri first described development in *Ascaris.* The existence of invariant cell lineage does not prove that lineage determines the developmental fate of a cell. The constancy may be present but not causal. Sulston and H. Robert Horvitz have made observations indicating that at least for some cells, lineage determines developmental fate. They

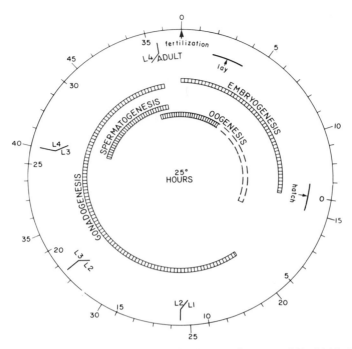

Figure 18-2.   Life cycle of *Caenorhabditis elegans.* Courtesy of David Hirsh, Department of Cell and Molecular Biology, University of Colorado, Boulder.

have killed specific embryonic cells using a microbeam laser. The structures normally derived from the killed cells are absent in the adult. Surviving cells do not replace the killed ones by, for example, undergoing additional cell divisions, even though they may physically occupy the position of a dead cell in the developing embryo. Thus, for these cells their position in the embryo does not determine their developmental fate.

In *Drosophila,* cells that give rise to adult structures are sequestered in larvae in small organs called *imaginal discs.*

The development of *Drosophila* contrasts strikingly with that of *C. elegans.* Following fertilization sequential, synchronous nuclear divisions occur, one division every ten minutes for about one hundred minutes, to produce, in the absence of cell division, an embryo composed of only one cell, but of 1024 (2^{10}) nuclei. The nuclei then migrate to the surface where they become enclosed in membranes to form a cell layer one cell thick. Now both nuclear and cell divisions occur, followed by cell and tissue migration that produce the internal and external structures that subsequently form the larva. Small ingrowths from the surface produce packets of cells called *imaginal discs.* The imaginal discs are the embryonic precursors of the

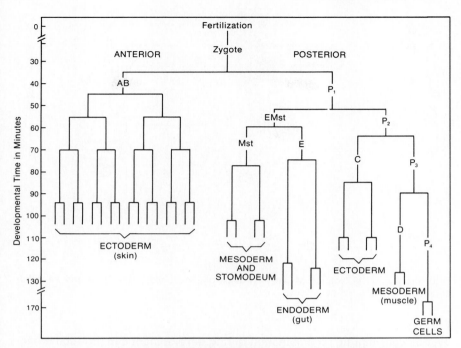

Figure 18-3. The cell lineage of *Caenorhabditis elegans* during early embryogenesis. After Deppe et al., *Proceedings of the National Academy of Sciences, U.S.,* v. 75, p. 376, 1978.

external structures of the adult insect, the *imago*. Following hatching, the larva feeds and grows and passes through three larval stages and then undergoes metamorphosis as we have described previously (pp. 473–474). The adult has eyes, wings, and legs and differs strikingly from the larva. The various adult external structures form from specific imaginal discs, eyes from eye discs, legs from leg discs, wings from wing discs (Figure 18-4). We know that the imaginal discs in the last larval stage have already received their instructions as to the structure into which they are to develop, but we do not know how these instructions are imparted. In *Drosophila* we believe that cells learn their developmental assignments by being in a particular place at a particular time.

Studies using somatic mosaics suggest that a cell's developmental fate in *Drosophila* is not determined by cell lineage.

The number of cells in the adult fly—well over one million—precludes tracing their lineage by direct observation. The use of somatic genetics, however, provides an opportunity to learn that the cell lineage patterns are not primarily responsible for developmental fate. We recall that in hetero-

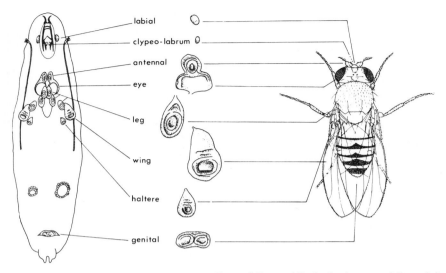

Figure 18-4.   The location of imaginal discs of *Drosophila* in the larva and the adult structures into which they develop.

zygous cells mitotic recombination during development can produce homozygous somatic cells whose clonal descendents can be identified as a patch of phenotypically distinct tissue on the surface of the adult (pp. 246–249, Figures 9-7 and 9-8). For example, cells that are homozygous *singed* (*sn/sn*) have *singed* bristles and are readily distinguished from heterozygous *sn/+* cells that appear normal.

If there is a constant lineal history, we expect neighboring bristles, each of which forms from a single cell, always to be either related or unrelated in their lineage, for example, both always being descended from a particular precursor cell. Assuming the clone derived from the recombinant cell is large enough and is appropriately placed, we expect to find that any mosaic patch that contains one bristle always contains the other bristle. What numerous *Drosophila* geneticists have found, dating back to Alfred Sturtevant in the 1920's, is that sometimes both bristles arise from cells derived by mitoses from the same precursor cell and sometimes from cells that belong to separate lineages. We can conclude from this that the cell lineage patterns leading to the production of the two bristles vary and that the formation of the bristles in their particular locations is not a result of their lineage.

A similar conclusion is drawn from a different kind of experiment by Peter Bryant, his co-workers, and others. These investigators found that if a piece of a wing disc is removed by surgery, under some circumstances the remaining wing tissue will grow and replace the missing tissue, that is, *regenerate.* Upon development all of the wing structures are found. Cells that were originally intended to form one set of structures within the wing can undergo additional cell divisions and replace missing structures. This result contrasts with the one described in *C. elegans* in which killing a single progenitor cell early in development eliminates structures normally derived from that progenitor cell.

Progressive restriction in developmental fate arises in increments during *Drosophila* development.

It is not true, however, that any structure in *Drosophila* can be replaced by any other structure or even that any region within a given tissue, for example, an imaginal wing disc, can be replaced by neighboring cells. Colin Murphy observed that when one of two wing discs was entirely removed from a larva, the fly that developed, even though there had been time for regeneration, always lacked one wing. None of the remaining tissues, including the other wing disc, was able to regenerate the missing wing. The developmental capacities of the remaining tissues are restricted. In multicellular animals developmental restrictions arise progressively. In some animals, such as *Caenorhabditis elegans,* the restrictions apparently arise early, the zygote being the only cell capable of producing any structure in the organism. In other organisms, the restrictions arise more slowly, many embryonic cells retaining the ability of forming any structure.

Observations originally made in the laboratory of Antonio Garcia-Bellido in Madrid suggested how developmental restrictions arise. According to the model put forth by Garcia-Bellido, at a particular time in development the cells in a specific limited region of tissue called a *compartment* become developmentally restricted in unison. Garcia-Bellido and his co-workers formulated their model by analysing clones from cells made homozygous by somatic recombination. Early in development any cell destined to form part of the wing could form *any* part of it. However, later cells could only give rise to structures either in the anterior half or compartment of the wing, or in the posterior half or compartment. The cells had become restricted in their developmental fates. Later cells within each compartment, for example, the anterior compartment, became further restricted, forming only structures on the top of the wing blade or on the bottom. Still later cells were so limited as to form structures within even smaller regions of the wing. Thus, as each smaller compartment is demarcated, cells become incrementally limited, restricted to forming ever more precisely limited regions of the adult. Each of these progressive limitations occurs to a group of cells simultaneously in a specific region of tissue at a particular time in development.

Progressive developmental decisions may involve dichotomous choices to repress or derepress groups of genes in response to geographic cues.

We may well wonder about the basis of the progressive restriction of developmental fate that occurs during *Drosophila* development. By examining a class of mutants called *homeotic* mutants, we can deduce some possibilities. Homeotic mutants all cause one region of the fly to develop structures characteristic of another region. For example, one mutant, *Antennapedia (Antp)*, is a homozygous lethal but as a heterozygote causes the

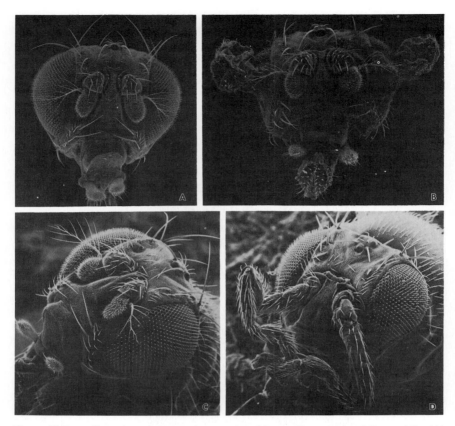

Figure 18-5.   Scanning electron micrographs of homeotic mutants of *Drosophila*. (A). Front view of normal head. (B) Front view of head of *opthalmoptera* mutant showing wings extending from eye regions. (C) Side view of normal head. (D) Side view of head of *Antennopedia* mutant showing legs replacing antennae. (A) and (B) courtesy of John Postlethwait, Department of Biology, University of Oregon, Eugene. (C) and (D) courtesy of Larry Salkoff, Department of Biology, Yale University, New Haven, Connecticut.

antenna disc to develop not into an antenna but into a leg like one of those in the second pair of legs on the fly's thorax. Thus, instead of two antennae on the head of a fly, there may be two legs (Figure 18-5). Another homeotic mutant, *Opthalmoptera,* causes wings to develop in the head (Figure 18-5). In still another, *tumorous head,* head structures are replaced by external genitalia. One important lesson we learn from homeotic mutants is that the failure of a single gene can switch the developmental fate of a tissue. We can view homeotic mutants as mutations in genes whose products, in an unknown manner, regulate the activity of other genes necessary for the development of particular tissues. Considering the development of the second pair of legs and the antennae, we speculate that normally the *Antp* gene is derepressed in second leg discs so that legs can form. In the head region in the embryo, the gene is repressed and its product is not synthesized so that antennae form. We assume that in the mutant the *Antp* product is synthesized in antennal discs. In the *Antp* heterozygote the presence of the *Antp* product causes the antennal discs to develop into legs. We deduce from properties of the mutant that alternative states of gene function regulate the development of antennae versus legs. In one state the gene is "on" and legs develop, in the other state the gene is "off" and antennae form. If the gene mutates so that it is on in antennal discs, legs automatically develop there.

By comparing the properties of homeotic mutants, that is, the developmental change that each mutant causes, and the positions of compartments during embryogenesis in *Drosophila,* Stuart Kauffman and his colleagues have proposed a scheme of developmental decision-making that depends on the activation of single genes or sets of genes. For early development, Kauffman proposes four pairs of alternative functional states that are either "on" or "off" at different times and places in the embryo. We can consider these gene sets to be either repressed, the 0 state, or derepressed, the 1 state. A gene is either repressed or derepressed because a cell is in a particular place at a particular time during development—ah, the American plan. The first compartment is formed between the anterior and the posterior of the embryo, with the anterior being identified as having one gene set turned on (1 0 0 0) and the posterior remaining off (0 0 0 0). The next compartment bisects each of the original two to produce four compartments each with a different pattern of gene function (Figure 18-6). Through another change in the activity of the functional elements these four compartments are divided into anterior and posterior halves, creating eight compartments each with its own activity profile. Another change in activity state divides these eight compartments into tops and bottoms to form sixteen compartments, each of which has a different activity profile for the four pairs of functional states. Kauffman interprets the effects of the different homeotic mutants as involving changes in activity of one of the pairs of alternative functional states, of *gene sets*, that produce the embryonic compartments. Thus, we see that any structure has a specific binary Zip Code, for example, the region that gives rise to the antenna is 1 0 1 0, that which produces the second leg, 1 1 1 0. For the antenna to be converted to a second leg a change in function of only one gene set is required. Right or wrong, the impressive aspect of Kauffman's model, like the model of DNA proposed by Watson and Crick, is its simple elegance. We can visualize the developmental progression from the zygote to the adult as a series of alternative decisions made one after another in response to different geographic cues. In the development of *Caenorhabditis elegans,* we can imagine that the basis for making the alternative decisions depends on alternatives created as a result of the lineage of particular cells, with each cell division during early development providing each alternative. In *Drosophila* development and possibly in that of other higher animals including humans, groups of cells in a compartment recognize in some manner their positions in a developing embryo and make in unison the appropriate response to the position.

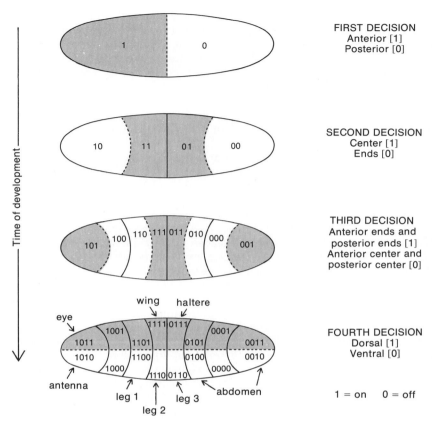

Figure 18-6.   Compartment formation during early *Drosophila* embryogensis resulting from a series of binary decisions so that each compartment can be assigned a "zip code." After Kaufman, S., et al., *Science,* v. 199, p. 259, 1978.

BEHAVIORAL GENETICS

Single gene changes affect behavior dramatically.

Courtship in *Drosophila* lacks many of the refinements found in its human counterpart. Males chase after females and go into a brief, stylized mating dance involving wing vibrations—in place of soft music—before mounting females. The success of their advance depends entirely on the female who may reject the male by literally kicking him away. We wonder what attracts males to females and vice versa, not only in *Drosophila* but in other species as well.

In *Drosophila melanogaster,* genetic manipulations provide some interesting answers. One approach is to construct flies that are genetically part male and part female. Such flies, called *gynandromorphs* or simply *gynanders,* are made through the use of a ring-shaped X chromosome. When heterozygous with a normal X chromosome, the ring X chromosome is occasionally lost during the first few mitotic divisions following fertilization. Two types of cells are produced, female ones with two X chromosomes and male ones with one X. Cells derived from these cells are, respectively, female and male. If the loss of the ring X chromosome occurs during the first mitotic division, the adult fly produced is half male and half female. Such gynanders are often bilaterally symmetric; all of one side being male, the other side being female. Sometimes the anterior of the fly is male and the posterior is female. If the head is male, despite the presence of a female reproductive system in the abdomen, the gynander behaves like a male and chases and courts females. If two such gynanders are placed together, they will chase each other around. As long as the cells in the fly's head are male—presumably including those of the brain—the fly behaves like a male.

The behavior of male flies is strikingly affected by a mutation originally isolated by Kalbir Gill and now studied by Jeff Hall and his associates. The recessive mutant, *fruitless (fru)*, maps to position 62 on chromosome three. Males, homozygous for *fru,* produce no offspring just as the name of the mutant indicates, whereas *fru* females are fertile. The *fru* condition does not involve any dysfunction in the male genitalia or sperm, which appear normal. Instead homozygous *fru* males behave abnormally. They court normal males and each other, lines of *fru* males following each other around in culture bottles. When placed with females exclusively, *fru* males will court the females, though with little vigor, but do not mount them. Finally, normal males will court *fru* males, whereas they seldom court normal males. Normal males also court isolated male *fru* abdomens, approaching the abdomen and doing their dance complete with wing vibrations. The behavior of the *fru* males does not attract normal males nor does the appearance of the *fru* male abdomens, which, to the human eye, appear normal. Apparently normal male flies are attracted to females by a chemical attractant, a sex *pheromone*. We believe that the *fru* males make the sex pheromone normally found in females. The basis for the abnormal production of the pheromone is not understood. Nevertheless, in *fru* males significant deviation from normal behavior results from the mutation of a single gene that, we can be confident, encodes a single polypeptide product.

Normal behavior depends upon both the correct function and interconnection of nerve cells.

If you step on a tack with a bare foot, you will jerk your foot back and at the same time make a short adjustment with your other leg to keep your balance. You might also respond with a vocal expletive. The first two of these responses are reflexes, but the vocal response depends on complex neural function. These behavioral responses depend on the correct connection between a very long sensory nerve running from the foot up the leg to the spinal cord and motor neurons running to muscles in both legs. The vocal response depends on neurons running to regions of the brain that control speech. Only the vocal response requires brain function; the others are mediated entirely by nerves located outside of the brain. The success of the responses depends on several conditions: The sensory and motor neurons must be connected correctly; the stimulus from the foot must be transmitted along the axon of the sensory neuron as an impulse of electricity resulting from the rapid movement of ions across nerve cell membranes; the connections between the sensory and motor neurons, *synapses*, across which chemicals called *neurotransmitters* are released, must function normally; an electrical impulse must travel down the motor neuron to the muscle and cause the release of transmitters across the neural muscular junction by which the nerve cell is connected to the muscle. These responses have adaptive value. The rapid removal of the foot protects against further damage; the expletive warns nearby members of our species about a real danger. Both are evolutionarily favored and are specified by the genome. The failure of these responses or of any behavioral activity can be a result of the failure of any of the components necessary for this reflex.

Mutants affect either the development or function of the nervous system.

In general we can divide adverse genetic effects on behavior into two categories: 1) Those that affect the nervous system by interfering with development of the neurons and their connections, that is, of the wiring of the nervous system. Included are those defects that eliminate nerve cells, for example, the *eyeless* mutant in *Drosophila.* 2) Those that affect the function of the nervous system, that is, interfere with the functional properties of nerve cells. Included in the second category is, probably, the Lesch-Nyhan syndrome, resulting from a deficiency in HPRT (p. 252) and causing severe behavioral abnormalities.

Developmental abnormalities. We will consider several mutants in the house mouse, *Mus musculus,* that affect the cerebellum. The cerebellum is the part of the brain that regulates muscular movement, and all of the mutants, carrying names like *swayer* and *leaner,* each an autosomal recessive, affect coordination and general muscle tone. Superficially at least, the phenotypes of three of the mutants, *weaver, staggerer,* and *reeler,* are similar. In the structure of the cerebellum, however, each of the three mutations has a unique effect, and each maps independently in the mouse genome.

The cerebellum is part of the hind brain and contains an enormous number of nerve cells, estimated in humans to be around 10^{10} to 10^{11}. Yet only five types of nerve cells are present, with each type residing in a specific layer in the cerebellum (Figure 18-7). The nuclei of the golgi and the granule cells are located in the granular layer, those of the purkinje cells in the purkinje layer. The basket cells sit between the purkinje layer and the molecular layer in which the stellate cells are found. The basket, golgi, granule, and stellate cells are all neurons that transmit or retransmit sensory input from sensory receptors located in muscles, skin, and joints. All of these cells act, directly or indirectly, on the purkinje cells, the only neurons in the

1 = MOLECULAR LAYER
2 = PURKINJE CELL LAYER
3 = GRANULAR LAYER

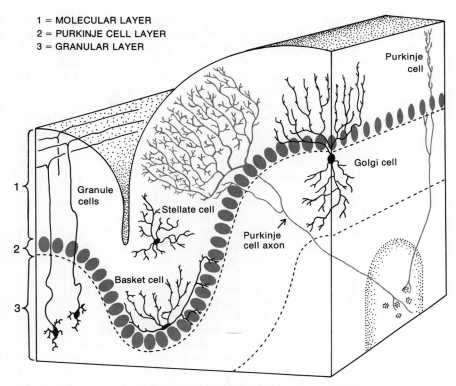

Figure 18-7.   The structure of mouse cerebellum. After Ramon y Cajal, S., *Histologie du systeme nerveux.* II. C.S.I.C., Madrid, 1955, and Palay and Chan-Palay, *Cerebellar Cortex,* Springer-Verlag, Berlin, 1974, and Kuffler, S. W., and J. Nicholls, *From Neuron to Brain,* Sinauer Associates, Inc., Sunderland, Mass., 1976.

cerebellum that provide output to peripheral structures such as muscles. The interconnections, the *synapses,* between the granule and the purkinje cells are particularly elaborate. Any mutation affecting the function of inter-connections of these cells will have dramatic consequences on coordination.

Three mutants, *weaver, staggerer,* and *reeler,* have been studied in some detail by Richard Sidman and his associates. In homozygous *weaver* mice the primary defect is the total absence at any stage of development of the granule cells. In the *staggerer* homozygote, granule cells are also deficient, but the development of the mutant phenotype differs. In *staggerer* mice the embryonic precursors to the granule cells are present, and appear to be normal but then degenerate. Also, the development of the purkinje cells is slow. Purkinje cells normally migrate to their position in the purkinje layer and then undergo development. In *staggerer,* the purkinje cells arrive at their normal station on schedule but then develop slowly. The slow development in the formation of contacts, that is, synapses, between the purkinje cells and the sensory cells in the cerebellum may be responsible for the degeneration of the granule cell precursors. Perhaps the most interesting mutant of the three is *reeler.* The purkinje cell precursors normally migrate to their positions in the purkinje layer and lie external to granule cells. In *reeler,* the purkinje cells migrate to an abnormal location, mainly internal to the granular cells. Many of the purkinje cells fail to send out processes to form synapses with the granule cells. The number of granule and purkinje cells is essentially normal, but the cells fail to establish a normal array of synapses. Thus, although the behavioral phenotypes of these three mouse mutants are similar, at the cellular level each results from a different developmental defect.

Abnormalities in neural function. In the *staggerer, reeler,* and *weaver* mutants the underlying basis of the defects stems from a loss of neurons or the failure of neurons to make normal connections. Therefore, the defects are all developmental failures that lead ultimately to abnormal neuronal wiring. Mutants also exist in which the function of nerve cells is defective but in which the wiring is apparently normal. We might guess that if neurons do not function properly, the mutation might be lethal. If, for example, nerve cells did not conduct electrical impulses along their axons, the organism could not move and would no doubt die. Conditional mutants are very useful in the study of these mutations. In *D. melanogaster* several mutants have been identified that are normal at low temperatures but are paralyzed at slightly elevated temperatures. Because the mutant flies behave normally at low temperatures, we can conclude that the wiring of the nerve cells is normal. The paralysis must then result from loss of function. Several temperature sensitive (*ts*) recessive mutants become paralyzed at elevated temperatures, including the X chromosome ones, *paralytic^{ts}, shibire^{ts} (shibire* is the Japanese word for "paralyzed"), and the second chromosome mutant *nap^{ts}.* The mutant name *nap,* in addition to describing the phenotype literally, is an acronym for **n**o **a**ction **p**otential. In *nap* flies, at low temperatures, when nerve cells are stimulated electrically, an impulse of electricity—an action potential—travels down the axon as a result of the flow of sodium ions across the cell membrane from inside to outside. In *nap* mutants at elevated temperatures, no impulse of electricity travels along the axon, suggesting that the flow of sodium ions across the cell membrane might be blocked. In contrast to the mutants in mice, *nap* and the other mutants noted here are functionally defective: Paralysis results from a physiological failure not a developmental one.

In higher vertebrates, development of normal behavior and nerve cell function depends on receipt of environmental stimuli.

In the behavioral mutants discussed, starting with *fruitless* in *Drosophila* and including the mouse mutants and the paralytic ones, we have shown that genetic changes result in striking behavioral changes. We do not wish, however, to leave the impression, particularly in respect to higher vertebrates, that there is a precise one-to-one relationship between behavior and genetic constitution, that someone is "crabby" merely because he has "crabby" genes, that complex behavioral patterns are direct inevitable results of genetic constitutions. A single example suffices. If a newborn kitten has a translucent cover placed over one eye that prevents the visualization of objects but not the transmission of light for six to eight weeks after birth, the kitten will become permanently functionally blind in that eye. If the cover is not applied until eight weeks after birth and is left on for eight weeks, upon removal the eye will function normally. The experiment indicates that the proper neurophysiological development of eye function during the first eight weeks after birth requires visual stimuli from discrete objects. Once that development has occurred, visual stimuli can be removed without causing functional damage.

No great imagination is required to extend this kind of observation to areas of human behavior such as speech, learning ability, intelligence, and personality and so realize that the absence of appropriate stimuli during the early life of children can have enormous, irreversible effects on the ultimate behavioral phenotype. This is not to say that mutational defects do not alter human behavior; witness the Lesch-Nyhan syndrome. The development of complex behavioral attributes results from complex interactions between genetic and environmental influences in which both play significant roles. We should keep these observations in mind when evaluating claims about the relative roles of genes and environment in determining behavioral characteristics of groups or of individuals exposed to different social and educational environments. For example, much has been written about the basis for the observation that, on average, blacks score somewhat lower on intelligence tests than whites in the United States. Whether these differences reflect genetic differences in neural function, environmental differences that affect neural function, inherent cultural biases in the tests, or some complex mixture of all three, remains an area of hot debate which, because of the inadequacy of available analytic tools and procedures, we feel is justifiably categorized as speculation and not science.

THE ORIGINS OF ANTIBODY DIVERSITY

In our discussion of red blood cell antigens (ABO, MN, rh, pp. 162–163) we noted that antigens are substances that elicit the formation of antibodies. Antibodies are proteins called *immunoglobulins* that combine with specific antigens, such combination eventually leading to the destruction of the antigen. Prior to exposure, antibodies specific for the antigen are not found in the blood. After exposure, over a period of several weeks, the level of an immunoglobulin specifically combining with the antigen increases as a result of increased synthesis.

We can ask how antibodies specific for different antigens are made. The question is a fascinating one. There are tens of thousands of different antibody molecules, all structurally similar but not identical. Mechanisms must exist for providing the right immunoglobulin to combine with a particular antigen and for a high level of synthesis of this immunoglobulin molecule in response to antigenic stimulation.

Immunoglobulins are composed of one pair of large polypeptides, the *heavy chains,* and one pair of small polypeptides, the *light chains.*

We have extensive knowledge of immunoglobin structure owing to efforts in many laboratories catalysed by the pioneering research from the laboratories of R. R. Porter in London and Gerald Edelman in New York City. Humans have several kinds of structurally similar immunoglobulins, called immunoglobulins G, A, M, D, and E and abbreviated IgG, IgA, IgM, IgD, and IgE. We shall be primarily concerned with IgG, which constitutes about 80 per cent of the total immunoglobulin found in blood. Immunoglobulins are composed of one pair of large polypeptides, each 53,000 D, called *heavy chains,* and one pair of small polypeptides, each 22,500 D, called *light chains.* The two heavy chains in IgG are called γ chains and are paired with a pair of either of two types of light chains, λ or κ (Figure 18-8). The regions of immunoglobulins that combine with an antigen are near the N-terminal ends of the two pairs of polypeptides, both light and heavy chains binding to an antigen. Each molecule of IgG has two sites that can combine with an antigen (Figure 18-8). Thus, it is possible to form networks of immunoglobulin and antigen molecules (Figure 18-8) that result in a precipitate.

Antigenic specificity of each antibody is determined by the amino acid sequence at the N-terminal end of both light and heavy chains.

What is the basis of antibody specificity? The answer is found in the primary structure of immunoglobulins. The amino acid sequences of the 330 C-terminal amino acids are the same in all IgG heavy chains. The amino acid sequences of the 105 C-terminal amino acids are the same in the light chains. The N-terminal regions of immunoglobulins have variable amino acid sequences in the first 110 amino acid residues of both heavy and light chains. Within the variable regions at the N-terminal ends are regions with highly

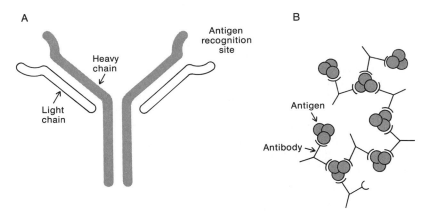

Figure 18-8.   (A) The structure of an immunoglobin (antibody) molecule. (B) Network formed between antibodies and antigens.

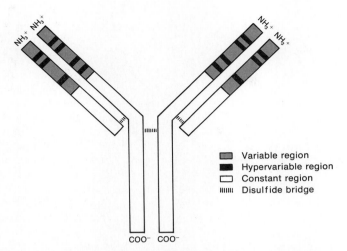

Figure 18-9.   The structure of immunoglobin showing constant, variable, and hyper-variable regions.

variable amino acid sequences (hypervariable regions), three in the light chains and four in the heavy chains (Figure 18-9). The variable regions of the light and heavy chains interact with antigens. The specificity of binding of antibody to antigen resides in the amino acid sequence in the variable region of each IgG molecule. The seemingly endless array of immunoglob-ulins specific for different antigens stems from an equally extensive array of amino acid sequences in the variable regions of IgG molecules.

Clonal selection: Increased synthesis of a specific immunoglobulin results from increased mitotic growth of cells producing that immunoglobulin.

The increase in the serum of an immunoglobulin specific for a particular antigen results from increased synthesis of that immunoglobulin. Immuno-globulins are synthesized by plasma cells, a form of white blood cell, with each plasma cell synthesizing only one type of IgG molecule. The plasma cells are produced by divisions of precursor cells called *plasmablasts* that stem from small lymphocytes that are located in the lymph nodes (Figure 18-10). There are many types of small lymphocytes, each of which gives rise through cell division to a clone of plasma cells that produce a specific IgG molecule. Small lymphocytes retain in their cell membranes the immuno-globulin molecules that they produce. According to the *clonal selection theory,* when an antigen recognized by the immunoglobulin on the cell surface binds to the surface, the small lymphocyte is stimulated to divide rapidly. The increased rate of division leads to a specific increase in a clone of cells responsible for synthesizing the type of IgG molecule that binds that particular antigen. The antibody synthesized by these cells is released into the serum to combine with the antigen. Thus, the increased synthesis of an

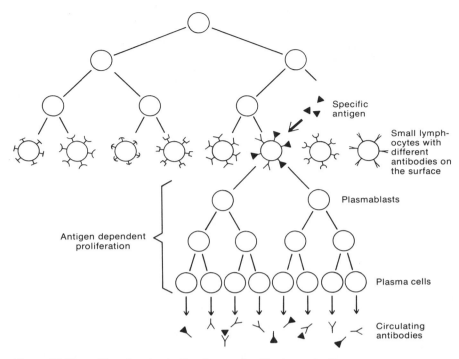

Figure 18-10.   The clonal selection theory of antibody production.

immunoglobulin for a specific antigen results from increased production of a clone of cells that make the immunoglobulin.

Based on our considerations of gene regulation we might suspect that the gene product encoded by a particular IgG gene is expressed in the responding lymphocyte. We might also predict an enormous number of IgG genes, each encoding a light or heavy chain with the same constant regions but differing in the variable regions. Such a belief is erroneous.

The number of immunoglobulin genes in embryonic cells is small.

The immunoglobulins of mice have structures similar to those of humans with constant carboxyl terminal regions and variable amino terminal regions. Because mice are far more amenable to experimentation, we know more about the immunoglobulin genes in mice than in humans. Mouse immunoglobulins are composed of light chains with three different types of constant regions, called κ, λ_I, and λ_{II}, with 95 per cent of the light chains in the blood having the κ sequence. We estimate, for example that plasma cells in mice synthesize light chains with about 1,000 different variable regions joined to the κ constant region. If there is a gene for each light chain, we expect to find about 1,000 genes each composed of a region encoding a particular variable region along with one of the three kinds of constant regions. In germ line and embryonic cells only 300 sequences exist in the DNA that encode the variable regions. Furthermore, and more astonishingly, only one coding sequence for the κ constant region is present per haploid genome. Two problems are evident. First, how are 300 copies of the coding sequences for the variable regions accommodated with only one copy of the coding sequence for the κ constant region? Second, how are 300 copies of coding sequences for the variable regions turned into the 1,000 genes encoding all of the different variable regions in plasma cell DNA, that is, what is the source of the great diversity in immunoglobulin structure?

Immunoglobulin genes become rearranged during development.

We can ask ourselves the question: How can two apparently separated coding regions, one encoding the constant portion of an immunoglobulin polypeptide and one a variable portion, produce a single polypeptide? Two answers come to mind. One is that the constant region coding sequence and the variable region are transcribed together to produce a single transcript, and the intervening sequence between the two coding regions is removed during processing to produce an mRNA with the fused coding sequences. A second answer we might consider is that a genetic rearrangement occurs during development of the lymphocytes so that the two coding regions in the DNA are fused. A region between two coding regions could be deleted, perhaps, as we saw with hemoglobin-Lepore, as a result of unequal crossing over. Alternatively, the DNA between the two coding regions could be removed by the formation of a stem-loop structure, like that envisioned for transposons (p. 284), and the excision of the loop. It is likely that both removal of an intervening sequence through processing of a pre-mRNA and a genetic rearrangement in somatic cells are necessary to produce the various immunoglobulin chains. A group of researchers in Basel, headed by Susumu Tonagawa, have with a cDNA clone to a specific light chain mRNA demonstrated that somatic rearrangement produces the light chain. These workers find that in embryonic DNA the variable and constant regions encoding the specific λ light chain are located in two separate chromosomal fragments created by digestion of total mouse embryonic DNA by the restriction endonuclease, *Eco*RI (Figure 18-11). In contrast, when plasma cell DNA is digested by *Eco*RI, in addition to the fragments found in embryonic DNA, there is a single fragment of DNA in which the coding sequences for both the constant and variable regions are found together. The left and right ends of this DNA fragment are identical to the left and right ends of the two separate fragments found in the embryonic DNA (Figure 18-11).

One plausible explanation for the production of the new gene arrangement in plasma cells is that it results from unequal crossing over during the development of a lymphocyte as depicted in Figure 18-12. More likely, the formation of a stem-loop structure and the excision of the loop with the joining of the cut ends could also bring the two coding regions together (Figure 18-13). As a result of rearrangement, one homolog has a rearranged

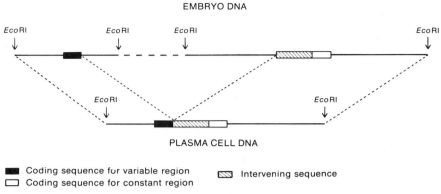

Figure 18-11.   Restriction fragments containing a specific mouse immunoglobin light chain produced by *Eco*RI digestion of embryonic (top) and plasma cell DNA (bottom). After Brack et al., *Cell,* v. 15, p. 1, 1978.

chromosome with the coding regions of the variable and constant regions close together, and the other chromosome has the two regions far apart. Endonuclease digestion of the DNA of such a cell will produce three kinds of *Eco*RI fragments containing light chain coding regions. Two are found in the embryonic DNA and one is found only in the plasma cell DNA. We note, however, in the plasma cell chromosome that the variable and constant coding regions are not fused; a non-coding region remains between them. We presume that the entire rearranged region is transcribed as a whole and that the transcript is then processed to remove the intervening sequence and produce an mRNA encoding one particular λ light chain.

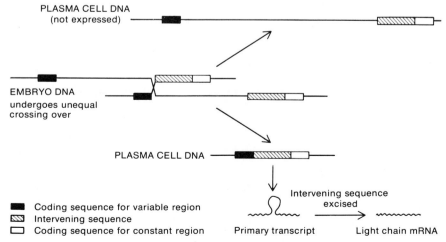

Figure 18-12.   The production of an immunoglobin light chain gene as a result of unequal crossing over.

EMBRYO DNA

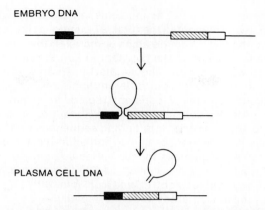

PLASMA CELL DNA

Figure 18-13.   An alternative method for rearranging the coding sequences for variable and constant regions. Inverted repeated sequences pair to form a "lollipop" which is then excised.

In this particular example we have demonstrated two possible ways in which a particular light chain can be generated from two coding sequences that in embryonic DNA are physically separate. Such rearrangements account not only for the generation of this particular λ chain but also for the juxtaposition of the constant region with a large number of alternative variable regions. We presume that the 300 coding sequences for the variable regions are arranged essentially as a series of tandem repeats linked to the coding sequence for the constant region. Through genetic rearrangement, the sequence encoding the constant region can be juxtaposed to any variable region sequence. For example, in one case rearrangement can place the constant region next to variable sequence 1, in another next to variable region 2, or 3, and so on. Rearrangements can occur in numerous lymphocytes and produce a group of cells, each of which can give rise to a clone of lymphocytes synthesizing a different light chain. Similar mechanisms can be invoked to produce different heavy chains. Upon antigenic stimulation, the cell that happens to manufacture an immunoglobulin that binds to the antigen is selected for growth. The resulting clone of cells then produces substantial quantities of antibody. The chromosomal rearrangements postulated to produce different plasma cells are examples of mitotically dividing somatic cells that are *not* genetically identical. We can speculate whether similar manipulations of genomes are involved in other developmental processes and to what extent our assumption is valid that mitotically derived cells in a developing organism are identical.

The generation of different forms of variable chains also involves genetic rearrangements during lymphocyte development.

We are still confronted with the problem of producing 1,000 different variable regions from a few hundred germ line or embryonic sequences. We can produce increased diversity in the variable regions by invoking, again, genetic rearrangements. The 110 amino acids in the κ chain variable regions are divided into two segments; the first 97 amino acids comprising the variable segment, the last 13 amino acids comprising the *joint* segment. The variable segment is attached to the constant region through the joint. Philip Leder and his co-workers have shown that in embryonic DNA there are separate coding sequences for four different joints residing in tandem near the coding sequence for the κ constant region. In plasma cell DNA the coding sequence for the constant region is joined, as a result of genetic rearrangement, to any one of the four coding sequences for the joints. These four constant plus joint sequences can be joined in plasma cell DNA to any one of the 300 coding sequences for the variable segments, creating 1,200 possible genes for κ light chains. Similarly, 1,000 or more heavy chains could be made and, if any pair of light chains can join with any pair of heavy chains, over one million different antibody molecules could be produced. Thus, genetic and molecular rearrangements can generate all of the diversity found in immunoglobulins.

In summary, the enormous number of immunoglobulins and their production in response to an antigenic challenge depends on an intriguing cooperative effort of genetics and cell biology. Because of genetic rearrangements, an enormous series of lymphocytes is produced that synthesize the different light and heavy chains comprising immunoglobulin molecules. An antigen binds to the cell surface of a particular lymphocyte because of the presence of immunoglobulins in the cell membrane. As a result of that binding, that particular lymphocyte is stimulated to divide, producing a clone of cells, all of which synthesize an immunoglobulin that binds that particular antigen. As a result of increased cell division of cells synthesizing a particular antibody, the level of that antibody increases in the serum where it functions to protect against an invading organism or to complicate tissue and organ transplants.

GENETIC ENGINEERING

A key to understanding the probable mechanisms by which antibody diversity arises during lymphocyte development was the use of a cDNA clone to the mRNA of a particular plasma cell light chain. This is but one more example of the use of recombinant DNA technology to gain critical insight into genetic and biological phenomena. Recombinant DNA technology often is referred to in lay journals, for example, in *Time* magazine, as *genetic engineering*. Genetic engineering has been hailed as providing new opportunities for understanding gene function and regulation, particularly in eukaryotes, and providing tools in the service of agricultural and medical sciences for crop and livestock improvement and cures for diseases. Yet "geneticists" have been practicing genetic engineering for millennia. Fossilized ears of corn dating from 5,000 to 400 years ago have been unearthed

Figure 18-14.   Fossil ears of corn recovered in the Tehuacan Valley in Mexico. The smallest is about 7,000 years old, the largest about 400 years old. From Manglesdorf et al., in *The Prehistory of the Tehuacan Valley,* Douglas S. Byers, Ed. University of Texas Press, 1967.

in Mexico (Figure 18-14). The size of the ears of corn increases strikingly from the earliest to the latest example, which, compared with modern hybrid corn, is still small. The increase in the size of the ears of corn and, presumably, in their nutritional value as well probably resulted from the selective recovery of progressively larger varieties, that is, from changes in the genetic constitution of the corn. The domestication of corn in the Western Hemisphere and of wheat in Egypt is a striking example of the success of early genetic engineers. We will look here, however, at the new genetic engineering, and some of the possible benefits and problems associated with its use.

Recombinant DNA technology: the dark side.

Ring around the rosy.   Red spots surrounded by rings of discoloration are symptoms of bubonic plague.

Pocket full of posy.   Herbs were believed to protect against the plague.

Ashes, ashes.   Ashes from fires lit to destroy corpses and plague-
                                  ridden areas of London.

All fall down.   Everybody dies.

What would happen if the gene or genes responsible for the pathogenicity of an organism such as *Pasteurella pestus,* the plague bacterium of the
Black Death of the Middle Ages immortalized in "Ring Around the Rosy,"
were to be moved into *Escherichia coli,* the common bacterium residing in
the intestines of all of us? What would happen if an *E. coli* strain carrying
a gene or genes from a cancer virus were let loose? Maybe even when
cloning a gene for a human hemoglobin, a sequence from a highly pathogenic human virus might be carried along as a hitch-hiker.

These prospects and many others have provoked a sense of caution in
the United States, in particular, and around the world about the use of
recombinant DNA technology and have produced a new federal bureaucracy.
Rules have been drawn up to regulate recombinant DNA research that are
monitored by the federal government. Special strains of bacterial cells are
required that are multiply auxotrophic and cannot survive outside of their
soup of enriched medium. The presence of *F* factors in these cells, which
might allow them to transfer a plasmid with its insert of foreign DNA to other
bacteria, is scrupulously avoided. A catalog of guidelines and rules has been
established with experiments being ranked at differing levels of risk. As a
function of increasing assessment of danger, facilities for conducting experiments become increasingly restrictive. Putting *Drosophila* genes into *E.
coli* requires little more than a closed room and the observance of routine
laboratory precautions, a so-called *P2* facility. Putting the genome of a
dangerous animal virus into *E. coli* requires the use of a highly contained
facility, a so-called *P4* facility.

Fort Dietrick in Maryland had long been a center in the United States
for the development of products to be used in chemical and biological
warfare, activities that ended in the late 1960's. The facilities at Fort Dietrick,
needless to say, are exceptional and allow the handling of highly poisonous
chemicals and pathogenic bacteria with the ultimate level of containment.
Fort Dietrick is now the site of continued research to determine how real are
asserted dangers inherent in the new genetic engineering, that is, in recombinant DNA technology.

A group under the direction of Wallace Rowe and Malcolm Martin, have
been engaged in research using the genome of polyoma virus, a virus similar
to SV40 that, like SV40, can induce tumors when injected into newborn mice.
When mice are infected with polyoma virus or with virus DNA and the virus
multiplies, the immune systems of the mice respond by making antibodies
against coat proteins of the virus. The presence of the antibodies against
viral proteins can be used as a sensitive test for a polyoma infection.

The group at Fort Dietrick inserted the entire polyoma chromosome into
the plasmid pBR322 and cloned the resulting recombinant plasmid in *E.
coli.* They also inserted the polyoma chromosome into λ WES (p. 345) either
as a single copy or as two copies in tandem, aligned head to tail (Figure

18-15). Bacteria carrying the recombinant plasmid or phage DNAs or purified recombinant plasmid or phage DNAs themselves were then injected into
or fed to mice. In no case did feeding produce a polyoma infection, that is,
result in the production of antibody for the viral proteins. Also, when the
plasmid DNA containing the polyoma insert or the λ DNA containing a single
copy of the polyoma chromosome was isolated and injected, no polyoma
infection resulted. Only when extracted DNA or λ phage carrying the dimeric
head-to-tail insert of polyoma chromosomes were injected did a polyoma
infection result. The head-to-tail arrangement of the inserted DNA makes
recombination possible between the two copies of the chromosome to produce a circular polyoma chromosome. This chromosome, like DNA extracted
from the virus itself, is expected to be infectious (Figure 18-15). Presumably,
the infections caused by these recombinant molecules resulted from such
recombination. Even then, the DNA was only one tenth as infective as DNA
prepared from the virus and over 10^8 times less infective than the virus itself.
Nevertheless, there is some danger from head-to-tail clones because of the
apparent generation of intact polyoma chromosomes by recombination.
Based on these experiments, there is no danger from cloned, single chromosomes and no danger from oral infection by any recombinant configuration.

It is impossible to conduct or even imagine all possible experiments to
evaluate all possible risks. The results from the polyoma experiments at Fort
Dietrick might reduce some fears, but they cannot eliminate all fears about
recombinant DNA technology. Many would argue that the need for government control of recombinant DNA research continues. It seems possible that
a modern molecular biologist Dr. Frankenstein could construct a form of an
organism that would be highly pathogenic and to which natural resistance
would be very low. The pathogenicity would start not with the organism but
with the scientist, and it is unlikely that any rules promulgated to control
recombinant DNA research would matter. Moreover, many geneticists and
molecular biologists advocate the abolition of rules regulating recombinant

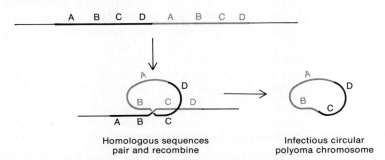

Figure 18-15.   Production of a normal, circular polyoma chromosome by crossing over
between two tandem polyoma chromosomes inserted into a λ chromosome.

DNA research. These people argue that possible dangers cannot be assessed quantitatively and that the assessment of different levels of risk is spurious. Moreover, there is much less danger working with the recombinant DNA cloned in *E. coli* than with the bacterium or virus itself. Indeed, James Watson, in an article published in the prestigious British journal, *Nature*, in 1979, urges abolition of government regulation of recombinant DNA research and argues that "recombinant DNA-induced diseases fall into the category of UFO's and witches" and that levels of risk assigned by different individuals reveal more about their neuroses than about real biological danger. Watson concludes by suggesting that there is no more reason to fear "recombinant DNA than there is to fear the Loch Ness Monster."

Recombinant DNA technology: the bright side.

Several human diseases with direct or indirect genetic bases respond dramatically to specific therapies, for example, phenylketonuria (p. 304). Another genetic disease is a form of pituitary dwarfism that results from the lack of the growth hormone, a polypeptide with 190 amino acids secreted by the pituitary gland. Children with identified pituitary dwarfism can be greatly helped by the proper injection of growth hormone. Growth hormones are species specific. The form of the hormone that functions in one species of mammals may not function in another distantly related mammalian species. Children with growth hormone deficiencies cannot be treated with hormones from distantly related animals such as cows and pigs, and the availability of human growth hormone is highly limited and, as we could guess, very expensive. There is a clear need for a reliable, cheap source of human growth hormone.

Another disease, insulin-dependent diabetes, always caused early death until the 1920's when Frederick Banting and Charles Best first isolated insulin and demonstrated that it could be used to control the disease. There is, fortunately, an extensive supply of insulin available from the pancreases of slaughtered animals, principally pigs and cattle. Insulins from these animals are highly active in humans and are used by diabetics. The amino acid sequences of pig and beef insulin, however, differ from human insulin, and a significant number of diabetics develop antibodies against the injected insulin that they must take to live, and severe complications can result. As with growth hormone, there is a ready need for human insulin. We do not look to animals or chemical synthesis for sources of these hormones but to bacteria carrying appropriate genes. Both human insulin and human growth hormone are now being synthesized in bacteria, a result of recombinant DNA technology. We will consider only the techniques involved in producing insulin.

Human insulin is synthesized by bacteria carrying cloned insulin genes.

Insulin is constructed of two polypeptides, called A and B, held together by disulfide bonds. The A chain contains 21 amino acids and the B chain contains 30 amino acids. Neither chain contains methionine, an important fact, as we shall see shortly. The synthesis of polynucleotides with known sequences continues to improve. H. Gobind Khorana and his co-workers produced in 1979 the first chemically synthesized gene, a 207 base pair sequence containing a coding region for the synthesis of a tRNA. Molecular geneticists at the City of Hope Medical Center in Duarte, California—Robert Crea, Adam Krasewski, Tadaaki Hirose, and Keiichi Itakura—have turned their attention to synthesizing polynucleotides that encode the amino acid sequences of the two chains of human insulin. They made two DNA's, one 104 nucleotides long for the B chain and one 77 nucleotides long for the A chain. The AUG codon for methionine is added to the end of each coding sequence, so that if the sequence were translated, the A and B polypeptides would both start with a methionine residue. In addition to double-helical coding regions the synthesized molecules have single-stranded tails at both ends. One tail has the sequence of an *Eco*R1 restriction site, the other of a *Bam*1 site.

Each synthetic fragment is inserted into a plasmid, pBR322, that has been cut open by first cleaving it with *Bam*1 and then with *Eco*R1 (Figure 18-16). The synthetic DNA carrying the coding sequence for an insulin chain is then inserted with a specific polarity by base pairing between the single stranded tails and is sealed into place with ligase. The site of insertion is a site corresponding to the C-terminal end of a β-galactosidase gene carried by the plasmid. Insertion occurs so that the orientation and reading frames of the insulin gene and the β-galactosidase gene are the same. The result is that bacteria carrying the recombinant plasmid with its insert make β-galac-

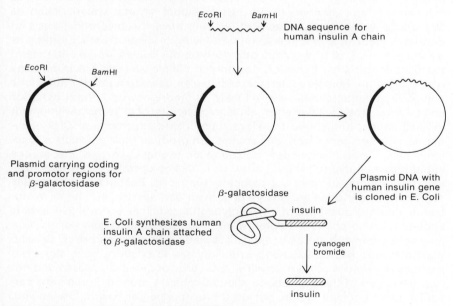

Figure 18-16.   Cloning the genes for human insulin.

tosidase to which an insulin polypeptide is covalently linked through the C-terminal end. It is at this stage that the presence of the methionine residue is critical. A reagent used by peptide chemists, cyanogen bromide (CNBr), cleaves a peptide bond immediately following a methionine residue. When the polypeptide, composed of an insulin polypeptide attached to the end of β-galactosidase through a methione residue, is treated with CNBr, the insulin polypeptide is split off. The released A and B chains, obtained from separate bacterial clones, are then purified and joined together through their disulfide bonds. The molecule that results is identical to human insulin. The yields of the A and B chains are high, about 10 milligrams each for every 24 grams of bacteria. The bacteria carrying the cloned human insulin coding sequences have now passed from the hands of the experimentalists in Duarte to those of industrialists, namely, Eli Lilly Company, which is the major supplier of insulin in the United States. Lilly has recently started pilot tests on the quality of the insulin produced, and we can look forward in the near future to the commercial availability of human insulin courtesy of *Escherichia coli.*

PROSPECTUS

We have described in this chapter the foundations for revolutionary applications of genetics. We are beginning to understand the genetic mechanisms by which different tissues arise during development. When these mechanisms are understood, not just in model genetic systems such as *Drosophila* but in humans as well, we will have the theoretical basis to develop a medical technology to replace failing or diseased organs, such as a kidney, a heart, or a liver, from other cells in our bodies. As we understand the genetic basis for behavior, not only in simple animals but in humans where there are complex interactions between developing neural functions and environmental stimuli, we will learn in theory how to control behavior and to direct it to specified ends. Most revolutionary of all is the development of techniques for genetic transformation. Using recombinant DNA techniques, specific genes from any source can produce their products in bacterial cells. These techniques, coupled, for example, to new discoveries about enzyme function, should allow us to design specific genes that manufacture enzymes to carry out particular functions. Bacteria may some day make human-designed enzymes that convert garbage to gasoline or manufacture copious quantities of brain polypeptides such as endorphin—literally endogenous morphine, the body's own drug for numbing pain—or particularly potent forms of this polypeptide, for medical or other purposes. Genetic transformation will be extended to eukaryotes to produce new, still more productive varieties of corn, to allow real cures of genetic diseases and not just alleviation of their effects, to allow parents to order children with particular genetic constitutions—not just the sex but height, weight, disposition, intelligence . . . . Whether we like it or not, the technology for a brave new world is in the wings.

IV

Population Genetics

The Search for Genetic Variation

We now turn to population genetics, which links classical and molecular genetics with evolutionary biology. Population genetics is an integral part of genetics; its perspectives, its insights, and the questions it addresses come, however, from evolutionary biology. The significance of population genetics lies in the contributions it makes to our understanding of evolutionary processes. The key to understanding population genetics is to think from the point of view of an evolutionary biologist.

For evolutionary biology in general and population genetics in particular, the starting point is the consideration of genetic variation. A casual look at one's friends and neighbors reveals many differences among individuals. A look at the families of those individuals reveals that many of the differences are to some extent shared by relatives. Inherited differences among individuals—genetic variations—are the basic material upon which evolutionary processes operate to change a species' make-up over time and space. Population genetics, in its simplest terms, is the study of genetic variation in groups of individuals. Population genetics is fundamentally concerned with the *extent* of variation within a population and the way in which evolutionary processes *change* and shape the genetic variation of a population over time. In other words, how extensive are genetic differences within a population of individuals, and how are such differences affected by natural selection, movement of individuals in and out of the population, and historical events?

Population genetics is a combination of empirical and theoretical studies of genetic variation.

Empirical population genetics is concerned with genetic differences in natural populations, the geographic and ecological distributions of such differences, and the significance of such differences in terms of the organism's biology. Theoretical population genetics is concerned with analytical models that describe the effects on genetic variation produced by evolutionary phenomena such as natural selection, mutation, chance events, and the exchange of individuals between populations. Some of the models are verbal, based upon qualitative concepts such as heterozygosis and adaptiveness. The majority are mathematical models that make use of the quantitative nature of such factors as Mendelian segregation ratios, linkage, mutation rates, biological fitnesses, and population size to deduce precise mathematical relationships. Since the classic works of R. A. Fisher, Sewall Wright, and J. B. S. Haldane, which appeared in the early 1930's and made the connection between Mendelian genetics and Darwinian evolution, theoretical population genetics and empirical population genetic studies by geneticists such as Theodosius Dobzhansky have formed the backbone of the modern theory of evolution.

Empirical studies of genetic variation are limited by our ability to detect differences in natural populations that can be attributed to single genes.

The extent of genetic variation within any given population can be assessed, in principle, by asking how many different alleles occur in the population at any particular gene locus and what the relative proportions are within the population of the various alleles at that locus. Any trait for which two or more forms commonly occur within a group of individuals is *polymorphic*. Conversely, a trait for which virtually all individuals exhibit the same form is *monomorphic*. A basic question of population genetics is to ask what fraction of the genome consists of loci that are polymorphic within a population and what fraction consists of loci that are monomorphic. The terms *polymorphic* and *monomorphic* apply to genes or traits within groups of individuals. At the individual level, one can ask what proportion of an individual's loci are, on the average, heterozygous rather than homozygous.

The general problem of the extent of genetic variation can be phrased in two diverse questions. Is it true that most loci are monomorphic, with most individuals being homozygous for the "wild-type" allele? Or, are most loci polymorphic and individuals heterozygous at many of the loci?

The fundamental problem for the student of genetic variation is the difficulty of detecting those variants that can be ascribed to genetic differences at single loci. The variants used in laboratory studies (eye-color mutants in *Drosophila* and nutritional mutants in bacteria and fungi, for example) are rare in natural populations. Consequently, empirical studies have had to be limited to a few *classes* of genetic variation for which techniques have been devised to resolve genetic differences among individuals. In this chapter we will survey briefly four classes that have become well documented over the past fifty years: morphological polymorphisms, gene arrangements, molecular variants, and fitness modifiers.

MORPHOLOGICAL POLYMORPHISMS

Differences in morphology provide obvious examples of differences among individuals. Think of the variety of sizes, shapes, and colors of the dogs one sees daily. Although morphological variations are usually inheritable to some extent—puppies tend to resemble their parents—many, if not most, morphological variants in natural populations fail to behave in simple Mendelian fashion, making it difficult to know just how much genetic variation they reflect. Cases for which the genetics of the polymorphisms have been determined have, however, provided some excellent material for empirical population genetics studies.

A famous example of genetic polymorphism occurs in the shell pattern of the land snail *Cepaea nemoralis.*

Snails of the species *Cepaea nemoralis* vary greatly in the patterns on their shells, which differ in the presence, number, and appearance of up to five bands (see Figure 19-1) and in general color, principally yellow, pink, and brown. Genetic studies have determined a number of loci for which specific alleles can be identified. One locus, for example, has alleles for dark brown, dark pink, pale pink, faint pink, dark yellow, and pale yellow shell color. Another locus governs the presence or absence of bands. A third modifies the number of bands where bands occur.

An enormous body of data exists on the frequencies with which different alleles occur in thousands of *Cepaea* populations throughout Europe and England. Consider the frequencies of the yellow-color form, as an example of the kinds of data available. Near the Mediterranean it is the predominant form. In England, it is about equal in frequency to other color forms although its frequency varies considerably from one population to another. Figure 19-2 shows the frequency of the yellow form in each of 18 populations along a three-kilometer stretch of the New Bedford River in England. In these New Bedford River populations the frequency varies from population to population with no obvious pattern at all. In other areas, around Oxford for one, the frequency of yellow shells correlates with the population's habitat; evidence implicates camouflage from bird predation as an important factor in the frequencies of the shell colors. Elsewhere, there are large areas, in-

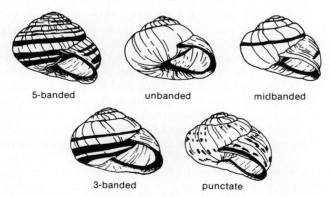

5-banded   unbanded   midbanded

3-banded   punctate

Figure 19-1   Examples of shell banding patterns in *Cepaea nemoralis*. After Jones, et al., *Ann. Rev. Ecol. Syst.* v. 8, pp. 109–144, 1977. Reproduced, with permission, from the *Annual Review of Ecology and Systematics*, Vol. 8. © 1977 by Annual Reviews Inc.

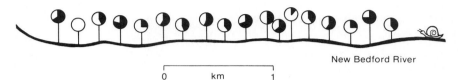

yellow ◑ non-yellow

New Bedford River

0   km   1

Figure 19-2   Frequency of the allele for yellow shell color in populations of *C. nemoralis* along an English river. Data from Jones, et al., *Annual Review of Ecology and Systematics,* v. 8, pp. 109–144, 1977. Reproduced, with permission, from the *Annual Review of Ecology and Systematics,* Vol. 8. © 1977 by Annual Reviews Inc.

corporating a variety of habitats, in which only the yellow form is found, while in adjacent areas with the same sorts of habitats, the yellow form is almost totally absent.

The extensive studies of *Cepaea* make it clear that no single explanation exists for the data. Almost every kind of evolutionary process has been implicated by one study or another as having a detectable influence upon the frequencies with which the various shell types occur.

Melanism in the peppered moth is a genetic trait that is famous as a demonstration of natural selection acting upon genetic variation.

The peppered moth, *Biston betularia,* in Great Britain gets its common name from the typical light peppered coloration on its body and wings (Figure 19-3). Dark melanic forms are, however, common in many populations. In fact, the black *carbonaria* form occurs at frequencies exceeding 90 per cent in some populations. An *insularia* form, intermediate in coloration, is also common in certain areas. The three forms are determined by a series of alleles at one locus, with the *carbonaria* form dominant to both the peppered and *insularia* forms. The *insularia* form, in turn, is dominant to the typical peppered form.

Frequencies of the three forms in different parts of Great Britain are shown in Figure 19-4. The pattern differs from that seen earlier for *Cepaea* in that most populations of *Biston* are composed either predominantly of the peppered form or predominantly of the melanic form. Peppered forms predominate in rural populations where the trees upon which the moths rest during the day are covered with lichens. Melanic forms predominate in areas where industrial pollution has killed the lichens and turned the tree trunks black.

The polymorphism is a recent development, concomitant with the rise of industrialization in England. The first black moth was found in 1848; by 1900, populations near Manchester were composed almost entirely of melanic individuals. Extensive field studies demonstrated that predation by

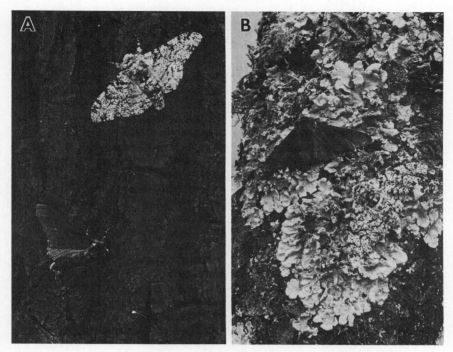

Figure 19-3   Typical and *carbonaria* forms of the peppered moth, *Biston betularia*. A) At rest on soot-covered oak trunk near Birmingham, England. B) At rest on lichened tree trunk in unpolluted countryside, Dorset, England. From the experiments of Dr. H. B. D. Kettlewell, University of Oxford, England.

birds creates a strong selective pressure against whichever form is conspicuous in a given area. Different forms predominate in different populations depending upon which form is best camouflaged. Although the species as a whole is polymorphic, most populations tend towards monomorphism for one form or the other.

The genes involved in known morphological polymorphisms constitute a minute fraction of the genome.

The polymorphisms in *Cepaea* and *Biston* are two of the best studied cases of morphological polymorphism; but examples are known in many species. In any one species morphological polymorphisms for which the genetic bases can be determined are comparatively few and involve genetic differences at only a small number of loci. Consequently, although such traits provide excellent material for studying particular evolutionary changes within populations and species, they tell us little about how much variation exists within the genome as a whole.

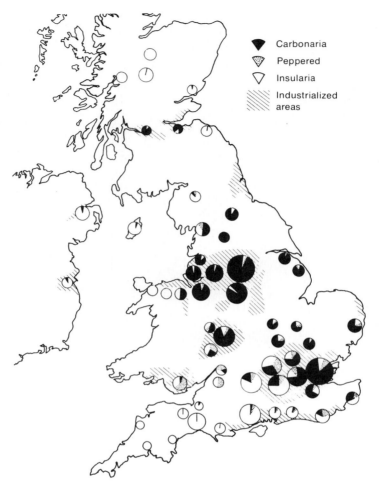

Figure 19-4   Frequencies of the three forms of *Biston betularia* in various populations in Great Britain. Hatching indicates areas of high population density and industrialization. After P. M. Sheppard, 1975, *Natural Selection and Heredity,* Hutchinson and Co., London.

GENE ARRANGEMENTS

A form of genetic variation at the chromosomal level that has received considerable attention concerns the order of genes along the chromosomes. In earlier chapters we have frequently discussed the polytene chromosomes of *Drosophila,* whose distinctive banding patterns allow differences between homologous chromosomes, reflected in the sequence of the bands, to be observed by light microscopy. Alternative banding patterns are commonly found in natural populations. In *Drosophila pseudoobscura,* for example, over twenty alternative arrangements are known for Chromosome III. Because genes can be mapped to particular bands, a particular sequence for a given chromosome is referred to as a particular *gene arrangement* and given a name or other designation. The alternative arrangements are related cytogenetically by a series of paracentric inversions (see Figure 8-9). Figure 19-5 shows the deduced relationships among the various arrangements in *Drosophila pseudoobscura* and *Drosophila persimilis.*

The cytogenetic relationships give insight into the historical relationships and origins of the various arrangements. The principal fact before us, however, is that different arrangements exist and are well established in natural populations. In *Drosophila pseudoobscura* polymorphism for gene arrangements is restricted, for the most part, to Chromosome III. In other

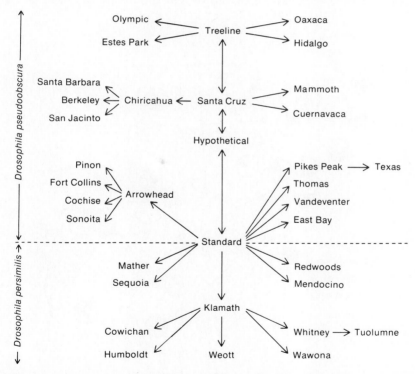

Figure 19-5   Cytogenetic relationships of gene arrangements in the third chromosome of *Drosophila pseudoobscura* and *Drosophila persimilis*. Each arrow denotes a difference in band sequence that can be accounted for by a single inversion. "Hypothetical" is a nonexistant sequence that must be postulated to complete the otherwise unbroken single-step connections. "Standard" is a sequence common to both species. After Anderson, et al., *Evolution,* v. 29, pp. 24-36, 1975.

species, such as *Drosophila willistoni* in South America and *Drosophila robusta* in eastern North America, all chromosomes are polymorphic for gene arrangements.

Most local populations of *Drosophila pseudoobscura* are polymorphic for gene arrangements on Chromosome III and can be grouped into major chromosomal races.

Figure 19-6 shows the frequencies of the major third chromosome gene arrangements found within a number of populations of *Drosophila pseudoobscura*. Several different arrangements generally occur within each population and individuals often differ, one from another, in the arrangements they carry. For example, in the Strawberry Canyon population in Berkeley, California, the Standard, Arrowhead, Chiricahua, and Tree Line arrangements are all common.

Obvious differences among populations are apparent. The populations shown in Figure 19-6 can be placed in four groups on the basis of gene

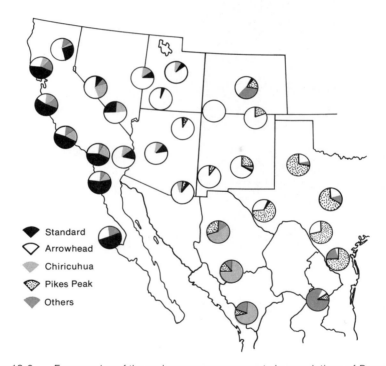

Standard
Arrowhead
Chiricuhua
Pikes Peak
Others

Figure 19-6   Frequencies of the major gene arrangements in populations of *Drosophila pseudoobscura*. Pacific Coast populations are distinguished by high frequencies of Standard; Great Basin populations by high frequencies of Arrowhead; Texas populations by Pikes Peak; and populations in the Chiricahua mountains of Mexico by Chiricahua. After Th. Dobzhansky, 1951, *Genetics and the Origin of Species,* Columbia University Press, New York.

arrangements, particularly on the basis of the locally predominant arrangement. These groupings are referred to as *chromosomal races.* Altogether, seven chromosomal races have been identified for *Drosophila pseudoobscura.* In addition to the four shown in Figure 19-6, three others occur farther south in Mexico and Central America. The principal characteristic of chromosomal races is that populations of the same race exhibit similar arrays of gene arrangements, whereas populations of two different races—even if geographically close together—exhibit distinct arrays. The boundaries between two races are usually sharp and are generally associated with obvious geographical and ecological barriers. For example, the east face of the Sierra Nevada mountain range separates the populations of the Pacific Coast race from those of the Great Basin.

The frequencies of gene arrangements exhibit well-documented clines and seasonal cycles.

Within chromosomal races, certain regular patterns of spatial and temporal variation in the frequencies of gene arrangements have been found. A situation in which the frequency of a particular allele, gene arrangement, or trait changes in a progressive, more or less uniform fashion across a geographical region is a *cline.* Figure 19-7 illustrates a cline in the frequencies

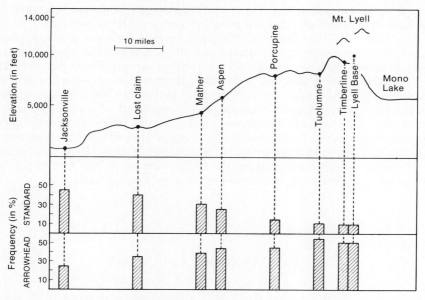

Figure 19-7   Clines in the frequencies of the Standard and Arrowhead gene arrangements in populations of *Drosophila pseudoobscura* along an altitudinal transect in the Yosemite region of the Sierra Nevada. Frequencies in local populations do not total 100 per cent because the frequencies for the other arrangements in the populations have been omitted. After Th. Dobzhansky, *Genetics,* v. 33, pp. 158–176, 1948.

of the Standard and Arrowhead arrangements along an altitudinal gradient on the west slope of the Sierras: the frequency of the Standard arrangement steadily decreases with increasing elevation. Analogous clines have been found for populations living along altitudinal gradients in the San Jacinto mountains in southern California. Populations of *Drosophila persimilis* show a north-south cline involving the Klamath and Mendocino arrangements in the coastal mountains of California, and a north-south cline involving the Klamath and Whitney arrangements is found in the Sierras.

In a number of instances the frequencies of gene arrangements within a population have been found to vary seasonally in a cyclical fashion. In a study of a population on Mt. San Jacinto, Standard had a frequency of about 50 per cent when the population became active in March. Several generations later, in June, its frequency had dropped to below 30 per cent. The frequency then began increasing until it once again was about 50 per cent in October when the population became inactive for the following winter.

Different gene arrangements apparently have different combinations of alleles at loci within the arrangement.

We shall see in Chapter 21 that gene arrangements are strongly affected by natural selection. Alternative arrangements are identical in terms of the loci present on the chromosomes: they differ merely in the physical order of those loci. Presumably, the significance of genetic variation for gene arrangements lies not so much in the arrangements themselves but rather in genetic differences at individual gene loci within the arrangements.

In individuals that are heterozygous for two different arrangements, however, the inversions that distinguish the arrangements greatly reduce recombination between loci within the inverted segments (see Figure 8-12). It is possible, therefore, that a different combination of alleles, involving particular genetic variants at a number of loci within the inverted region, might become more or less permanently associated with each particular gene arrangement in the population. In other words, polymorphism for gene arrangements allows for the possibility of polymorphism for particular combinations of alleles that are protected from recombination. Even so, nothing can be said about either the nature or the number of loci that might be involved in any such allelic differences between arrangements.

MOLECULAR VARIANTS

Differences in amino acid sequences of particular proteins indicate genetic variation. A number of biochemical techniques are used to find such differences. The most direct approach is comparison of amino acid sequences. Technical considerations prohibit large-scale surveys involving many different individuals, and in general, comparisons of amino acid sequences are limited to a few well-known molecules such as hemoglobin, cytochrome c, insulin, and fibrinopeptide and involve comparisons between typical sequences for different species. The sort of data provided by this method is abbreviated in Table 19-1 for the first 15 sites of the hemoglobin α chains of five vertebrate species.

Two comparisons can be made directly from sequence data. First, comparison by pairs shows the fraction of amino acid sites at which two sequences differ. For example, human and dog sequences differ at 4 of the 15 sites shown in Table 19-1 (numbers 10, 12, 13, 15); human and carp sequences differ at 7 of the 15 sites. Second, the data as a whole can be used to classify each site according to the number of different amino acids found within the assemblage of sequences under comparison. For the data in Table 19-1, at five of the 15 sites the same amino acid is found in all five species; at four sites two different amino acids are found; at two sites three amino acids are found; and at four sites four amino acids are found. There is no site at which all five sequences are different from each other.

Obtaining full-sequence data is laborious. An indirect method for comparing two species is based upon immunological techniques. If, say, rabbit antibodies are made to human albumin, the degree to which the antibodies react with, say, dog albumin gives a measure of the sequence similarity between the human and dog albumins. For certain molecules such as albumin the "immunological distance" has been sufficiently calibrated that it can be translated into numbers of amino acid differences. Extensive comparisons between species have been made with this method. The technique is insensitive, however, to the small differences in sequence that occur between members of the same species. Like the direct sequence data, immunological distances are of principal value for determining genetic differences between distantly related organisms rather than among populations of a single species. We shall return to sequence data in the final chapter when we discuss genetic changes that occur during the evolutionary divergence of different lineages.

Table 19-1   Amino Acid Sequences for the First Fifteen Sites of the Hemoglobin α Chain.

Site	1	2	3	4	5	6	7	8	9	10	11	12	13	14	15
Human	Val-	Leu-	Ser-	Pro-	Ala-	Asp-	Lys-	Thr-	Asn-	Val-	Lys-	Ala-	Ala-	Trp-	Gly-
Dog	Val-	Leu-	Ser-	Pro-	Ala-	Asp-	Lys-	Thr-	Asn-	Ile-	Lys-	Ser-	Thr-	Trp-	Asp-
Gray Kangaroo	Val-	Leu-	Ser-	Ala-	Ala-	Asp-	Lys-	Gly-	His-	Val-	Lys-	Ala-	Ile-	Trp-	Gly-
Chicken	Val-	Leu-	Ser-	Asn-	Ala-	Asp-	Lys-	Asn-	Asn-	Val-	Lys-	Gly-	Ile-	Phe-	Thr-
Carp	Ser-	Leu-	Ser-	Asp-	Lys-	Asp-	Lys-	Ala-	Ala-	Val-	Lys-	Ile-	Ala-	Trp-	Ala-
Number of Different Amino Acids	2	1	1	4	2	1	1	4	3	2	1	4	3	2	4

From Dayhoff et al., 1972, *Atlas of Protein Sequence and Structure*, National Biomedical Research Foundation. Washington, D.C.

Certain kinds of molecular differences among individuals can be detected by gel electrophoresis.

Assessing molecular variation in populations requires techniques that can be easily repeated for a large number of individuals so as to permit broad surveys of variation within and between populations. Electrophoresis, coupled with enzyme staining techniques, provides such a capability. In Chapter 11, we saw how sickle-cell hemoglobin can be distinguished from normal hemoglobin using electrophoresis.

For survey work a crude extract of soluble proteins is obtained by grinding a piece of tissue in a small amount of buffer. Extracts from different individuals are placed side by side on slab gels and subjected to electrophoresis, following which the gels are subjected to one or another of a variety of staining procedures, each of which is specific for a particular enzyme. As the stains develop, bands appear indicating the position to which the enzyme has migrated during electrophoresis. Molecular variants of a given protein that affect electrophoretic mobility will be reflected by differences in their relative positions on the gel. An example is seen in Figure 19-8.

The distance a protein migrates during electrophoresis is determined mainly by its charge and conformation. Not all genetic differences in a protein necessarily produce differences in electrophoretic mobility. Electrophoresis, therefore, detects a subset of the molecular variation that may exist among the individuals studied. Allelic forms of a protein that can be distinguished by electrophoresis are generally called *allozymes.* In practice, surveys of allozyme variation are restricted to a class of soluble enzymes that can be obtained in crude extracts and for which there exist biochemical staining techniques that allow the enzymes' locations to be viewed on gels. About 30 commonly used enzymes meet these criteria. Many are coded for by single genes; others are multimers involving gene products of two or more loci. Typical allozyme studies survey electrophoretically detectable genetic variation at 20 to 40 loci.

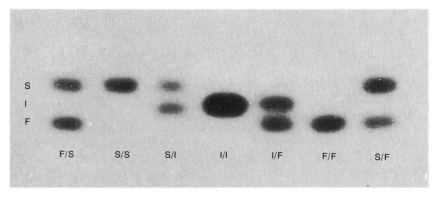

Figure 19-8   Starch gel showing electrophoretically detectable variation for the enzyme phosphoglucomutase (PGM) in frogs of the genus *Rana* from Travis County, Texas. Homozygotes and heterozygotes for the three allozyme alleles F, I, and S can be distinguished. Courtesy of Dr. R. Sage, University of California, Berkeley.

Allozyme variation within populations is widespread throughout the living world.

Allozyme variation in natural populations can be characterized in several ways. One is to determine the fraction of the various allozyme loci that are polymorphic in a given population. For example, in one study of *Drosophila pseudoobscura* a total of 24 loci were tested for allozyme variation; 10 of the 24 loci (42 per cent) proved to be polymorphic; no variation was found at the other 14 loci. An alternative, related method is to determine the fraction of the loci at which individuals are, on the average, heterozygous. At monomorphic loci all individuals are necessarily homozygous for the one and only allele; at polymorphic loci, however, individuals may be either homozygous or heterozygous. In the *Drosophila pseudoobscura* study, individuals were found, on the average, to be heterozygous at 12.3 per cent of the 24 loci studied. Put another way, on the average an individual was heterozygous at 3 of the 10 polymorphic loci; which of the 10 were heterozygous varied from one individual to another.

The proportion of polymorphic loci per population and the average heterozygosity per individual are both defined for an entire population on the basis of the variation within the particular collection of allozyme loci in a given study. Different enzymes do different things, and we might expect the variation for a given enzyme to be influenced by its particular biochemical and physiological circumstances. Indeed, some enzymes appear to be consistently more variable than others. Consequently, average measures of allozyme variation may be affected to some extent by the choice of enzymes in each particular study.

A wealth of allozyme data has been obtained from a wide variety of species, including many that were previously not amenable to genetic investigation. Overall average measures of allozyme variation for vertebrates, invertebrates, and vascular plants are given in Table 19-2. In all three groups there is substantial variation from species to species. Species exhibiting average heterozygosities anywhere from zero to 10 per cent are commonly found in all three groups. Among the vertebrates, species with average heterozygosities greater than 10 per cent are uncommon; among the inver-

Table 19-2   Average Measures of Allozyme Variation.

	Number of species	Proportion polymorphic loci per population	Average proportion heterozygous loci per individual
Vertebrates	135	.173	.049
Invertebrates	93	.397	.112
Vascular Plants	113	.368	.141

Data for animal species are from E. Nevo, 1978, *Theoretical Population Biology* 13: 121–177. Of the 93 species of invertebrates, 43 are members of the genus *Drosophila*. Data for plant species are from J. Hamrick et al., 1979, *Annual Review of Ecology and Systematics* 10: 173–200. The plant data include a large number of studies that involved only a few loci (e.g., fewer than 10) and may, therefore, be somewhat biased in favor of polymorphic loci.

tebrates and vascular plants, however, typical averages range as high as 25 per cent or more.

For most species, allozyme frequencies are more or less the same throughout the species' range.

Allele frequencies at polymorphic allozyme loci can be compared among populations in the same fashion as gene arrangements. Frequencies at each of four loci in three populations of *Drosophila pseudoobscura* are given in Table 19-3. The convention used in this table for naming the alleles at a

Table 19-3

	Pt-8 Chromosome II		
Allele	Strawberry Canyon	Mesa Verde	Austin
.80	.014	.009	.011
.81	.472	.410	.441
.83	.514	.576	.512
.85	—	.005	.035

	Xanthine dehydrogenase Chromosome II		
Allele	Strawberry Canyon	Mesa Verde	Austin
.90	.053	.016	.018
.92	.074	.073	.036
.99	.263	.300	.232
1.00	.600	.580	.661
1.02	.010	.032	.053

	Larval acid phosphatase-4 X Chromosome		
Allele	Strawberry Canyon	Mesa Verde	Austin
.93	—	—	.028
1.00	1.0	1.0	.860
1.05	—	—	.112

	Pt-10 Chromosome III		
Allele	Strawberry Canyon	Mesa Verde	Austin
1.02	.005	.021	.010
1.04	.615	.970	.935
1.06	.380	.008	.054

Allele frequencies at four allozyme loci in three geographically distant populations of *Drosophila pseudoobscura* in Strawberry Canyon, California; Mesa Verde, Colorado; and Austin, Texas.

From Prakash, Lewontin, and Hubby, 1969, *Genetics* 61: 841–858.

locus involves arbitrarily assigning one allele the designation *1.00*. The other alleles are then designated by the distance of their electrophoretic bands from the origin *relative* to the band for the allele *1.00*. For example, the protein coded for by the XDH allele *1.02* migrates 2 per cent farther than the protein coded for by the allele *1.00*.

Although most of the loci in Table 19-3 exhibit substantial variation *within* populations, there is little difference *among* populations. All three populations have similar arrays of allele frequencies except for the Pt-10 locus. For example, at the XDH locus, allele *1.00* is the most common allele in all three populations, with its frequency varying only from .600 in Strawberry Canyon to .580 in Mesa Verde to .661 in Austin. Because each of these populations is in a different chromosomal race, the pattern of variation between populations for allozymes contrasts sharply with the pattern for gene arrangements. In particular, allozyme variation does not reflect the geographical differentiation that gene arrangements show.

The exception at the Pt-10 locus can be ascribed to associations between allozyme alleles and gene arrangements of Chromosome III. In particular, the Pt-10 locus is located on Chromosome III, and it happens that virtually all third chromosomes with Standard, Arrowhead, or Pikes Peak gene arrangements have the allozyme allele *1.04* at the Pt-10 locus, whereas those with Chiricahua or Tree Line arrangements generally have the allele *1.06*. The frequencies of allele *1.06* in the three populations simply relect the frequencies of the Chiricahua and Tree Line arrangements.

As a rule the pattern in Table 19-3 is typical of allozyme variation. For many species, including *Drosophila,* humans, and *Neurospora,* the array of alleles in any one local population tends to be similar to the arrays in other populations of the species. In most species an allele that has a high frequency in one population is rarely absent in other populations. There are, however, some well-documented exceptions to this general pattern. We shall look further at allozyme differences among populations when, in Chapter 25, we are in a better position to put the general topic of genetic differentiation among populations into an evolutionary perspective.

MODIFIERS OF VIABILITY

We shall see in Chapter 21 that natural selection occurs when genetic variations result in differences among individuals with respect to their relative abilities to survive and reproduce. The genetic variants that we have so far considered produce discrete morphological, cytological, or biochemical differences among individuals. Determination of the connections, if any, between such phenotypic differences by which genetic variants are detected and differences in the organisms' abilities to survive and reproduce are separate matters to which we shall devote further attention in Chapter 21.

There is, however, one extensively studied class of genetic variation that is directly related to differences in viability. We now turn to variation that can be detected by measuring the effects upon viability that result from making individual *Drosophila* homozygous for an entire chromosome.

Natural populations of *Drosophila* contain enormous amounts of concealed variation that reduce viability when made homozygous.

The origins of empirical population genetics are usually traced to S. S. Chetverikov, who in 1926 suggested that diploid organisms ought, on the basis of Mendelian genetics, to contain a significant number of hidden, recessive, detrimental mutants. We saw in Chapter 14 that roughly one out of three chromosomes in natural populations of *Drosophila melanogaster* carry at least one gene that is lethal when homozygous. Such recessive lethals involve drastic reductions in viability. It is also possible to assay for the presence of genes on a chromosome that have less severe effects.

Suppose, for example, that we are looking for alleles on Chromosome II of *Drosophila pseudoobscura* that modify viability. By using the technique shown in Figure 14-2 a *single* male fly taken from a natural population can be crossed to a female from a balanced marker stock for Chromosome II. In the F_2 generation a stock is produced that consists of both males and females with the genotype $+/M$, where $+$ denotes a chromosome that is identical with one of the two second chromosomes carried by the original wild male and M denotes a marked balancer chromosome that, like the Curly-Plum and ClB chromosomes, causes a visible phenotype when heterozygous, is lethal when homozygous, and has inversions that suppress recombination. Repeating the procedure with different wild males, we can produce a large number of different stocks, each of which has a different, possibly unique, second chromosome that was originally present in a natural population. We can designate these different chromosomes as $+_1$, $+_2$, $+_3$, et cetera.

Suppose we cross males and females from the i-th stock. Both parents have the genotype $+_i/M$. Zygote genotypes should be $+_i/+_i$, $+_i/M$, M/M in 1:2:1 proportions. The M/M zygotes die; the $+_i/M$ individuals are phenotypically marked by the M chromosome; the $+_i/+_i$ are homozygous for the $+_i$ chromosome and as adults will be wild type. We expect, therefore, one third of the adult flies to be phenotypically wild type. But if we allow more eggs to be laid than can possibly survive to adulthood on the amount of available food, the ratio of wild-type adults to marked adults may deviate from 1:2, depending upon the relative likelihoods of the two kinds of individuals surviving from egg to adult. In fact, the fraction of wild-type adults produced by such a cross provides a measure of viability for individuals that are homozygous for the $+_i$ chromosome relative to the $+_i/M$ heterozygotes.

Performing this cross for each stock yields a viability measure for each of the different wild-type second chromosomes isolated from natural populations. Figure 19-9A gives results from making 208 different stocks homozygous for second chromosomes obtained from a subspecies of *Drosophila pseudoobscura* near Bogota, Colombia. The histogram, which is typical of

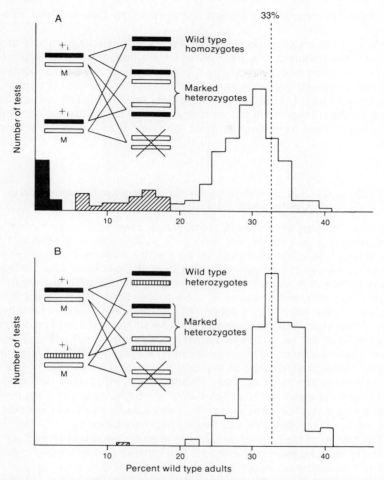

Figure 19-9   Distribution of the relative viabilities of wild-type second chromosomes obtained from natural populations of *Drosophila pseudoobscura*. Relative viability is measured by the percentage of wild-type flies that survive to adulthood in the indicated crosses. A. Tests involving individuals homozygous for each of 208 different wild-type second chromosomes. B. Tests involving individuals heterozygous for 209 different pairs of wild-type second chromosomes. Data from Dobzhansky et al., *Genetics*, v. 48, 361–373, 1963.

such experiments, is bimodal. The left mode contains 15 crosses that produced no wild-type flies, indicating that the wild-type chromosomes involved in these crosses carry recessive lethal alleles at one or more gene loci. In these particular data the frequency of such chromosomes happens to be less than the more common frequencies shown in Table 14-3. Another 25 to 30 crosses, indicated by the hatched area of Figure 19-9A, revealed chromosomes, known as *semi-lethals,* that cause drastic reductions in viability when made homozygous although some wild-type flies survived to adulthood.

Most of the crosses, however, produced appreciable numbers of wild-type flies. Nevertheless, the right mode of the histogram is somewhat less than the expected frequency of one third, indicating that most second chromosomes in natural populations carry alleles that, when made homozygous, cause at least slight reductions in the ability to survive. The chromosomes carried by stocks that fall in the righthand peak of the histogram are referred to as *quasi-normal* to distinguish them from lethal and semi-lethal chromosomes.

Most individuals that are heterozygous for different wild chromosomes show little, if any, reduction in viability.

If, say, females from the i-th stock are crossed with males from the j-th stock, the array of zygotes will be $+_i/+_j$, $+_i/M$, $+_j/M$, and M/M in 1:1:1:1 proportions. Again, the M/M zygotes die and the expected ratio of wild type to marked flies is 1:2. However, in this case the wild-type flies are heterozygous for second chromosomes obtained from two different wild males. Figure 19-9B shows the results of crosses involving 210 different pairs of the same stocks as in Figure 19-9A. The frequencies of wild-type heterozygotes in the majority of the crosses are centered about the expected number, indicating that the reduction in viability for chromosomal homozygotes is, for the most part, masked in chromosomal heterozygotes. The histogram is, however, somewhat skewed to the left, suggesting that some of the wild-type heterozygotes have reduced viability. This result may be caused by some chromosomal heterozygotes being homozygous at the gene level for recessive variants that reduce viability. It may also be caused by the presence of alleles that exert deleterious effects when heterozygous.

The nature of the genetic variation that underlies the observed variation in viability must be indirectly inferred. The crosses just described involve an entire chromosome on which there are many gene loci. How many of those loci have variants that affect the results for a given test is not fully known. The lethality of any particular chromosome appears to be caused by only one lethal gene, or at most a few such genes, on the chromosome. The absence of the lethal class in Figure 19-9B indicates that chromosomal heterozygosity masks the lethal effects of the 15 lethal chromosomes that are seen in Figure 19-9A. The lethal chromosomes of different stocks presumably carry different, non-allelic lethal genes so that wild-type heterozygotes composed of two different lethal chromosomes are not lethal. As a rule, two lethal chromosomes from a natural population prove to have their lethality caused by allelic genes approximately 0.1 per cent of the time.

In any event, we can see that Chetverikov's prediction of hidden, detrimental variation in natural populations is amply borne out. Moreover, such variation is constantly being introduced by mutation. Estimates by Terumi Mukai and his colleagues for the second chromosome of *Drosophila melan-*

ogaster place the mutation rate at about 0.006 per chromosome per generation for mutations that have drastic effects when homozygous. They estimate the rate per chromosome is 10 to 20 times higher for mutants that produce small reductions in viability.

The salient fact of empirical population genetics is the widespread existence of genetic variation in natural populations.

The kinds of genetic variation that we have discussed in this chapter involve genes that collectively represent a small fraction of an organism's genome. Nevertheless, within this small sampling alone we see evidence of plentiful variation. The theoretical basis of Darwinian evolution is the assumption of genetic variability among the individuals of a species. Darwin's greatness lies in his shifting of our attention from the similarities of individuals to their differences. In the *Genetic Basis of Evolutionary Change* (1974), Richard Lewontin puts it thus:

> He [Darwin] called attention to the *actual* variation among *actual* organisms as the most essential and illuminating fact of nature. Rather than regarding the variation among members of the same species as an annoying distraction, as a shimmering of the air that distorts our view of the essential object, he made that variation the cornerstone of his theory.

The accumulated data of empirical population genetics leave no doubt that actual genetic variation among actual organisms is a reality of nature. The raw material for evolution exists.

The high level of genetic variability in natural populations raises a significant philosophical point. Consider a locus such as the Pt-8 allozyme locus for *Drosophila pseudoobscura*. If 47 per cent of the genes at that locus code for one allele and 51 per cent for another allele, what then is the "wild type" or "normal" allele? Clearly the notion of wild type is on shaky ground. Its usage as a convenient term is certainly justifiable in some contexts, such as when distinguishing clearly deleterious alleles or nonfunctional mutants from otherwise normally functioning alleles. As a biological concept, however, the notion of wild type is the antithesis of variation. The term implies genetic uniformity among individuals, and we know that genetic uniformity is precisely not the case at many loci. In fact, if we consider an individual's entire genome as a unit, genetic uniqueness is the case. For example, taking the known data for polymorphic allozyme loci for humans and taking the known data for blood groups, such as ABO and Rh, it is possible to compute the probability that two unrelated individuals have exactly the same alleles at every one of these loci—a probability of less than 10^{-20}. In other words, there are far more possible, different genotypes than there are individuals. The conclusion, reached long ago by poets and philosophers, is unmistakable: excepting, possibly, identical twins, each and every one of us is genetically unique.

SUMMARY

Genetic variation is the central issue of population genetics. Studies of genetic variation in natural populations have dealt primarily with a few kinds of variation for which genetic differences among individuals can be readily detected. In a few cases, polymorphisms for such morphological characters as the color of snail shells can be attributed to allelic differences at one or a few gene loci. In *Drosophila* widespread polymorphism exists with respect to the physical order of genes along their chromosomes, which can be determined from the sequence of bands in polytene chromosomes. In populations of *Drosophila pseudoobscura,* for example, alternative *gene arrangements* of the third chromosome are common.

For a few well-known genes, such as the hemoglobin genes, genetic differences between species can be assessed from direct comparisons of the amino acid sequences in their gene products. For about 30 to 40 gene loci that code for biochemically known soluble enzymes, gel electrophoresis can detect genetic variation that affects electrophoretic mobility. *Allozyme* variation is common in a wide variety of vertebrate, invertebrate, and plant species. Depending upon the species, individuals are on the average heterozygous at roughly 5 to 20 per cent of these allozyme loci.

For many diploid species such as *Drosophila, viability studies* reveal that most individuals carry chromosomes with one or more recessive alleles that, upon being made homozygous, reduce the organism's viability. In many instances these detrimental alleles are lethal when homozygous; in other instances they result in only a slight reduction in viability relative to that of heterozygotes.

The genes for which genetic variation in natural populations can be accurately described encompass a minute portion of the total genome. Nevertheless, enough genetic variation is revealed by those genes to support the conclusion that most individuals are genetically unique.

20

The Mendelian Genetics of Populations

The last chapter gave us an inkling of the vast reservoir of genetic variation in natural populations and a feeling for some of the forms in which variation is manifested. Our discussion, however, has given a *static* view of variation, and the central characteristic of evolution is change. In the chapters ahead we want to develop a *dynamic* view of genetic variation.

The dynamics of genetic variation is the principal concern of population genetic theory. The goal of any quantitative theory is the prediction of future conditions, but biological systems are complex—too complex, in fact, for population genetics to have reached the level of a predictive science. Given our inability to construct a predictive theory, the value of population genetic theory lies in the understanding it provides, through quantitative analyses, of the effects that particular biological processes (such as natural selection) have upon genetic variation.

Population genetic theory begins, so to speak, in the middle of things, with expressions for changes in the genetic composition of a population from one generation to the next in a given set of circumstances. The theory with which we shall be dealing is mostly concerned with genetic variation at a single, arbitrary, gene locus. Although it should not be imagined for a moment that any genes exist independently of other genes in the genome, only individual loci are, or are not, polymorphic. If genetic variation is to be explained, it must be explained in terms of processes that affect the alleles at individual loci.

For any gene locus the fundamental quantities dealt with in population genetics are the frequencies of alternative alleles and genotypes within a population.

Before addressing the dynamics of genetic change, we first need to develop a mathematical expression of genetic variation within natural populations. For any particular locus, we are primarily interested in the array of alleles carried by the individuals of a given population, and particularly in the frequencies of all existing alleles. Recall, for example, the allozyme data in Table 19-3. Allele frequencies are always defined in terms of the alleles at

531

a single locus. By convention, they are expressed as fractions and must have values of not less than zero nor more than unity. The combined frequencies of all the alleles at a locus must add up to unity—each haploid genome in the population must contain one or the other of the alleles at the locus in question. Traditionally, allele frequencies have been called *gene frequencies,* but the term is a misnomer because the entities whose frequencies are under consideration are alternative alleles at one gene locus and not alternative genes in the genome.

In much of what follows we shall consider an arbitrary gene locus with two alleles, A and a. For concreteness the A-locus may be thought of as an allozyme locus, with, say, the allele A coding for the faster-migrating form and the allele a coding for the slower form, or as a blood group locus such as the MN locus. We shall use the letters p and q to denote the frequencies of A and a respectively. Because we have only two alleles at the A-locus, $p + q = 1$—a fact that should be kept in mind whenever we do any algebraic manipulations with allele frequencies.

Most of population genetic theory deals with populations of diploid individuals, which are classified with respect to a particular locus on the basis of genotype. Genotype frequencies, like allele frequencies, are expressed as fractions. Table 20-1 gives, as an example, data for a phosphoglucomutase locus that is polymorphic for two alleles, PGM^1 and PGM^2, in a Chinese population of 419 individuals. Allele frequencies follow directly from an enumeration of individuals according to their diploid genotypes. Because each individual carries two copies of the PGM gene, there is a total in the population of 838 copies of the gene, of which 480 are PGM^1 alleles carried by homozygous individuals and 154 are PGM^1 alleles carried by heterozygotes, giving an allele frequency of $634/838 = .757$ for PGM^1.

For our generalized A-locus the genotypes are AA, Aa, and aa. We denote their frequencies as G_{AA}, G_{Aa}, and G_{aa} ($G_{AA} + G_{Aa} + G_{aa} = 1$). All copies of the A-gene that are carried by AA individuals are A alleles, and half of the copies that are carried by Aa individuals are A alleles. The frequency of the allele A is, therefore, equal to the frequency of the AA homozygotes plus half of the frequency of the Aa heterozygotes. Symbolically, $p = G_{AA} + \frac{1}{2}G_{Aa}$.

When a population's composition is given in terms of genotype frequencies for a given locus, the corresponding allele frequencies are directly obtainable. The converse is generally not true. Two populations with very different genotype frequencies may have the same allele frequencies. For

Table 20-1   Genotypes and Alleles at a Phosphoglucomutase (PGM) Locus in a Chinese Population.

Genotype	PGM^1/PGM^1	PGM^1/PGM^2	PGM^2/PGM^2	Total
Number of Individuals	240	154	25	419
Frequency	.573	.367	.060	1.0
Number of PGM^1 Alleles	480	154	0	634
Number of PGM^2 Alleles	0	154	50	204

Data from Bodmer and Cavalli-Sforza, *Genetics, Evolution and Man,* W. H. Freeman & Co., 1976.

example, suppose that the frequencies of AA, Aa and aa are .5, .2, .3 in one population and are .36, .48, .16 in another one. In both populations the allele frequency for A is $p = .6$.

HARDY-WEINBERG PROPORTIONS

In a population of N diploid individuals, there are $2N$ copies of each gene in the genome. The total collection of all copies of genes present in a population is the population's *gene pool*. Although individuals carry genes and individuals collectively constitute a population, individuals live only for one generation. Genes, however, are passed from parent to offspring and can persist, in principle, indefinitely. It is convenient, therefore, to think in terms of the collection of all copies of the A-gene that are present in the gene pool from generation to generation.

In order to follow a gene pool through time we must relate the composition of the gene pool in a parental generation to its subsequent composition in the offspring generation. Although it is not possible to do so in a universally applicable way, the British mathematician G. H. Hardy and the German physician W. Weinberg, working independently of each other in 1908, developed a significant mathematical expression of quantitative relations. Given one major assumption, we can thereby arrive at the *genotype frequencies of the offspring* generation in terms of the *allele frequencies of their parents*.

Whenever mating is random with respect to genotype, the genotype frequencies of the resulting zygotes will be in Hardy-Weinberg proportions.

If, following our convention, p and q denote the frequencies of A and a ($p + q = 1$), then the particular set of genotype frequencies $G_{AA} = p^2$, $G_{Aa} = 2pq$, and $G_{aa} = q^2$ is known as the *Hardy-Weinberg proportions*. Notice that $G_{AA} + \frac{1}{2}G_{Aa} = p^2 + pq = p(p + q) = p$, so there is no inconsistency between the allele frequencies being p and q and the genotype frequencies being p^2, $2pq$, and q^2. Moreover, for each value of p there is a single, unique, set of Hardy-Weinberg proportions. Figure 20-1 illustrates the Hardy-Weinberg proportions that correspond to various allele frequencies.

The relation found by Hardy and Weinberg applies to any *autosomal* (that is, not sex-linked) locus at which the allele frequencies are the same in males and females. Their result depends upon the union of gametes (eggs and sperm; eggs and pollen) within the population to form zygotes *randomly with respect to which allele each carries*. Given the random union of gametes, if p is the frequency of the allele A in the *parental gene pool* at the time of mating, then the genotype frequencies for AA, Aa, and aa *among newly formed zygotes* will be in the Hardy-Weinberg proportions p^2, $2pq$, q^2.

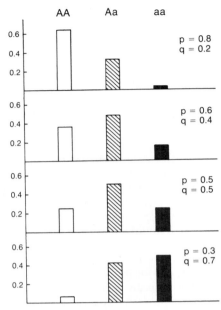

Figure 20-1   Genotype frequencies in the Hardy-Weinberg proportions $p^2 : 2pq : q^2$ for various allele frequencies p and q. Any one of the three genotypes may be the most common, depending upon the allele frequencies. The maximum Hardy-Weinberg frequency of Aa is .5.

This implies the frequency of the allele A in the gene pool of the zygotes is the same as in the gene pool of their parents at mating.

Hardy-Weinberg proportions follow directly from the mathematical definition of random union. Let us think of a species, such as sea urchins, that sheds large numbers of eggs and sperm into a tide pool in which fertilization occurs through random collisions of eggs and sperm. Random union implies that the frequency with which any union involves an egg with a particular allele (A or a) and a sperm with a particular allele (A or a) is equal to the product of the frequency with which such an egg occurs among all eggs times the frequency with which such a sperm occurs among all sperm. If p and q are the allele frequencies for A and a among both eggs and sperm, then the frequencies of the possible kinds of zygotes formed by random union of gametes are obtained from the 2 × 2 checkerboard in Table 20-2. Because the unions aA and Aa are indistinguishable, we have three different kinds of zygotes, Aa, Aa, aa, with genotype frequencies of p^2, $2pq$, q^2.

In many species, including humans and *Drosophila,* gametes unite by the indirect route of matings between diploid individuals, and the concept of the parental gene pool may seem a bit artificial and unreliable. Random mating with respect to diploid genotypes turns out to be equivalent to random union of gametes. We shall illustrate this point with a numerical example and see explicitly how Mendelian genetics enters into population genetics.

Table 20-2   Genotype Frequencies of
Zygotes Produced by Randomly Uniting
Gametes.

		Sperm	
		Allele A Freq. p	Allele a Freq. q
Eggs	Allele A Freq. p	AA p^2	Aa pq
	Allele a Freq. q	aA pq	aa q^2

Suppose the genotype frequencies at the A-locus are $G_{AA} = .5$, $G_{Aa} = .2$, and $G_{aa} = .3$. Notice that these frequencies are far from being in Hardy-Weinberg proportions. Three steps are required to determine the genotype frequencies of offspring produced by random matings. First, we find the frequency of each possible kind of mating. Second, for each kind of mating we apply Mendelian segregation ratios to determine the relative proportions (within the total population) of the different genotypes produced by such a mating. Third, we combine the relative proportions of all offspring with the same genotype.

The first step is done by means of the 3 × 3 checkerboard in Table 20-3. The condition of random mating implies that the frequency with which a particular type of mating—defined by a particular genotype for the mother and a particular genotype for the father—occurs is simply the product of the respective genotype frequencies for females and males. For example, the frequency of matings that involve a female with genotype aa and a male with genotype Aa is .3 × .2 = .06.

For the second step, shown in Table 20-4, the frequencies of reciprocal matings may be combined. The procedure consists of: (1) assuming that the frequency of each kind of mating equals the fraction of the offspring generation contributed by such matings, and (2) using Mendelian segregation ratios to distribute the contribution of each kind of mating according to the resulting offspring genotypes. For example, Table 20-4 shows that hetero-

Table 20-3   Frequencies of the Possible Types of Matings
in a Randomly Mating Population in Which the Genotype
Frequencies Are .5 AA, .2 Aa, and .3 aa.

			Paternal Genotypes		
			AA	Aa	aa
Maternal Genotypes	AA	.5	AA × AA .25	AA × Aa .10	AA × aa .15
	Aa	.2	AA × AA .10	Aa × Aa .04	Aa × aa .06
	aa	.3	aa × AA .15	aa × Aa .06	aa × aa .09

Table 20-4   Frequencies in the Offspring
Generation for Individuals Classified According to
Both Their Own Genotype and Their Parents'
Genotypes.

Parental Genotypes		Offspring Genotypes		
Type of Mating	Freq.	AA	Aa	aa
AA × AA	.25	.25		
AA × Aa Aa × AA	.20	.10	.10	
AA × aa aa × AA	.30		.30	
Aa × Aa	.04	.01	.02	.01
Aa × aa aa × Aa	.12		.06	.06
aa × aa	.09			.09
Total	1.00	.36	.48	.16

Frequencies for the different types of matings are from Table 20-3 with reciprocal matings combined.

zygote-by-heterozygote matings (Aa × Aa) contribute 4 per cent of the offspring generation. That 4 per cent is distributed among the three genotypic classes in the proportions 1 : 2 : 1. Consequently, 1 per cent of the offspring, *in toto,* consists of individuals produced by Aa × Aa matings *and* have the genotype AA; 2 per cent are Aa individuals produced by Aa × Aa matings; 1 per cent are aa individuals produced by Aa × Aa matings.

The third step involves simply adding up each column of offspring genotypes in Table 20-4. A quick check will show that the genotype frequencies of .36, .48, and .16 are in Hardy-Weinberg proportions of $(.6)^2$, $2(.6)(.4)$, and $(.4)^2$, and that .6 and .4 are the allele frequencies for A and a in both the parental and offspring generations. This example illustrates the general conclusion that, whatever the parental genotype frequencies are, random mating immediately produces offspring genotype frequencies that are in the Hardy-Weinberg proportions appropriate to the parental allele frequencies.

We have so far assumed for simplicity that the A-locus has only two alleles. Hardy-Weinberg proportions for a locus with more than two alleles are a direct extension of the two-allele case. Our notation, however, must be modified. Suppose the A-locus has n different alleles denoted as $A_1, A_2, \ldots, A_n$. Let the respective allele frequencies be denoted as $p_1, p_2, \ldots, p_n$. The Hardy-Weinberg proportions for homozygotes are p_i^2 for A_iA_i ($i = 1, 2, \ldots, n$), and for heterozygotes they are $2p_ip_j$ for A_iA_j ($i < j$). The condition $i < j$ for heterozygotes means that we combine reciprocal heterozygotes in the usual fashion for Aa and aA. These Hardy-Weinberg proportions can be verified for any particular value of n by writing out the $n \times n$ checkerboard, analogous to Table 20-2, for randomly uniting male and female gametes.

Allele frequencies are not changed by random matings.

The calculations we have made deal with *allele* frequencies in a parental pool of randomly uniting gametes and with *genotype* frequencies in the resulting array of zygotes. A corollary to our analysis is that the zygotic allele frequencies are the same as the allele frequencies in the parental gamete pool. In other words, allele frequencies in a randomly mating population remain unchanged during the transmission of alleles from parents to offspring.

During the course of a generation's development from zygotes to reproducing adults, genotype frequencies may depart from their initial Hardy-Weinberg proportions in response to a variety of factors, for example, natural selection in the form of viability differences like the ones discussed in the previous chapter. Whether or not changes in genotype frequencies occur, genotype frequencies for the next generation initially will be in the Hardy-Weinberg proportions that correspond to the allele frequencies in the gene pool of the present generation at the time of random mating. Figure 20-2 gives a hypothetical illustration of genotype frequencies that change over three generations. Regardless of any changes in genotype (and allele) frequencies that occur during each generation, at the beginning of each generation we see random mating causing the genotype frequencies to assume the Hardy-Weinberg proportions that are appropriate to the allele frequencies that exist at the time of random mating.

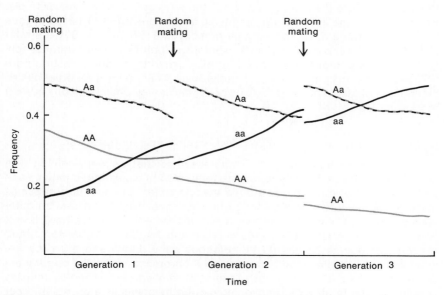

Figure 20-2   Plots of genotype frequencies over three successive generations in a hypothetical population. During each generation the frequencies of AA and Aa are presumed to decrease by 20 per cent as a result of natural selection and other biological processes. Because of random mating, each generation begins in Hardy-Weinberg proportions that are appropriate to the allele frequencies at the end of the previous generation.

In our discussion so far the critical condition has been that mating be random with respect to parental genotypes. This condition is not as stringent as it might at first appear; it requires randomness only with respect to the alternative genotypes at the locus in question. The requirement may be met for some loci and not for others. For example, it is well known that humans do not mate randomly with respect to height—matings between individuals of similar height occur more frequently than random mating would predict. On the other hand, there is evidence that humans often mate randomly with respect to such genetic differences as blood types. Numerous examples of local populations exhibiting Hardy-Weinberg proportions have been found. In the majority of species for which allozyme data exist allozyme loci typically show Hardy-Weinberg proportions. Gene arrangements in natural populations of *Drosophila pseudoobscura* have been found to be in Hardy-Weinberg proportions during the egg stage (that is, very early within a generation).

With random mating *rare alleles* occur mostly in heterozygotes.

The common occurrence of Hardy-Weinberg proportions has significance for recessive alleles. If the allele a is recessive so that its phenotypic effects are discernible only in aa homozygotes, we can ask what is the frequency of heterozygous (Aa) individuals that carry, but do not express, the a allele relative to homozygous aa individuals. Using Hardy-Weinberg proportions, the ratio of the frequencies of Aa to aa can be expressed as $2pq/q^2 = 2p/q = 2(1 - q)/q = 2/q - 2$. This ratio is inversely related to q. If the a allele is rare (i.e., if q is near zero) the ratio can be large, for example, when $q = .01$ the ratio is 198, implying that there are 198 heterozygous "carriers" of the a allele for each aa homozygote. When $q = .001$ the ratio is 1998. These simple calculations exemplify a point of major significance. Rare alleles (whether recessive or not) are carried in randomly mating populations primarily by heterozygous individuals. Thus, recessive lethal alleles can remain hidden, with most copies being carried by phenotypically normal heterozygotes.

Genotype frequencies at sex-linked loci require separate treatment.

In the preceding discussion it was assumed that we were dealing with an autosomal locus. If, instead, we are interested in a sex-linked locus, such as the white-eye locus in *Drosophila melanogaster,* the situation becomes more complex. Suppose we are dealing with heterogametic sex chromosomes such that females have two X chromosomes and males have one X and one Y. Suppose that the A-gene is located on the X chromosome, and the Y chromosome carries no homologous gene. Let p and q be the frequencies of the alleles A and a among all X chromosomes in the population. Suppose that p and q apply both to all X chromosomes carried by females as a group and to all X chromosomes carried by males as a group. In other words, the allele frequencies for A and a are the same in both sexes.

To determine the genotype frequencies that result from randomly uniting gametes we must consider male and female offspring separately. Because females receive an X chromosome from each parent, the checkerboard in Table 20-2 remains appropriate. Consequently, the genotypes among female offspring are AA, Aa, and aa in Hardy-Weinberg proportions. Males, on the other hand, receive an X chromosome only from their mothers. Their possible genotypes are AY and aY. The frequencies of these two alternatives are simply the frequencies p and q with which the alleles A and a occur in the eggs of the maternal generation.

A consequence of these genotype frequencies is the more common occurrence of *recessive sex-linked traits* among males than among females in a randomly mating population. Suppose the allele a is responsible for such a trait. With random mating the trait occurs among females with a frequency of q^2 (genotype aa) and among males with a frequency of q (genotype aY). As long as q is less than unity, q^2 is less than q. Color blindness in humans is such a trait. It is sex-linked and occurs among men with a frequency of about 8 per cent ($q = .08$). Among women it is much rarer, having an observed frequency of about .005, which is reasonably close to the value of $(.08)^2 = .0064$ predicted by Hardy-Weinberg proportions. When recessive alleles are very rare, the few cases in which such traits occur are almost entirely restricted to males. For example, the genetic disease hemophilia A is a sex-linked trait that occurs in about one in every 10,000 men ($q = .0001$). Among women the trait is virtually unknown, as is to be expected—$q^2 = .00000001$.

INBREEDING

The requirement that pervades discussions of Hardy-Weinberg proportions is random mating. We now consider a common form of non-random mating known as inbreeding. We shall see that the principal effect of inbreeding is an increase in the frequencies of homozygous genotypes, relative to Hardy-Weinberg proportions, without any change in the allele frequencies.

Inbreeding refers to matings between related individuals and, in its most extreme form, self-fertilization. It occurs commonly in a variety of situations. In small, isolated human populations inbreeding has often had pronounced effects with respect to the occurrence of recessive genetic defects. Even in large human populations the occurrence of inbreeding is not necessarily negligible. For example, in a study of 33,000 marriages in Nagasaki, Japan, 8 per cent involved partners that were second cousins or closer. In a rural area of India, one third of all marriages were found to be between first cousins. Inbreeding is often a factor in breeding programs with domesticated animals and crops. Such programs generally use only a few choice individuals as parents, and within a very few generations all individuals in the breeding populations become related. Inbreeding is also a regular feature of many plant species, particularly those species in which individual plants engage in self-fertilization.

The two homologous copies of any particular gene of an inbred individual have a finite probability of being identical by descent.

An individual whose parents are related is said to be *inbred*: the term applies to a single individual but is based upon a relationship between his father and mother. Since related individuals tend to be genetically more alike than randomly chosen individuals, the alleles that an inbred individual receives from his parents ought to be alike more often than are any two randomly chosen alleles.

Two different copies of a gene are said to be *identical by descent* if both are derived through direct lines of gene replication from the same single copy of the gene that was carried by some ancestor. Identity by descent is a statement about a *historical* relationship between two copies of a gene. It is distinct from any statement regarding the nucelotide sequences of the gene copies. Two copies of a gene that both code for the same allele are said to be *identical in allelic state,* which is a statement about the biochemical and physiological properties of their gene products.

Clearly, homologous copies of a gene carried by two different individuals can be identical by descent only if the individuals have at least one ancestor in common. Brothers and sisters, for example, have both parents in common; first cousins have two of their four grandparents in common. Consequently, for any particular gene locus, the copy that an inbred individual receives from his mother has a certain probability of being identical by descent with the copy received from his father. This probability is called the *inbreeding coefficient, F.* For individual I the inbreeding coefficient is $F_I =$ the probability that two homologous copies of any given gene carried by individual I are identical by descent. For a non-inbred individual, whose parents are unrelated, the inbreeding coefficient is zero.

We can more fully appreciate the concepts of identity by descent and inbreeding coefficients by considering a specific pedigree. Figure 20-3 illustrates the pedigree for an individual whose parents are half-sibs. Earlier pedigrees (e.g., Figures 6-9 and 6-10) were shown in the "genealogical form," which clearly shows relationships between individuals; Figure 20-3 is in the "genetic form," which is better suited for analysis. In the figure, the inbred individual I is identical by descent (at a given locus) for two copies of a gene that are both direct copies of the shaded copy carried by the common grandparent. We can easily calculate the probability of such an event.

The probability is one-half that the father receives from the grandfather a shaded copy instead of an unshaded copy. The same is true for the mother. So, the probability is $(\frac{1}{2})(\frac{1}{2}) = \frac{1}{4}$ that both the father and the mother receive shaded copies. The probability is now one-half that the father transmits to individual I a shaded copy (instead of the dotted copy that came from the paternal grandmother), and one-half that the mother transmits to individual I a solid copy rather than a dotted copy. The probability that individual I receives two dark copies of the gene—one each from his father and his mother—is $(\frac{1}{2})^4 = \frac{1}{16}$.

The probability of $\frac{1}{16}$ just calculated is only part of the inbreeding coefficient F_I for the inbred individual in Figure 20-3. There is also, for example,

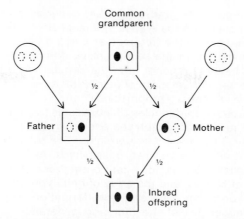

Common
grandparent

Father

Mother

Inbred
offspring

Figure 20-3   Pedigree for a half-sib mating shown in the "genetic form." I is the inbred individual. His father and mother are half-sibs, having the same father but different mothers. In the genetic form of a pedigree each individual is connected by arrows with each of his two parents and with his offspring. The direction of the arrow indicates the direction in which genes are passed from parent to offspring.

a probability of one-sixteenth that individual I receives two copies that are identical by descent from the unshaded copy carried by the grandfather. The inbreeding coefficient, F_I, must take into account both of these two possibilities, because the two homologous copies carried by individual I will be identical by descent in either case. The inbreeding coefficient is concerned only with the state of being identical by descent without regard to the means by which this occurs. General formulas are available for calculation of inbreeding coefficients from specific pedigrees. We need here only recognize that in populations in which inbreeding occurs, there is a finite probability that the homologous copies of any gene carried by a randomly chosen individual may be identical by descent.

In a population in which the average inbreeding coefficient is F, the genotype frequencies for AA, Aa, and aa are initially $p^2 + Fpq$, $2pq(1 - F)$, and $q^2 + Fpq$.

We can now return to determining the deviations from Hardy-Weinberg proportions that occur in a population in which inbred individuals exist but in which mating is otherwise random. Let F (without any subscript) denote the *average* of the inbreeding coefficients (many of which may be zero) of all individuals in the population. For a randomly chosen individual, therefore, the probability of being identical by descent at any given locus is simply F. We can divide individuals into two groups with respect to the A-locus: 1) on the average a fraction F of the individuals carry alleles that are identical by descent, and 2) the remaining fraction $1-F$ carry unrelated alleles. Alleles that are identical by descent, being copies of a single ancestral allele, are necessarily identical in allelic state, so those individuals in the first category are homozygous—some AA and others aa, depending upon whether the ancestral allele happened to be A or a. The respective probabilities of these two possibilities are simply the frequencies p and q of the alleles A and a in the ancestral generation. In the second case, unrelated alleles can be regarded as being drawn at random from the population's gene pool, so that individuals in the second category should exhibit the three genotypes AA, Aa, and aa in Hardy-Weinberg proportions.

Table 20-5 gives the frequencies of individuals (within the population as a whole) both in respect to genotypes and to whether their alleles at the A-locus are identical by descent or unrelated. The total frequency for each genotype is the sum of the last two possibilities. As with Hardy-Weinberg proportions, the allele frequencies p and q are those of the parental generation at the time of mating, and the array of genotype frequencies is for the offspring generation at the time of zygote formation. We can easily verify that the allele frequencies of the offspring generation are unaffected by inbreeding. For example, the frequency of the allele A is: $(p^2 + Fpq) + \frac{1}{2}(2pq - 2Fpq) = p^2 + Fpq + pq - Fpq = p^2 + pq = p(p + q) = p$.

We see from Table 20-5 that these genotype frequencies, with the arbitrary inbreeding coefficient F, are a generalized version of Hardy-Weinberg proportions, which corresponds to the special case in which $F = 0$. Inbreeding increases the frequency of each homozygote above Hardy-Weinberg proportions by the amount Fpq, while the frequency of heterozygotes is correspondingly reduced by the amount $2Fpq$. Because the two homozygotes are increased by the same amount (Fpq), the increase is proportionately greater for the rarer homozygote. The effect is especially notable with respect to recessive traits, such as many human genetic diseases, that are caused by very rare alleles. Suppose, for example, that the frequency of such an allele is $q = .001$. In a randomly mating population the frequency of affected homozygotes is $q^2 = 1 \times 10^{-6}$, whereas in a population with an average inbreeding coefficient of $F = .01$, the frequency is $q^2 + Fq(1 - q) = 10.99 \times 10^{-6}$, indicating a more than ten-fold increase in the incidence of the trait. With $F = .1$ the frequency of affected homozygotes is 100.9×10^{-6}!

Table 20-5   Genotype Frequencies in a Population with an Average Inbreeding Coefficient F.

	Genotype			
	AA	Aa	aa	Fraction
Identical by Descent	$F \cdot p$	0	$F \cdot q$	F
Unrelated	$(1 - F)p^2$	$(1 - F)2pq$	$(1 - F)q^2$	$1 - F$
Total	$Fp + (1 - F)p^2$ $p^2 + Fp(1 - p)$	$(1 - F)2pq$ $2pq - 2Fpq$	$Fq + (1 - F)q^2$ $q^2 + Fq(1 - q)$	1

A fraction F of the population has alleles that are identical by descent, and a fraction $1 - F$ has alleles that are randomly drawn from the population's gene pool. The two expressions for the total frequency of each genotype are simply algebraic identities; e.g., $Fp + (1 - F)p^2 = Fp + p^2 - Fp^2 = p^2 + F(p - p^2) = p^2 + Fp(1 - p)$. For a locus, such as this one, with only two alleles we can write the frequency in the form of $p^2 + Fpq$.

In a population that repeatedly engages in inbreeding, the inbreeding coefficient increases with time, and the frequency of heterozygotes in the population declines.

Many plant species reproduce almost entirely by self-fertilization; many agricultural herds are routinely maintained through matings of closely related individuals; many lab strains of mice and *Drosophila* are perpetuated by repeated matings between sibs. We might therefore expect the inbreeding coefficient to increase from generation to generation. Such is in fact the case. Where inbreeding occurs in a systematic manner generation after generation, it is possible to explicitly calculate the increase in F over successive generations.

The simplest instance is repeated self-fertilization. To see the pattern that develops, let us suppose a large population that is initially in Hardy-Weinberg proportions with respect to genotypes at the A-locus. The frequency of heterozygotes (H) in this initial, non-inbred population is $H_0 = 2pq$. Suppose now that the population reproduces solely by self-fertilization. Homozygotes breed true under self-fertilization, giving rise only to like homozygotes. The heterozygotes, however, segregate in 1:2:1 ratios, thereby reducing the frequency of heterozygotes by one half each generation. After one generation the frequency of heterozygotes is $H_1 = \frac{1}{2}H_0 = pq$. After two generations it is $H_2 = \frac{1}{2}H_1 = (\frac{1}{2})(\frac{1}{2})H_0 = (\frac{1}{2})^2 H_0$. After three generations, $H_3 = (\frac{1}{2})^3 H_0$. After t generations, $H_t = (\frac{1}{2})^t H_0$. We can see that the frequency of heterozygotes is rapidly reduced and soon becomes, for all practical purposes, zero. This process is illustrated graphically in Figure 20-4.

Because Mendelian segregation in self-fertilized heterozygotes yields equal numbers of the two kinds of homozygotes, half of the original fraction of heterozygotes in the population will be converted eventually into AA homozygotes and half into aa homozygotes, giving genotype frequencies that approach $p^2 + pq = p$ AA, zero Aa, and $q^2 + pq = q$ aa. Comparison with Table 20-5 shows that these ultimate genotype frequencies correspond to an inbreeding coefficient of $F = 1$. In general, for any generation t, the amount of heterozygosity H_t can be expressed as $H_t = 2pq(1 - F_t)$, where F_t is the inbreeding coefficient for that generation. Inasmuch as $H_0 = 2pq$ and $H_t = (\frac{1}{2})^t H_0$, we have $F_t = 1 - (\frac{1}{2})^t$.

What all this shows, therefore, is that with repeated self-fertilization after just a few generations almost the entire population has become homozygous. The same result occurs, at a slower rate, with repeated matings between sibs, first cousins, or any other related individuals. Repeated sib mating, for example, is routinely used to produce strains of laboratory animals that are homozygous for most of their genome.

Within a population of individuals, the increase in homozygosity is the result of each individual possessing alleles that are identical by descent; it is not the result of changes in allele frequencies within the population. Some individuals are homozygous AA and others homozygous aa. The alleles remain present at the same frequencies as they would with no inbreeding. In terms of allele frequencies, the population has lost none of its genetic variation as a result of inbreeding. If the entire population were to mate completely at random, so that uniting gametes become associated at random

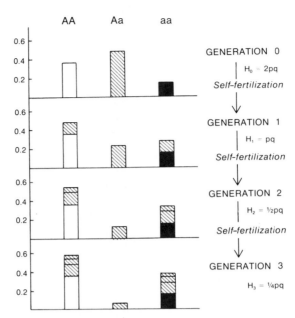

Figure 20-4   Systematic changes in genotype frequencies produced by repeated self-fertilization in a population with allele frequencies of $p = .6$ and $q = .4$. Each generation, the frequency of heterozygotes is reduced by 50 per cent, with the frequencies of both homozygotes being increased by equal amounts. The allele frequencies remain unchanged from one generation to the next.

in the population, Hardy-Weinberg proportions would be immediately restored. In short, *the effect of inbreeding is solely upon genotype frequencies*—increasing the frequencies of homozygotes and reducing the frequencies of heterozygotes—without any change in allele frequencies.

Plant species that reproduce by a mixed strategy of self-fertilization and outcrossing give us a simple model to illustrate the dynamic interplay of inbreeding and random mating.

The reproductive biology of numerous plant species, of which the slender wild oat *Avena barbata* is typical, is such that each generation a certain fraction, say S, of the ovules are self-fertilized by pollen from the same plant as the ovule, and the remaining ovules are fertilized by pollen from other plants. The fraction S may be quite large; in the case of *Avena barbata* it is over 90 per cent. A single population consists of a mixture of individuals with differing pedigrees that we can group systematically. Some individuals are produced by random unions of ovules and pollen; other individuals will be the result of a single generation of self-fertilization that was preceded by random mating; others by two generations of self-fertilization; others by three generations of self-fertilization; and so on.

For those individuals produced by random mating, genotype frequencies should be in Hardy-Weinberg proportions. Among the remainder, the likelihood of heterozygosity decreases in accordance with the number of successive generations of self-fertilization in each pedigree. So, in the population as a whole, the frequency of heterozygotes should be the *average* of the values $2pq$, pq, $\frac{1}{2}pq$, $\frac{1}{4}pq$,..., weighted according to the fractions of individuals that are the products of exactly 0, 1, 2, 3, ... successive generations of self-fertilization. We can predict what these fractions should be on the basis of chance. If, in each generation, self-fertilization occurs on a random basis with probability S, then these fractions should be $1-S$, $(1 - S)S$, $(1 - S)S^2$, $(1 - S)S^3$, ..., respectively. The frequency of heterozygotes is then,

$$H = \sum_{i=0}^{\infty} [2pq(\tfrac{1}{2})^i][(1 - S)S^i] = 2pq(1 - S)\sum_{i=0}^{\infty} \left(\frac{S}{2}\right)^i.$$

The infinite sum in the last term is a geometric series that is equal to $2/(2 - S)$, so, on the average, the amount of heterozygosity is

$$H = 2pq \left[\frac{2 - 2S}{2 - S}\right].$$

Or, using $H = 2pq(1 - F)$, the inbreeding coefficient is $F = S/(2 - S)$, which ranges from $F = 0$ when $S = 0$ to $F = 1$ when $S = 1$. The greater the likelihood of self-fertilization, the greater the inbreeding coefficient. But as long as there is some random mating, the population cannot become fully inbred. In other words, any random mating at all insures that at least some of the alleles in the population are randomly associated and that all possible genotypic combinations are maintained.

The image we get from this example is one of a constant turnover of pedigrees. Each generation a fraction S of the pedigrees are extended for one more generation of self-fertilization with a concomitant loss of heterozygosity. At the same time, in a randomly chosen fraction $1 - S$ of the population, outcrossing reassorts alleles at random, thereby undoing the effects of inbreeding in those particular pedigrees. The larger S is, the greater the effects of inbreeding because individual pedigrees will, on the average, involve a greater number of generations of self-fertilization between episodes of random mating.

This example of mixed self-fertilization and random mating is an extreme case because loss of heterozygosity occurs more rapidly with repeated self-fertilization than with other forms of inbreeding. It is also unusual in that it is a sufficiently simple case that we are able to construct a quantitative model to predict the heterozygosity (or inbreeding coefficient) for known fractions of the population and thereby come up with a particular expression for the average probability of identity by descent for individuals in the population. In other respects, however, this example is characteristic of any population, of which *Avena barbata* is only one of many examples, in which the pattern of mating involves a combination of random mating and various degrees of inbreeding. In any such case we can expect genotype frequencies to depart from Hardy-Weinberg proportions to some extent, depending upon how extensive the average level of inbreeding is.

JOINT GENOTYPE FREQUENCIES AT TWO LOCI

In spite of our attention to a single locus at a time, it is unreasonable to imagine that genes are unaffected by other genes in the organism's genome. In this section we will look at genotype frequencies in a randomly mating population in which we simultaneously consider an individual's genetic constitution at each of two different loci. The purpose is not to see whether our results obtained for a single locus become invalid when two loci are considered simultaneously—we shall see, for example, that Hardy-Weinberg proportions still hold for each locus. Rather, we want to see if there are phenomena peculiar to the joint consideration of two loci that cannot be predicted from the consideration of each locus by itself. There is one such phenomenon, and it involves the frequencies with which particular combinations of alleles at the two loci occur in gametes.

Let us call the two loci in which we are interested the A-locus and the B-locus, with alleles A and a and B and b. Changing our notation slightly, let p_A and p_a denote allele frequencies at the A-locus, and p_B and p_b the frequencies at the B-locus. Because allele frequencies are defined separately for each locus, $p_A + p_a = 1$ and $p_B + p_b = 1$.

In a randomly mating population, genotype frequencies among newly formed zygotes are determined by the frequencies of the *gamete types* (g) produced by the parental generation.

For our two loci there are four types of haploid gametes, AB, Ab, aB, and ab, whose frequencies we denote as g_{AB}, g_{Ab}, g_{aB}, and g_{ab}. Relationships between gamete frequencies and allele frequencies are illustrated in Table 20-6. Each allele frequency is the sum of the frequencies of the two gamete types that contain the allele in question. That is, $p_A = g_{AB} + g_{Ab}$; $p_a = g_{aB} + g_{ab}$; $p_B = g_{AB} + g_{aB}$; and $p_b = g_{Ab} + g_{ab}$. The sum of all four gamete frequencies is unity: $g_{AB} + g_{Ab} + g_{aB} + g_{ab} = 1$.

For a randomly mating population, frequencies of the joint genotypes among offspring can be obtained by our standard checkerboard method for randomly uniting gametes. In this case we use a 4 × 4 checkerboard with

Table 20-6   Gamete and Allele Frequencies for Two Loci with Two Alleles Each.

A-Locus		B-Locus		
		B	b	
	A	g_{AB}	g_{Ab}	p_A
	a	g_{aB}	g_{ab}	p_a
		p_B	p_b	1

Table 20-7   Genotype Frequencies in a Randomly Mating Population for Two Loci with Two Alleles Each in Terms of the Frequencies of Randomly Uniting Gametes.

Genotype		Frequency	Gamete Types Produced
AB/AB	(AABB)	g_{AB}^2	All AB
AB/Ab	(AABb)	$2g_{AB}g_{Ab}$	.5 AB & .5 Ab
Ab/Ab	(AAbb)	g_{Ab}^2	All Ab
AB/aB	(AaBB)	$2g_{AB}g_{aB}$	.5 AB & .5 aB
AB/ab	(AaBb)	$2g_{AB}g_{ab}$	$\begin{cases}(.5)(1-R)\ AB\ \&\ (.5)R\ Ab \\ (.5)(1-R)\ ab\ \&\ (.5)R\ aB\end{cases}$
Ab/aB	(AaBb)	$2g_{Ab}g_{aB}$	$\begin{cases}(.5)(1-R)\ Ab\ \&\ (.5)R\ AB \\ (.5)(1-R)\ aB\ \&\ (.5)R\ ab\end{cases}$
Ab/ab	(Aabb)	$2g_{Ab}g_{ab}$	.5 Ab & .5 ab
aB/aB	(aaBB)	g_{aB}^2	All aB
aB/ab	(aaBb)	$2g_{aB}g_{ab}$	.5 aB & .5 ab
ab/ab	(aabb)	g_{ab}^2	All ab

The doubly heterozygous genotype (AaBb) is divided into two classes.

the frequencies of the four gamete types among eggs and sperm. Combining reciprocal classes gives the genotype frequencies listed in Table 20-7. The doubly heterozygous genotype, AaBb, is divided into two classes that depend upon whether the A and B alleles came together in the cis or trans configuration. From these frequencies we can show that the single locus genotypes are in Hardy-Weinberg proportions. For example, the frequency of AA is the sum of the first three rows:

$$G_{AA} = G_{AABB} + G_{AABb} + G_{AAbb} = g_{AB}^2 + 2g_{AB}g_{Ab} + g_{Ab}^2 = (g_{AB} + g_{Ab})^2 = p_A^2.$$

The question we ask next is whether we can express the two locus gentoype frequencies in terms of the Hardy-Weinberg proportions at each locus. Is it true, for example, that the frequency of AABB is $p_A^2 \times p_B^2$? The answer is "yes" only if an individual's genotype at the A-locus is *random* with respect to the genotype at the B-locus. From our earlier discussion we know that random mating results in the alleles at a single locus being randomly associated in diploid individuals (Hardy-Weinberg proportions). That does not, however, give us any reason to suppose that the genotypes at any one locus should be random with respect to genotypes at some other locus. On biological grounds there are reasons to suppose that the genotypes at different loci may not necessarily be randomly associated. Many traits involve interactions between a number of different genes; we can imagine that in some cases only certain combinations of genotypes at the different gene loci might result in efficiently functioning organisms. On the other hand, genetic recombination between loci reassorts the alleles at different loci and breaks up particular combinations that may have come together in diploid individuals, thereby tending to randomize any associations between alleles at different loci.

Non-random associations between alleles at different loci are measured by the "coefficient of linkage disequilibrium."

Our first task is to develop mathematical relationships for the concept of non-random association between loci. Because genotype frequencies are determined by gamete frequencies, we need only ask whether the frequencies of the four gamete types can be found in terms of the allele frequencies. In general, the mathematical answer is no.

The essence of the problem can be seen in two examples. 1) Suppose the gamete frequencies are $g_{AB} = .36$, $g_{Ab} = .24$, $g_{aB} = .24$, $g_{ab} = .16$. 2) Suppose the same gamete frequencies are .6, 0, 0, .4. In both examples the allele frequencies are $p_A = .6$, $p_a = .4$, $p_B = .6$, and $p_b = .4$, yet the gamete frequencies are quite different. In the first case gamete frequencies are the products of the appropriate allele frequencies in the rows and columns of Table 20-6: we say that the two alleles at the A-locus are randomly associated with each of the two alleles at the B-locus. In the second case there is a complete association between the alleles at the two loci: the allele A occurs only in gametes with the allele B, and the allele a occurs only in gametes with the allele b.

To describe the differences between these examples mathematically, we need a measure of non-random association between alleles at two different loci. Consider the gamete type AB. The criterion for random association is that $g_{AB} = p_A p_B$. Expressing $p_A p_B$ in terms of gamete frequencies gives:

$$p_A p_B = (g_{AB} + g_{Ab})(g_{AB} + g_{aB})$$
$$= g_{AB}^2 + g_{AB}g_{Ab} + g_{AB}g_{aB} + g_{Ab}g_{aB}$$
$$= g_{AB}(g_{AB} + g_{Ab} + g_{aB}) + g_{Ab}g_{aB}$$
$$= g_{AB}(1 - g_{ab}) + g_{Ab}g_{aB}$$
$$= g_{AB} - (g_{AB}g_{ab} - g_{Ab}g_{aB}).$$

We make the definition:

$$D = g_{AB} - p_A p_B = g_{AB}g_{ab} - g_{Ab}g_{aB}. \qquad (20\text{-}1)$$

D is the *coefficient of linkage disequilibrium*—something of a misnomer because non-random associations do not require that the two loci be genetically linked on the same chromosome. D is a function of the particular gamete frequencies within a population and is a measure of how the alleles at two loci are associated within the population. We can see that the alleles A and B are randomly associated only if $D = 0$. D may be either positive or negative, depending upon which of the two kinds is more common. When D is positive we speak of the gamete types AB and ab as being "in excess," and Ab and aB are in excess when D is negative.

Repeating analogous calculations for the other three gamete types gives us the following relations.

$$g_{AB} = p_A p_B + D \qquad g_{Ab} = p_A p_b - D$$
$$g_{aB} = p_a p_B - D \qquad g_{ab} = p_a p_b + D \qquad (20\text{-}2)$$

Table 20-8   Genotype Frequencies in a Randomly Mating Population for Two Loci with Two Alleles Each Expressed in Terms of the Allele Frequencies and the Coefficient of Linkage Disequilibrium, D.

		Genotype at the B-Locus		
		BB (p_B^2)	Bb $(2p_Bp_b)$	bb (p_b^2)
	AA (p_A^2)	$p_A^2p_B^2$ $+$ $2p_Ap_BD$ $+$ D^2	$2p_A^2p_Bp_b$ $+$ $2p_A(p_b - p_B)D$ $+$ $-2D^2$	$p_A^2p_b^2$ $+$ $-2p_Ap_bD$ $+$ D^2
Genotype at the A-Locus	Aa $(2p_Ap_a)$	$2p_Ap_ap_B^2$ $+$ $2p_B(p_a - p_A)D$ $+$ $-2D^2$	$4p_Ap_ap_Bp_b$ $+$ $2(p_a - p_A)(p_b - p_B)D$ $+$ $4D^2$	$2p_Ap_ap_b^2$ $+$ $-2p_b(p_a - p_A)D$ $+$ $-2D^2$
	aa (p_a^2)	$p_a^2p_B^2$ $+$ $-2p_ap_BD$ $+$ D^2	$2p_a^2p_Bp_b$ $+$ $-2p_a(p_b - p_B)D$ $+$ $-2D^2$	$p_a^2p_b^2$ $+$ $2p_ap_bD$ $+$ D^2

Frequencies for the two doubly heterozygous genotypes (AB/ab and Ab/aB) are combined. D is defined by equation 20-1.

To return to joint genotype frequencies, we substitute Equations 20-2 into Table 20-7. The results of so doing are given in Table 20-8, from which we can see that the joint genotype frequencies are products of the separate Hardy-Weinberg proportions if, and only if, D is zero. Whenever D is not zero, the joint genotype frequencies are involved functions of D as well as the allele frequencies. Random mating alone is an insufficient basis for predicting the joint genotype frequencies at two (or more) loci solely in terms of the allele frequencies at the individual loci. Knowing or making some assumption about the extent of associations between alleles at the loci is also necessary.

Recombination consistently acts to randomize the alleles at different loci.

Several factors can promote or reduce linkage disequilibrium between loci. A consistent, although not always the predominant, factor is genetic recombination in the form of *either* crossing over or random assortment. Because of recombination between two loci, a doubly heterozygous individual can produce different gamete types from those that originally united to form the individual. The result is that recombination may cause a change in the frequencies of the gamete types within the population as a whole.

Consider again Table 20-7. Let R denote the frequency of recombination between A and B loci. ($0 \leq R \leq .5$; if the loci are on different chromosomes, $R = .5$.) The right column indicates the kinds and proportions of gamete types produced by each genotype. Let us suppose that the genotype frequencies of the parental generation are in the random mating proportions given in the table. The frequencies of the gamete types produced by this parental generation can be determined by simple enumeration from Table 20-7. The details are set forth in Box 20-1, where a prime denotes the new frequency for a gamete type.

With the definition of Equation 20-1 we can see from Table 20-7 that D is proportional to the difference between the frequencies of the two kinds of double heterozygotes (AB/ab and Ab/aB). We see in Box 20-1 that when the two kinds of double heterozygotes are not equal in frequency (i.e., $D \neq 0$), recombination changes the gamete frequencies by $\pm RD$. Table 20-7 and Box 20-1 indicate that these changes depend directly upon recombination frequency R and upon the fraction of individuals in the population that are heterozygous at both loci.

Recombination also changes the value of D. The new gamete frequencies in Equation 20-1 yield a coefficient of linkage disequilibrium of $D' = (1 - R)D$ for the offspring generation. (See Box 20-1.) In the absence of other factors, recombination reduces the amount of linkage disequilibrium in a population by a factor of R per generation. For unlinked loci $R = .5$, so D is halved each generation; for tightly linked loci the effect is much weaker. In either case, linkage disequilibrium is gradually, but not immediately, eliminated by recombination.

Various factors can promote linkage disequilibrium but are difficult to describe in a general quantitative fashion.

Although recombination always acts in the direction of randomizing the alleles at different loci, the quantitative result just obtained depends upon the assumption that the parental genotypes contribute gametes to the offspring generation in proportion to the frequencies in Table 20-7 with which the parental generation began as zygotes. Such an assumption is not always justified. Natural selection, for example, might cause the parental genotype frequencies at the time of reproduction to differ from their original random mating proportions. If so, the calculations in Box 20-1 are inappropriate. In particular, if changes in genotype frequencies that happen to occur during the course of a generation are of such a nature that one complementary pair of gamete types (say, AB and ab) increase in frequency within the population while the alternative (Ab and aB) decrease, the linkage disequilibrium present at the start of the subsequent offspring generation may be greater than what was present at the start of the parental generation in spite of the randomizing effect of recombination.

Non-random associations within a population are, of necessity, caused by factors that perturb genotype frequencies in just such a fashion. One of the principal factors is natural selection. However, mathematical descriptions of the conditions under which natural selection will promote linkage disequilibrium and also maintain both loci in a polymorphic state are incompletely worked out and, in their present form, are beyond the scope of this

BOX 20-1.

$$g'_{AB} = g^2_{AB} + g_{AB}g_{Ab} + g_{AB}g_{aB} + (1 - R)g_{AB}g_{ab} + Rg_{Ab}g_{aB}$$
$$= g_{AB}(g_{AB} + g_{Ab} + g_{aB} + g_{ab}) - R(g_{AB}g_{ab} - g_{Ab}g_{aB})$$
$$= g_{AB} - RD.$$

$$g'_{Ab} = g_{AB}g_{Ab} + g^2_{Ab} + g_{Ab}g_{ab} + (1 - R)g_{Ab}g_{aB} + Rg_{AB}g_{ab}$$
$$= g_{Ab}(g_{AB} + g_{Ab} + g_{ab} + g_{aB}) + R(g_{AB}g_{ab} - g_{Ab}g_{aB})$$
$$= g_{Ab} + RD.$$

$$g'_{aB} = g_{AB}g_{aB} + g^2_{aB} + g_{aB}g_{ab} + (1 - R)g_{Ab}g_{aB} + Rg_{AB}g_{ab}$$
$$= g_{aB}(g_{AB} + g_{aB} + g_{ab} + g_{Ab}) + R(g_{AB}g_{ab} - g_{Ab}g_{aB})$$
$$= g_{aB} + RD.$$

$$g'_{ab} = g_{Ab}g_{ab} + g_{aB}g_{ab} + g^2_{ab} + (1 - R)g_{AB}g_{ab} + Rg_{Ab}g_{aB}$$
$$= g_{ab}(g_{Ab} + g_{aB} + g_{ab} + g_{AB}) - R(g_{AB}g_{ab} - g_{Ab}g_{aB})$$
$$= g_{ab} - RD.$$

$$D' = g'_{AB}g'_{ab} - g'_{Ab}g'_{aB} = (g_{AB} - RD)(g_{ab} - RD) - (g_{Ab} + RD)(g_{aB} + RD)$$
$$= g_{AB}g_{ab} - RD(g_{AB} + g_{ab}) + R^2D^2 - g_{Ab}g_{aB} - RD(g_{Ab} + g_{aB}) - R^2D^2$$
$$= (g_{AB}g_{ab} - g_{Ab}g_{aB}) - RD(g_{AB} + g_{Ab} + g_{aB} + g_{ab})$$
$$= D - RD = (1 - R)D.$$

book. We shall see in Table 21-2 a case in which natural selection will promote non-random associations between the two loci; but permanent polymorphism at the two loci is not to be expected in that particular example.

The mixing of genetically dissimilar populations can also create linkage disequilibrium. As an extreme example, think of one population that is monomorphic for the genotype AABB and another that is monomorphic for aabb. The only gamete types present in these two populations are AB and ab (Ab and aB are absent). Admixture of the two populations will create a population in which there is complete association between the alleles at the A and B loci. If the occurrence of such admixture is a single event in the history of the population, with random mating and in the absence of other factors, recombination will cause this linkage disequilibrium to decay in accordance with the relation $D' = (1 - R)D$. How long the linkage disequilibrium will persist at an appreciable level depends upon the amount of recombination between the two loci.

Examples of linkage disequilibria are not uncommon; the extent and causes of linkage disequilibria in natural populations are still, however, open questions.

We saw in Chapter 19 (p. 525) that the allozyme alleles at the Pt-10 locus of *Drosophila pseudoobscura* were non-randomly associated with gene arrangements of the third chromosome. Linkage disequilibrium between allozyme alleles and gene arrangements is a common occurrence in *Drosophila* populations although there are also many examples of allozyme loci whose alleles show random associations with gene arrangements of the chromosomes on which they are carried.

Disequilibrium between pairs of allozyme loci has proved to be harder to find. In a North Carolina population of *Drosophila melanogaster* none was found for four allozyme loci spaced at roughly 20 map unit intervals along Chromosome II. Two of these loci did, however, show weak associations with gene arrangements. A study of three *Drosophila melanogaster* populations from Massachusetts and New York revealed four cases of non-random associations out of 30 pairwise comparisons of five allozyme loci distributed over a total interval of 20 map units on Chromosome III. Different pairs of loci were involved in each of the four cases, and in only one of the cases was the association between adjacent loci.

A particularly good case was found in Colorado populations of *Drosophila montana* for four esterase loci that map within a span of one map unit. Each locus is polymorphic for an *active* allele, which stains the gels, and a *null* allele, which produces no bands. Strong disequilibria were found between loci 1 and 2 and between loci 3 and 4. Data are shown in Table 20-9; in each pair the associations involve an excess of gamete types that have one active and one null allele. No disequilibrium was found between other pairs of the four loci. These particular associations were found not to change over five years' time and were also found to exist in nearby but genetically distinct populations.

Table 20-9   Gamete Frequencies for Two Pairs of Esterase Loci in a 1970 Population of *Drosophila montana* at Gothic, Colorado.

| | Locus 2 | | | | Locus 4 | | |
	Active	Null			Active	Null	
Locus 1 Null / Active	Obs. 31 Exp. 82.8	Obs. 99 Exp. 47.2	130	Locus 3 Null / Active	Obs. 16 Exp. 93.1	Obs. 253 Exp. 175.9	269
	Obs. 271 Exp. 219.2	Obs. 73 Exp. 124.8	344		Obs. 148 Exp. 70.9	Obs. 57 Exp. 134.1	205
	302	172	474		164	310	474
	$D = -.1093.$				$D = -.1626.$		

Expected numbers are based upon the assumption of random association between the active and null alleles at each locus. Data are from W. K. Baker, 1975, *Proceedings of the National Academy of Science, U.S.* 72: 4095–4099.

Another example involves linkage disequilibrium between alleles at the xanthine dehydrogenase (XDH) and aldehyde oxidase (AO) loci in two populations of *Drosophila subobscura,* one from near Athens, Greece, the other from the island of Crete. Despite their geographical separation, the two populations had similar values for *D.* Although these two loci are five to ten map units apart, their enzymes have a close physiological relationship: They share a common cofactor and the activities of both enzymes can be modified by mutants at at least two other loci. On the other hand, XDH and AO are two of the loci for which no linkage disequilibrium was found in the New England populations of *Drosophila melanogaster.*

Findings of linkage disequilibrium are not limited to *Drosophila.* The most extensive disequilibria yet found occur in Californian populations of the slender wild oat, *Avena barbata,* and involve five allozyme loci and a morphological polymorphism. The case is notable because this species reproduces almost entirely by self-fertilization. We saw in the previous section that inbreeding reduces the frequency of heterozygotes. Consequently, in *Avena barbata* the effectiveness of genetic recombination is greatly diminished—a fact that may explain the persistence of linkage disequilibrium but not its existence.

Two esterase loci of unknown map positions were found to exhibit similar linkage disequilibria in four of seventeen populations of the salamander *Plethodon cinereus*: in the other populations the two loci were randomly associated. In the snail *Cepaea nemoralis* the gene for shell color is linked to the gene that controls the presence or absence of bands (see Figure 19-1). The allele for dark brown shells is almost completely associated with the allele for unbanded shells. Yellow shells and pink shells may be either banded or unbanded. In some populations associations have been found between the pink and yellow alleles and the alleles for bandedness; the direction of association, however, varies from area to area. One population that recolonized an English river bank following a flood had an excess of the unbanded-pink and banded-yellow combinations. Sixteen years later the linkage disequilibrium had disappeared although the allele frequencies at the two loci had not changed.

The cases just described illustrate empirical evidence of linkage disequilibria in natural populations. For none of these cases can we say anything definite about the causes although in some the evidence suggests plausible explanations. For example, there is reason to suspect that natural selection is primarily responsible for the situation in *Drosophila subobscura* because 1) the same non-random associations between XDH and AO alleles are found in two well-separated populations; 2) the known biochemical relationships between the two genes make it conceivable that certain allele combinations might happen to function together better than other combinations; and 3) the comparatively loose linkage between the two loci reduces the likelihood that the observed disequilibrium in the two populations is a residual effect of some historical event such as admixture. On the other hand, the example of the *Cepaea nemoralis* population that recolonized the English river bank is an apparent example of linkage disequilibrium arising from simple admixture and decaying over time as a result of random mating and genetic recombination. The points to note are that linkage disequilibria may arise for any one of a number of reasons, and, having arisen for whatever reason, are most likely to persist when the effectiveness of recombination is reduced by tight linkage or extensive inbreeding.

In our earlier discussion we saw that linkage disequilibrium, if present in a population, affects the frequencies of multiple-locus genotypes in a way that cannot be described on the basis of single-locus theory alone. We also saw that genetic recombination enters into population genetics entirely within the context of linkage disequilibrium. Recombination can change gamete frequencies—and thereby alter a population's genetic organization—only if D is not zero. Any discussion of the evolutionary significance of recombination necessarily involves consideration of linkage disequilibria.

In general, any interactions between different genes that cannot be accounted for by single-locus theory are expressed in their effect upon linkage disequilibrium. The extent to which single-locus theory is inadequate for describing the evolutionary behavior of genes within the genome as a whole is principally determined by the extent to which linkage disequilibrium exists in natural populations. At present the data are inadequate to permit general conclusions concerning the commonness of linkage disequilibria or the kinds of genes most often involved in actual instances of linkage disequilibria. Both questions remain open for research.

SUMMARY

The basic quantities of theoretical population genetics are the allele frequencies and the genotype frequencies for a given gene locus. Allele frequencies are determined by genotype frequencies, but the converse is not, in general, true. In a *randomly mating population,* the genotype frequencies of offspring are given by the *Hardy-Weinberg proportions* that correspond to the allele frequencies of their parents. For a sex-linked gene the genotype frequencies for daughters in a randomly mating population are in Hardy-Weinberg proportions (provided the allele frequencies are the same in both sexes); however, for sons, the (hemizygous) genotype frequencies are the same as the allele frequencies.

When inbreeding occurs, there is a finite probability that the two homologous copies of any gene carried by an inbred individual may be *identical by descent.* This probability, the *inbreeding coefficient,* is determined by the individual's particular pedigree. In a population of individuals whose average inbreeding coefficient is F, the genotype frequencies of homozygotes is increased above Hardy-Weinberg proportions by an amount Fpq, and the frequency of heterozygotes is reduced by $2Fpq$. With repeated inbreeding, the average inbreeding coefficient may steadily increase until the entire population becomes homozygous. Allele frequencies, however, are unaffected by inbreeding, and random mating can undo increases in homozygosity caused by inbreeding.

When simultaneously considering genotype frequencies at each of two different gene loci, it is necessary to take into account the possibility of *non-random associations* between the alleles at the two loci. The extent of such non-random associations is expressed quantitatively by the *coefficient of linkage disequilibrium*. When linkage disequilibrium is present in a randomly mating population, the joint two-locus genotype frequencies cannot be determined solely from the single-locus Hardy-Weinberg proportions. The role of *genetic recombination* in population genetics is to randomize any associations between alleles at different loci and, thereby, to reduce any linkage disequilibrium that may exist in the population.

21

Natural Selection

Charles Darwin did not originate the idea of biological evolution, but his enunciation of natural selection as the mechanism of evolutionary change made his work monumental. We shall see in subsequent chapters that although changes manifest in genetic variation depend upon more than natural selection, it continues to stand at the very center of evolutionary biology.

Darwin's understanding of natural selection comprises four points: 1) Not all individuals survive and reproduce. 2) Because individuals differ, some have better chances of surviving and reproducing than do others. 3) Because offspring resemble their parents, they will tend to have those traits that enabled their parents to survive and reproduce. 4) Consequently, those traits that increase an individual's chances of survival and reproduction will increase in frequency from generation to generation.

The fourth point means that natural selection involves change in the frequency of some genetically determined trait or characteristic. As the animal breeder selects some quality—milk production, coat color, running speed—so, too, natural selection acts upon some polymorphic trait or, in the basic genetic sense, upon some polymorphic gene.

We can rephrase Darwin's description of natural selection in more genetic terms by saying that natural selection upon a trait occurs when three requirements are met: 1) Differences in viability and fertility exist among individuals. 2) Genetic differences in the trait exist among individuals—that is, the trait is polymorphic. 3) There is a correlation between the genetic differences and differences in viability and fertility. To come to a good understanding of natural selection we need to take a close look at these three points.

Differences in viability and fertility create the opportunity for natural selection.

The failure of many individuals to live to reproductive age and the differences among parents in the number of offspring per parent are basic facts of biology. Some plants, for example, produce large numbers of seeds, any one of which has a very low probability of survival. Even though many mammals have small numbers of offspring and devote large amounts of parental care to each individual, appreciable differences in mammalian viability and fertility nevertheless exist.

Prior to modern medicine human life was a risky business. In primitive societies infant mortality rates of 50 per cent are still not uncommon. Over one third of white Americans born in 1840 failed to reach age 15. Table 21-1 gives some contemporary data on differences among parents in numbers of children. The data for white American women imply that one third of the women produce over half of the children. Among the Xavantes, an Indian tribe in the interior of Brazil, differences are even more pronounced, especially with respect to a few males who father a disproportionate number of children.

The existence of differences in viability and fertility is not, however, in itself natural selection, nor does it even imply that natural selection is necessarily occurring. Rather, it provides what we can call the "opportunity for natural selection," which, together with the existence of genetic variation, is the necessary prerequisite for the action of natural selection upon a trait.

Natural selection occurs when there is a connection between genetic variation and differences in viability and fertility among individuals in a population.

The key to natural selection acting upon a particular locus is the third of our three points. Consider our A-locus with alleles A and a. The individuals in a population can be divided into three genotypic classes according to whether their genotype at the A-locus is AA, Aa, or aa. The critical question is whether differences in viability and fertility among individuals *within each of these three classes* differ from such differences among individuals within the population as a whole. In other words, is there any relationship between the genotypes at the locus and the likelihood of survival and reproduction?

We can see a striking example in melanism in the peppered moth. Individuals that are either homozygous or heterozygous for the *carbonaria* allele are darkly pigmented, and individuals homozygous for the typical allele

Table 21-1   Percentage of Parents Who Had Completed Families Consisting of a Given Number of Children.

	Number of Children						
	0	1	2	3	4	5	6 or more
White American Women	4%	7%	26%	28%	21%	7%	7%
Xavantes Women	2%	16%	16%	16%	16%	16%	18%
Xavantes Men	6%	19%	23%	11%	11%	10%	19%

Included in the class of "6 or More" is the Xavantes Chief who had 23 children. Data are from Kirk, *Proceedings of the National Academy of Sciences,* U.S., v. 59, p. 662, 1968 and from Salzano et al., *American Journal of Human Genetics,* v. 19, p. 463, 1967.

are peppered. In forests with lichen-covered trees the dark melanic forms are far more conspicuous (see Figure 19-3) and are more likely to be eaten by birds than are the peppered forms. Consequently, a direct connection between mortality and the genotypes at the *carbonaria* locus is created by bird predation and the different degrees to which the melanic and peppered forms are camouflaged when resting upon lichen-covered trees. The *result* is natural selection upon the alleles at the *carbonaria* locus.

BIOLOGICAL FITNESSES

The concept of *biological fitness* is used to characterize the connections between genotypes and differences among individuals in viability and fertility. The fitness of each genotype is the relative contribution to the next generation that is made, on the average, by an individual with the given genotype. Fitnesses are defined for genotypic classes as *averages* per individual. A single individual cannot half survive to adulthood nor have 1.83 offspring. It is possible, however, that half of all AA individuals, for example, die before adulthood; it is possible that AA individuals as a group have, on the average, 1.83 offspring per individual.

Because population genetics is concerned with frequencies rather than absolute numbers, fitnesses need not measure the actual numbers of offspring. They need only measure the average contribution of an individual *relative* to the contributions of individuals with different genotypes at the locus in question. A fitness may be zero in the case of a lethal genotype or a genotype that causes total sterility; otherwise, fitnesses are positive numbers.

Fitnesses are complex properties of the genotypes, involving other genes and the environment of the species.

Fitnesses are based on differential contributions to the next generation that are ascribable to differences in the genotypes of individuals. Whether genotypic differences cause differential contributions depends not only upon the direct effects of the genotypes but also upon the effects of other genes in the genome and upon the environment in which the species lives.

Consider again the peppered moth. The genotypes at the *carbonaria* locus determine wing and body coloration. The relative fitnesses, which are based upon camouflage, result from a combination of 1) phenotypic differences in wing coloration, 2) coloration of the tree trunks upon which the moths rest, and 3) bird predation. The role of the environment is readily apparent. Depending upon whether the moths live where trees are covered with lichens or are darkened by soot, the fitnesses of the genotypes at the *carbonaria* locus can be very different.

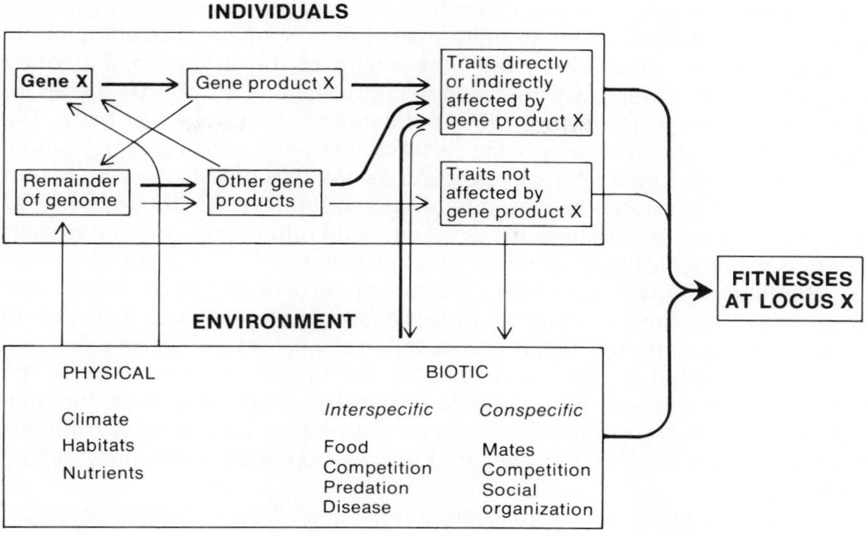

INDIVIDUALS

Figure 21-1   Interactions of potential importance in determining the fitnesses of geno-types at gene locus X. Heavy lines indicate direct connections of major importance. Light lines indicate indirect connections that may be important in some cases. The list of environmental factors is suggestive rather than exhaustive.

Figure 21-1 illustrates the principal factors that combine to determine the fitnesses associated with an arbitrary locus X. A given gene has the immediate effect of producing some gene product that, in conjunction with the products of other genes, may directly or indirectly affect a number of phenotypic traits. Those traits, in combination with environmental circumstances, influence the organism's ability to survive and reproduce. With respect to natural selection, the net result of importance is whether different fitnesses are in some way connected with different genotypes at the X locus. We can illustrate a part of the biological complexity reflected by Figure 21-1 with a few examples.

Phenotypic differences associated with genetic differences at a locus are primary factors affecting relative fitnesses.

Beginning with a specific locus and its product, such as allozyme variants, one can characterize biochemical differences between the products of the different alleles. For example, allozyme variants of alcohol dehydrogenase in *Drosophila melanogaster* differ in catalytic efficiency, sensitivity to changes in temperature and pH, and substrate specificity, as well as in electrophoretic mobility. These characteristics, like the genes themselves, are a far cry from the overall phenotype of the organism. Connections between genetic differences at the biochemical level and differences in fitness are generally indirect. For some enzymes the organism's physiology may be sufficiently buffered that small differences in biochemical properties might have virtually no effect upon the functioning of the organism. For other enzymes, small differences in biochemical functions at critical stages of development might produce profound differences in a variety of traits. The biological complexity is simply too great to warrant a broad generalization.

On the other hand, for genes, such as the *carbonaria* gene in *Biston betularia,* that produce obvious differences at the level of the organism as a whole, connections between their phenotypic differences and differences in fitnesses in a particular environment are generally more direct.

When dealing with genetic differences at the level of morphological and behavioral traits, the concept of *adaptiveness* is closely related to that of fitness. An *adaptation* is a trait that is especially suited for accomplishing a particular function or task in the life of the organism. For example, the peppered coloration of wings in *Biston betularia* is an adaptation for camouflage on lichen-covered trees. A hand with opposing thumb and fingers is an adaptation for grasping. Nectar in flowers is an adaptation for attracting pollinators.

Adaptiveness is not an inherent property of a trait. The concepts of adaptiveness and adaptation must be set within a context of a trait's being adaptive *for something.* To the extent that that "something" is important to the organism's survival and reproduction, differences in degrees of adaptiveness imply differences in fitnesses. A particular trait, however, may be adaptive for one thing and detrimental for another. For example, the structure of the human pelvis is an adaptation for erect posture; it also results in difficult childbirth. During human history the advantages of efficient erect locomotion with the hands freed for other tasks apparently outweighed the concomitant risks to mother and child.

A more subtle example occurs in the snail *Cepaea nemoralis.* Evidence links differences in shell coloration and banding patterns with differences in camouflage from bird predation in various habitats. Other evidence links differences in shell darkness with differences in heat absorption. Relative fitnesses for shell color and banding patterns are presumably affected by associated differences in both camouflage and thermoregulation. However, the way and extent to which either of these two factors affects fitnesses vary from population to population, depending upon such things as the severity of predation, whether local habitat coloration is dark or light, whether the climate is such that overheating is a problem, and whether the ability to rapidly increase body temperature in the early morning is a decided benefit to an individual.

Fitnesses can be affected by environmental factors intrinsic to the population.

The environmental factors that affect fitnesses in the examples discussed so far are extrinsic to the population. Figure 21-1 indicates that there may also be factors intrinsic to the population. One such factor is intraspecific competition among the members of the population. In particular, if individuals compete for some sort of resource—food, nutrients, space—and *if* individuals with different genotypes (at the locus in question) utilize slightly different resources, then the degree of competition that an individual experiences in acquiring necessary resources may depend upon how many other individuals in the population have the same genotype as the individual in question. If so, then individuals with rare genotypes may have higher fitness than individuals with common genotypes simply by virtue of facing less severe competition.

Any situation in which the fitness at a locus, X, is affected by the frequencies of the locus X genotypes is known as *frequency-dependent selection*. Experimental studies with different varieties of barley and wheat have shown that, in many cases, an individual plant has a higher yield if it is surrounded by plants of a different variety of the same species than if it is surrounded by plants of its own variety.

Mating behavior experiments with *Drosophila* have provided examples of an interesting phenomenon known as the "minority-male effect." When females are given the opportunity to choose a mate from among a group of males with two different genotypes, the females often preferentially mate with males who have the rarer of the two genotypes. Repeating the experiment using different proportions of the two kinds of males shows that the preferential mating is, in fact, with males of the rarer genotype, regardless of which genotype is rarer.

Many species have traits, such as the colorful plumage of male peacocks and bright spots on the sides of male guppies, that derive adaptive significance from their role in competition for mates. Any situation in which fitnesses differ between alternative genotypes because of associated differences in ability to attract mates is referred to as *sexual selection*. Such situations undoubtedly involve complex intraspecific interactions, because the attractiveness of such traits is determined by the responses of the opposite sex. The responses as well as the traits themselves are undoubtedly subject to genetic variation, most likely at different sets of loci.

Interactions with other loci can have important effects upon fitnesses.

Although we are concerned with fitnesses at a single locus, X, no locus exists in a vacuum; genes act within the highly organized milieu of the whole organism. One can expect those traits that act as intermediaries between the genotypes at locus X and their fitnesses to be affected by genetic differences at other loci. The way genotypic differences at locus X interact with genotypic differences at other loci can have a profound effect on fitnesses at locus X.

The principles involved are easily illustrated by a hypothetical example. Suppose two loci, A and B, have major effects upon wing coloration in a butterfly. Suppose individuals with genotypes AABB, AABb, AaBB, or AaBb have dull, inconspicuous coloration, but individuals that are homozygous aa have bright wings and individuals that are homozygous bb have strong dark spots. Finally, suppose that aabb individuals, which have bright wings with dark spots, mimic another butterfly species that is unpalatable and avoided by predators. The example is summarized in Table 21-2. Of the four pheno-typic classes, two have high fitness by virtue of avoiding predation by being either inconspicuous or a good mimic. The other two classes are simply conspicuous and have low fitness.

Now, consider the relative fitnesses at the A-locus. Does the genotype aa have high or low fitness? Clearly the answer depends upon the B-locus: if the B-locus genotype is BB or Bb, then the aa genotype has low fitness because it is simply conspicuous (aaBB or aaBb), but it has high fitness if the B-locus genotype is bb because it is a mimic (aabb). We speak of the genotypes aa and bb as being *coadapted* in that together they combine to produce an adaptation—mimicry, in this case.

Recall that fitnesses are *averages* for individuals grouped according to their genotype at the locus in question. The fitness of aa, for example, is an average between the relatively low values for conspicuous bright-winged individuals (aaBB and aaBb) and the high values for mimetic individuals (aabb). The average depends both upon the genotype frequencies at the B-locus and upon how the various genotypes at the A and B loci are associated within the population. If, for example, the genotype aa always occurs with bb, the fitness of aa will be higher than if the aa genotype occurs with bb on a purely random basis. Furthermore, in the second instance, the fitness of aa will be higher if bb has a frequency of .9 (in which case 90 per cent of all

Table 21-2   Fitnesses Affected by Interactions Between Two Loci.
Genotypes at Locus A

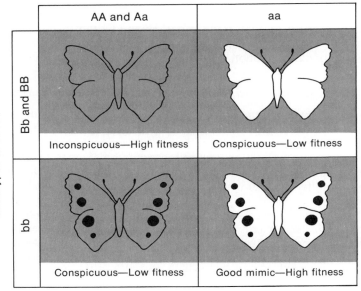

aa individuals will be mimics) than if bb has a frequency of only .2. Although this example is artificial, the principles it illustrates are fundamental to the concept of fitness. Complications caused by differences in *genetic background*, that is, differences at loci other than the ones under consideration, are well known in genetic research.

In general, a trait that is not affected by a given locus (X) has no influence on fitnesses at that locus, because there is no connection between differences in the trait and different genotypes at the X locus. However, if differences in such a trait are *non-randomly associated* with different genotypes at locus X, the fitnesses at locus X will be affected by the differences in the trait. For example, let us suppose that some allozyme locus in peppered moths has three genotypes FF, FS, and SS, which have nothing whatsoever to do with *carbonaria*. Suppose further that 80 per cent of the SS individuals happen to be melanic, whereas only 10 per cent of the FF and FS individuals are. In an environment of lichen-covered trees, SS individuals will suffer proportionately more from predation than will their FF and FS peers, and the fitness of the SS genotype will be correspondingly lower than the fitnesses of the FF and FS genotypes. Differences in fitnesses that are caused solely by such non-random associations are known colloquially as "hitchhiking" effects. In general, whenever there is linkage disequilibrium between two loci (see Chapter 20), the fitnesses at each locus are affected to some extent by the fitnesses at the other locus.

GENETIC CONSEQUENCES OF NATURAL SELECTION

Natural selection upon a particular locus occurs whenever the relative fitnesses of the various genotypes at the locus differ. We have elaborated upon various biological factors that can produce differences among the fitnesses at a locus. Such differences are the *cause* of natural selection. We turn now to the *effects* that natural selection has upon genetic variation at a locus.

Consider our A-locus with alleles A and a. The starting point is a particular set of fitnesses for the three genotypes. Suppose that the average contribution to the next generation by an individual with genotype AA relative to an individual with genotype Aa is in the ratio of $W_{AA}:W_{Aa}$ and in the ratio of $W_{AA}:W_{aa}$ relative to an individual with genotype aa. The quantities W_{AA}, W_{Aa} and W_{aa} are then the relative fitnesses of the genotypes AA, Aa, and aa. They are given quantities that characterize the *selection regime* at the A-locus. For purposes of analysis we shall regard these fitnesses as constant from one generation to the next, even though this assumption is inappropriate for a number of important situations such as frequency-dependent selection and cases involving major interactions between loci as in our hypothetical butterfly example.

The most important factor determining natural selection's effect on a locus is the fitness of the heterozygotes relative to the homozygotes.

The consequences that natural selection has upon a locus are readily described in general terms. Before deriving quantitiative results we shall describe qualitatively the changes in allele frequencies produced by natural selection. For a single locus with two alleles in a randomly mating population, selection regimes fall into three or four classes, depending upon whether the fitness of heterozygotes is greater than, intermediate between, or less than the fitnesses of the two homozygotes. The class of heterozygote intermediacy can be divided into two sub-classes on the basis of which homozygote has higher fitness.

In Figure 21-2 the effect of natural selection upon allele frequencies in each of these classes is illustrated graphically by plotting characteristic curves that show the value of the frequency p of the allele A as it changes over time. The curves are representative of the qualitative behavior in each

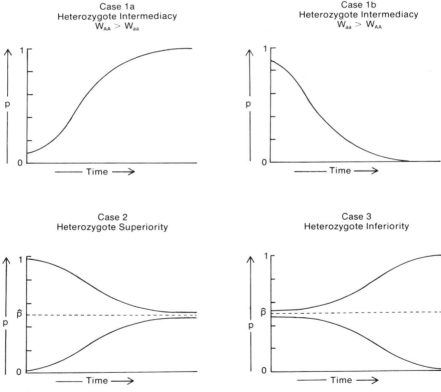

Figure 21-2   Changes in the frequency, p, of the allele A caused by natural selection. Four general patterns occur, depending upon the fitness of the heterozygote Aa relative to the fitnesses of the homozygotes AA and aa.

case. The precise shapes of the curves depend upon the particular values of the relative fitnesses and the initial allele frequency.

Heterozygote intermediacy. *Case 1a* encompasses all selection regimes in which the order of fitnesses is $W_{aa} \leq W_{Aa} \leq W_{AA}$; in *Case 1b* the order is $W_{AA} \leq W_{Aa} \leq W_{aa}$. (These cases include situations in which the heterozygote's fitness is equal to one or the other of the homozygotes' fitnesses.) In the first case, the allele A is favored by natural selection regardless of its initial frequency. Over time the frequency of A steadily increases until the allele A is *fixed* as the only allele at the locus. *Case 1b* is the converse of *Case 1a*: the frequency of A steadily decreases until A is ultimately *lost* from the population. In both cases, natural selection leads to *monomorphism,* fixing the population for the favored allele.

Heterozygote superiority. *Case 2*, which is also known as *overdominance,* encompasses those situations in which $W_{Aa} > W_{AA}$ *and* $W_{Aa} > W_{aa}$. Neither A nor a is strictly favored by natural selection because the genotype with greatest fitness, Aa, has both. For each overdominant selection regime there is a particular allele frequency between zero and unity that is denoted by the symbol $\hat{p}$ (read "*p* hat") and known as the *overdominant equilibrium point.* The actual value of $\hat{p}$, which is defined mathematically in the next section, depends upon the particular values of the relative fitnesses. Natural selection causes the frequency of A to steadily converge on the overdominant equilibrium point. If the initial frequency of A is greater than $\hat{p}$, it steadily decreases towards $\hat{p}$; if the initial frequency is less than $\hat{p}$, it steadily increases towards $\hat{p}$. Of all the cases in Figure 21-2, *Case 2* is the only one in which natural selection will maintain the population in a polymorphic condition.

Heterozygote inferiority. For selection regimes in *Case 3* $W_{Aa} < W_{AA}$ *and* $W_{Aa} < W_{aa}$. Again there is a particular allele frequency denoted by $\hat{p}$. This case is the converse of overdominance, however, in that natural selection causes the frequency of A to move away from $\hat{p}$. If the initial frequency happens to be greater than $\hat{p}$, A will increase steadily until it becomes fixed. If it happens to be less than $\hat{p}$, natural selection leads to the loss of A. In either case the result is monomorphism.

The primary genetic event of evolutionary change is an allele substitution.

Evolutionary change can be defined in part as change in the composition of a population's gene pool. In its simplest form such change involves shifts in allele frequencies. The replacement of the predominant allele within a population by some other, initially rare, allele is an *allele substitution.* An allele substitution constitutes complete genetic change at its locus. It might occur following a mutation that produces an allele with fitness greater than the fitnesses of the existing alleles. Alternatively, it might occur following a change in environmental conditions that alters the selection regime, making a rare allele more favorable than the previously predominant allele.

The replacement of the typical peppered by the melanic form within populations of *Biston betularia* in industrial areas of England is a clear example of an allele substitution. The *carbonaria* allele was non-existent

before 1848, but, following the environmental changes produced by industrialization, it had increased to frequencies of over 95 per cent in most of
the industrialized regions by 1900.

The critical aspect of natural selection in allele substitutions is the effect
of selection upon *rare* alleles. The graphs in Figure 21-2 fall in two classes
in this respect. In *Cases 1b* and 3 selection removes the A allele when it is
rare. In both cases we expect the A allele, if initially rare, to remain rare at
best, unless some other evolutionary process such as genetic drift (see
Chapter 24) should establish it in spite of natural selection. On the other
hand, in *Cases 1a* and 2 selection causes the A allele to increase in frequency
when rare. Rare alleles in either of the latter two cases are prime candidates
for producing allele substitutions.

**The various effects of natural selection are illustrated by a variety of
examples.**

Many of the common laboratory mutants, such as eye-color mutants in
Drosophila, when kept for a number of generations in laboratory populations
that contain both mutant and wild-type flies show changes in the frequencies
of the mutant alleles that closely follow the pattern of *Case 1b.* Many genetic
birth defects of humans are caused by single genes whose deleterious alleles
exhibit heterozygote intermediacy.

One of the best known examples is sickle cell anemia, the genetics of
which we discussed in Chapter 11. Homozygous Hb^S/Hb^S individuals suffer
severe anemia caused by sickling and premature destruction of their red
blood cells. The anemia is usually lethal and leads to fitness near zero for
the genotype. Although Hb^S/Hb^A heterozygotes have physiological difficulties
in some circumstances, in the majority of the world their relative fitness is
essentially the same as that of Hb^A/Hb^A homozygotes. Consequently, the
relative fitnesses for Hb^S/Hb^S, Hb^S/Hb^A, and Hb^A/Hb^A are approximately 0:1:1.
Such a selection regime is characteristic of a recessive lethal allele and falls
in Case 1. The observed frequency of approximately .0001 for the Hb^S allele
is consistent with the expectation for a situation in which natural selection
removes all Hb^S genes other than a residue attributable to recurrent mutations.

In certain parts of the world where malarial infections are common the
situation is different. The physiological component of fitness that leads to
lethality, or near lethality, of Hb^S/Hb^S individuals is the same in malarial
environments as in the rest of the world. The difference is that heterozygous
Hb^S/Hb^A individuals are more likely to resist malarial infection than are
Hb^A/Hb^A individuals. In particular, Hb^A/Hb^A homozygotes are about 85 per
cent as likely to survive infection as heterozygotes. In addition, there is
evidence that in malarial areas heterozygous individuals have slightly higher
fertility than Hb^A/Hb^A homozygotes, possibly because of malarial infection of
the placenta leading to a higher spontaneous abortion rate among Hb^A/Hb^A
mothers. Consequently, in areas where malaria is prevalent the relative fitnesses for Hb^S/Hb^S, Hb^S/Hb^A, and Hb^A/Hb^A are roughly 0:1:.8.

This particular selection regime is quite different from that in environments free of malaria. In particular, in malarial environments the Hb^A and

Hb^S alleles are subject to an overdominant selection regime in which, at the overdominant equilibrium point, the advantage of the Hb^S allele among heterozygous individuals balances its disadvantage among homozygous Hb^S/Hb^S individuals who are afflicted with the sickling disease. The observed frequency of 15 per cent to 20 per cent Hb^S in malarial regions is consistent with the expected overdominant equilibrium point for the relative fitnesses given.

A parallel example is found in the Norway rat, *Rattus norvegicus,* which is a common agricultural pest in Britain. The poison warfarin is used in some areas to control rat infestations in barns. In the early 1960's resistance to warfarin arose in rat populations on the England–Wales border, where it climbed to a frequency of about 45 per cent, at which it subsequently remained more or less constant. The resistance is attributed to a single gene with a resistant allele R that is dominant to the usual sensitive allele S so that the genotypes RR and RS are resistant to warfarin. The biochemical action of warfarin is related to blood coagulation and vitamin K metabolism. Warfarin-resistant rats turn out to be susceptible to vitamin K deficiency. Compared to SS individuals, RS heterozygotes require 2–3 times as much vitamin K, and RR homozygotes require nearly 20 times as much. The apparent situation, therefore, is an overdominant selection regime within populations exposed to warfarin: SS homozygotes are sensitive to warfarin poisoning, and RR homozygotes suffer severe vitamin K deficiency; the added vitamin K requirement of RS heterozygotes is the least of the three evils.

On the other hand, in the absence of warfarin, the poisoning factor is removed, and the vitamin K deficiency associated with the R allele leads to a selection regime exhibiting heterozygote intermediacy. When the use of warfarin was discontinued in two isolated valleys, the frequency of resistance declined substantially. We have here another example of fitnesses—and therefore, the effect of natural selection—being dramatically different in different environments.

Heterozygote superiority is a major factor in the gene arrangement polymorphisms of *Drosophila.*

The effects of natural selection upon gene arrangements of *Drosophila* have been extensively studied in laboratory populations. The populations are established with two or more arrangements present at particular frequencies and allowed to reproduce at random for many generations. The frequencies of the arrangements are monitored at regular intervals. Figure 21-3 shows data obtained from two laboratory populations of *Drosophila pseudoobscura* with flies carrying Standard (ST) and Arrowhead (AR) arrangements from Mather, California. The frequencies follow the pattern characteristic of overdominant selection, converging from both above and below to an apparent overdominant equilibrium point.

On the basis of many such experiments, certain generalizations emerge regarding gene arrangements in *Drosophila.* First, the results are sensitive to temperature. The experiment in Figure 21-3 was conducted at 25°C; at 16°C the frequencies of the gene arrangements tend to remain at whatever

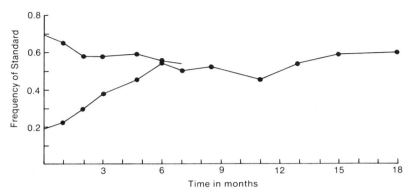

Figure 21-3   Frequency of the Standard (ST) gene arrangement is plotted over time for two laboratory populations of *Drosophila pseudoobscura* that were founded with a mixture of Standard and Arrowhead gene arrangements obtained from a natural population at Mather, California. From Cage Nos. 29 and 32 in Dobzhansky, *Genetics,* v. 33, pp. 588–602, 1948.

the frequencies were at the beginning. (For *Drosophila persimilis,* a species that lives in cooler localities than *Drosophila pseudoobscura,* such experiments show overdominance at 16°C and no change at 25°C.)

Second, the results depend upon the geographic source of the chromosomes used in the experiment. As long as all the chromosomes are obtained from a single natural population, the results typically show heterozygote superiority and are repeatable, although the particular results obtained with, say, flies from Mather differ from the results for the same gene arrangments from a different locality such as Pinon Flats, California. On the other hand, if the gene arrangements with which the experimental population is founded are from different natural populations (for example, AR from Mather and ST from Pinon Flats), the results often do not show overdominance and are not repeatable.

Further evidence concerning selection upon gene arrangements comes from comparing genotype frequencies with respect to Hardy-Weinberg proportions in natural populations. In particular, if a census is made at the egg stage, the gene arrangements generally conform to Hardy-Weinberg proportions, whereas a census of adults typically shows an excess of heterozygotes, suggesting that heterozygotes have a higher probability of surviving to adulthood in natural populations.

Chromosomal variants are, in general, likely candiates for selection regimes involving heterozygote inferiority.

The gene arrangements in *Drosophila* are atypical of chromosomal variants in that they rarely lead to inviable offspring as a result of meiosis. Such is not generally the case. Many forms of chromosomal rearrangements, especially translocations, result in meiotic difficulties in individuals hetero-

zygous for a normal chromosome complement and a rearranged complement. (See Figures 8-9 and 5-23.) In such individuals meiosis often leads to
an appreciable fraction of gametes with duplications or deficiencies, which,
in turn, lead to inviable offspring, causing a reduction of fitness for heterozygous individuals. The production of inviable offspring is, of course, only
one component of fitness, but if not compensated for by other, favorable
components, it leads us to expect heterozygote inferiority.

In general, chromosomal variants are rare in natural populations. When
a species is polymorphic for chromosomal rearrangements, usually either
the polymorphism is confined to different races within which populations
tend to be monomorphic or some sort of cytological mechanism allows
heterozygous individuals to avoid the meiotic difficulties that would otherwise lead to inviable offspring.

QUANTITATIVE ANALYSIS OF NATURAL SELECTION

We turn now to a mathematical treatment of natural selection. Our task,
so to speak, is to take the mathematical bull by the horns in order to
document the qualitative results of selection already described and to obtain
a few basic equations that we can use subsequently. Our analysis has two
phases: 1) determination of changes in allele frequencies over a single
generation and 2) determination of equilibrium allele frequencies that will
be attained if selection should continue to operate in the same form for an
indefinite length of time. The first phase involves calculation of the allele
frequencies in an offspring generation in terms of the genotype frequencies
and relative fitnesses of the parental generation. The principles are easily
illustrated by a hypothetical example.

Suppose we have 1,000 plants consisting of 250 AA plants, 500 Aa, and
250 aa. (The plants are in Hardy-Weinberg proportions with $p = q = .5$.) Suppose the relative fitnesses of the three genotypes are 4:3:2. The relative
contribution made by the three parental genotypic classes to the offspring
generation can be calculated in the form of Table 21-3.

Relative fitnesses are averages *per individual,* so the *relative* contribution
of each genotypic class is obtained by multiplying each relative fitness times
the number of individuals with that genotype. Because we work with fre-

Table 21-3.

Genotypes	AA	Aa	aa	Total
Relative Fitnesses	4	3	2	—
Number of Plants	250	500	250	1000
Relative Contributions to Offspring Generation	1000	1500	500	3000
Fractional Contributions	$\frac{1000}{3000}$	$\frac{1500}{3000}$	$\frac{500}{3000}$	1.0

quencies and because fitnesses are relative rather than absolute numbers, we need to convert the relative contributions into *fractional* contributions that sum to unity. We do so by dividing by the total. In this example, therefore, one third ($\frac{1000}{3000}$) of the genes in the offspring generation are contributed by AA parents; one half ($\frac{1500}{3000}$) by Aa parents; and one sixth ($\frac{500}{3000}$) by aa parents.

All alleles from AA parents are necessarily A: half the alleles from Aa parents are A; none from aa parents is A. It follows from the fractional contributions that the frequency of the A allele among offspring is $1/3 + 1/4 = 7/12 = .58333$. Selection, therefore, increases the frequency of A from 0.5 in the parental generation to 0.58333 in the offspring generation. The genotype frequencies in the offspring generation are a separate matter. If the gametes produced by the parental generation (in accord with their fractional contributions) randomly unite, then the new genotype frequencies will be in Hardy-Weinberg proportions for allele frequencies of 7/12 and 5/12.

General equations are stated in terms of *selection coefficients*.

For algebraic analyses it is convenient to assign one genotype the relative fitness of unity and to express the other fitnesses in relation to the fitness of the reference genotype. It is also more convenient to work in terms of the *differences* between the relative fitnesses, which are known as *selection coefficients*. Any genotype may be taken as the reference. We shall adopt the convention of using the AA homozygote as the reference, although an alternative model that uses the heterozygote as the reference is also employed.

Let us suppose we are given a set of relative fitnesses W_{AA}, W_{Aa}, W_{aa}. As long as W_{AA} is not zero, we can divide all three fitnesses by W_{AA} without altering their ratios. We can then express the relative fitnesses as $1:1 - hs$: $1 - s$, where the selection coefficients s and h are defined by the relations $1 - s = W_{aa}/W_{AA}$ and $1 - hs = W_{Aa}/W_{AA}$, from which we get $s = 1 - W_{aa}/W_{AA} = (W_{AA} - W_{aa})/W_{AA}$ and $h = (1/s)(1 - W_{Aa}/W_{AA}) = (W_{AA} - W_{Aa})/(W_{AA} - W_{aa})$.

Six numerical examples are given in Table 21-4. The selection coefficients may be either positive or negative, depending upon the order of the relative fitnesses, and are well defined as long as $W_{AA} \neq 0$ and $W_{AA} \neq W_{aa}$. If

Table 21-4   Selection Coefficients for Six Arbitrary Sets of Relative Fitnesses. The Case Number Listed in the Right Column Corresponds to Figure 21-2. $s = 1 - W_{aa}/W_{AA}$; $h = (1/s)[1 - W_{Aa}/W_{AA}]$.

	Relative Fitnesses			Fitnesses Relative to W_{AA}	Selection Coefficients		
Example	W_{AA}	W_{Aa}	W_{aa}		s	h	Case
1	4	3	2	$1 : \frac{3}{4} : \frac{1}{2}$	$\frac{1}{2}$	$\frac{1}{2}$	1a
2	2	2	0	$1 : 1 : 0$	1	0	1a
3	1	3	4	$1 : 3 : 4$	-3	$\frac{2}{3}$	1b
4	3	4	2	$1 : \frac{4}{3} : \frac{2}{3}$	$\frac{1}{3}$	-1	2
5	2	4	3	$1 : 2 : \frac{3}{2}$	$-\frac{1}{2}$	2	2
6	4	2	3	$1 : \frac{1}{2} : \frac{3}{4}$	$\frac{1}{4}$	2	3

the relative fitnesses are all the same ($W_{AA} = W_{Aa} = W_{aa}$) there is no natural selection acting on the locus; mathematically, $s = h = 0$. For those cases in which the two homozygotes happen to have equal fitness but are different from the heterozygote, this particular algebraic model is unsuitable because $s = 0$ and $hs \neq 0$, making h undefined: a different model must be adopted to deal explicitly with such cases. We shall use $s = 0$ to denote the case in which there is no selection upon the locus and otherwise assume that the homozygotes differ at least slightly in their relative fitnesses.

The selection coefficient h is equal to the difference in fitness between the heterozygote and the reference homozygote *relative* to the difference between the two homozygotes: $h = (W_{AA} - W_{Aa})/(W_{AA} - W_{aa})$. For a case involving heterozygote intermediacy, h has a value between 0 and 1 and immediately tells us which of the homozygotes the heterozygote is more like in fitness. For example, $h = 0$ implies that the heterozygote is identical to AA in relative fitness; $h = 1/2$ implies that the heterozygote is precisely intermediate between the two homozygotes; $h = 1$ implies $W_{Aa} = W_{aa}$. In the first case we speak of the A allele as being dominant with respect to fitness and the a allele as being recessive. When h is strictly greater than zero and less than unity we speak of both alleles as being *partially dominant*. When $h = 1/2$ the fitnesses are said to be *additive*, because two A alleles in a genotype add twice as much to the relative fitness as a single A allele.

If h is either negative or greater than unity, the fitness of the heterozygote is more extreme than the fitnesses of both homozygotes. Whether the heterozygote is superior or inferior to the homozygotes depends upon the sign of hs, which, in turn, depends upon both the sign of h and the sign of s; that is, upon which homozygote has the greater fitness. See examples 4, 5 and 6 in Table 21-4 and also Table 21-5.

For a randomly mating population, equations for changes in allele frequencies are easily derived.

Algebraic analyses for a single generation of selection are presented in Box 21-1. The given quantities are the relative fitnesses in the form $1 : 1 - hs : 1 - s$, and the parental genotype frequencies. For simplicity we restrict attention to randomly mating situations in which each generation *begins* in Hardy-Weinberg proportions; subsequent deviations caused by differences in viability—along with all other factors affecting fitness—are accounted for by the relative fitnesses.

The logic of the analyses parallels the numerical example in Table 21-3. The relative contribution of each genotypic class is obtained by multiplying its frequency times its relative fitness. Each relative contribution is converted to a fractional contribution by dividing by the sum of all three relative contributions (denoted by $\overline{W}$). The frequency of the allele A in the new generation (denoted by p') is equal to the entire contribution of the AA class plus half the contribution of the Aa class. $\overline{W}$ enters the algebra simply as the factor needed to normalize the contributions of the three genotypes so that they sum to unity. $\overline{W}$ is, in addition, the *average* of the relative fitnesses, with each weighted according to its frequency in the population.

BOX 21-1

Genotype	AA	Aa	aa	Total
Relative Fitness	1	$1 - hs$	$1 - s$	—
Frequency	p^2	$2pq$	q^2	1
Relative Contribution	p^2	$2pq(1 - hs)$	$q^2(1 - s)$	$\overline{W}$
Fractional Contribution	$\dfrac{p^2}{\overline{W}}$	$\dfrac{2pq(1 - hs)}{\overline{W}}$	$\dfrac{q^2(1 - s)}{\overline{W}}$	1

$\overline{W} = p^2 + 2pq(1 - hs) + q^2(1 - s) = p^2 + 2pq + q^2 - 2pqhs - q^2s$
$= 1 - 2pqhs - q^2s.$

$p' = p^2/\overline{W} + \frac{1}{2}[2pq(1 - hs)]/\overline{W} = [p^2 + pq(1 - hs)]/\overline{W}$
$= p(p + q - qhs)/\overline{W} = p(1 - qhs)/\overline{W}.$

$\Delta p = p' - p = \dfrac{p(1 - qhs)}{\overline{W}} - p = [p(1 - qhs) - p\overline{W}]/\overline{W}$

$\quad = p[(1 - qhs) - (1 - 2pqhs - q^2s)]/\overline{W}$
$\quad = p(2pqhs - qhs + q^2s)/\overline{W} = pqs[(2p - 1)h + q]/\overline{W}$
$\quad = pqs[(p - q)h + q]/\overline{W} = pqs[ph + q(1 - h)]/\overline{W}.$
$\quad$ (Notice that $2p - 1 = p - q.$)

The *change* in the frequency of the allele A is denoted by Δp. The new frequency equals the old plus the change, so $p' = p + \Delta p$ and $\Delta p = p' - p$. The calculations in Box 21-1 give the basic result

$$\Delta p = \frac{pqs}{\overline{W}} [ph + q(1 - h)]. \tag{21-1}$$

Equilibrium allele frequencies can be determined from the equation for Δp.

We now have a general equation for a single generation's change in the frequency of A in terms of its present frequency and the selection coefficients. As long as the fitnesses do not change from one generation to another, Equation 21-1 can be used to determine the equilibrium conditions that selection will lead to if given sufficient time. Analyses of equilibria in general have three aspects: 1) determination of *stationary points* at which no change occurs, 2) determination of which stationary points the system moves towards and which ones it moves away from, and 3) determination of the rate at which the system moves toward a stationary point.

For a single locus with two alleles we can find the *locations* of stationary allele frequencies simply by solving Equation 21-1 for the values of p at which $\Delta p = 0$. If s is not zero there are three possibilities: 1) $p = 0$, 2) $q = 0$ and therefore $p = 1$, and 3) $ph + q(1 - h) = 0$. The first two possibilities are the monomorphic states in which the allele A is either lost or fixed in the population and there is no genetic variation at the locus upon which natural selection can operate. To address the third possibility we substitute $q = 1 - p$ into the equation $ph + q(1 - h) = 0$ and solve for p with the result that,

$$\hat{p} = \frac{1 - h}{1 - 2h} . \tag{21-2}$$

The symbol $\hat{p}$ is employed to emphasize that we are referring to a particular, mathematically determined, *equilibrium point* for the allele frequency p. The $\hat{p}$ of Equation 21-2 is in fact the same $\hat{p}$ that we encountered in Figure 21-2.

Our simple analysis of Equation 21-1 proves that natural selection will cause no change in the frequency of A—the system will be at an equilibrium—if the frequency of A should happen to equal 0, 1, or $\hat{p}$. The equilibrium point $\hat{p}$ merits further attention. An interesting feature of Equation 21-2 is the dependence of $\hat{p}$ entirely upon the selection coefficient h: the value of $\hat{p}$ is independent of s. What matters is simply the difference in fitness between AA and Aa *relative* to the difference between AA and aa.

Figure 21-4 gives a graphical plot of $\hat{p}$ as a function of h. Allele frequencies must always be between zero and unity; as an equilibrium point, therefore, $\hat{p}$ is biologically important only when its value is between zero and unity, which is denoted by the shaded area in Figure 21-4. We see that the curve for $\hat{p}$ lies in the shaded area except when the selection coefficient h is between zero and one (indicated by dark shading). We know that the condition $0 \le h \le 1$ corresponds to the condition that the fitness of the heterozygote is intermediate to the fitnesses of the homozygotes. So, what we see is that with heterozygote intermediacy there are only two equilibrium points—0 and 1—while with heterozygote superiority or inferiority there are three—0, 1, and $\hat{p}$. Of the three possible equilibrium points, only $\hat{p}$ is a polymorphic equilibrium point.

The equilibrium point towards which allele frequencies move depends upon the order of the relative fitnesses.

If the frequency of the allele A does not happen to be 0, 1, or $\hat{p}$, the locus is not at equilibrium and natural selection will change the frequency of A in the direction of one or the other equilibrium points. We can determine the *direction* of change from the sign of Δp. Because p, q, and $\overline{W}$ are nonnegative, the sign Δp is determined by the product $s \cdot [ph + q(1 - h)]$ and may be either positive or negative depending upon the sign of s and the values of h and p.

We first consider the quantity $ph + q(1 - h)$ for which we can distinguish three cases that can be analyzed algebraically. 1) If $0 \le h \le 1$, this quantity is always positive. 2) if h is negative, $ph + q(1 - h)$ is positive when p is less than $\hat{p}$ and negative when p is greater than $\hat{p}$. 3) if h is greater than one, the reverse of 2) is true: the quantity is negative for p less than $\hat{p}$ and positive

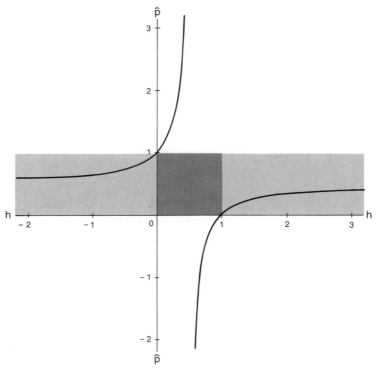

Figure 21-4   Plot of $\hat{p} = (1 - h)/(1 - 2h)$. The polymorphic equilibrium point $\hat{p}$ is biologically important only when its curve lies in the shaded area, which does not happen when there is heterozygote intermediacy—dark shading. When h is negative, $\hat{p}$ is between .5 and 1. When h is greater than one, $\hat{p}$ is between zero and .5.

for p greater than $\hat{p}$. If we now take the sign of s into consideration we have six cases, which are enumerated in Table 21-5 and which correspond to the six possible orders that the relative fitnesses may have.

Because a positive Δp implies that the frequency of A will be greater in the next generation than it is in the present generation and a negative Δp implies the reverse, the results in Table 21-5 confirm the qualitative descriptions made in connection with Figure 21-2. In particular, with heterozygote intermediacy, the frequency of A either steadily increases or steadily decreases depending upon which homozygote has the highest fitness; with heterozygote superiority, the frequency of A always moves toward $\hat{p}$, whether from above or from below; with heterozygote inferiority, the frequency of A always moves away from $\hat{p}$.

On the basis of this analysis we can specify the equilibrium point (0, 1, or $\hat{p}$) that will be attained under a given selection regime. In doing so, however, we must assume that the relative fitnesses remain the same over successive generations. In a number of interesting situations, such as fre-

Table 21-5   The Direction of Change in the Frequency of Allele A According to the Value of p and the Six Possible Orders of the Relative Fitnesses W_{AA}, W_{Aa}, and W_{aa}.

Order of Fitnesses	Selection Coefficients		Change in p
1. Heterozygote Intermediacy			
a. $W_{aa} \leq W_{Aa} \leq W_{AA}$	$0 \leq h \leq 1$	s pos.	Δp pos.
b. $W_{AA} \leq W_{Aa} \leq W_{aa}$	$0 \leq h \leq 1$	s neg.	Δp neg.
2. Heterozygote Superiority			
a. $W_{aa} < W_{AA} < W_{Aa}$	$h < 0$	s pos.	Δp pos. If $p < \hat{p}$
b. $W_{AA} < W_{aa} < W_{Aa}$	$h > 1$	s neg.	Δp neg. If $p > \hat{p}$
3. Heterozygote Inferiority			
a. $W_{Aa} < W_{aa} < W_{AA}$	$h > 1$	s pos.	Δp neg. If $p < \hat{p}$
b. $W_{Aa} < W_{AA} < W_{aa}$	$h < 0$	s neg.	Δp pos. If $p > \hat{p}$

quency-dependent selection regimes and cases where the fitnesses at the A-locus depend upon genotype frequencies at other loci, the relative fitnesses themselves may change from one generation to the next. In such cases nothing can be said about changes in allele frequencies over more than one generation unless there is some way to know what the fitnesses will be in subsequent generations. Nevertheless, Equation 21-1 is appropriate for a single generation at any particular time provided that selection coefficients appropriate for that particular time are used. For any selection regime, at equilibrium it must be the case that $\Delta p = 0$ and either $p = 0$, $p = 1$, or $ph + q(1 - h) = 0$ where h is the selection coefficient that obtains at that particular equilibrium.

RATES OF CHANGE

Given our knowledge of equilibrium conditions we need to assess the likelihood that allele frequencies at any particular locus are at, or near the equilibrium point appropriate for the selection regime to which the locus is subject. Sewall Wright suggested the possibility that evolution "may consist of an unswerving pursuit of an equilibrium point, which is itself continually on the move because of changing conditions." If his suggestion is correct, much of the genetic variation we observe may be in a transient state, heading towards, but not yet near, equilibrium conditions appropriate to the selection regime of the moment. The likelihood of a population's being near equilibrium depends upon how long the present selection regime has been in effect and upon the length of time required for a population to attain, at least approximately, the equilibrium frequencies that are associated with the particular selection regime.

In a mathematical analysis, allele frequencies attain the precise equilibrium values only after an infinite amount of time. Consequently, we cannot simply ask how long it takes to get to equilibrium. Rather, we must phrase the problem in terms of the number of generations required to change an allele frequency from some initial value, p_0, to some other value, p_t, which may be arbitrarily close to the equilibrium value.

The strength of selection is the principal factor determining the time required for an allele substitution.

Rates of change in allele frequencies can be illustrated by the general case of an allele substitution that is the result of selection involving heterozygote intermediacy. Suppose that A is an initially rare, favored allele with the relative fitnesses for AA, Aa, and aa represented as $1 : 1 - hs : 1 - s$. Equation 21-1 gives us a feeling for the magnitude of the change that occurs in one generation. Exact expressions for the *rate* of change in the frequency of A, however, are difficult to obtain. We can, nevertheless, gain insight into the dynamics of the matter from numerical examples. For various values of s and h, Table 21-6 gives the number of generations required for the frequency of A to change from $p_0 = .01$ to $p_t = .99$. As we should expect, weak selection takes longer than strong selection to effect an allele substitution. The time required is approximately proportional to $1/s$. For example, the time length with $s = .1$ is close to twice the time with $s = .2$. If selection is quite weak (say, $s = .01$, or less) an allele substitution can extend over a substantial number of generations. Evolutionary fine tuning—replacing a functional allele by one that is only slightly better—can be quite slow.

When a *recessive* allele is rare the rate of change becomes very slow.

For a given value of s, the number of generations is smallest when fitnesses are additive ($h = .5$). If h is near zero or unity, the total length of time is greatly extended. The increase in time when either allele is recessive, or nearly so, is attributable to the weak effect of selection when the recessive allele (whether A or a) is rare. Consider as a specific example an allele with a 1 per cent selective advantage (i.e., $s = .01$). In Table 21-7 its progress is broken down into stages for three cases with different values of h. The progress is shown graphically in Figure 21-5. If the advantageous allele is recessive ($h = 1$) its initial increase within the population is very slow. Over 75 per cent of the total time is taken up by the change from .01 to .05.

Table 21-6   The Number of Generations (t) Required to Change the Frequency of an Allele from an Initial Value of $p_0 = .01$ to a Subsequent Value of $p_t = .99$ Are Given for Various Values of the Selection Coefficients s and h.

h	.2	.1	.05	.01	.005
			s		
.5	83	175	359	1,830	3,700
.1 or .9	155	317	640	3,225	6,500
0 or 1	534	1,075	2,157	10,800	21,600

Table 21-7   The Progress of a Favored Allele, A, with Selection
Coefficient $s = .01$ Is Broken Down into Stages Between an Initial
Frequency of $p_0 = .01$ and a Subsequent Frequency of $p_t = .99$. The
Number of Generations Required for Each Stage Is Given for Three
Cases with Different Values of h.

Change in p		Number of Generations		
		A recessive	Additive Fitnesses	A dominant
From	To	$(h = 1)$	$(h = .5)$	$(h = 0)$
.01	.05	8,164	330	169
.05	.1	1,075	150	81
.1	.2	581	160	95
.2	.8	652	550	652
.8	.9	95	160	581
.9	.95	81	150	1,075
.95	.99	169	330	8,164
		10,817	1,830	10,817

Conversely, if the advantageous allele is dominant ($h = 0$), it becomes established within the population quickly, but the rate of change is exceedingly
slow when the allele approaches fixation. Only when a recessive allele has
a substantial frequency is its rate of change faster than the corresponding
rate of change for an allele with additive fitnesses.

We can appreciate the biology of this result by recalling that, with
random mating, rare alleles occur mostly in the heterozygous condition.
When a dominant or partially dominant allele is rare, the effect of selection
depends primarily upon the difference between the relative fitnesses of the
heterozygote and the common homozygote. When a recessive allele is rare,
most copies of the allele, being in the heterozygous condition, are unaffected
by selection, and the effect of selection, which is wholly dependent upon
the infrequent occurrence of recessive homozygotes, is very small.

The preceding examples all involve heterozygote intermediacy. Similar
dynamics govern the rates at which overdominant alleles become estab-

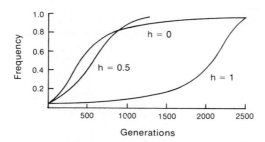

Figure 21-5   Plots of the changes over time for the allele substitutions given in Table
21-7. To keep the three curves more or less commensurate, plots are for changes from a
frequency of .05 to a frequency of .95.

lished and approach their equilibrium frequencies. For the reasons just discussed, if A is a rare overdominant allele, its initial rate of increase in the population is determined primarily by the difference between the fitnesses of the heterozygote Aa and the common homozygote aa.

SUMMARY

The *cause* of natural selection acting upon a gene is a connection between genotypic differences at the locus and differences in viability and fertility in a population. Such connections are characterized by the concept of *relative fitness*. Differences in relative fitnesses are the result of phenotypic differences produced by alternative genotypes and the effects that such phenotypic differences have upon survival and reproduction of organisms in a particular population in a particular physical and biological environment. The *carbonaria* locus in the peppered moth superbly illustrates the roles of phenotypic and environmental factors in determining relative fitnesses for a gene locus.

The *results* of natural selection are systematic changes in allele frequencies. Quantitative analyses of changes caused by natural selection can be made in terms of the *selection coefficients* for a given set of relative fitnesses. The long-term effects of natural selection depend upon the fitness of the heterozygote relative to the fitnesses of the two homozygotes. With heterozygote intermediacy, selection leads to fixation of the allele that has the larger fitness as a homozygote. With heterozygote superiority or inferiority there exists an overdominant equilibrium point, $\hat{p}$. With heterozygote superiority, selection causes the allele frequency to move toward $\hat{p}$, whereas with heterozygote inferiority, selection moves the allele frequency away from $\hat{p}$ and toward fixation of one or the other of the two alleles.

Fixation of a previously rare allele is known as an *allele substitution* and constitutes the fundamental event of evolutionary change within a gene pool. The time required for natural selection to cause an allele substitution is inversely proportional to the selection coefficient s and is less when the fitnesses are additive than when either allele is recessive or nearly so. The rate of change in allele frequencies is slow whenever a recessive allele is rare.

22

Continuous Variation and Natural Selection

We have made heavy use of industrial melanism in the peppered moth as a model for the means by which natural selection acts upon genetic variation. This example is especially illuminating because the peppered and melanic forms of *Biston betularia* constitute two discrete phenotypes that are determined by simple genetic differences upon which natural selection can act. There are many traits—such as body shape, size, and weight—for which differences among the individuals of a population do not fall into discrete classes but instead show a continuous spectrum of variation from short to tall, light to heavy, etc. For more than a half century following Darwin's publication of *The Origin of Species*, a controversy raged over the relative importance of continuous versus discrete variations in the course of evolutionary change. Darwin favored the notion that evolution proceeds by gradual change in which the mean value of a trait showing continuous variation is steadily shifted by natural selection. Others, including Thomas Henry Huxley who was Darwin's chief advocate, held the view that major evolutionary change involves discontinuous jumps akin to the change from peppered to melanic forms in industrial melanism.

The rediscovery of Mendel's work initially served to intensify the feud. The use of traits that exhibit discrete variants—e.g., green versus yellow cotyledons—was one of the keys to Mendel's success. In the first decade of the twentieth century, Mendelian geneticists, led by William Bateson in England, focused their attention upon discrete variants and resolutely held to a discontinuous view of evolution. They regarded continuous variation in characters to be mere random fluctuations about the wild-type norm and to have neither genetic basis nor evolutionary significance. Moreover, they put little stock in Darwin's view of evolution through natural selection, which they saw entirely in the context of eliminating deleterious variants from a population.

On the other side of the debate was a group, led by W. F. R. Weldon and Karl Pearson, known as the biometricians. They were firmly convinced of the correctness of Darwin's views about continuous variation and gradual evolution through natural selection. Well before the rediscovery of Mendelism they were gathering voluminous data on continuous variations and developing statistical methods for determining the degree to which offspring resemble their parents.

Human stature was one of the traits to which the biometricians devoted considerable attention and which will serve to illustrate the nature of continuous variation. Figure 22-1 shows the distribution of 174 college men according to their heights. The figure is typical of data for many traits that exhibit continuous variation, conforming moderately well to a smooth bell-shaped curve that is centered upon the mean value in the group.

The primary characteristic of inheritance is a greater resemblance between related individuals than between randomly chosen individuals. In other words, there should be a correlation between parents and their offspring. Figure 22-2 illustrates the correlation between mother and son with respect to height. From the scatter plot we can see that the majority of the points fall in either the top-right or bottom-left quadrants, which implies that taller-than-average mothers tend to have taller-than-average sons and shorter mothers tend to have shorter sons. Notice that the criterion for whether a particular mother is tall or short is in relation to the mean height of the mothers, and the criterion for the sons is in relation to the mean of the sons. A correlation between parents and their offspring for a given trait suggests that differences for the trait among individuals reflect to some extent genetic variation among those individuals. Such a conclusion does not, however, automatically follow. Cultural inheritance, in particular, also may lead to correlations between parents and offspring. For example, most individuals speak the same language that their parents speak.

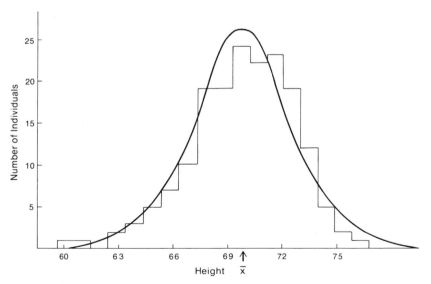

Figure 22-1   Frequency distribution for the heights (in inches) of 174 college men studying general genetics at Berkeley. Mean height is 70.2 inches.

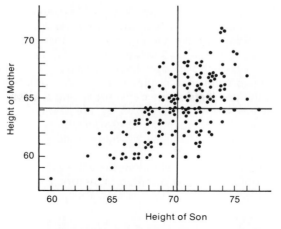

Figure 22-2   Scatter plot of the heights of mothers and sons for 174 college men. The cross lines indicate the mean heights for the sons and their mothers.

The biometricians were, nevertheless, correct in their conviction that such traits as height that vary in a continuous fashion may indeed be based upon genetic variation upon which natural selection can act. They were stymied by their inability to explain adequately the genetic basis or mode of inheritance for continuous variation. Resolution of the matter came in 1918 when R. A. Fisher developed a genetic model capable of interpreting continuous variation on the basis of the Mendelian principles of discrete, particulate inheritance, thereby achieving a synthesis of Darwinism, Mendelism, and biometry.

A GENETIC MODEL FOR CONTINUOUS VARIATION

A trait that is affected to a small degree by allelic differences at each of a large number of polymorphic loci can be expected to show continuous variation.

To see how discrete Mendelian genetic differences might result in a continuously varying array of phenotypes like those shown in Figure 22-1, let us consider a hypothetical example. Let us suppose that a trait, say, wing length, is affected by a number of polymorphic genes (A, B, C, . . .) with alleles A and a, B and b, C and c, etc. Suppose that, for each of these genes, being homozygous for the upper-case allele increases wing length by some small amount relative to the wing length of individuals that are homozygous for the lower-case allele. That is, individuals with the genotype BB, say, have wings that are, on the average, one unit longer than individuals with the genotype bb. For numerical simplicity, let us assume that the frequencies of the upper-case alleles at all the loci are .4 and the frequencies for lower-case alleles are .6. With random mating, therefore, the genotype frequencies for each locus are .16 for AA, .48 for Aa, and .36 for aa, and so on for BB, Bb, bb, etc. Let us further suppose that gentotypes at the various loci are randomly associated within the population so that multiple-locus genotype frequencies are the products of the constituent single-locus genotypes: the frequency of AAbb is $(.16)(.36) = .0576$; the frequency of AaBbcc is $(.48)(.48)(.36) \doteq .083$; the frequency of AaBbCcDdEe is $(.48)^5 \doteq .0255$; etc. In terms of our discussion in Chapter 20, there is no linkage disequilibrium among the loci involved.

The distribution of relative wing lengths among individuals of the population depends upon a number of different factors. Figure 22-3 shows four

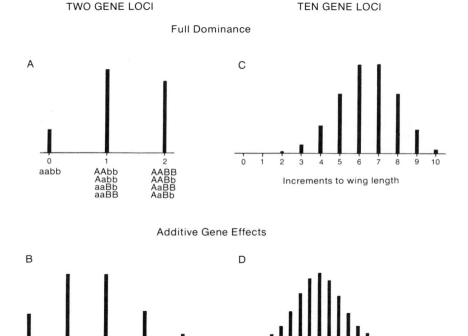

Figure 22-3   Four hypothetical examples of frequency distributions for the phenotypes of individuals in a population with polymorphic loci that affect wing length. The height of each vertical bar indicates the frequency of individuals whose genotype results in a given contribution to wing length. Genotype frequencies are based upon random mating and frequencies of .4 and .6 at each locus. A and B: two polymorphic genes; C and D: ten polymorphic genes. A and C: full dominance at each locus; B and D: additive gene effects at each locus.

examples. In (A) and (B) there are two genes that affect wing length. The heights of the vertical bars indicate the combined frequencies of all genotypes that contribute a given amount to relative wing length. In (C) and (D) there are ten loci, each of which has a small effect upon wing length. With only a small number of polymorphic loci affecting a trait, the distribution of multiple-locus genotypes leads to a more or less continuous distribution of phenotypes.

The differences between (A) and (B) and between (C) and (D) are the result of differences in the kind of dominance shown by the alleles at each locus. In (A) and (C) the upper-case alleles are fully dominant: a heterozygous locus makes the same contribution to wing length as a locus homozygous for the upper-case allele. In (B) and (D), however, a heterozygous locus makes half the contribution that a locus homozygous for the upper-case allele makes. In (B) and (D) it is as if each upper-case allele in the genotype independently adds a half increment to wing length: such a situation exhibits *additive gene effects.* Although the kind of dominance makes a quantitative difference in the phenotypic distributions—the locations of the means in (C) and (D), for example—the qualitative result of a more or less continuous distribution of phenotypes can be seen in both examples.

In all four examples the heights of the vertical bars indicate genotypic frequencies and, therefore, depend upon the allele frequencies at the loci that affect the trait. For numerical simplicity we used frequencies of .4 and .6 at each locus. We can imagine a combination of polymorphic loci with various allele frequencies and with different degrees of dominance resulting in a continuous, roughly bell-shaped, distribution of phenotypes.

If we further suppose that the phenotypes of individuals who have identical genotypes randomly fluctuate about the average value for their genotype, we can imagine the gaps between adjacent bars in Figure 22-3 smoothed out to give a continuous phenotypic distribution. Such fluctuations may occur for a variety of reasons. Traits like body size and stem length can be affected by a host of environmental factors, among them the kinds and quantities of food and nutrients available, climate, and the incidence of disease. The larger such environmentally induced differences are relative to the genotypic differences illustrated in Figure 22-3, the more the population's overall phenotypic distribution will be spread out.

In spite of its simplicity, the polygene model for continuous variation is consistent with many known examples from breeding experiments.

The genetic model we have just described, in which continuous variation in a trait is ascribed to random, environmentally induced fluctuations plus accumulated genetic effects caused by allelic differences at a number of gene loci, is known as the *polygene* or *multiple-factor* model of inheritance. The various polymorphic gene loci that affect the trait are rather loosely referred to as *polygenes*. In the form just presented, the polygene model gives a simplistic view of gene action and the biological basis for quantitative traits such as height or wing length. Nevertheless, the model has proved to be a reasonable means for understanding the results of numerous breeding experiments.

A classic example is afforded by the work of the early geneticist E. M. East with the tobacco plant *Nicotiana longiflora*. His experiments are illustrated in Figure 22-4. Like Mendel, East began with two, highly inbred, pure breeding lines of plants. The two lines differed greatly in the size of their flowers; in one line the average length of the corolla was 40.5 millimeters, and in the other line it was 93.3 millimeters. Both lines were presumably homozygous at most loci because of having been repeatedly self-fertilized, and the differences between the lines were attributed to their being homozygous for different alleles at one or more gene loci (e.g., AABBCC. . . for the tall line and aabbcc. . . for the short line). From the histograms of Figure 22-4 we can see some variation among the individuals of each line, which presumably reflects random fluctuations about their respective means.

The F_1 generation, produced by crossing the two parental lines, consists entirely of individuals that are heterozygous at each of the loci responsible for the difference between the two parental lines. The intermediate corolla lengths of F_1 individuals suggest that the gene effects at these loci are additive. Because all F_1 individuals are genetically identical, the variation in corolla length among them is again attributable to random fluctuations.

At this stage in the experiment there is no difference between the observed results and what would be expected for a simple Mendelian trait caused by a single gene for which heterozygotes have a phenotype that is intermediate to the phenotypes of homozygotes. It is in the F_2 generation, produced by self-fertilization of F_1 individuals, that the polygenic nature of corolla length is observed. For a simple Mendelian trait we would expect a 1:2:1 distribution of F_2 individuals, with one fourth being like each of the two parental lines. We see nothing of the sort in Figure 22-4. Instead we see an array of phenotypes among the F_2 that looks similar to Figure 22-3D.

Two characteristics of the F_2 generation are important to our interpretation of the results. First, none of the F_2 individuals is as extreme as either of the parental lines. This indicates that a number of different gene loci are involved in the inheritance of corolla length: If n independently assorting genes are involved, the probability of an F_2 individual being homozygous for the short allele at every locus (aabbcc. . . like the short parental line) is $(\frac{1}{4})^n$. If n is, say, five or more, the likelihood of finding such extreme individuals in the F_2 generation is very small.

Second, phenotypic variation among the individuals of the F_2 generation is substantially broader than among individuals of the F_1 generation, indicating that differences between F_2 individuals are caused by more than environmentally induced fluctuations. Indeed, although each individual in the F_2 generation is expected to have, on the average, 50 per cent short alleles and 50 per cent long alleles, on the basis of Mendelian segregation and chance deviations we expect individuals in the F_2 generation to be genetically different, some having more than 50 per cent short alleles and some having more than 50 per cent long alleles.

Further evidence of genotypic differences among the F_2 generation is provided by self-fertilizing three phenotypically different F_2 individuals. Each

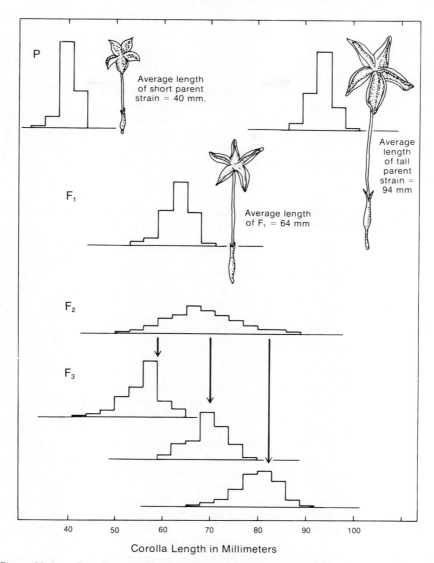

Average length
of short parent
strain = 40 mm.

Average
length
of tall
parent
strain =
94 mm

Average length
of F₁ = 64 mm

Corolla Length in Millimeters

Figure 22-4   Breeding experiments with two highly inbred lines of *Nicotiana longiflora*, illustrating the polygenic inheritance of corolla length. Data of E. M. East, reprinted with permission of Macmillan Publishing Co., Inc. from *Genetics* by M. W. Strickberger. Copyright © 1976 by Monroe W. Strickberger.

of the F_3 generations shown in Figure 22-4 exhibits variation that can be attributed to further segregation of heterozygous loci as well as to environmentally induced fluctuations. Even so, in each case mean corolla length is more similar to the length of the parent chosen from the F_2 generation than to the mean of the F_2 as a whole.

These data on corolla length show that segregation in the F_2 and F_3 generations does not conform to the Mendelian pattern expected for differences that result from a single gene. Instead, segregation is consistent with our polygenic model of phenotypic variation based upon Mendelian segregation at a number of loci. In other words, we have a set of results for which a simple single-locus genetic model is inadequate and the polygenic model is adequate.

Human red blood cell acid phosphatase activity shows continuous variation that can be subdivided on the basis of electrophoretic variants.

We get a slightly different perspective on the genetic basis of continuous variation from data on human red blood cell acid phosphatase activity. Figure 22-5 shows the distribution of enzyme activity measured for red blood cell acid phosphatase obtained from 275 individuals in an English population. The distribution has the typical shape of a trait that varies continuously within a population. The structural gene for this enzyme is polymorphic for three electrophoretically detectable alleles within the English population. The three alleles, which we shall simply denote as A_1, A_2, and A_3, have frequencies of .36, .57, and .07 among the 275 individuals sampled. Genotype frequencies are close to Hardy-Weinberg proportions. The enzyme activities can be grouped according to the allozyme genotypes of the individuals. Figure 22-5 shows the distributions of enzyme activities for each genotype.

The overall distribution of enzyme activity is strongly affected by three biologically distinct factors: 1) differences in mean activity for the different allozyme genotypes; 2) variation among individuals having the same allozyme genotype; 3) the frequencies within the population of the different genotypes. In this particular case comparatively large differences in enzyme activity are associated with the allozyme genotypes. For example, individuals with the A_1A_1 genotype have on the average 122 units of enzyme activity and individuals with the A_2A_2 genotype have 188 units on the average. Such differences are primarily determined by the biochemical and physiological properties of the enzymes specified by the different genes. In this particular example we can see that the model of additive gene action is reasonably applicable to the A_1 and A_2 alleles: the mean activity of 154 for A_1A_2 heterozygotes is close to being precisely intermediate between the mean activities of 122 for A_1A_1 homozygotes and 188 for A_2A_2 homozygotes.

We can also see from Figure 22-5 that two individuals with the same allozyme genotype can have quite different enzyme activities. The distribution for the A_2A_2 homozygotes, for example, ranges from 140 to 240. Differences among individuals with the same allozyme genotype may be attributed to two different factors. First, the individuals may have experienced different

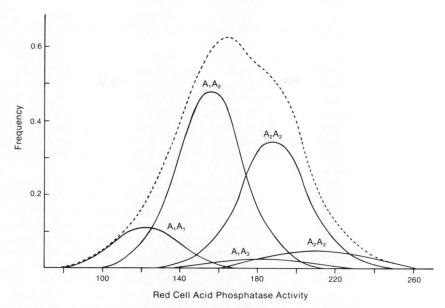

Red Cell Acid Phosphatase Activity

Figure 22-5   Distribution of red blood cell acid phosphatase activity for an English population. The dotted line shows the distribution for the overall population. Solid curves indicate the activity distributions for individuals grouped according to five electrophoretically distinguishable genotypes. The curve for the overall population is the sum of the constituent distributions. The genotype A_3A_3 is too rare to be included. Drawn after H. Harris, *The Principles of Human Biochemical Genetics,* North Holland/American Elsevier, 1970.

environmental conditions that in some way affect the individuals' levels of enzyme activity. Second, two individuals with the same allozyme genotype may have different genotypes at other polymorphic loci that also affect the activity level of red blood cell acid phosphatase. We can imagine the curve for each allozyme genotype in Figure 22-5 being further broken down into constituent distributions on the basis of genotypes at some other locus in the same fashion that the distribution for the total population has been subdivided on the basis of allozyme genotypes.

Finally, Figure 22-5 illustrates the effects of genotype frequencies upon the population's overall distribution of enzyme activities. Because the overall distribution is the sum of the constituent distributions for the different genotypes, its curve is dependent upon the number of individuals with each of the genotypes. If, for example, the allele A_1 were common enough so that A_1A_1 became the most common genotype, the distribution for the population would be shifted substantially to the left. Recognition of the role played by allele frequencies in determining the distribution of a continuously varying trait is crucial to our understanding of the effects of selection upon such a trait.

Continuous variation can be represented as the net result of differences among individuals with respect to their genotypes and their environments.

In the parental and F_1 generations of Figure 22-4 we see small but distinct environmentally caused variation among individuals of like genotypes. In the F_2 and F_3 generations, environmentally caused variation is confounded with genotypic differences among individuals caused by imprecisely known amounts of genetic segregation. In Figure 22-5 we see substantial variation among individuals with the same allozyme genotype, but we must leave unresolved the question of how much of that variation is caused by genotypic differences at other loci and how much is attributable to environment differences.

We incorporate environmentally caused variation into the polygenic model of continuous variation by representing each individual's phenotype as the sum of a value associated with the individual's particular genotype and a deviation attributed to the specific environmental circumstances that the individual experienced. $P = G + E$. For example, with reference to Figure 22-3B, consider three hypothetical individuals. The first has genotype AAbb and a relative wing length of 1.2 increments on the scale of the figure; the second also has a phenotype of 1.2 but genotype AABb; the third has genotype AABb like the second but a phenotype of 1.5. The three individuals' phenotypes are (1) $1.2 = 1.0 + .2$, (2) $1.2 = 1.5 - .3$, and (3) $1.5 = 1.5 + 0$.

The first two individuals have identical phenotypes but different genotypes. We might suppose that the first individual received, say, a greater than average amount of food as a juvenile, while the second was afflicted with a mild parasite that stunted growth. Their identical phenotypes arise from the first individual's being above average for its genotype while the second is below average for its. With respect to long wings, however, the second individual has the better genotype and will, on the average, have longer-winged offspring than the first individual. The second and third individuals, on the other hand, are genetically identical in spite of their different phenotypes. The environmental deviation of zero for the third individual means the individual experienced average environmental conditions; it does not mean that the individual had no environment.

When analyzing continuous variation, one of the principal tasks is the sorting out of the relative degrees to which phenotypic differences between individuals are caused by 1) the individuals' having different genotypes, and 2) environmentally induced differences. To do so requires some means for quantifying the amount by which individuals within a population differ from each other. Two related quantities, the *variance* and the *standard deviation,* are commonly used.

The variance of a population (denoted by σ^2) is the average squared deviation from the mean within the population. For empirical situations, a sample of measurements from a population can be used to estimate the value of the population's variance by the calculations shown in Box 22-1. The square root of the variance is the standard deviation (σ) and is a direct measure of the extent to which individuals in the population vary about the mean. For bell-shaped distributions—those in Figure 21-1 and 21-5, for ex-

BOX 22-1
Sample Mean and Variance.

Suppose that we have a sample of n measurements for some trait—say the heights of 150 women. Let us denote the individual measurements by X_1, X_2, X_3,..., X_n. The sample mean, denoted by $\overline{X}$, is calculated by

$$\overline{X} = \frac{X_1 + X_2 + \cdots + X_n}{n} = \frac{1}{n} \sum_{i=1}^{n} X_i.$$

The sample variance, denoted by V to distinguish it from the population's true variance, is calculated by

$$V = \frac{(X_1 - \overline{X})^2 + (X_2 - \overline{X})^2 + \cdots + (X_n - \overline{X})^2}{n - 1}$$

$$= \frac{1}{n - 1} \sum_{i=1}^{n} (X_i - \overline{X})^2.$$

We divide by $n - 1$ (instead of n) when calculating the sample variance because we are estimating the true value of a population's variance with calculations based on only a sample of individuals from the population.

The standard deviation is the square root of the variance. Bell-shaped curves have the property that approximately two out of every three individuals are within one standard deviation of the mean and 95 per cent are within two standard deviations of the mean. Diagramatically:

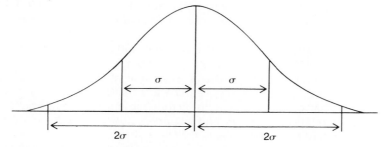

ample—the standard deviation has the graphical interpretation shown in Box 22-1.

We can characterize the phenotypic variation of any trait within a given population by its mean and variance. We let σ_P^2 denote the variance of phenotypic values among the individuals of the population. We have just seen, however, that each individual's phenotypic value can be broken down into the sum $P = G + E$. Depending upon their genotypes, different individuals may have different values of G. The distributions in Figure 22-3 are, for example, distributions of genotypic values. Genotypic differences within a population can be characterized by the variance, σ_G^2, of the values of G among the individuals of the population. Similarly, depending upon the array of environmental conditions experienced by the individuals within the population, different individuals may have different values for the environmental deviation, E. The variety of effects upon the trait that are attributable to environmental differences is characterized by the variance, σ_E^2, of the values of E among the individuals of the population.

Variances have an important property that was noted by R. A. Fisher in his 1918 paper. If one quantity is the sum of two or more uncorrelated quantities, then the variance of the sum is simply the sum of the variances. So, if the environmental deviations are unrelated to genotype, we get the relation $\sigma_P^2 = \sigma_G^2 + \sigma_E^2$. It is not necessarily the case, however, that genotype and environment are uncorrelated. For example, sometimes dairy farmers allot feed to their cows in accordance with the amount of milk each cow gives, so that the better milk producers get more feed. The systematic differences in feed allotment accentuate any genetic differences in milk production, and, as a result, σ_P^2 is *greater* than $\sigma_G^2 + \sigma_E^2$.

A measure of the degree to which phenotypic differences are inheritable within a population is given by the genotypic variance relative to the total phenotypic variance.

The key point is that we can explicitly measure *differences* among the members of a particular group of individuals. For any trait in a particular population, we define its *heritability* by $H = \sigma_G^2/\sigma_P^2$. Because σ_G^2 is a component of σ_P^2, the value of H is between zero and unity and gives a measure of the extent to which phenotypic differences among individuals are the result of genotypic differences and can, to some extent, be passed on to their offspring.

Determining actual values for H is generally difficult because only phenotypic variation is directly observable. Direct determination of σ_G^2 requires knowing what genes are involved and what their allele frequencies are. In practice, estimates of heritability are generally made by statistical and genetic models that predict observable phenotypic correlations between various kinds of relatives (parents and offspring, full-sibs, half-sibs, etc.) in terms of the population's heritability for the trait in question.

Our definition of heritability is known as *broad-sense heritability*. We shall see in the next section that genotypic differences among individuals are not always entirely inheritable. Using σ_G^2 as the measure of genetic variation in a trait overestimates the trait's heritability unless—as in Figure 22-3D, for example—gene effects are strictly additive and there is no linkage disequilibrium. Nevertheless, we gain a general appreciation of heritability from our broad definition.

One important property is readily apparent: the heritability of any trait is specific to a particular population. With even the simplest cases in which

$H = \sigma_G^2/(\sigma_G^2 + \sigma_E^2)$ we can see that the value of H depends upon the circumstances peculiar to each population. In particular, distributions of genotypic values—such as in Figure 22-3—depend upon the genotype frequencies in the population. So, too, therefore, does the value of σ_G^2 and, in consequence, the value of H. In Figure 22-5, we can see that the genotypic variance would be substantially smaller if the frequency of allele A_2 were, say, .99.

The environmental variance, σ_E^2, also necessarily depends upon the array of environmental conditions to which the individuals of the population are exposed. A nurseryman who is sloppy in his application of fertilizer can expect to have more variability in his plants than his meticulous, even-handed colleague will have. Animal and plant breeders go to great effort to make environmental conditions as uniform as possible for the express purpose of increasing the heritability of the trait they are interested in.

In short, heritability, like allele frequencies, describes a particular state of affairs *within* a particular population at a particular time. Estimates of heritability obtained from one population cannot be relied upon for other populations. Nor can heritability be used to say anything about differences between populations. The inappropriateness of attempting to do so can be seen from two examples. The two inbred parental populations in Figure 22-4 are very different from each other genetically, but within each population all the phenotypic variation is environmental. As a result, the heritabilities within the populations are zero, yet the differences between the populations are predominantly genetic. Alternatively, we can think of two identical but genetically heterogenous plant populations that are grown in very different but uniform environments—one, say, with liberal water and fertilizer, the other on an arid, unfertilized plot. The populations will probably be very different, yet variation within each population may be predominantly genetic. In other words, differences between the populations may be almost entirely the result of their different environments even though heritability within each population may be high.

When kept in its proper context, however, heritability is an important concept. Breeders want to know that the better-than-average individuals chosen to be parents will have better-than-average offspring. Biologists want to know that variations in the traits that they study are "genetic." But its primary importance is that evolutionary change in a trait and the efficiency with which selection can modify a phenotypic trait are directly related to the trait's heritability.

SELECTION ON TRAITS EXHIBITING CONTINUOUS VARIATION

What we have accomplished so far is to construct a model that can account for a continuous distribution of phenotypes on the basis of Mendelian segregation at a number of gene loci. We now want to look at our model with the purpose of making inferences about the effects of selection upon a trait that shows a spectrum of continuous variation among the individuals of a given population. Because our model is founded upon the principles of Mendelian genetics, we begin by analyzing the effect of selection upon a single gene locus that contributes to the overall phenotypic distribution in the population.

Selection changes a population's distribution for a continuously varying trait by changing allele frequencies at loci that affect the trait.

We saw in the last chapter that natural selection upon a polymorphic locus produces a change in allele frequencies. Selection upon a continuously varying trait is no different. We can get a feeling for some of the major factors involved from hypothetical examples patterned after the situation in Figure 22-5 for red blood cell acid phosphatase activity. Suppose that Figure 22-6A represents a population's frequency distribution for some continuously varying trait that is strongly affected by the polymorphic A-locus. Suppose that the frequency of the allele A is $p = \frac{2}{3}$, and the genotype frequencies are in Hardy-Weinberg proportions. The mean value of the trait in the population is $M = 125$. The means for the three genotypes are $M_{AA} = 135$, $M_{Aa} = 120$, and $M_{aa} = 105$. Within each genotype, however, there is considerable variation.

Now, suppose that only those individuals whose phenotypic value is greater than 125 are allowed to reproduce. Compared to the population as a whole, the genotype AA will be overrepresented among the reproducing individuals, and the genotype aa will be underrepresented. In fact, the shaded and hatched areas above the value of 125 in Figure 22-6A are roughly in the proportions of 37:16:1 for the genotypes AA, Aa, and aa. So, among the reproducing individuals, the frequency of the allele A is approximately $\frac{5}{6}$ compared to its frequency of $\frac{2}{3}$ in the population as a whole.

If the reproducing individuals mate randomly among themselves, the resulting offspring will be in Hardy-Weinberg proportions with allele frequencies of $p = \frac{5}{6}$ and $q = \frac{1}{6}$. Figure 22-6B shows the expected frequency distribution for our trait among the offspring. The genotypic means, M_{AA}, M_{Aa}, and M_{aa}, are considered to be the same as before. So is the amount of variation among individuals with the same genotype. Nevertheless, the overall distribution of phenotypic values changes in the offspring generation. The mean value for the population has increased from 125 to 130. The change is entirely the result of different frequencies of the alleles A and a in the two generations.

With additive gene action the amount of change caused by selection is proportional to the difference between the genotypic means relative to the overall variation in the trait.

We can see from Figure 22-6A what is going on. Because the mean values of the three genotypes (M_{AA}, M_{Aa}, M_{aa}) are different, there is a correlation between higher-than-average phenotypes and the genotypes at the A-

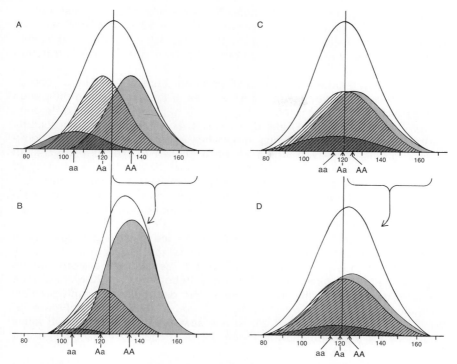

Figure 22-6   Hypothetical examples of frequency distributions for a continuously vary-ing trait, subdivided according to genotype at the A-locus. AA—light shading; Aa—hatching; aa—dark shading. The horizontal axis is numbered in units appropriate to the trait—e.g., length, weight, enzyme activity, etc. Gene action at the A-locus is additive in its effect upon the trait. The distributions in B and D are the results of allowing only those individuals in A and C, respectively, whose phenotypic value is greater than the popu-lation's mean, to mate at random among themselves. The frequency of the allele A is p = 2/3 in A and C. In B, p = 5/6; in D, $p \doteq$ 5/7. Genotype frequencies are in Hardy-Weinberg proportions.

locus. This correlation and the extent to which selection will change the allele frequencies depend upon several factors. One obvious factor is the amount by which the genotypic means differ relative to the amount of vari-ation between individuals in the population as a whole.

To illustrate this point, compare Figures 22-6A and C. The total amount of variation is about the same in both distributions. In the latter, the genotypic means are closer together and there is more variation among individuals with the same genotype. The overlap of the constituent distributions for the three genotypes is considerably greater. As a result the correlation between genotype and phenotype is weaker. If, as before, only those individuals in Figure 22-6C whose phenotypic values are larger than the population's mean are allowed to mate, the frequency of the allele A among mating individuals will be roughly $\frac{5}{7} = \frac{15}{21}$, which is only slightly larger than the frequency of

$\frac{2}{3} = \frac{14}{21}$ in the population as a whole. The expected offspring distribution, comparable to Figure 22-6B, is shown in Figure 22-6D. We can see that selection on the distribution in Figure 22-6C has less effect upon the allele frequencies than does the same amount of selection upon the distribution in Figure 22-6A.

In both figures the gene action at the A-locus is additive. The difference between the genotypic means for AA and Aa is the same as the difference between Aa and aa. That is, $M_{AA} - M_{Aa} = M_{Aa} - M_{aa}$. Let us call this difference α (i.e., $\alpha = M_{AA} - M_{AA}$). The actual value of α is not by itself particularly informative. Among other things, it depends upon the units—inches, centimeters, grams, etc.—with which we have chosen to measure the trait in question. The value of α relative to the amount of variation in phenotypic values in the total population is, however, important. We saw in Box 22-1 that the standard deviation σ_P characterizes the amount by which the phenotypic values of the individuals within the population differ from the mean value of the population.

By comparing Figures 22-6A and C we can see that the change in allele frequencies caused by selection is proportional to α/σ_P. Both α and σ_P are in the units with which the trait is measured; their ratio, therefore, has no units—i.e., α/σ_P is a dimensionless quantity. In both Figures 22-6A and C the standard deviation for the overall phenotypic distribution is approximately 20 units. In Figure 22-6A, $\alpha = 15$ units and $\alpha/\sigma_P = .75$, whereas in Figure 22-6C, $\alpha = 5$ units and $\alpha/\sigma_P = .25$. As we saw earlier, the change in allele frequency for the population in Figure 22-6A is roughly three times larger than the change for Figure 22-6C.

Changes in allele frequencies also depend upon the intensity and efficiency of selection upon the trait.

In the two examples we have considered we assumed that only the half of the population that had phenotypic values above the mean served as parents of the offspring generation. Obviously, if a larger or smaller fraction of the population served as parents, the change in the allele frequencies would be different. For example, if all the individuals in the population served as parents, there would be no change at all. Or, at the other extreme, if only those few individuals in Figure 22-6A whose phenotypic value is greater than 150 units served as parents, the frequency of the allele A would be nearly 100 per cent in the offspring generation.

Furthermore, in our examples we have assumed that the demarcation between those individuals that do or do not serve as parents is drawn sharply on the basis of a particular phenotypic value (for example, the population's mean). For artificial selection in plant or animal breeding programs, sharp demarcations on the basis of phenotype are possible. They are, however, unlikely in the case of natural selection upon a continuously varying trait. Natural selection will be less efficient than artificial selection in that the connection between an individual's phenotype and the individual's contribution to the next generation is a more probabilistic matter.

With respect to both the intensity and efficiency of selection, the thing that counts is the extent to which the phenotypic values of those individuals that contribute to the next generation differ from the phenotypic values in their own generation as a whole. Suppose we let M_S denote the mean phenotypic value of those individuals that actually contribute to the offspring generation, where the phenotypic values of individual parents are weighted according to the number of offspring each one produces. If we let M_P denote the mean phenotypic value in the parental generation as a whole, then the difference between the average phenotype of selected parents and the average of their generation as a whole is $M_S - M_P$. Once again, it is not the actual difference that counts but, rather, the difference relative to the amount of variation in the population. The quantity $i = (M_S - M_P)/\sigma_P$ is known as the *intensity of selection*. It summarizes the effect wrought by selection at the phenotypic level of the parental generation.

Our examples based upon Figure 22-6 have so far shown us two major factors that influence the extent to which selection upon our trait in question will lead to changes in allele frequency at the A-locus. The effects we have seen are expressed quantitatively in the following equation, which can be rigorously derived by using the statistical techniques of regression analysis:

$$\Delta p = i \cdot pq \cdot \frac{\alpha}{\sigma_P}. \tag{22-1}$$

The equation is written as the product of three terms. The term i we know is a measure of the intensity and efficiency of selection upon the trait in the parental generation. The term α/σ_P can be thought of as a measure of the extent to which phenotypic differences within the population can be attributed to genetic differences at the A-locus. The term pq measures the amount of genetic variation at the A-locus in the population: it is largest when both alleles are common, is small when either allele is rare, and is zero when the locus is monomorphic.

If the gene action is not additive, differences between genotypic means are not entirely inheritable.

Now let us consider Figure 22-7. Superficially, the distribution looks the same as that in Figure 22-6A, but the distributions for the genotypes AA and Aa are reversed. In Figure 22-7 there is overdominant gene action: The Aa heterozygotes have, on the average, the highest phenotypic values. If, as before, only those individuals whose phenotypic value is greater than 125 units are chosen as parents, the frequency of the allele A among the parents turns out to be $\frac{2}{3}$, just as in the population as a whole. Consequently, such selection produces no change in allele frequencies. Because heterozygotes, who segregate both A and a alleles, tend to have the highest phenotypic values in this example, the population is still polymorphic but selection for high phenotypic values no longer changes the population's phenotypic distribution. In short, we have an example of an overdominant equilibrium for selection upon a trait showing continuous variation.

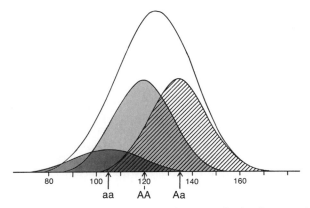

Figure 22-7   Hypothetical example of a continuous trait showing overdominant gene action at the A locus. It is identical with Figure 22-6A except that the genotypic means for AA and Aa are reversed.

In a sexually reproducing population, alleles rather than genotypes are passed from parents to offspring. If gene action is additive we can think of each allele as adding or subtracting a certain fixed increment to or from an individual's phenotype independently of the other allele in the diploid genotype. On the other hand, if the heterozygote is not precisely intermediate between the two homozygotes, diploid genotypic means cannot be expressed solely in terms of independent effects of the two alleles. In the second instance there exist interactions between the alleles that cannot be attributed directly to either allele alone and, therefore, cannot be regarded as being inheritable.

We can determine the differences between expected phenotypic values for the alleles A and a.

Regardless of the form of gene action, if the genotypic means are different, in general, there ought to be some effect that can be ascribed to the alleles involved. Our problem is to assign to each allele a phenotypic value that we can regard as being passed from parent to offspring.

Suppose we know that a gamete, which randomly unites with another gamete, has the allele A (rather than the allele a). What can we say about the expected phenotypic value of the resulting diploid individual? The answer depends upon which allele the other gamete has. With probability p, the other gamete also has the allele A, and the resulting genotype is AA with a mean phenotypic value of M_{AA}. Alternatively, with probability q the other gamete has the allele a, and the expected phenotypic value is M_{Aa}. Multiplying each of these possible phenotypic values by their respective probabilities and adding the two possibilities together gives an expected phenotypic value of $pM_{AA} + qM_{Aa}$ for the gamete that is known to have the allele A. A similar argument for a gamete known to have the allele a gives an expected phenotypic value of $pM_{Aa} + qM_{aa}$. Both arguments are set forth in tabular form in Table 22-1.

What difference does having the allele A instead of the allele a make for the diploid individual's phenotypic value? The answer to this question gives us a direct means for describing an effect that can be attributed to the alleles themselves. A quantitative answer is given by the difference between the expected phenotypic values for the two alleles. In particular, let $\alpha = (pM_{AA} + qM_{Aa}) - (pM_{Aa} + qM_{aa})$. Notice that there is a direction to the question: We asked what difference having the allele A instead of the allele a makes. We could just as well ask the question the other way around. The sign of α depends upon the direction in which the question is asked.

Our present expression for α is not particularly illuminating. By employing a suitable change of variables, however, we can express α in a fashion that we can directly relate to additive gene action and departures therefrom. We define the variables a and d in terms of the following differences between genotypic means: $M_{AA} - M_{aa} = 2a$ and $M_{Aa} - M_{aa} = a + d$. The definitions of a and d are illustrated graphically in Figure 22-8. Both a and d may be either positive or negative. $M_{aa} + a$ is the midpoint between the genotypic means of the two homozygotes. Any situation in which $d = 0$ is a case of additive gene action. If we substitute $M_{AA} = M_{aa} + 2a$ and $M_{Aa} = M_{aa} + a + d$ into our expression for α we get, by means of the algebra shown in Box 22-2, the result $\alpha = a + (q - p)d$ as an expression of the difference between the expected values of the two alleles.

If the gene action at the A-locus is additive that is, ($d = 0$), then $\alpha = a = M_{AA} - M_{Aa} = M_{Aa} - M_{aa}$, so our present definition of α agrees with the earlier definition in conjunction with Equation 22-1. In fact, although we introduced that equation solely in the context of additive gene action, with $\alpha = a + (p - q)d$, it is appropriate whether or not the gene action is additive. The term α/σ_P in Equation 22-1, therefore, is the fraction of phenotypic variation within the population that can be attributed to differences between the alleles at the A-locus.

If d is not zero, which is to say that the gene action is not additive, the value of α depends upon the allele frequencies p and q and will change as p and q change. In other words, it depends upon the composition of the population and changes as the population changes. The value of α may even be zero. In Figure 22-7, $a = 7.5$ units and $d = 22.5$ units; with allele frequencies of $p = \frac{2}{3}$ and $q = \frac{1}{3}$ we have, therefore, $\alpha = 7.5 + (-\frac{1}{3})(22.5) = 0$, which confirms on the basis of Equation 22-1 our previous finding of no change in

Table 22-1   Average Effects of Alleles.

Allele	A		a	
Randomly unites with:	A	a	A	a
Probability of occurrence:	p	q	p	q
Resulting genotype:	AA	Aa	Aa	aa
Genotypic values:	M_{AA}	M_{Aa}	M_{Aa}	M_{aa}
Average value of allele:	$pM_{AA} + qM_{Aa}$		$pM_{Aa} + qM_{aa}$	

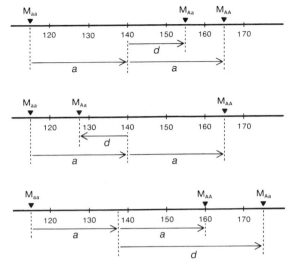

Figure 22-8   Three graphical examples of the variables *a* and *d*. The mean genotype values M_{AA}, M_{Aa}, and M_{aa} are plotted along the scale of phenotypic values. The arrows indicate the values for *a* and *d*. In B the arrow indicates a negative value for *d*. In C gene action is overdominant, and *d* is greater than *a*.

allele frequency for the population in Figure 22-7. We have here an example of the general condition that $\alpha = 0$ at an overdominant equilibrium point.

Selection changes allele frequencies simultaneously at all loci that affect the trait.

We have concentrated in this section on the contribution of a single locus to a continuously varying trait. Even so, we have seen that such traits are affected by a number of loci, each of which should be affected by selection upon the trait. A simple example of two loci, each with additive gene action, is shown in Figure 22-9. The figure is a two-locus version of Figure 22-6A, except that, for clarity, the frequency of each genotype is represented by a bar rather than a bell-shaped distribution.

Starting with the phenotypic values and genotype frequencies given by Figure 22-9 for the two-locus genotypes, we can determine genotypic means for each locus separately by averaging over the other locus in the fashion shown in Table 22-2. For example, the A-locus genotype aa occurs in the three darkly shaded cases (aabb, aaBb, aaBB) with phenotypic values of 95, 105, and 115. Because in this example the frequency of the allele B is .5 and there is no linkage disequilibrium between the A and B loci, the frequencies of these three cases are in relative proportions of 1:2:1, and the mean value for the aa genotype is $(\frac{1}{4})(95) + (\frac{1}{2})(105) + (\frac{1}{4})(115) = 105$. Similarly, at the B-locus the bb genotype is found in three cases (aabb, Aabb, AAbb) in relative proportions of 1:4:4. Its mean is 115. The single-locus means determine the

BOX 22-2

Let: $M_{AA} - M_{aa} = 2a$ and $M_{Aa} - M_{aa} = a + d$.

So that: $M_{AA} = M_{aa} + 2a$ and $M_{Aa} = M_{aa} + a + d$.

Define: α $= (pM_{AA} + qM_{Aa}) - (pM_{Aa} + qM_{aa})$.

Then: α $= (pM_{aa} + 2pa + qM_{aa} + qa + qd)$
$- (pM_{aa} + pa + pd + qM_{aa})$
$= M_{aa} + (2p + q)a + qd - M_{aa} - pa - pd$.
$= (p + q)a + (q - p)d$
$= a + (q - p)d$.

values of a, d, and α for each locus. For example, both loci in Table 22-2 have additive gene action: At the A-locus (which is the same as in Figure 22-6A) $\alpha = a = 15$ and $d = 0$; at the B-locus $\alpha = a = 10$ and $d = 0$.

For each locus Equation 22-1, with the values of p, q, and α appropriate for that locus, indicates the amount of genetic change wrought by selection upon the trait. The intensity of selection, i, and σ_P are, of course, the same for all loci. We get a picture, therefore, of allele frequencies simultaneously changing at all loci that affect the trait with changes at each locus having some effect upon the multiple-locus genotype frequencies that will occur in the next generation and, therefore, upon the distribution of phenotypic values among the offspring.

The calculations of Table 22-2 explicitly illustrate the important property that in polygenic situations the means for the single-locus genotypes depend

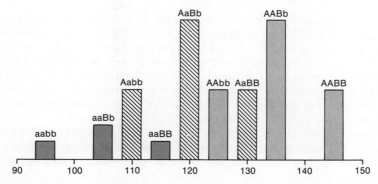

Figure 22-9   Mean phenotypic values for nine genotypes involving two loci with two alleles each. Heights of the bars are proportional to the genotype frequencies. Allele frequencies for A and a are ⅔ and ⅓. Allele frequencies for B and b are both ½. Both loci are in Hardy-Weinberg proportions, and the joint two-locus genotype frequencies are the products of the constituent single-locus frequencies. The shading and the genotypic means for the A-locus are the same as in Figure 22-6A. Both loci act additively.

Table 22-2   Single-Locus Genotypic Means for Figure 22-9.

$$
\begin{aligned}
M_{aa} &= G_{bb}M_{aabb} + G_{Bb}M_{aaBb} + G_{BB}M_{aaBB} \\
&= (\tfrac{1}{4})(95) \quad + (\tfrac{1}{2})(105) \quad + (\tfrac{1}{4})(115) \quad = 105 \\
M_{Aa} &= G_{bb}M_{Aabb} + G_{Bb}M_{AaBb} + G_{BB}M_{AaBB} \\
&= (\tfrac{1}{4})(110) \quad + (\tfrac{1}{2})(120) \quad + (\tfrac{1}{4})(130) \quad = 120 \\
M_{AA} &= G_{bb}M_{AAbb} + G_{Bb}M_{AABb} + G_{BB}M_{AABB} \\
&= (\tfrac{1}{4})(125) \quad + (\tfrac{1}{2})(135) \quad + (\tfrac{1}{4})(145) \quad = 135
\end{aligned}
$$

$$
\begin{aligned}
M_{bb} &= G_{aa}M_{aabb} + G_{Aa}M_{Aabb} + G_{AA}M_{AAbb} \\
&= (\tfrac{1}{9})(95) \quad + (\tfrac{4}{9})(110) \quad + (\tfrac{4}{9})(125) \quad = 115 \\
M_{Bb} &= G_{aa}M_{aaBb} + G_{Aa}M_{AaBb} + G_{AA}M_{AABb} \\
&= (\tfrac{1}{9})(105) \quad + (\tfrac{4}{9})(120) \quad + (\tfrac{4}{9})(135) \quad = 125 \\
M_{BB} &= G_{aa}M_{aaBB} + G_{Aa}M_{AaBB} + G_{AA}M_{AABB} \\
&= (\tfrac{1}{9})(115) \quad + (\tfrac{4}{9})(130) \quad + (\tfrac{4}{9})(145) \quad = 135
\end{aligned}
$$

Because the genotypes at the A and B loci are randomly associated within the population, the single-locus genotypic means (M_{aa}, etc.) for each locus are obtained by averaging the two-locus genotypic means (M_{aabb}, etc.) weighted by the Hardy-Weinberg proportions at the other locus (G_{aa}, G_{bb}, etc.).

upon the genotype frequencies at the other loci and can be expected to change if allele frequencies change at the other loci. The picture is a dynamic one with changes over time at the various loci being interdependent. An exact description of the changes over a number of generations in the population's distribution of phenotypic values may be quite complex. Nevertheless, in any generation the single-locus genotypic means determine a value of α for each locus, and Equation 22-1 tells us that any selection upon the trait (that is, $i \neq 0$) will result in genetic change at each locus in proportion to the value of $pq\alpha$ for the locus.

Continual selection upon a trait will reduce the amount of its inheritable genetic variation within the population.

In our discussion of heritability we said that using the genotypic variance, σ_G^2, overestimates the heritability of the trait if gene action is not additive. For a single locus, R. A. Fisher defined $2pq\alpha^2$ as the locus's inheritable contribution to variation within the population. Adding over all loci he defined the quantity $\sigma_A^2 = \Sigma\, 2pq\alpha^2$ as the *additive genetic variance* of the trait in the population. He defined heritability in a strict sense by the fraction σ_A^2/σ_P^2, which uses the additive genetic variance in place of the genotypic variance that we used in our broad definition.

Fisher arrived at his definition after building a statistical and genetic model for the inheritance of continuous variation. From our discussions in this chapter we can get a feeling for what it is that the additive genetic variance measures. Consider the quantity $2pq\alpha^2$ for any one locus. $2pq$, in addition to being the Hardy-Weinberg heterozygosity, is a simple measure of the amount of allelic variation at the locus. α^2 gives a non-negative measure of the inheritable effect of the locus upon the trait. The product, $2pq\alpha^2$, therefore, takes into account both the amount of allelic variation and the inheritable effect of the variation. It will be greater than zero unless either the locus has no effect upon this trait ($\alpha = 0$) or is monomorphic.

We have seen that natural selection produces no genetic change at a locus only if $pq\alpha = 0$. We might expect that continual selection upon a trait ought to lead eventually to some sort of equilibrium, and, if so, the inheritable variation at each locus ($2pq\alpha^2$) ought to become close to zero. Such is the case. One of Fisher's most famous results says that selection reduces the additive genetic variance of any trait upon which selection is acting. For a trait that has been subject to strong selection we can, therefore, expect most phenotypic variation within a population to be attributable to non-inheritable causes such as environmental variation and overdominance. This statement, startling as it may seem, when thought about will come to be fully consistent with the descriptions in Chapter 21 of the long-term effects of natural selection upon individual gene loci.

SUMMARY

In this chapter we have taken a detour to build a model that explicitly brings into the realm of Mendelian genetics such traits as height and wing length that show continuous distributions of phenotypes. It has been an important detour because many of the traits that are of greatest interest to evolutionary biologists exhibit continuous variation in natural populations. The inheritance of traits that show continuous phenotypic variation can be explained, in part, by Mendelian segregation at a number of gene loci, each of which has allelic differences that have small effects upon the trait. In addition, many continuously varying traits are affected by environmental conditions, so that a continuous distribution of phenotypes among the population is the net result of differences among individuals with respect to both their genotypes and the environmental conditions to which they were subject.

Beginning from the point of view of selection acting upon phenotypic variation in a particular trait, we have made, in the form of Equation 22-1, a quantitative connection between selection acting on the trait and genetic change at individual gene loci that contribute to phenotypic variation in the trait. In so doing, we have further elaborated upon the relationships between genes, traits, and relative fitnesses implied in Figure 21-1. Selection upon a continuously varying trait produces a change in the population's phenotypic distribution by simultaneously changing allele frequencies at all loci that have inheritable effects upon the trait. At each such locus, changes in allele frequencies are proportional to 1) the intensity and efficiency of selection upon the trait, 2) the amount of genetic variation at the locus, and 3) the fraction of phenotypic variation within the population that can be attributed to differences between the alleles at the locus.

It is important to realize that, whether we approach natural selection in terms of traits that have inheritable genetic variation or in terms of polymorphic loci with different relative fitnesses for the alternative genotypes, we are in fact dealing with two sides of the same coin. Any evolutionary phenomena that are explicitly found by one approach are, of necessity, implicitly involved in the other approach.

23

Mutation and Gene Flow

In addition to natural selection, three other processes affect the allele frequencies of natural populations. In this chapter we shall look at the effects of mutation and gene flow. In the next chapter we shall consider random, or chance, changes in allele frequencies. Each of these four factors is independent of the others in that for each we can make a separate quantitative analysis of its effect upon genetic variation and determine the results attributable to it. Even so, in any natural population all four factors can be expected to be in operation, to lesser or greater extents, at any time.

Although the basic effect of each factor is qualitatively unchanged by the presence of the others, the changes in allele frequencies are the result of the simultaneous influences of all four processes which may act either in concert or antagonistically. So, although we shall see that the effect of gene flow is to average the allele frequencies of two or more populations, natural selection may act to maintain different allele frequencies in the different populations and obscure the averaging effect of gene flow. Each process is only one component in a complex, dynamic biological system.

MUTATION

Chapter 14 dealt extensively with various kinds of mutations, their causes and rates of occurrence. We now turn to the population genetic and evolutionary consequences of mutations. As we already know, mutations are inheritable alterations in the genetic material, producing physical and chemical changes in the allelic state of a gene. In the sense that mutation is the process responsible for the creation of new alleles, it is the ultimate source of all genetic variation and supplies the raw material for evolutionary change.

The quantitative effect of mutation is in terms of the *rate* of mutational events per gene per generation.

To represent the quantitative effect of mutation upon a gene pool, let us suppose that the allele A mutates to the allele a at a rate of u mutations per gene per generation. For our purposes we can think of the allele a in either of two ways. We can regard it as a particular alternative form of the A-gene in the sense in which we normally think of A and a alleles, or as all non-A alleles, so that the "a allele" is a possibly heterogenous class of distinct alleles, each of which differs from the allele A. For example, we might think of the allele A as the normal form of the gene and the allele a as all forms that code for non-functional gene products. Or, we might think of A denoting the typical chromosomal arrangement while a denotes all rearrangements. Because chromosomal aberrations tend to be unique, it does not make sense to speak of the rate at which a *particular* rearrangement is produced by mutation, but we can speak of rearrangements of a given chromosome in general occurring at rate u.

The mutation rate, u, is the probability that any one A allele will mutate to an a allele during the course of a single generation. The probability of its not mutating is $1 - u$. If the frequency of the allele A in any one generation is p, its frequency in the next generation, when the probability of mutation is taken into account, will be $p' = p(1 - u) = p - up$. The change in the frequency of allele A caused by A alleles mutating to a alleles is:

$$\Delta p = p' - p = -up.$$

If a alleles mutate to A alleles at a rate v, then the net change in the frequency of the allele A is:

$$\Delta p = -up + vq.$$

Mutation rates for a number genes in a variety of organisms were given in Table 14-1. The rates vary but are all low. The highest are on the order of 10^{-5}. Changes in allele frequencies from mutation are, therefore, very small. Mutation alone has little effect upon allele frequencies within a population. On the other hand, in large populations the absolute numbers of new mutants are not necessarily negligible. For example, infantile amaurotic idiocy, Tay-Sachs disease, is determined by a recessive allele at a single autosomal locus. The IAI allele has an estimated mutation rate of 11×10^{-6}. In 1968, there were roughly 3.5 million births in the United States. Because each birth involved two copies of the IAI gene, we can figure that, on the average, $2 \times (3.5 \times 10^{6})(11 \times 10^{-6}) = 77$ *new* IAI-producing alleles were introduced into the U.S. population in that one year. Nevertheless, the significance of mutation is not in the magnitude of the changes it causes but rather in its persistent, continuing introduction of new alleles at a very low but steady rate into a species' gene pool.

Mutation is non-directional in an evolutionary sense.

Mutation is a non-directional process: mutants arise regardless of whether they improve the adaptiveness of the individuals who carry them.

Indeed, the vast majority of mutants are in all likelihood detrimental. Although mutation is the primary process of evolution in that it provides the raw material for evolutionary change and precedes all other processes, evolutionary change involves more than the mere introduction of a new variant into one or two individuals. A previously rare or nonexistent allele must become *the* common form within an entire population. The primary event of evolution is an allele substitution as described in Chapter 21. Mutation is no more responsible for evolutionary change than the lumberjack cutting trees is responsible for the house built from those trees.

The evolutionary fate of a new mutant depends upon the relative fitnesses of the genotypes that result from the mutant in combination with existing alleles. Mutants that result in overdominance or have superior fitness as homozygotes are presumably rare but, nevertheless, are prime candidates for allele substitutions. Deterministic selection theory of the sort we have considered so far predicts that such mutants will be steadily incorporated into the gene pool. To get a true view of the fate of such mutants, however, we must wait until the next chapter when we take into account random fluctuations in allele frequencies.

There is a paucity of information concerning the distribution of fitness values associated with randomly occurring mutants. The general presumption—based upon extensive knowledge of genetic birth defects, viability studies of *Drosophila,* and the theory that a random change in something that works will most likely produce something that does not work—is that the vast majority of mutants are deleterious, having reduced fitnesses as homozygotes and fitnesses as heterozygotes that are intermediate between the fitnesses of homozygotes. Natural selection will eliminate such mutants from the gene pool; if, however, deleterious mutants recur at a constant rate we can expect a reservoir of deleterious alleles within the gene pool at some low frequency. The purpose of this section is to develop quantitative predictions of the frequencies at which deleterious alleles will be found when the production of new mutants is exactly balanced by their elimination by natural selection.

The frequency of a deleterious allele depends upon its mutation rate and its relative fitnesses, especially its fitness as a heterozygote.

Let us suppose that the allele A mutates to the allele a at rate u. Let the relative fitnesses of the genotypes AA, Aa, and aa be $1:1 - hs:1 - s$. We shall restrict ourselves to the case where the allele a is deleterious in the sense that s and h are both between zero and unity. Because we expect the allele a to be rare we shall ignore the possibility of its mutating to the allele A. This analytic model is presumably applicable to the majority of alleles that produce human birth defects and genetic diseases and to the recessive lethals of *Drosophila* we discussed in Chapter 19.

Over one generation in a randomly mating population, natural selection changes the frequency of the allele A from p to $p(1 - qhs)/\overline{W}$, where $\overline{W} = 1 - 2pqhs - q^2s$ (see Box 21-1). On the average, a fraction u of these A alleles will mutate to a alleles. Consequently, the joint effect of selection and mutation is to change the frequency of the allele A from p to $p' = p[(1 - qhs)/\overline{W}]$ $(1 - u)$.

Our goal is to find an equilibrium value for q. At equilibrium allele frequencies do not change, and $p' = p$. Setting $p' = p$ and multiplying both sides by $\overline{W}$ give the following condition for equilibrium allele frequencies:

$$\hat{p}(1 - 2\hat{p}\hat{q}hs - \hat{q}^2s) = \hat{p}(1 - \hat{q}hs)(1 - u).$$

We use the "hats" to show that we are now dealing with equilibrium values. Substituting $\hat{p} = 1 - \hat{q}$ and doing some algebraic rearranging yield an implicit equation for the equilibrium frequency of the deleterious allele a.

$$u = \hat{q}hs + \hat{q}^2s(1 - 2h) + \hat{q}u \cdot hs. \tag{23-1}$$

We saw in Chapter 21 that the selection coefficient h indicates the degree to which the deleterious effects of the allele a are expressed in Aa heterozygotes. The value of h has a strong effect upon $\hat{q}$. In particular, if the deleterious effects of the allele a are fully recessive ($h = 0$), Equation 23-1 simplifies to $u = \hat{q}^2s$, implying that the equilibrium allele frequency is

$$\hat{q} = \sqrt{\frac{u}{s}}. \tag{23-2}$$

On the other hand, if h is substantially larger than both $\hat{q}$ and u, the term $\hat{q}hs$ dominates Equation 23-1 and the very small terms involving $\hat{q}^2$ and $\hat{q}u$ can be ignored, giving an approximate equation $u = \hat{q}hs$. The equilibrium allele frequency is given approximately by

$$\hat{q} \doteq \frac{u}{hs}. \tag{23-3}$$

This mathematical approximation amounts to ignoring selection against aa homozygotes on the grounds that they are very rare compared to Aa heterozygotes. In other words, if selection acts, even weakly, on individuals who are heterozygous for deleterious alleles, the equilibrium allele frequencies in a randomly mating population will be determined primarily by the selection pressure on such heterozygous individuals.

Equation 23-2 gives substantially higher values for the equilibrium allele frequency than does Equation 23-3. Suppose that aa homozygotes are lethal (i.e., $s = 1$) and the mutation rate is 10^{-6}. For $h = 0$, Equation 23-2 gives $\hat{q} = .001$, while for $h = .1$, Equation 23-3 gives $\hat{q} = .00001$, a hundredfold difference. The biology of the situation is relatively easy to understand. As long as h is not zero, every a allele in the population is affected by natural selection, and natural selection is comparatively efficient at holding the frequency of the allele a near the mutation rate at which new mutant alleles are introduced into the population each generation. On the other hand, when h is near zero, the majority of the a alleles in a randomly mating population occur in heterozygotes and are protected from the effects of selection. Consequently, a alleles accumulate in the population until the frequency of aa homozygotes (q^2) is sufficiently high that selection acting against the homozygotes removes as many a alleles each generation as mutation introduces into the population. The result is an allele frequency that is low on an absolute scale but substantially higher than the mutation rate.

Random mating is a key ingredient in both the qualitative picture just described and the quantitative result of Equation 23-2. In a highly inbred population the *equilibrium* allele frequency for recessive alleles is substantially lower than the value given by Equation 23-2 because inbreeding increases the fraction of a alleles that occur as homozygotes and are exposed to natural selection.

However, an inbred population recently derived from a large, randomly mating group is likely not to be at equilibrium. Its allele frequencies are likely to be similar to the equilibrium frequencies of the randomly mating group. In such a case, inbreeding results in a comparatively high incidence of individuals with detrimental genotypes. Only after inbreeding has been in effect for a substantial length of time can we expect the frequencies of deleterious alleles to be reduced to equilibrium frequencies consistent with the population's degree of inbreeding.

Recurrent mutation maintains a reservoir of nonadaptive genetic variation at low frequency within randomly mating populations.

The theoretical results just stated are fully consistent with the empirical evidence, discussed in Chapter 19, concerning viability differences in *Drosophila.* Both argue for the existence at many gene loci of deleterious alleles, most commonly recessive ones, at low frequencies in randomly mating populations. The widespread existence of nonadaptive variations—albeit at low frequencies—implies that every generation a few individuals carrying detrimental genotypes are produced and subjected to the afflictions associated with such genotypes. The recurrent, unavoidable introduction of deleterious mutants into a population represents a chronic decay of genetic organization, checked by the continual action of natural selection, and results in what has been termed a continuing mutational load on the population.

On the positive side, the existence of a reservoir of genetic variations at low frequency offers possibilities for future evolutionary changes. Although selection currently acts against them, some of the alleles that are being maintained at mutation-selection equilibria might become adaptive under some future environmental conditions, thereby providing the material for allele substitutions. For example, the Lederbergs' experiment described in Chapter 14 showed genes conferring resistance to phage infection were present at low frequency in laboratory populations of *E. coli* prior to exposure to phage infection. A population's immediate response to sudden changes in environmental conditions appears to be generally based upon existing variation within the gene pool. An allele already present at a frequency of, say, .001 has a definite head start—should it prove later to be favored by natural selection—over a newly arisen mutant.

GENE FLOW

The ranges of most species greatly exceed the range of a single population. Except in rare instances, biological species are comprised of many local populations. Gene flow is the process by which genetic variation is exchanged between populations, thereby engendering genetic connectedness among sundry populations of a species. The local population is the arena wherein evolutionary changes wrought by natural selection, mutation, and genetic drift must take place. But local populations are not independent of each other. Gene flow is capable of spreading genetic innovations throughout a species' entire gene pool and tends to equalize, or average, allele frequencies among populations.

Gene flow is a consequence of dispersal by individuals and the ability of conspecific individuals to engage in sexual reproduction.

The biological basis of gene flow is dispersal. Dispersal is a complex trait that varies greatly among organisms. It is obviously dependent upon an organism's vagility: fruit flies are capable of traveling far greater distances than are snails. In addition, dispersal may be strongly influenced by behavioral and ecological factors. An illustrative example has been found in studies of the butterfly *Euphydryas editha.* In three populations on a ridge near Stanford University the movements of adult butterflies amounted to a radius of less than 100–200 meters, so that populations only 500 meters apart exchanged no individuals. On the other hand, individuals of the same species living at another site 50 miles away routinely flew distances of nearly 1,000 meters. The apparent explanation for these differences is found in the spatial distributions of the two different plant species used by *Euphydryas editha* for egg-laying sites and for feeding by adults. At the first site both plant species occur together in clusters, with a different butterfly population inhabiting each cluster. At the second site, the two kinds of plants are scattered about at separate locations, so that the adult butterflies must move over a much wider range with a single population inhabiting an area of the same size as that occupied by three populations at the first site.

Obtaining quantitative data on individual dispersal is a difficult task. The nemesis of field work on dispersal is the disappearance of individuals from a study site: it is impossible to distinguish individuals that disappeared in death from those that successfully migrated to unknown destinations. The situation is particularly difficult for the many species that have relatively stable adult populations but highly vagile juveniles. For example, pocket gophers of the genus *Thomomys* are adapted to a fossorial habit. Movement of adults is almost totally confined to their burrows. Juveniles, however, are known to move above ground, but almost nothing is known about how far they move. Many species of marine invertebrates have sessile adult stages and free-floating larva. How much intermixing of larva from different adult populations occurs is unknown. Most plant species disperse by pollen, seeds, and spores.

In general, available data indicate *potential* for dispersal more than quantitative estimates of the amount of dispersal that actually occurs. For example, large quantities of pine pollen have been collected aboard a research ship 40 miles at sea. Fungal spores have been collected in the upper atmosphere. Isolated gopher mounds have been found in obviously temporary habitats, many miles from the nearest population.

Patterns of dispersal are characterized by a preponderance of individuals that move comparatively short distances and a small percentage that go enormous distances.

Data concerning dispersal and gene flow are probably more accurate and abundant for human populations than for any other species. The distances between the birthplaces of husbands and wives provide a convenient measure of gene flow. Data for three human groups are shown in Figure 23-1. Although these groups differ, especially with respect to the average distance between birthplaces, all three histograms illustrate a major generalization that holds for almost all studies on dispersal. Namely, dispersal distances differ in two ways from the pattern that would be expected if movement of individuals were as purely random as diffusion of a gas. 1) The proportion of "home-bodies" who do not disperse at all is substantially greater than would be expected on the basis of random dispersal. Notice the large fraction of marriages between individuals born in the same village— that is, within five kilometers of each other. 2) A small, but appreciable, fraction of individuals are "wanderers" who disperse much further than would be expected on the basis of random dispersal. In each of the histograms there is a distinct tail, well removed from the rest of the population.

This generalization concerning "wanderers" and "home-bodies" applies to most species, including those for which dispersal is essentially a passive affair. Studies of fungal species estimated that 99 per cent of all spores land within 100 meters of their point of release, with the majority travelling less than ten meters. The 1 per cent that go farther may go enormous distances. Spores of *Puccinia graminis,* known to have been released from infected wheat fields in Texas, were recovered in substantial numbers across the Great Plains as far as Wisconsin.

Gene flow requires more than mere movement of individuals.

Gene flow is based, in part, upon dispersal; it is not, however, synonymous with dispersal. Gene flow is also based upon the ability of conspecific individuals to mate with each other. Mere movement of individuals from one place to another does not constitute gene flow. Immigrant individuals must contribute their genes to the gene pool of future generations of the recipient population if there are to be any genetic consequences. Salmon provide an extreme example of extensive movement with little gene flow; they hatch in shallow fresh-water streams, spend several years roaming vast distances in

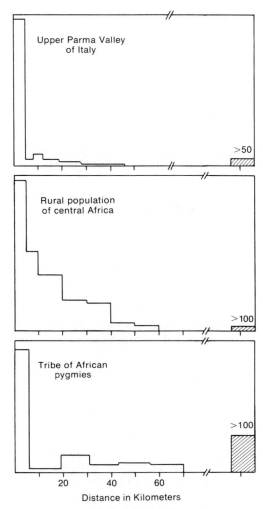

Figure 23-1   Distributions of the distances between birthplaces of husbands and wives in three human populations. A. Upper Parma Valley of Italy. B. Rural population of central Africa. C. Tribe of African Pygmies. Population density and the average village size are highest for the Parma Valley and lowest for the African Pygmies. Data from Cavalli-Sforza and Bodmer, *The Genetics of Human Populations,* Freeman, 1971.

the ocean, and then return to the stream in which they were hatched in order to mate and reproduce. Gene flow in salmon is a function of the extent to which fish return for breeding to streams other than the ones in which they were hatched. The distances travelled in the ocean and the other salmon encountered there are irrelevant. Gene flow involves the mixing of not just individuals but of gene pools.

ANALYSIS OF GENE FLOW

Theoretical analysis of gene flow is reasonably straightforward. Quantitative results for a collection of populations as a whole depend upon the pattern of migration of individuals among the constituent populations. Changes will be quantitatively different if migration occurs exclusively between adjacent populations from the changes if all populations exchange migrant individuals each generation. Regardless of the pattern of migration, the effect of gene flow upon allele frequencies *within a given population* can easily be described.

Within each population allele frequencies are averages of the frequencies among residents and immigrants.

Let us consider a particular population that, each generation, exchanges migrants with other populations. For our A-locus let p and q denote the frequencies of the alleles A and a within the population before we take migration into account. Emigrant individuals will have no effect upon the allele frequency of A unless there is a preferential tendency of individuals with A alleles to either emigrate or not. In general, we assume emigration is random with respect to the alleles A and a unless there is evidence to the contrary. On the other hand, immigrants will affect the frequency of A within the population if the allele frequency among the immigrants differs from the frequency among the residents. Let p_I denote the frequency of the allele A among immigrants. The frequency of the allele a among immigrants is $q_I = 1 - p_I$. The value of p_I depends upon where the immigrants have come from.

We let m denote the fraction of breeding individuals who are new immigrants to our particular population after migration. That is, if we should take a census of the breeding individuals we would find a fraction $1 - m$ of the individuals were born within the population, and a fraction m were born elsewhere. The value of m, therefore, measures the amount of immigration. Because among residents the frequency of the allele A is p and among immigrants it is p_I, the overall frequency of the allele A following immigration, p', is obtained by averaging these two values according to the respective fractions of residents and immigrants in the population. That is,

$$p' = (1 - m)p + mp_I = p + m(p_I - p).$$

The change in the frequency of the allele A that we can attribute directly to immigration is

$$\Delta p = p' - p = m(p_I - p).$$

This equation indicates that the change in allele frequency is directly proportional to the difference between the frequencies among immigrants (p_I) and among residents (p) and to the fraction of immigrant individuals entering the population (m). The change is zero only if there is no gene flow ($m = 0$) or if $p = p_I$. Inasmuch as p' is a weighted average of p and p_I, it is intermediate between p and p_I, and the frequency of the allele A within the population is always closer to p_I after immigration than it was before.

Among an ensemble of populations gene flow is a dynamic affair with each population being affected by the others.

The value of p_I is extrinsic to the recipient population and can be treated as a constant for purposes of the change in the allele frequency within the population during a single generation. For a particular population within an ensemble of populations, the value of p_I depends upon where immigrants came from and what the allele frequencies were within the source populations. Those source populations may themselves be affected by gene flow—from our original population and elsewhere—so that their own allele frequencies change and thereby cause the value of p_I (for immigrants into our original population) to change from generation to generation. The image we should have, therefore, is of gene flow shifting allele frequencies within each population toward the frequencies of those other populations from whom it receives immigrants. All else being equal, gene flow will, in the long run, average out allele frequencies among the populations in the group. The rate at which it does so obviously depends upon the amount and pattern of migration among the populations.

GENE FLOW AND SELECTION

In biology all else is rarely equal. Several factors may prevent gene flow from equalizing allele frequencies in different populations. We turn now to a joint consideration of natural selection and gene flow. We have seen that relative fitnesses depend upon the environment in which an organism lives. Most species are distributed over diverse environmental conditions. Conceivably, varying environmental conditions might result in different selective pressures upon individuals in different environments.

The critical factor is whether individuals living in one environment face selective pressures favoring one allele at the particular locus in question while individuals in a different environment face pressure favoring another allele. The peppered form of the peppered moth is favored by selection in rural lichen-covered forests whereas the melanic form is favored in soot-blackened forests surrounding industrial centers. The mere existence, however, of environmental heterogeneity does not necessarily imply contradictory selection pressures upon a particular locus. Moreover, environmental differences that affect relative fitnesses at one locus may be irrelevant to genetic differences at another locus.

When heterogeneous environmental conditions do affect the relative fitnesses at a given locus, dispersal of individuals from one area to another results in gene flow that tends to average allele frequencies throughout the species' range. The critical factor is the degree to which the disruptive effect of contradictory selection pressures is offset by the averaging effect of gene flow, which ultimately depends upon the magnitude of the selection coefficients involved and the amount of gene flow between populations inhabiting different environments.

Gene flow in a heterogeneous environment depends upon the patchiness of the environment relative to the average dispersal of individuals.

Let us think of situations in which a species' range is a mosaic of, say, two different environmental conditions (lichen-covered versus soot-blackened trees) that result in contradictory selection pressures upon a given locus. The primary factor determining the amount of gene flow is the geographical size of the various patches of environmental conditions relative to average dispersal distances. Figure 23-2 schematically illustrates three sit-

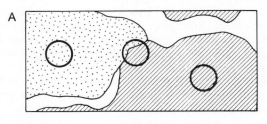

A   Large patches of uniform environmental conditions. High correlation between parents and offspring with respect to environmental conditions.

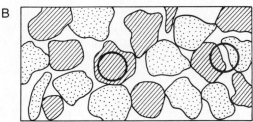

B   Patch sizes similar to average dispersal distances. Moderate correlation between conditions experienced by parents and their offspring.

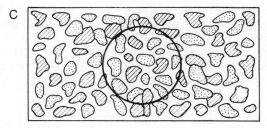

C   Fine grained environmental conditions relative to average dispersal distances. Conditions experienced by offspring are random with respect to those experienced by their parents.

Figure 23-2   Environmental patch size relative to average dispersal distances. Stippling and hatching denote the geographical distributions of two different environmental types. Clear areas denote uninhabitable regions across which individuals can disperse. The species is presumed to be distributed throughout all inhabitable areas of the range. Circles indicate the average radius of individual dispersal over a generation's time.

uations in a continuum of possibilities for a mosaic consisting of two alternative kinds of patches. The possibilities for gene flow are low in A, moderate in B, and high in C. Naturally occurring examples of these three situations are easy to imagine.

Industrial melanism in the peppered moth serves as an example of a situation involving extremely large patch size. Soot-blackened trees extend over large geographical areas surrounding and downwind of industrial areas of England (Figure 19-4). Very few moths emerging in soot-blackened forests will have offspring that live among lichen-covered trees and vice versa.

As a second example, consider plant species inhabiting regions in which serpentine soil is common. Serpentine soil is famous among plant ecologists because of the ecological differences that can be observed among plants living on it. Patches of serpentine soil are generally of moderate size so that an entire population can be almost entirely on serpentine soil or not. Depending upon the species' mechanisms for pollen and seed dispersal, there may be substantial gene flow from areas of one soil type to the other.

In many insect species, females deposit eggs upon particular plants. Development of the eggs and their subsequent larva is confined to the plants upon which the eggs are deposited. Emerging adults, however, randomly mate and lay eggs throughout the range of the entire population. Suppose that a population has two species of host plants scattered throughout its range and that the relative fitnesses at a particular locus during the larval stage are different for the two species of host plants. Such a situation is an example of a mosaic environment with small patches among which a thorough mixing of individuals occurs each generation.

Implicit in these examples is the principle that gene flow is directly related to the likelihood that offspring live in environments that are different from the ones in which their parents lived. Maximum gene flow occurs when the environmental conditions experienced by offspring are random with respect to the conditions experienced by their parents. A high correlation between the conditions experienced by parents and offspring implies little gene flow.

In addition to geographical size, behavioral characteristics can affect the amount of gene flow between different environmental patches. If dispersing individuals seek out environmental patches similar to the ones in which they were born, the correlation between parent and offspring will be increased and the effectiveness of dispersal reduced. The amount of gene flow between different kinds of patches in a mosaic environment is the joint result of 1) the sheer geographical size of individual patches, 2) the dispersal distances of which the species is capable, and 3) any tendencies on the part of dispersing individuals to select one kind of environment over another.

The effects on allele frequencies are qualitatively different for the three situations illustrated in Figure 23-2.

Our main interest in studying the joint behavior of gene flow and natural selection is to determine the effect upon allele frequencies. How much genetic polymorphism within local populations can occur as a result of gene flow between areas in which different alleles are favored by selection? It is vital to know whether the effectiveness of gene flow is substantially less than, approximately equal to, or substantially greater than the selection pressures upon the locus in question. Mathematical analyses are complicated and not necessary for our purposes here. Using Figure 23-2 as a model we can describe the qualitative consequences of this trichotomy.

Let us suppose that in the stippled patches the relative fitnesses for AA, Aa, and aa are 1.14:1:.9, and in hatched patches they are .9:1:1.05. So, the allele A is favored in stippled areas and selected against in hatched areas.

Very Large Patches. If, as in Figure 23-2A, the patches are much larger than average dispersal distances, allele frequencies in each local population will be governed primarily by the relative fitnesses of the patch in which they are located. Populations in stippled patches, for example, become essentially fixed for the allele A and those in hatched patches for the allele a. Unless the relative fitnesses happen to be overdominant, there should be little polymorphism within the patches except near the boundaries where mixing of individuals from both patches creates local polymorphism. The spread of alleles from one region into the other is most pronounced at the boundary and progressively decreases away from the boundary. Consequently, we can expect the joint action of gene flow and natural selection to produce a cline in allele frequencies across the boundary.

The width of such a cline depends upon two factors: the amount of gene flow across the boundary and the magnitude of the selection coefficients within the two patches. Mathematical analyses suggest that the width of transition zones is given approximately by $\ell/\sqrt{s}$, where ℓ is the average dispersal distance and s is the difference in relative fitnesses between the locally favored genotype and the genotype of immigrant individuals from the adjoining patch. Strong gene flow produces wider transition zones than weak gene flow. Strong selection results in narrow zones.

Intermediate-Sized Patches. As the size of the patches becomes smaller, as in Figure 23-2B, the effectiveness of gene flow increases. Even though allele frequencies may still be predominantly influenced by the relative fitnesses of their respective patches, gene flow will result in substantial mixing of alleles between patches. A substantial amount of polymorphism may then exist in local populations as a result of the opposing effects of selection and gene flow.

Very Small Patches. When the size of individual patches is substantially less than average dispersal distances, as in Figure 23-2C, gene flow, for all practical purposes, obscures the effects of the relative fitnesses in the individual patches. Each population encompasses a mixture of both types of patches, and allele frequencies are influenced primarily by the relative fitnesses of each genotype *averaged over the different patches.* These averages depend upon the relative fitnesses in the individual patches and also upon the relative proportions of the two kinds of patches in the area as a whole. For example, if the stippled and hatched areas in our example occur in a 50:50 mixture, the average fitnesses for AA, Aa, and aa are 1.02:1:.975. In a 40:60 mixture of stippled and hatched patches, however, the average fitnesses are .996:1:.99. In a 30:70 mixture the average fitnesses are

.972:1:1.005. Thus, in a 50:50 mixture we can expect the allele a to be lost from the population, while in a 30:70 mixture the allele A will be lost. But in a 40:60 mixture, the relative fitnesses are, on average, overdominant, and we can expect both alleles to be maintained.

From our discussion of Figure 23-2 we see that the likelihood of gene flow causing an allele to become established in the face of hostile selection pressures is remote except over short geographical distances of the sort in Figures 23-2B and 23-2C. The primary significance of gene flow, therefore, lies in its ability to spread new genetic variants throughout the range of a species to all those regions where the variant is in fact favorable. In a metaphorical sense, gene flow achieves significance not as an adversary but as the handmaiden of natural selection.

SUMMARY

Mutation is the primary cause (in the sense of preceding all others) of genetic variation and evolution. The quantitative effect of mutation upon a gene pool is given in terms of its rate per gene per generation. Mutation is a non-directional process. The majority of mutations are detrimental and, because of natural selection, are of no consequence to evolution. The recurrence of detrimental mutations does, however, lead to a reservoir of rare, selectively disadvantageous, alleles within each gene pool. In randomly mating populations, mutations whose deleterious effects are fully recessive can attain frequencies that are small on an absolute scale but are, nevertheless, substantially greater than their mutation rates.

Dispersal and reproduction within a new population are the biological basis of gene flow. Gene flow allows for the spread of genetic variation throughout a species' range and tends to *average* allele frequencies among the various populations of the species. If environmental heterogeneity exists such that natural selection favors different alleles in various areas of a species' range, the extent to which the effects of natural selection are influenced by gene flow depends upon the patchiness of the environment relative to the average dispersal between patches. Along the spectrum of possible situations, three qualitatively different results can be distinguished.

24

The Weighted Odds of
Evolutionary Change

Life is a chancy business. Although we have treated natural selection, gene flow, and mutation as systematic processes that lead to specific genetic changes, there are few sure bets in evolution. Genetic changes may often take any one of a variety of forms, so that, in some respects, the most accurate descriptions of evolutionary change are as probabilities for the various alternatives rather than as predictions of particular outcomes.

In many biological situations, specific results are quite predictable. There are other cases, however, in which chance events may significantly affect the evolution of a population. In this chapter we will look at the role of random events in evolutionary processes, identify conditions under which they may be significant, and develop a view of natural selection and genetic change in which possible results are described by probabilities rather than by specific predictions.

GENETIC DRIFT

All populations, regardless of size, consist of a finite number of individuals. Consequently, the alleles in any generation are a finite sample of the alleles from the previous generation. Finite samples generally deviate randomly from expectations. Random fluctuations in allele frequencies as a result of the finite size of natural populations are known as *genetic drift*.

Genetic drift differs in an important way from the random deviations observed in flipping a coin 100 times, when the expected ratio of "heads" to "tails" is always 50:50 (assuming the coin is fair). In genetic drift, the expected allele frequencies among offspring are based upon the allele frequencies in the parental gene pool. As a result, random deviations may *accumulate* over successive generations, leading to far greater deviations than the kind we dealt with when discussing χ^2 tests.

Chance fluctuations depend upon population size, being least significant in large populations and most important in small populations.

The magnitude of chance deviations within any sample depends upon the size of the sample. A population of N diploid individuals contains $2N$ copies of each gene, which we can regard as a sample of the previous generation's gene pool. The effects of genetic drift are illustrated graphically in Figure 24-1, which shows computer simulations of two populations of very different size. In both, the allele A is a dominant allele strongly favored by natural selection ($s = .2$). In the large population ($2N = 10,000$), the effects of genetic drift are negligible: The frequency of the allele A closely follows the predictions of selection theory. Random fluctuations show as small squiggles in a smooth curve. In the small population ($2N = 100$), however, the random fluctuations are large and dramatic. At times they overwhelm and obscure the systematic effects of natural selection. After 100 generations the allele A is still a long way from fixation.

Even though computer simulation produced the two graphs, their patterns are characteristic of the effects expected from genetic drift. In large populations random fluctuations in allele frequencies from one generation to the next tend to be small or even negligible, whereas in small populations substantial random fluctuations will probably occur in spite of the systematic effects of natural selection.

The strength of genetic drift in a natural population is characterized by the effective population size.

Although genetic drift is directly related to the number of individuals in a population, a number of biological factors affect the magnitude of random fluctuations. For example, genetic drift in a dioecious population may differ substantially from that in a monoecious population of exactly the same size. Cumulative effects of genetic drift over a number of generations may be strongly affected by fluctuations in the population size from one generation to another.

To compensate for such variation, we can define an "effective population size," denoted by N_e, such that two populations of the same effective size are directly comparable in respect to genetic drift. As our standard of reference we take a population of randomly mating monecious individuals that is of constant size from one generation to the next. For a given population that differs from our reference population in some way, we need an expression that gives the effective population size N_e in terms of the actual number of individuals. Without going into derivations, we can illustrate the problem with two cases.

Separate sexes. In a dioecious population the effective population size depends upon the sex ratio as well as upon the total number of individuals. Suppose that a population of N individuals consists of N_f females and N_m males ($N_f + N_m = N$). Let $r = N_f/N$ denote the fraction of female individuals. The effective population size is given by

$$N_e = \frac{4N_fN_m}{N_f + N_m} = 4Nr(1 - r).$$

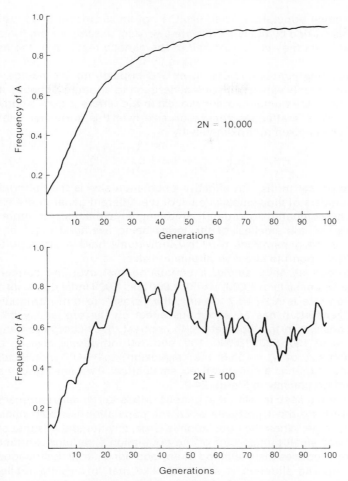

Figure 24-1   Computer simulations showing the effects of genetic drift and natural selection on the frequency of the allele A over 100 generations. In both cases the initial frequency of A is .1 and the relative fitnesses are $W_{AA} = W_{Aa} = 1.0$ and $W_{aa} = .8$. A) Population size is $N = 5,000$. B) Population size is $N = 50$. Simulations courtesy of Dr. F. Awbrey, San Diego State University.

Because $r(1 - r)$ is largest when $r = .5$, the effective population size is greatest when there is the same number of males and females. In such a case, $N_f = N_m$ and the effective population size is $N_e = N$, indicating that dioeciousness, as such, has no effect.

On the other hand, the effective population size is substantially reduced if one of the two sexes is rare (or if only a few individuals of one sex act as parents). As an extreme example, consider a dairy herd of 98 cows and 2 bulls. The effective population size is $N_e = 7.84$, which is quite small for a

100-individual population. This result is not so surprising if we realize that the alleles carried by the two bulls are a very small sample from the previous generation, yet they constitute *half* of the genetic material of the next generation.

Fluctuating population size. In natural populations the number of individuals commonly varies from one generation to another. Suppose that over *t* generations the numbers of individuals in the different generations are N_1, N_2, . . . , N_t. The effective population size over this particular interval of *t* generations is given approximately by,

$$\frac{1}{N_e} = \frac{1}{t}\left[\frac{1}{N_1} + \frac{1}{N_2} + \ldots + \frac{1}{N_t}\right] = \frac{1}{t}\sum_{i=1}^{t}\frac{1}{N_i}.$$

In mathematical terms, the effective population size is the harmonic mean of the numbers of individuals in each of the different generations. Harmonic means, as compared to ordinary arithmetical averages, are more heavily influenced by the smallest of the terms being averaged together. Consequently, those generations with few individuals have a disproportionately large effect upon the effective population size.

Consider, as an example, a population that over the course of five generations consists of 1,000, 10, 50, 1,000 and 1,000 individuals. Its effective population size is $N_e = 40.7$. Clearly, the "crash" to ten individuals in the second generation has an enormous effect on the probability of random fluctuations within this population. In contrast, five successive generations consisting of 100, 500, 10,000, 100, and 100 individuals give an effective population size of $N_e = 155.8$: the "explosion" to 10,000 individuals in the third generation has a comparatively small effect. Two analogous cases are illustrated graphically in Figure 24-2.

We should keep in mind that genetic drift is a *cumulative* process. Major fluctuations are most probable when the population is small; minor ones, when it is large. When the population is large, any fluctuations that occurred when it was small remain part of the population's heritage; an increase in size merely reduces the likelihood of any additional large fluctuations for the time being. The situation is somewhat like that in an automobile that is alternately gunned and coasted. When coasting, not much fuel is consumed, but that does not compensate for the excessive consumption during hard acceleration.

Severe reductions in population size are known as *bottlenecks* and may have major effects on a population's gene pool.

We have just seen that a few generations of small population size can have a strong effect—in terms of genetic drift—upon a population that generally is made up of a large number of individuals. In some species bottlenecks in population size occur on a regular basis. Populations of *Drosophila melanogaster* in New England, which are generally quite large during the summer and fall, are yearly reduced to those few individuals that manage to survive the rigors of winter. Field mice of the genus *Microtus* are famous for

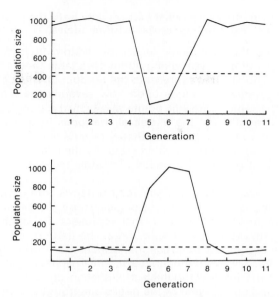

Figure 24-2   Effective population size when the actual population size fluctuates from one generation to another. Dotted lines show the effective sizes that correspond to the actual population sizes shown with solid lines. Temporary reductions in population size have a far greater effect on the effective population size than do brief bursts of high population density.

population "crashes" every few years, at which time their densities are reduced from several hundred individuals per acre to only a few. Other species may experience bottlenecks on a less regular basis, such as when occasional catastrophes, such as forest fires, strike.

Bottlenecks provide an opportunity for unusually large changes as a result of genetic drift. Allele frequencies after a bottleneck may be different. In particular, if sufficiently severe, bottlenecks may eliminate a part of a population's genetic variation. Common alleles will generally persist in the gene pool, but many of the rarer alleles present before a bottleneck may not be represented among the survivors. A population's reservoir of rare alleles may be greatly depleted by a severe bottleneck.

The evolutionary consequences of a bottleneck depend heavily upon what happens in the generations following the crash. If the population remains very small through subsequent generations, the likelihood of alleles being lost from the gene pool becomes even greater. On the other hand, if the population subsequently receives substantial immigration from other populations, alleles lost during the bottleneck may be restored to the gene pool. Because the biology of natural populations is a dynamic, multifaceted affair, the whole range of circumstances peculiar to the history of each population strongly affects its evolution.

Random changes in allele frequencies accumulate until the population eventually becomes fixed for one of the existing alleles.

To appreciate the cumulative nature of random fluctuations in allele frequencies, think of a marble rolling about on a table that is being shaken back and forth. The shaking represents the random effects of genetic drift. The marble's movement depends upon how severely the table is shaken. The edges of the table represent points at which the population becomes fixed for one or another of the alleles at the locus in question when the marble rolls off the table. Probability theory tells us that sooner or later the marble is bound to fall off the shaking table. By the same token, probability theory tells us that genetic drift leads ultimately to each allele becoming either fixed or lost.

In principle, this statement of loss or fixation is true even if the population is large and the locus is under strong selection. We can imagine ridges and troughs on the table's surface that represent the effects of natural selection and gene flow. The troughs direct the natural roll of the marble toward an edge in the case of directional selection or toward the interior of the table in the case of overdominant selection. The steepness of the table's topography and the force with which it is shaken determine the extent to which the marble is constrained by the ridges and troughs. Even steep ridges around the entire edge of the table may not keep the marble from rolling off the table. By the same token, even with overdominant selection, ultimate loss and fixation are to be expected.

PROBABILITY OF ULTIMATE FIXATION

The elementary event of evolutionary change is an allele substitution. We saw in Chapter 21 that allele substitutions are one of the major results of natural selection. Our analogy of a marble rolling off the edge of a shaking table suggests the possibility of allele substitutions occurring as a result of genetic drift. In our treatment of natural selection, allele substitutions proceed in relentless, deterministic fashion. In our genetic drift analogy, they are the fortuitous results of random fluctuations in allele frequencies. In reality, both of these statements apply, to a lesser or greater extent to allele substitutions. Our goal is to combine the chance events of genetic drift with the systematic pressures of natural selection so as to understand the probabilities involved in allele substitutions.

Genetic drift theory tells us that, with or without natural selection, a polymorphic locus will sooner or later become fixed for one of the existing alleles, which leads to the obvious question: Which allele will become fixed? This question, however, often does not have an explicit answer. We must settle for the *probability* that a particular allele will or will not become fixed.

For a given allele, we define its *probability of ultimate fixation,* denoted by $U(p)$, as the probability that the allele will become fixed given that its current frequency is p. The alternative to ultimate fixation is loss, and $1 - U(p)$ is the probability that the allele in question will be lost and some other allele at the locus fixed. An allele's probability of ultimate fixation is a direct measure of its potential for evolutionary success. Indeed, what better indicator of future success is there than the probability that the allele will take over the gene pool?

In the absence of natural selection the probability of ultimate fixation for an allele is equal to its frequency.

We expect natural selection to have a major influence on which allele becomes fixed. We need, therefore, to determine the relationship between the relative fitnesses and the probabilities of ultimate fixation for the alleles at a given locus.

Suppose for the moment that the relative fitnesses of AA, Aa, and aa are equal. In other words, suppose that no selection is acting upon the A-locus. In such a case we speak of the alleles A and a as being *selectively neutral,* or, simply, *neutral.* If selection gives no intrinsic advantage to one allele over another, any copy of the gene should have the same chance as any other copy of being passed on to succeeding generations. So, if the allele A is represented by a fraction p of all the copies of the A-gene in the population and the allele a by a fraction q, then their respective probabilities of ultimate fixation should be simply p and q. That is, in the absence of selection an allele's probability of ultimate fixation is equal to its frequency ($U(p) = p$).

Natural selection weights the odds of ultimate fixation in favor of advantageous alleles.

On the other hand, if the relative fitnesses of the three genotypes are unequal, we should expect the allele that is favored by selection to have an increased likelihood of becoming fixed. To see a specific example of the influence of natural selection upon the probability of ultimate fixation, let us suppose that the relative fitnesses for AA, Aa, and aa are in the form $1 : 1 - \frac{1}{2}s : 1 - s$. Suppose that p is the frequency of the allele A in a population of effective size N_e.

If selection is not too strong (if s is small), changes in allele frequencies under the joint effects of natural selection and genetic drift can be analyzed by mathematical methods similar to those used to study diffusion of gases. Such methods give the following probability of ultimate fixation for the allele A in this situation:

$$U(p) = \frac{1 - e^{-2N_e s p}}{1 - e^{-2N_e s}}, \qquad (24\text{-}1)$$

where $e = 2.718$ is the base of the natural logarithms.

The biological significance of this equation is revealed by the way in which the value of $U(p)$ is influenced by each of the three parameters s, N_e, and p. Six examples are given in Table 24-1. For each example we can see the effect of natural selection by comparing the given value of $U(p)$ with the probability of ultimate fixation $U(p) = p$ that would be the case if natural selection did not influence the outcome.

We can also see the relative effects of the allele frequency p, the selection coefficient s, and the effective population size N_e upon the probability of ultimate fixation. The first example involves moderately strong selection in a medium-sized population. The value of $U(p)$ indicates that the favored allele A is virtually certain of being fixed in such a case unless it happens to be rare (example 2). In the other cases it is not a sure bet that the allele A will be fixed. The odds diminish if the population is small (example 3) or if selection is weak (example 4). It is not necessary that s be positive; a selectively disadvantageous allele might become fixed as a result of genetic drift. Example 6 shows that in a small population there is a small, but not necessarily negligible, probability of such an event. Finally, for a given population size and selection coefficient, the likelihood of fixation is less for rare alleles than for common ones (examples 1 and 2; 4 and 5).

In one sense, the probabilities of ultimate fixation in Table 24-1 are measures of whether random fluctuations caused by finite population size will ultimately obscure the deterministic effect of natural selection. In this sense we are dealing with the same phenomenon that we saw from a different perspective in the graphs of Figure 24-1. In both instances the critical factors are the relative strengths of natural selection and genetic drift. Looking at Equation 24-1 we see that these two factors enter together in the form of the product $2N_e s$: The probability of ultimate fixation in a population of size $N_e = 500$ with $s = .02$ is exactly the same as in a population with $N_e = 1000$ and $s = .01$. The quantity $2N_e s$, therefore, gives us a single composite measure for selection and drift.

Rare alleles are likely to be lost. Our criterion for rareness changes with the relative strengths of natural selection and genetic drift.

In addition to the value of $2N_e s$, the probability of ultimate fixation, as given by Equation 24-1, depends upon the allele's current frequency p. The

Table 24-1   Values for the Probability of Ultimate Fixation, as Given by Equation 24-1, for the Allele A Are Given for Various Values of the Current Allele Frequency, p, the Selection Coefficient, s, and the Effective Population Size, N_e. Relative Fitnesses of the Genotypes AA, Aa, and aa are $1 : 1 - \frac{1}{2}s : 1 - s$.

Example	p	s	N_e	$U(p)$
1	.10	.1	1000	.999999998
2	.001	.1	1000	.8647
3	.10	.1	100	.2096
4	.10	.01	1000	.8647
5	.01	.01	1000	.1813
6	.01	-.01	100	.0032

effects of p and $2N_e s$ are illustrated graphically in Figure 24-3, which gives plots of $U(p)$ versus p for four values of $2N_e s$. For $2N_e s = 100$ (say, $s = .1$ and $N_e = 500$ or $s = .01$ and $N_e = 5,000$), natural selection pretty much guarantees fixation of the favored allele once its frequency is more than .02 or .03. In contrast, for $2N_e s = 5$ (say, $s = .1$ and $N_e = 25$ or $s = .01$ and $N_e = 250$), there is a moderate likelihood that the favored allele will be ultimately lost even if it is rather common. Loss of a favored allele implies fixation of the alternative, selectively disadvantageous, allele. The curve for $2N_e s = -5$ explicitly shows the probability of ultimate fixation for the selectively disadvantageous allele in such a case. Finally, when $2N_e s$ is distinctly less than unity, as in the case of $2N_e s = .1$, the value of $U(p)$ is close to p, which is the probability that a neutral allele becomes fixed purely on the basis of genetic drift. In such a case we can consider the alleles to be, for all practical purposes, selectively neutral, even though the selection coefficient is not necessarily all that small (for example, $s = .01$ and $N_e = 5$).

Figure 24-3 reveals that, regardless of the value of $2N_e s$, the probability of ultimate fixation is small for sufficiently rare alleles. If $2N_e s$ is large, the criterion for what is sufficiently rare is rather narrow; nevertheless, if p is very small, the allele will probably be lost even if it is strongly favored by natural selection. Recall that changes wrought by natural selection are small when one allele is rare (see, for example, Table 21-6 and Figure 21-5) and that the likelihood of loss through random chance is highest when the allele is rare—a marble near the edge of a shaking table is likely to roll off the edge.

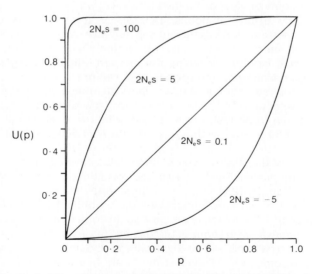

Figure 24-3   Probability of ultimate fixation, $U(p)$, plotted against the allele frequency p for four different values of $2N_e s$. The curve for $2N_e s = .1$ is almost indistinguishable from the line $U(p) = p$, which is the case for neutral alleles. For $2N_e s = 100$, fixation is virtually assured (i.e., $U(p) \doteq 1$) unless the allele is rare. For $2N_e s = -5$ the selection coefficient s is negative, implying that selection is against the allele in question.

We can determine the probability that a new mutant will lead to an allele substitution.

Rare alleles—for which ultimate fixation is uncertain even when they are favored by selection—are precisely the salient alleles with respect to the occurrence of new allele substitutions. Of particular interest is a new allele that has just been introduced into a gene pool as a single copy. The probability of ultimate fixation for such a new allele tells us what the chances are that its introduction into the population will lead to an allele substitution.

In a population of N individuals, the frequency of an allele that exists as a single copy is $p = 1/2N$. If there is no inherent selective advantage or disadvantage to the new allele, its probability of ultimate fixation is simply $U(1/2N) = 1/2N$. On the other hand, letting the allele A be the new allele, if the fitnesses of the three possible genotypes AA, Aa, aa are in the form $1 : 1 - \frac{1}{2}s : 1 - s$ and if we assume for simplicity that $N = N_e$, then substitution of $p = 1/2N$ into Equation 24-1 gives $U(1/2N) = (1 - e^{-s})/(1 - e^{-2Ns})$. If $2Ns$ is not too small (say, $2Ns > 10$) and if s is not too large (say, $s < .1$), then this equation can be approximated by the simple relation $U(1/2N) \doteq s$. For example, with $s = .01$ and $N = 500$, the approximate probability of .01 is in close agreement with the value of .00995 given by the exact equation.

The relationships between a new allele's probability of ultimate fixation, its selective advantage or disadvantage, and population size are shown graphically in Figure 24-4. $U(1/2N)$ depends upon both s and N; however, by scaling the axes in terms of $2Ns$ and $1/N$, the curve shown in Figure 24-4 is approximately correct for populations of almost any size. With larger populations the approximation $U(1/2N) = s$ is even better than in the figure.

From the point of view of relative fitnesses, we think of the allele A as being selectively disadvantageous, neutral, or advantageous depending upon whether the selection coefficient s is negative, zero, or positive. Figure 24-4 shows that we can make a similar trichotomy with respect to the probability of ultimate fixation; however, the division—illustrated by shading in Figure 24-4—is based upon the product $2Ns$ rather than upon s alone.

In the unshaded regions we can think of ultimate fixation as being dominated by selection. To the left, new, selectively disadvantageous alleles are almost certain to be lost. For example, with $s = -.01$ and $N = 500$, we get $U(1/2N) = 4.56 \times 10^{-7}$; with $s = -.01$ and $N = 5000$, we get $U(1/2N) = 3.74 \times 10^{-46}$!

To the right of the shaded region, new alleles have a selective advantage over the existing alleles. Their probabilities of ultimate fixation are close to the approximation $U(1/2N) \doteq s$. In a large population, a small selective advantage (say, $s = .02$) greatly increases the probability of ultimate fixation above the value of $1/2N$, which is the probability for a neutral allele. However, even with a selective advantage, $U(1/2N)$ is not large, and the most probable event is that the new allele will be lost from the population. Until, and unless, a new allele becomes established at moderate frequency, any intrinsic selective advantage is of limited help in preventing accidental loss by genetic drift. For every new allele that becomes an allele substitution, we can expect many others to enter the gene pool only to be lost, most likely within a few generations' time.

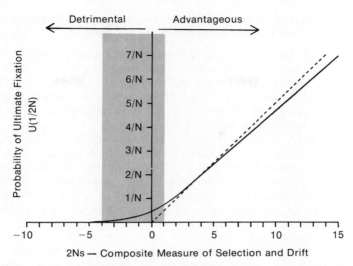

Figure 24-4   The probability of ultimate fixation $U(1/2N) = (1 - e^{-s})/(1 - e^{-2Ns})$ is plotted against the composite measure of selection and genetic drift, $2Ns$. Negative values of $2Ns$ imply that the allele in question is selectively disadvantageous. The vertical axis is scaled in inverse proportion to the population size. The dashed line shows the approximation $U(1/2N) \doteq s$. The curve shown is based upon $N = 50$.

The shading denotes the region in which eventual fixation of the new allele, should it occur, is primarily the result of random fluctuations in allele frequencies leading, on the basis of chance alone, to the spread of the new allele throughout the gene pool. In other words, fixation is a result of genetic drift. In the shaded region, the new allele's probability of ultimate fixation is small—less than $1/N$—but not necessarily negligible. Included within this region are some new alleles that have a selective disadvantage. For example, with $s = -.01$ and $N = 50$, we have $2Ns = -1$ and $U(1/2N) = .0037$.

The shaded region has greater biological significance in small populations than in large ones for two reasons. First, the probabilities of ultimate fixation in the shaded region are inversely proportional to population size and, in terms of absolute value, are therefore larger in small populations than in large ones. Second, in terms of the selection coefficient s, the width of the shaded region is greater in small populations. For example, we have just seen examples in which an allele with a 1 per cent selective disadvantage ($s = -.01$) falls within the shaded region in a population of 50 individuals but is outside the region in a population of 500 individuals and well to the left of the region in a population of 5,000 individuals.

In short, fixation of a new allele—especially one with a selective disadvantage—by genetic drift alone is most likely to occur in small populations, where the effect of genetic drift is strongest.

We can make qualitative generalizations concerning the establishment of a new allele for which the relative fitnesses are not additive.

Our explicit results concerning the effect of natural selection upon an allele's probability of ultimate fixation deal only with cases in which relative fitnesses are additive. We expect probabilities of ultimate fixation to be affected by differences in dominance. Nevertheless, we can use Figure 24-4 as a basis for qualitative statements about the likelihood of any new allele becoming well established within its population.

The probability of ultimate fixation for any new allele primarily reflects events that are likely to occur while the allele is rare. Because a rare allele occurs mostly in the heterozygous condition, the effect of natural selection is mostly determined by the difference between the fitnesses of the heterozygote and the predominant homozygote (i.e., $W_{Aa} - W_{aa}$, for the allele A), which is $s/2$ with additive fitnesses in Equation 24-1 but differs if fitnesses are not additive.

To apply Figure 24-4 loosely to an arbitrary new allele we can ask two questions. First, is selection initially acting for or against the allele? Advantageous and overdominant alleles are favored by selection when rare; disadvantageous alleles and alleles having heterozygote inferiority are disfavored by selection when rare (see Figure 21-2). Second, is the magnitude of the difference $W_{Aa} - W_{aa}$ sufficient to put the case in an unshaded region of Figure 24-4, or is it small enough to fall in the shaded region? The answer must take population size into account.

For example, suppose that the advantageous new allele A is fully recessive. It will be effectively neutral until its frequency is great enough for an appreciable number of AA homozygotes to be present in the population. We can, therefore, expect its probability of ultimate fixation to have a value closer to $U(1/2N) = 1/2N$ for a neutral allele than to $U(1/2N) = s$ for an allele with additive fitnesses.

The point is that the trichotomy of Figure 24-4 holds with respect to any newly introduced allele. Quantitative differences for alleles that do not have additive fitnesses have to do with the precise values of $U(1/2N)$ and with where along the horizontal axis of Figure 24-4 the demarcations between shaded and unshaded regions fall.

RECURRENT ALLELE SUBSTITUTIONS

Evolutionary change in a species' genetic repertory is a two-step process, involving 1) the introduction of a new variant into the genome of one, or a few, individuals and 2) subsequent establishment of the new variant within the gene pool as a whole. The probability of ultimate fixation gives us half the story. The other half comes from a consideration of mutation rates and—in order to use probabilities of ultimate fixation—the relative fitnesses of the resulting mutants.

Most mutations are irrelevant to evolutionary change.

Mutation occurs without regard to adaptiveness. However, the occurrence of new mutants is only the first of two steps. The vast majority of new mutants are probably detrimental, falling in the unshaded region on the left of Figure 24-4 where the probability that such mutants go on to complete the second step is negligible. Their probabilities of ultimate fixation are, for all practical purposes, zero. Natural selection directs evolutionary change by providing, so to speak, a ratchet mechanism that prevents establishment of deleterious alleles. Although most new mutants contribute to the reservoir of deleterious alleles in a population's gene pool, evolutionary change is restricted to the small minority of mutants (in, or to the right of, the shaded region in Figure 24-4) that may, with finite probability, go on to become allele substitutions. Many are called, but few are chosen. To the chosen few we now turn our attention.

For neutral alleles the rate of substitutions is equal to the mutation rate. The rate for selectively favored substitutions is more problematic.

We first consider the case of neutral or nearly neutral mutations for which $2Ns$ is near zero. Let u_0 denote the rate (per gene copy per generation) at which such mutations occur. In a population of N diploid individuals we can expect, on the average, $2Nu_0$ such mutants to be produced each generation. Because they are selectively neutral, the probability that any of these new mutants will be ultimately fixed is $1/2N$. So, the rate at which new mutants that are destined to lead to allele substitutions arise within a population is:

$$K_0 = (2Nu_0)\,(1/2N) = u_0.$$

K_0 is the rate per *population* per generation at which new selectively neutral allele substitutions commence. It happens to be equal to the mutation rate for neutral mutations per gene copy per generation. Therefore, once every $1/u_0$ generations, a neutral allele substitution will begin to take place within the gene pool.

A similar argument can be made for allele substitutions that involve selectively favored alleles for which the relative fitnesses are $1 : 1 - \tfrac{1}{2}s : 1 - s$. If we let u_s denote the mutation rate at which such alleles arise, then using $U(1/2N) = s$ as the probability of ultimate fixation for such an allele gives a rate of allele substitution of $K_s = 2Nsu_s$ per population per generation.

We can only speculate as to what value u_s might have. Presumably, it is exceedingly low. However, mutation rates may not be the critical factor in the case of selective allele substitutions. Quite possibly, such substitutions are initiated by changes in environmental conditions that cause some already existing, rare allele to become selectively favored over the existing predominant allele. The question, then, is how often do such changes in selection regimes occur? Every 100 generations? 500 generations? 1,000? 10,000? 100,000 generations?

One special class of initially disadvantageous genetic variations has had a significant rate of substitution during the history of biological evolution.

This is the class of chromosomal rearrangements, such as translocations, that result in meiotic difficulties when heterozygous. The mutation rate for chromosomal rearrangements as a class is generally considered to be moderately high as mutation rates go. However, because of reduced fitness associated with heterozygotes, the probability of ultimate fixation for a newly arisen rearrangement of this sort is qualitatively similar to those cases in Figure 24-4 for which $2Ns$ is negative. For a new rearrangement to become established, it must fall in the shaded region of the figure, which requires that $2Ns$ be close to zero. Consequently, if the reduction in fitness for heterozygotes is not negligible, the population size must be very small.

Evolutionary changes in the number and structure of chromosomes have most often been associated with populations that have probably gone through severe reductions. Furthermore, the rate of such changes appears to have been higher in organisms that typically live as small, semi-isolated, groups of individuals than in organisms in which a relatively large amount of gene flow between populations can be expected to occur.

An allele substitution generally requires substantial time.

The picture we are drawing is one of successive allele substitutions at varying rates, depending upon changes in environmental conditions, population sizes, relative fitnesses, and mutation rates for the alleles involved. Allele substitutions are not instantaneous. In cases such as the establishment of industrial melanism in the peppered moth, allele substitutions may occur rapidly if driven by strong selection. More generally, however, a substantial time elapses between the introduction of a new allele into a population and the time at which it becomes fixed. In neutral allele substitutions, the average number of generations required is approximately $4N_e$ *generations,* where N_e is the effective population size of any one generation. For large populations we are, therefore, dealing with long periods of time. The periods are so long, in fact, that several complications arise.

First, the number of generations is so great that mutations may alter the allelic states of some, perhaps many, of the copies of the allele that is becoming fixed. As a result, when discussing allele substitutions we must alter our notion of fixation. Rather than thinking of fixation in terms of a population's becoming monomorphic for a particular allele, we must instead think of the population's becoming comprised entirely of *descendents* of the particular allele in question. In other words, in the context of allele substitutions, fixation is a matter of identity by descent rather than a matter of monomorphism. The population may in fact be polymorphic throughout the entire evolutionary process.

Second, we need to consider the biological unit within which ultimate fixation occurs. For simplicity we have been speaking in terms of a single population. This simplified view is appropriate only if the population is a closed unit, isolated from other populations, for the duration of the allele substitution. There may be cases of small, isolated populations for which this simple view is appropriate, especially during episodes that lead to speciation (see Chapter 25). More generally, however, gene flow and the likelihood that individual populations will lose their individual identity during the long lengths of time involved suggest that the appropriate biological unit is more likely to be the entire species or at least a major segment of the species. That is, in any generation the various copies of, say, the A gene within a population may trace their respective lines of descent back to a number of different populations that existed in earlier generations, and fixation within a population may not occur until it occurs in most or all of the species. In such a case the relevant population sizes—and, therefore, the times involved—are very large indeed.

Depending upon their rate of occurrence and duration, several allele substitutions may occur simultaneously.

We are now in a position to make a rough sketch of the dynamic effects of successive allele substitutions within a species' gene pool (Figure 24-5). Suppose that the population is initially monomorphic for the allele A_0 and that at time t_1 a new allele A_1 enters the gene pool and proceeds toward fixation. The hatched area indicates the fraction of all alleles that are descendents from the original allele, A_1 at any subsequent time. At time t_3 the allele A_1 has become fixed: all alleles in the gene pool after time t_3 are descended from A_1. The period from t_1 to t_3 is the time, denoted by t_d, required for the substitution to take place. At time t_2 we suppose that a second, ultimately-to-become-fixed allele A_2 arises *among the descendents* of A_1. The shaded area indicates the alleles that are descended from A_2; necessarily they are also descendents of A_1 but have a different allelic state.

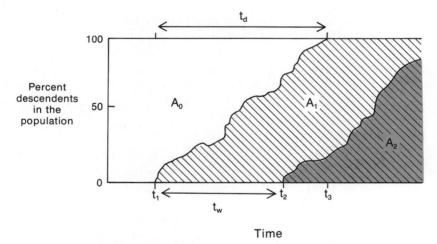

Figure 24-5   Successive allele substitutions within a gene pool. Time is measured on the horizontal axis. The vertical axis measures the fraction of alleles in the gene pool that are descendents of the allele progressing toward fixation. t_d is the duration time from the beginning to completion of an allele substitution. t_w is the waiting time between the commencement of successive substitutions.

The period from t_1 to t_2, denoted as t_w, is the time between the commencement of two successive allele substitutions.

We can use the time lengths t_w and t_d to make crude estimates of the number of allele substitutions likely to be in progress at any particular time. If the duration of each substitution is long compared to the waiting time between the commencement of successive substitutions ($t_d > t_w$), we can expect a number of substitutions to occur simultaneously. On the other hand, if the duration of a substitution is short compared to the waiting time between substitutions, there will be times when no substitution is occurring. These two cases are illustrated in Figure 24-6. On the average, t_d/t_w different allele substitutions will be occurring at any randomly chosen time.

The problem is to estimate the values of t_d and t_w. For substitutions of selectively favored alleles we have already seen that the waiting time, t_w, is open to speculation and most likely determined by environmental changes that affect the locus in question. Table 21-5 shows that the time required for such a substitution is relatively short if the selection pressures are strong. On the other hand, if selection is weak, say, $s < .01$, t_d is on the order of several thousand generations, and it is quite possible that such allele substitutions may be observed while in progress.

For neutral allele substitutions, it is possible to make better predictions. A new neutral allele substitution is expected to begin once every $1/u_0$ generations. That is, $t_w = 1/u_0$. The time required for fixation of a new allele by genetic drift alone is approximately $4N_e$ generations. So, for neutral allele substitutions $t_d = 4N_e$, and, as a rough average, the number of simultaneously occurring allele substitutions is $t_d/t_w = 4N_e u_0$.

Simultaneous allele substitutions may result in detectable transient polymorphism within a species' gene pool.

If each allele substitution involves an allele that is distinguishable from all other alleles in the gene pool, we can use the number of simultaneously occurring allele substitutions to estimate the amount of transient genetic

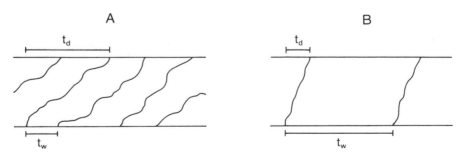

Figure 24-6   Differences in the average number of simultaneously occurring allele substitutions, depending upon whether the duration time is long (A) or short (B) with respect to the waiting time between successive substitutions.

variation (that is, the number of different alleles) expected at a randomly chosen time.

One allele must correspond to the last completed allele substitution. If, in addition, there is a distinguishable allele associated with each of the t_d/t_w substitutions in progress, there will be, on the average, $n_e = 1 + t_d/t_w$ different alleles in the gene pool. n_e is known as the *effective number of alleles* and is a loose average. Nevertheless, it gives us a general, quantitative indication of the amount of transient variation.

We can only speculate about values of t_d/t_w for allele substitutions in which natural selection plays a significant role. For neutral allele substitutions, however, our previous results give an effective number of alleles $n_e = 1 + 4N_e u_0$. If we make the further assumption that, on the average, the n_e different alleles present at any one time are equally frequent, then the average amount of heterozygosity is $\overline{H} = 1 - \Sigma(1/n_e)^2 = 1 - (1/n_e)$, which becomes

$$\overline{H} = \frac{4N_e u_0}{1 + 4N_e u_0}. \qquad (24\text{-}2)$$

These estimates show that the transient genetic variation that might exist as a result of recurrent neutral allele substitutions is determined by the product of the effective population size and the mutation rate for selectively neutral alleles. Table 24-2 gives values for the average amount of heterozygosity expected on the basis of Equation 24-2 for various values of $4N_e u_0$.

Unfortunately, it is virtually impossible to know with any degree of precision what sorts of values for N_e and u_0 exist in natural populations for particular genes. We know that u_0 is small, but whether it is nearer to 10^{-6} or 10^{-9} is a matter of speculation. The rate probably is different for different genes. Estimates for N_e are even more difficult: we discussed earlier the fact that, because of gene flow, the biological units in which ultimate fixation and neutral allele substitutions occur involve more than just local populations. Presumably, therefore, the appropriate values for N_e are large. The product $4N_e u_0$ is likely to fall anywhere within a range of values that surpasses the range in Table 24-2.

In spite of these difficulties and in spite of the assumptions that we had to make in order to predict the amount of transient polymorphism, we are, nevertheless, left with a picture of successive allele substitutions throughout the evolutionary history of any group of organisms. The preceding discussion has roughly indicated the conditions under which natural populations will

Table 24-2   Average Amounts of Heterozygosity, $\overline{H}$, Predicted by Equation 24-2 for Transient Variation Caused by Successive Substitutions of Selectively Neutral Alleles.

$4N_e u_0$	$\overline{H}$	$4N_e u_0$	$\overline{H}$	$4N_e u_0$	$\overline{H}$
.001	.0010	.05	.0476	1	.5000
.005	.0050	.1	.0909	5	.8333
.01	.0099	.2	.1667	10	.9090

exhibit significant amounts of transient variation as the by-product of such substitutions.

In the last few chapters we have glimpsed the dynamic nature of evolutionary change. On a random, non-directional basis, mutation incessantly introduces new variants. Gene flow provides genetic connectedness among the gene pools of local populations, allowing for the potential spread of genetic innovations throughout the species as a whole. The finite size of local populations adds an element of chance fluctuations and makes historical events significant factors in a population's evolution. Through it all, natural selection directs the otherwise random events of genetic change, stacking the odds against deleterious variations and weighting the odds in favor of those variations that increase adaptation or in any other way confer increased fitness upon the individuals who carry them.

In focusing upon the probability of ultimate fixation as the measure of a genetic variant's evolutionary potential, we come directly to grips with the probabilistic nature of evolutionary change. However, while recognizing the indeterminacy associated with making predictions in terms of probabilities, we must be careful not to attribute the results of organic evolution to pure capriciousness any more than we would attribute the profits of gambling houses in Las Vegas or Monte Carlo to capriciousness. Just as the weighted odds employed by a casino assure it of a tidy income, so, too, the weighted odds of natural selection exert a dominant influence upon evolutionary change.

SUMMARY

Chance fluctuations in allele frequencies—genetic drift—are a result of the finite number of individuals in any population. The significance of genetic drift depends upon the *effective population size.* In small populations, random fluctuations may be large and dramatic; in large populations, they produce only slight deviations from expected frequencies. Whether small or large, random fluctuations *accumulate* over time, with the result that sooner or later any particular allele in the gene pool will become either lost or fixed.

For a given allele, the probability that it becomes fixed (rather than lost) is known as the probability of ultimate fixation. For neutral alleles the probabilities of ultimate fixation are equal to their current allele frequencies; natural selection upon a locus, however, weights the odds of ultimate fixation in favor of advantageous alleles. In either case, the probability of ultimate fixation depends upon (1) the allele frequency p and (2) the product of $2N_e s$ (effective population size times the selective advantage of the allele). Whether favored by selection or not, an allele that is sufficiently rare—for example, present as a single copy in the population—has a low probability of ultimate fixation. For a rare, selectively disadvantageous allele, the probability of ultimate fixation is infinitesimal unless the product $2N_e s$ is close to zero.

By combining probabilities of ultimate fixation with mutation rates we can discuss the frequency with which allele substitutions can be expected. The picture that emerges is one of a continual succession, over evolutionary time, of allele substitutions. Depending upon the waiting time between the commencement of successive allele substitutions and the duration of each substitution, which can be quite long if selection is not a factor in the substitution, we can make general statements concerning the amount of transient genetic polymorphism that can be expected at any time as a result of on-going allele substitutions.

25

Genetic Divergence and Speciation

Biological evolution can be simply defined as change in the diversity and genetic composition of populations of organisms. Change in the composition of a particular gene pool constitutes what we can call "vertical evolution." The basic event of vertical evolution is the two-step process of an allele substitution, which we have discussed in great detail. Changes in the numbers and kinds of organisms that exist at any one time constitute what we can call "horizontal evolution." Horizontal evolution is a compound process, involving the splitting of a single gene pool into two or more separate gene pools, coupled with allele substitutions in one or all of the new gene pools leading to the establishment of new forms of organisms. Extinction of gene pools is also a component of horizontal evolution.

Population genetics generally focuses upon vertical evolution, particularly allele substitutions, as did the previous five chapters of this book. Many, if not most, evolutionary biologists are preoccupied with horizontal evolution. As the title of *The Origin of Species* suggests, Charles Darwin was primarily concerned with horizontal evolution. Because allele substitutions are fundamental to both horizontal and vertical evolution, the theory of population genetics that we have set forth in this book is, of course, pertinent to both. In this final chapter we shall discuss the theoretical and empirical aspects of population genetics that are relevant to the origin of species and changes in the diversity of organisms. We begin with genetic differences among populations of the same species.

Adaptation to local environmental conditions is a major source of genetic differences between populations.

We have repeatedly emphasized the role that local environmental conditions play in determining relative fitnesses and the course of natural selection in a population. We saw in Chapter 23 that large-scale environmental heterogeneity as shown in Figure 23-2A will lead to genetic differentiation between populations, each of which responds to its own local selective pressures.

Industrial melanism in *Biston betularia* (Figure 19-3) provides a well-documented illustration of genetic differentiation among populations as a result of adaptation to different local environments. Populations of pocket gophers differ strikingly with respect to pelt coloration and properties of their digging feet according to the color and graininess of the soil in which they live. Populations of Douglas fir in northern California, Oregon, and Washington differ with respect to onset and breakage of dormancy. Populations living in more severe winter climates have longer periods of dormancy than do populations in more moderate climates. (The length of dormancy is known to have a genetic basis because the trait is expressed in offspring that are experimentally grown in climates different from that of their parents.)

Genetic drift, especially in the form of *founder effects,* is a second major cause of genetic variation among populations.

Even if two populations are expected to have, on the average, identical allele frequencies, chance fluctuations of genetic drift may cause allele frequencies to increase in one population and decrease in another. However, such chance differences between populations should be comparatively minor unless the populations are small or have been genetically isolated from each other for a long time so that random fluctuations have *accumulated independently* in the different populations.

A special case of genetic drift is frequently credited with a major role in horizontal evolution. *Founder effects* are associated with the establishment of a new population by a small group of colonizers who emigrate from a large parental population. Because the colonizers represent a small sample of the parental population, allele frequencies in the new population may differ profoundly from those of the parental population (and from the allele frequencies in other colonies founded by different colonizers). The number of colonizing individuals is the critical factor. Conceivably, extreme cases might involve a self-fertilizing plant population founded by a single seed, a fungal colony established by a single spore, or an insect population started by a single inseminated female. Even if a new population grows rapidly, if it receives no further immigrants from the parental population, its gene pool will continue to reflect the heritage from the founding individuals. If the new population remains small, genetic drift in the following generations may further accentuate the population's genetic individuality.

An example of presumed founder effects is shown in Table 25-1, which lists the eight most common surnames in each of three Amish populations. Surnames are inherited with the Y-chromosome and have unusual mutational properties. Nevertheless, they have proved useful in human population genetic studies. The three populations in Table 25-1, which have remained isolated from each other and the rest of the world, came from the same region of Europe in separate waves of emigration between roughly 1720 and 1850. Although the lists account for approximately 80 per cent of all surnames in each population, there is little overlap in the three lists: a striking illustration of founder effects.

Table 25-1   Lists of the Eight Most Common Surnames in Three
Separate Populations of Old Order Amish.

Lancaster Co., Pa.		Holmes Co., O.		Mifflin Co., Pa.	
Stolzfus*	23%	Miller	26%	Yoder	28%
King	12%	Yoder	17%	Peachey	19%
Fisher	12%	Troyer	11%	Hostetler	13%
Beiler	12%	Hershberger	5%	Byler	6%
Lapp	7%	Raber	5%	Zook	6%
Zook	6%	Schlabach	5%	Spiecher	5%
Esh**	6%	Weaver	4%	Kanagy	4%
Glick	3%	Mast	4%	Swarey	4%
	81%		77%		85%

Totals:

1,106 families, 1957	1,611 families, 1960	238 families, 1951

* Including Stolzfoos.
** Including Esch.

Each population traces its descent to a separate group of emigrants from the same area of Europe. An apparent instance of an "allele substitution" (Beiler–Byler) can be seen.

McKusick et al., 1964, *Cold Spring Harbor Symposium on Quantitative Biology*, 29:99–114.

Figure 19-2 illustrated obvious genetic variation among populations of *Cepaea nemoralis* with respect to the frequency of yellow shells. Among the New Bedford River populations in Figure 19-2, there were no apparent relationships between the frequencies of yellow shells and the local environments, which were remarkably uniform along the entire length of the river bank. Moreover, these particular populations were established by rapid recolonization following a flood four years earlier that had wiped out existing snail populations. The haphazard pattern of frequencies in Figure 19-2 is presumed to be the result of founder effects.

Founder effects in horizontal evolution are comparable to bottlenecks in vertical evolution. In particular, strong founder effects may cause the derived populations to lack many rare variants present in the parental population. At the same time, any rare variant of the parental population that is foruitously carried along by colonizers may suddenly become common within the derived population. For example, each of the three Amish populations in Table 25-1 has at comparatively high frequency a normally rare allele that is responsible for a recessive birth defect. A different gene (i.e., a different birth defect) is involved in each of the three populations. Like bottlenecks, founder effects are individual events in the dynamic history of a species. The evolutionary significance of any one case is subject to the same considerations that we discussed in Chapter 24 with respect to bottlenecks (p. 621).

Genetic differentiation among populations is a potential prelude to horizontal evolution but is not by itself a sufficient cause.

Genetic divergence among populations is obviously a major ingredient in the diversification of organisms. Nevertheless, the mere existence of genetic differences carries no guarantee of evolutionary significance. Too much depends upon the subsequent histories of the populations. Differences caused solely by the vicissitudes of genetic drift depend upon isolation for long-term continuation. Population merging can readily undo genetic differences between the populations. Many, if not most, of the differences among populations may best be regarded as horizontal noise in the vertical evolution of the species.

The point being made here can be seen in extreme form in a study of house mice on a British farm. Before the advent of combine harvesting, unthreshed corn was commonly piled in stacks for the winter months. The corn ricks, as they were called, would be colonized by mice from the fields, which would then go through a number of rapid generations—each rick having a separate population—before the ricks were threshed in the spring and the mice scattered back into the fields and hedgerows. In 1960 the mouse populations from 15 ricks on one farm were captured when the ricks were broken down for threshing. Genetic heterogeneity among the rick populations took the form of markedly different frequencies—up to 20 or 30 per cent—for a number of skeletal variants (such as the presence or absence of certain small holes in the skull and vertebra). When, however, the mice from all 15 ricks were treated as a single group, they were found to have effectively the same frequencies of skeletal variants as occurred in house mouse populations throughout Britain. The local heterogeneity caused by founder effects within the ricks is reshuffled each year and has no apparent effect upon the genetic variation within the farm population as a whole.

This example is atypical in that a rick population is of regularly limited duration—and we know it. In more typical cases the length of time that a small population maintains an independent identity is irregular, uncertain, and difficult to measure. Nevertheless, the founder effects in rick populations and their apparent lack of evolutionary significance should be an important influence on the way we view microgeographic variation among populations.

Although gene flow and mixing of populations can readily undo founder effects and other forms of genetic drift, genetic differences based upon adaptations to local environments are another matter. We know from Chapter 23 that in a coarse-grained environment natural selection can maintain distinct locally adapted varieties even though there is a continuum of gene flow through the species. However, if differences between populations are caused by only a few genes, the species may nevertheless continue to evolve as a unit with respect to the rest of its genome. For example, might not an allozyme variant spread throughout the range of Douglas fir in spite of the differences in timing of dormancy among populations? As long as populations remain interfertile, gene flow and recombination will allow genetic variations not related to dormancy to be exchanged between populations although selection against immigrants of the locally disadvantageous variety may effectively slow down the rate of gene flow at which alleles at other loci become incorporated into the local gene pool.

The evolutionary significance of genetic differentiation among populations depends upon whether the populations continue sharing a common evolutionary fate or go separate ways to become distinct kinds of organisms. *Speciation* distinguishes these two alternatives and is, therefore, the fundamental event of horizontal evolution.

SPECIATION

Speciation is a compound process entailing the establishment of genetically determined isolation between biologically differentiated populations.

Speciation has two components, one relating to vertical evolution and one relating to horizontal evolution. On the one hand, speciation requires that two groups of what had previously been a single species become genetically isolated through the fixation within at least one of the groups of some inheritable trait that either prevents sexual reproduction between individuals who are members of the different groups or, if such matings can occur, leads to offspring who are incapable of contributing to the gene pool. The study of speciation is, in part, a study of the genetics and evolution of reproductive isolating mechanisms. It is also a study of the diversification of organisms. In terms of horizontal evolution, speciation requires such genetic divergence between closely related populations that what had previously been a single kind of organism becomes two distinct kinds of organisms. The study of speciation is also, in part, a study of the genetics and evolution of shifts in adaptation and ecological niches. In short, speciation requires at least two different kinds of genetic changes: those that directly or indirectly result in the establishment of reproductive isolating mechanisms and those that produce adaptive shifts to new ecological niches and biological differentiation of the populations.

Biological differentiation is limited by the "unity of the genotype."

The processes and events that lead to the differentiation of populations into distinct species differ only in degree, rather than in kind, from those already discussed in conjunction with genetic differentiation among populations in general. Active or passive dispersal of individuals continually expands the ranges of organisms both geographically and ecologically. Expansion into new territories opens the possibility of significant founder effects, new environments, new selection regimes, and new adaptations. Such "ecological experiments" play a major role in horizontal evolution. There are obvious limits, however, to the extent and speed with which biological differentiation can occur. No organism can live everywhere or do everything. For any particular organism at any one time, biological differentiation is limited by the genetic variation in its gene pool and, in addition, by what Ernst Mayr calls the "unity of the genotype."

The unity of the genotype is a concept based upon the requirement that all the genes in an individual's genome must be able to function as a single, coordinated unit. Darwin, for example, knew from his experiences with artificial selection that selection for any particular trait frequently results in modifications in other, seemingly unrelated, traits. Genetic variation is organized into balanced "co-adapted" complexes of interacting genes within which variants of any one gene may affect a number of different aspects of the organism's phenotype and variation in any one aspect of the phenotype is affected by variants at a number of gene loci. The phenotypic effects and relative fitnesses of particular alleles depend upon the entire gene complex.

According to this view, natural selection operates upon gene complexes as a whole to produce coordinated groups of alleles that function as a unit. Gene pools will be resistant to changes at individual loci that are not compatible with other genes in the complex: Major biological differentiation may require coordinated allele substitutions within entire complexes. The likelihood of coordinated allele substitutions occurring simultaneously at a large number of loci must be miniscule. The unity of the genotype is best thought of as a conservative factor in biological evolution that creates a barrier to major reorganization of the genome and limits biological differentiation, at any time, to changes that can be accomplished without any drastic or prolonged disruption of the organism's genetic balance. The high degree of biological similarity between closely related species, the persistence of rudimentary organs, the existence of "relic" species that have remained unchanged for millions of years, and the multitude of species that became extinct rather than adapt to changing environments are all indications of the evolutionary inertia that results from the unity of the genotype.

Intrinsic reproductive isolation between populations can be achieved in a variety of ways.

That particular kind of genetic differentiation that causes reproductive isolation of populations may involve any one of a diverse array of biological mechanisms (see Table 25-2). A basic distinction is made according to whether an isolating mechanism prevents fertilization (is *prezygotic*) or whether it prevents hybrid individuals from contributing to the gene pool of future generations (is *postzygotic*). Postzygotic mechanisms often waste reproductive effort; prezygotic mechanisms do not. Inability to produce viable, fertile hybrids is not an absolute requirement of isolating mechanisms: Ecologically separated species, for example, may be fully fertile when crossed in the laboratory. The critical aspect of reproductive isolating mechanisms is their *effect*—genetic isolation of gene pools—rather than the means by which the effect is achieved.

The amount of genetic change involved in reproductive isolation can vary greatly depending on the particular isolating mechanism. For example, many parasitic insects are highly specific with respect to the hosts used as rendezvous sites for courtship and mating: Relatively minor genetic changes

Table 25-2   Major Categories of Reproductive Isolating Mechanisms

A. PREZYGOTIC MECHANISMS. Factors that prevent fertilization and zygote formation.
  1. *Ecological.*
      a. *Habitat.* Interpopulational matings fail to occur because individuals of the two populations occupy or mate in different habitats within the same region.
      b. *Temporal.* Matings fail to occur because individuals in the two populations become sexually mature at different times.
  2. *Ethological.* Matings between individuals of the two populations are prevented by incompatible differences in their mating behaviors.
  3. *Mechanical.* Matings between individuals of the two populations are prevented or restricted by incompatible differences in the structure of their reproductive organs.
  4. *Physiological* (gamete incompatibility). Gametes fail to survive in alien reproductive tracts.

B. POSTZYGOTIC MECHANISMS. Factors that operate after fertilization takes place and result in inviable, weak, or sterile hybrid individuals.
  1. *Hybrid inviability or weakness.*
  2. *Hybrid sterility.*
      a. *Developmental.* Gonads fail to develop normally, or meiosis breaks down before completion.
      b. *Segregational.* Abnormal meiotic segregation of chromosomes, chromosomal segments, or combinations of genes results in the production of inviable F_2 individuals.
  3. *F_2 breakdown.* F_1 hybrids are normal and fertile, but F_2 individuals are weak or sterile.

may be sufficient for a population to shift to a new kind of host and thereby become ecologically isolated from other populations that continue to use the original host. Populations of the hawthorn fly, *Rhagoletis pamonella,* became isolated rather abruptly in the mid-1800's when some flies shifted from hawthorns to apples. In contrast, many common forms of reproductive isolation result from critical changes in, for example, highly programmed courtship patterns or morphology of genitalia. In the Hawaiian species *Drosophila pilimana* males court behind the females, only occasionally circling in front of them. In the closely related species *Drosophila aglaia,* a shift in the program caused males to adopt a ritual carried on in front of the females. We can only speculate about the amount of genetic change responsible for such revision of the courtship program.

The amount of genetic change responsible for postzygotic mechanisms covers a particularly broad spectrum. For example, differences in the karyotypes of two populations produced by comparatively simple genetic changes such as chromosomal translocations often result in meiotic difficulties for hybrid individuals (see Figure 8-9) and can cause segregational hybrid sterility. In other cases, reproductive isolation may be the by-product of extensive biological differentiation, involving a large number of genes, that results in incompatible genomes and hybrid inviability.

A certain amount of physical isolation between populations is usually necessary for speciation.

The first step towards speciation is usually some sort of temporary isolation between populations. Such isolation is the result of extrinsic factors—typically, simple geographical separation—and is distinct from the intrinsic, genetically determined, isolation of reproductive isolating mechanisms. Because genetic divergence requires that one or more allele substitutions occur in some populations and not in other populations, isolation of the gene pools is generally essential while the initial allele substitutions are in progress.

In a number of known examples, major geographical changes, such as expanding glaciers or shifts in a coastline, apparently caused the continuous distribution of a single species to be split into two large, discontinuous portions. Gross changes in climate have frequently caused major changes in vegetation, producing geographical discontinuities in the distributions of both the plants themselves and the animals that inhabit them. In many cases, however, physical isolation appears to have been the result of a small group of individuals becoming separated from the species as a whole, especially as a consequence of colonization of new territories. For example, among the Hawaiian *Drosophila,* fortuitous movements from one island to another have had a major role in speciation. Physical isolation need not necessarily be on so grand a scale. In some cases, temporary cessation of gene flow between discrete populations (for example, mice in corn ricks) for a few generations may allow a critical allele substitution to occur.

There is great variation in terms of how long physical isolation must last in order for speciation to occur. The more *genetic changes* required to establish reproductive isolation and biological differentiation, the longer the necessary period of time. In the case of a pathogen that shifts to a new host, the entire process may take place in little more than the time span of a single founder effect, whereas the gradual acquisition of hybrid sterility as a by-product of extensive biological differentiation requires a comparatively long period of geographical isolation. Furthermore, even extensive isolation is no guarantee that speciation will occur, because there is no assurance that the necessary allele substitutions will ever occur.

The major events of speciation can be illustrated with a generalized scenario.

A composite, hypothetical scenario of genetic divergence and reproductive isolation is presented in Figure 25-1. Stage I depicts a single species that consists of a number of populations connected by gene flow. In Stage II some of the populations (A-B, G, and H) are temporarily isolated from the main body of the species (C-D-E-F). The groups A-B and G have acquired genetic changes that are, therefore, not passed on to the other populations. In other words, a certain amount of genetic divergence has occurred within the species. In Stage III, G and H have become fixed for additional genetic

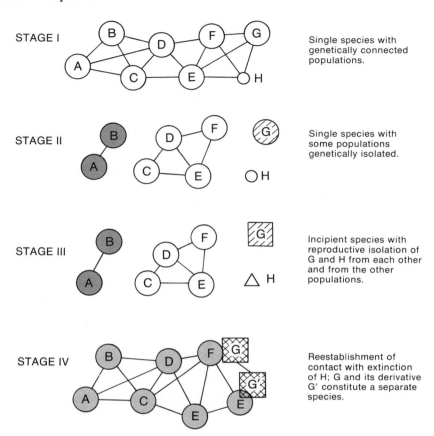

STAGE I — Single species with genetically connected populations.

STAGE II — Single species with some populations genetically isolated.

STAGE III — Incipient species with reproductive isolation of G and H from each other and from the other populations.

STAGE IV — Reestablishment of contact with extinction of H; G and its derivative G' constitute a separate species.

Figure 25-1   Composite diagram of possible events associated with genetic divergence and speciation. Lines connecting populations indicate genetic connectedness through gene flow. Populations with different shapes are separated by reproductive isolating mechanisms. Stippling, cross hatching, and double hatching denote genetic changes that have become incorporated in the gene pools.

changes that result in reproductive isolation for these groups. Population H, however, unlike G, has not differentiated from the main body of the species in any other way.

In Stage IV of our hypothetical scenario the physical isolation among populations of Stages II and III has ended. Because no intrinsic isolation exists between A-B and the main body of the species, the resumption of gene flow allows the genetic changes established in A-B to spread throughout the species' range. Population H, because of its reproductive isolation, cannot be reassimilated into the gene pool of the species: It is driven to extinction by virtue of its small size and competition with the expanding main body of the species. Population G, however, which has acquired additional genetic changes and expanded its range, has now achieved both reproductive isolation and biological differentiation and with its colony G' constitutes a new, distinct species.

The form, sequence, and likelihood of events leading to speciation are strongly influenced by the organisms involved.

The scenario just described illustrates the major themes of speciation. Many variations are possible in the order in which reproductive isolation and biological differentiation are established, the length of time spent in any one stage, and the relative sizes of the physically isolated subgroups. We see from Table 25-2 that the genetic changes responsible for speciation may take a variety of forms. The likelihood of any particular reproductive isolation mechanisms becoming established depends first of all upon the genetic variation present in the species' gene pool and also upon the totality of the organism's biological characteristics. Especially important characteristics are the organism's reproductive biology, mating behavior, ecological niche, vagility, and social structure. For example, habitat isolation is more likely to arise within a group that is already highly specific for a narrow ecological niche than within a vagile, highly outcrossing species that occupies a broad niche. Ethological isolation can most easily occur in groups that already have highly ritualized mating behavior.

Perhaps the most critical variation among scenarios for speciation is the sequence in which reproductive isolation and biological differentiation arise. Let us consider the consequences if temporary physical isolation ends before both of these requirements are fulfilled. If only biological differentiation has occurred, the scenario of populations A-B in Figure 25-1 is likely to be followed, even though temporary genetic differentiation may persist for some time. In other words, unless there is some reproductive isolation, resumption of gene flow will, in general, simply undo whatever steps towards speciation have occurred.

In the case of reproductive isolation without biological differentiation we have a different state of affairs. Although reproductive isolation is only half of the speciation story, its establishment commits the populations to separate lines of vertical evolution and is, therefore, a point of no return in the speciation process. Faced with intergroup competition, reproductively isolated groups must become biologically differentiated or face the prospect that one of the groups will be driven to extinction.

The likelihood that biological differentiation and eventual speciation will occur before extinction depends upon the amount of competitive interaction between the two groups, which, in turn, is related to the vagility, ecology, and behavior of the organisms involved. For example, within a highly vagile species like *Drosophila pseudoobscura,* a small reproductively isolated population is likely to be swamped almost immediately upon the loss of physical isolation from the main body of the species. On the other hand, among such organisms as the flightless grasshoppers of Australia, low vagility and patchy distribution of individuals may allow a reproductively isolated population to maintain its existence and integrity for a significant time even though in close geographical proximity to other populations with which it is in ecological competition. In other words, if dispersal is limited—either by low vagility or high niche specificity—the mere establishment of reproductive isolation (say, in conjunction with a brief founder effect) may prove sufficient to initiate successful speciation, whereas with higher levels of dispersal a certain amount of biological differentiation before physical isolation ends is generally necessary to avoid extinction.

The upshot of these various considerations is that speciation can occur in so many ways that we can prescribe neither necessary nor sufficient conditions for its occurrence. The form and rate of events leading to speciation are affected by a great variety of biological circumstances, and some modes of speciation are more likely to occur in certain kinds of organisms than in others. Within any group of organisms, however, speciation may occur in conjunction with major discontinuities in the species' distribution caused by such events as major geological changes or the establishment of remote colonies.

ALLOZYME CHANGES ACCOMPANYING SPECIATION

Genetic change is the principal theme of population genetics. In our discussion of speciation we have emphasized two kinds of genetic changes: 1) those that produce reproductive isolation and 2) those that achieve biological differentiation of the groups into distinct kinds of organisms. Because reproductive isolation commits the gene pools to separate evolutionary paths, we can expect a third kind of genetic change: independent vertical evolution that has no direct role in the speciation process. Evolutionary changes in allozyme variation appear to reflect primarily genetic changes of the third type. By examining patterns of allozyme variation that occur in conjunction with speciation, we can appreciate the varied scenarios that the events of speciation are capable of following.

Nei's genetic identity is a measure of the similarity of two populations with respect to their allele frequencies at a given locus.

In order to compare the arrays of allozyme alleles in two populations we need a means of measuring genetic similarity. A number of measures have been devised for this purpose. We shall employ the one developed by Masatoshi Nei, which is relatively simple and has gained increasingly common usage by evolutionary biologists. The measure is calculated from a locus by locus comparison of the allele frequencies within each of a pair of populations.

To describe the measure we need to introduce some notation. Let x and y denote the two populations being compared. For a particular locus, suppose that n different alleles, which we number 1 to n, are found in one or the other (or both) of the populations. Let $p_{1x}, p_{2x}, \ldots, p_{nx}$ denote the allele frequencies in population x, and let $p_{1y}, p_{2y}, \ldots, p_{ny}$ denote the frequencies in population y. For the particular locus under consideration we define the following quantities:

$$j_x = \sum_{i=1}^{n} p_{ix}^2, \quad j_y = \sum_{i=1}^{n} p_{iy}^2, \quad j_{xy} = \sum_{i=1}^{n} p_{ix}p_{iy}.$$

As our measure of genetic similarity we define,

$$I = \frac{j_{xy}}{\sqrt{j_x j_y}},$$

which is referred to as the *genetic identity* between x and y at the locus in question.

Possible values for I range from zero to unity. The maximum value of $I = 1$ occurs when the two populations contain the same alleles with identical frequencies, in which case $j_x = j_y = j_{xy}$. The minimum value of $I = 0$ occurs when the two populations have no alleles in common, in which case $j_{xy} = 0$. These properties are true regardless of whether the locus is polymorphic or monomorphic within the populations. In other words, I measures the genetic similarity *between* the populations independently of the genetic variability *within* the populations.

As an example, consider the genetic identity between the Strawberry Canyon and Austin populations of *Drosophila pseudoobscura* with respect to their allele frequencies at the Pt-8 locus (Table 19-3). The calculations are set out in tabular form in Table 25-3. The value of $I = .997$ indicates that these two populations are very similar with respect to genetic variation at the Pt-8 locus.

Since I is defined for a single locus and a single pair of populations, for a given pair of populations we can calculate a different I for each locus for which data are available from both populations. If we have data from a number of populations, we can calculate values of I for each possible pair of populations. Consequently, many values can be obtained from allozyme studies of natural populations. As an overall measure of genetic similarity

Table 25-3   Calculation of Nei's Genetic Identity for the Strawberry Canyon, California, and the Austin, Texas, Populations of *Drosophila Pseudoobscura* with Respect to Their Allele Frequencies at the Pt-8 Allozyme Locus.

Allele (i)	Strawberry Canyon (p_{ix})	Austin (p_{iy})	p_{ix}^2	p_{iy}^2	$p_{ix}p_{iy}$
.80 1	.014	.011	.000	.000	.000
.81 2	.472	.441	.223	.195	.208
.83 3	.514	.512	.264	.262	.263
.85 4	0.0	.035	0.0	.001	0.0
	1.0	.999	.487 (j_x)	.458 (j_y)	.471 (j_{xy})

$$I = (.471) / \sqrt{(.487)(.458)} \doteq .997.$$

among populations of a given species, we can take the average of the values obtained for the various loci and for all pairs of populations that belong to the species.

For most allozyme loci, genetic identity between a pair of conspecific populations is commonly close to unity.

The left column in Table 25-4 gives a sample of available data. The average genetic identity between two populations belonging to the same species is generally high. For most allozyme loci, a pair of conspecific populations tends to be either monomorphic for the same allele or polymorphic with similar allele frequencies in the two populations like the example in Table 25-3. Some differences, however, are noticeable. Tahitian snails and salamanders in the genera *Plethodon* and *Batrachoseps* have lower average values, which reflect the existence of genetic variation among populations with respect to alloyzme variation. These animals are notable for their comparatively low vagility. Although vagile organisms like *Drosophila* generally do not show any regional differentiation with respect to allozyme

Table 25-4   Average Values of Nei's Genetic Identity for Various Comparisons Between Populations Belonging to Closely Related Groups. See Text for Discussion of Some Major Exceptions to the General Patterns Illustrated by These Data.

| | Comparisons of Populations Belonging To: | | |
Group	Same Species	Different Subspecies	Closely Related Species
INVERTEBRATES			
Drosophila			
D. willistoni complex	.97	.80	.52
D. athabasca complex	.99	.90	—
D. mulleri—D. aldrichi	.998	—	.88
D. silvestris—D. heteroneura	.955	—	.94
Speyeria—Butterflies	.99	.98	.83
Partula—Tahitian land snails	.91	—	.91
VERTEBRATES			
Peromyscus—field mice	.97	.95	.72
Anolis (roquet group)—lizards	.996	—	.67
Salamanders			
Taricha	.98	.84	.75
Plethodon cinereus—P. serratus	.91	—	.66
P. dorsalis—P. welleri	.88	.76	.66
Batrachoseps pacificus complex	.86	.63	—
Lepomis—sunfish	.98	.84	.54
ANNUAL PLANTS			
Lupinus—legume	.97	—	.35
Hymenopappus—composite	.96	—	.90
Stephanomeria—composite	.98	—	.95

data (review Table 19-3), regional differentiation can sometimes be found among the populations of less vagile organisms. For example, among the populations of *Batrachoseps pacificus* that inhabit drainages on the western slopes of the Sierra Nevada mountains, increasing geographical separation is reflected by decreasing genetic identity.

A striking exception to the pattern of intraspecific variation in Table 25-4 is found in Californian populations of the slender wild oat, *Avena barbata*, for which the average genetic identity between pairs of populations was found to be $\bar{I} = .714$. This low value is the result of individuals in this highly self-fertilizing species having one or the other of two sharply differentiated genotypes whose distributions are strongly correlated with climatic conditions. In particular, in the dry environments of the central valley of California and southern California, populations are fixed for the same alleles at each of a number of allozyme loci; whereas in wetter environments an alternative allele at each of these loci is either fixed or very common. So, the genetic identity between a pair of populations living in the same environment is high, and the identity between populations in different environments is quite low.

For allozyme loci the genetic similarity between populations belonging to closely related species is highly variable.

Now that we have some idea of the allozyme similarity between populations of the same species, we can determine the similarity of two populations of two different, closely related, species. In cases such as that of the *Drosophila willistoni* complex, which consists of about 15 closely related species that live in the tropics of Central and South America, we can make a series of comparisons between groups with varying degrees of evolutionary separation.

Figure 25-2 shows a series of histograms based upon about thirty allozyme loci and several hundred populations in the *Drosophila willistoni* complex. Each value of I is based upon allele frequencies at a single locus within a particular pair of populations. Figure 25-2A shows a histogram of the various values of I obtained from comparisons between any two populations of the same species. Figure 25-2B shows the distribution of genetic identities between pairs of populations in which the two populations belong to different subspecies between which only partial reproductive isolation exists. Figure 25-2C shows genetic identities between populations that belong to different, reproductively isolated, but closely related, species. Figure 25-2D shows the results of comparisons in which one population belongs to the *Drosophila willistoni* complex and the other population belongs to *Drosophila nebulosa,* which is outside of the *Drosophila willistoni* complex.

These data show that the more distantly related two populations are, the lower their average genetic identity, $\bar{I}$, is likely to be. However, at all four levels of evolutionary relatedness shown in Figure 25-2, the majority of the values for I are either close to unity or close to zero, which means that for most genes, two populations (regardless of their relationship) tend either to have the same alleles at similar frequencies or to have completely different

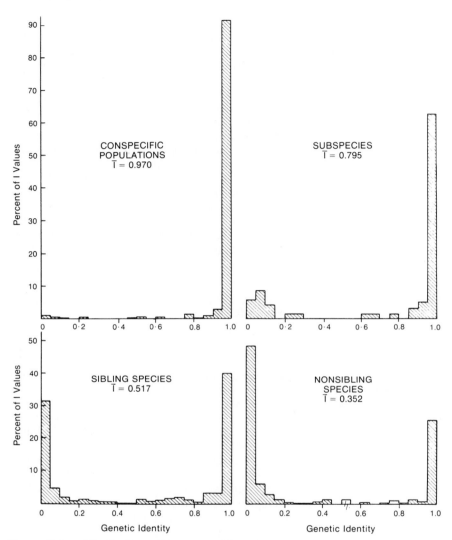

Figure 25-2   Distributions of the values of Nei's genetic identity, *I*, obtained from a number of allozyme loci within natural populations of the *Drosophila willistoni* complex. Each value of *I* is based upon allele frequencies at a single allozyme locus in a pair of populations. Data are plotted according to the taxonomic status of the pair of populations for which the value of *I* was calculated. After Ayala et al., *Genetics*, v. 77, pp. 343–384, 1974.

alleles. The more distantly related the populations are, the greater the percentage of loci that fall in the latter category. The pattern that emerges, therefore, is one of allele substitutions at more and more loci over time, so that populations belonging to two different lineages become genetically different at more and more loci as their lineages grow farther apart taxonomically.

Table 25-4 gives average genetic identities between populations of closely related species for a variety of organisms. The numbers in the second and third columns indicate considerable variation from case to case. To some extent these differences appear to be related to the presumed mode of speciation. The taxonomic relationships within the *Drosophila willistoni* complex, for example, conform to a pattern of progressive, gradual differentiation in conjunction with long-term geographical isolation. The allozyme data in Figure 25-2 are consistent with the varying lengths of genetic isolation between lineages within this complex. However, Table 25-4 shows that even among *Drosophila* speciation does not necessarily entail great reductions in allozyme similarities. Especially notable is the case of the Hawaiian species *Drosophila silvestris* and *Drosophila heteroneura,* which are easily distinguished morphologically and are reproductively islated in nature by a strong ethological mechanism (see Table 25-2). Their allozyme similarity is, nevertheless unusually high, and when hybrids are obtained in the laboratory, no loss of viability or fertility is detected.

In contrast to the *Drosophila willistoni* complex is a case of apparent abrupt speciation in the plant genus *Stephanomeria. Stephanomeria malheurensis* consists of a single known population that lives interspersed among a peripheral population of *Stephanomeria exigua* ssp. *coronaria,* which is the widespread and ecologically diverse taxon from which *Stephanomeria malheurensis* was derived. *Stephanomeria malheurensis* is highly self-fertilizing and is reproductively isolated from the obligate outcrossing population of ssp. *coronaria* by mating system and by a degree of F_1 and F_2 inviability in hybrid crosses. *Stephanomeria malheurensis* has diverged biologically in a number of growth characteristics. It has heavier roots and shoots and fewer but much heavier seeds. In terms of allozyme variation, however, genetic similarity between these two groups is high ($\bar{I} = .95$). In fact, the allozyme alleles found in *S. malheurensis* are, for the most part, simply a subset of the alleles of ssp. *coronaria,* containing the most common allele at each locus in the population of ssp. *coronaria.*

Genetic differentiation at allozyme loci is primarily related to the time that gene pools are isolated from each other.

Pocket gophers of the genus *Thomomys* provide interesting examples of speciation and allozyme differentiation. One group, which lives in western North America, is known as the *Thomomys talpoides* complex. This complex is subdivided into eight or more different "chromosomal types" that differ from each other by virtue of chromosomal rearrangements of a form that results in meiotic difficulties for hybrid individuals. There is little or no known interbreeding between the types, which, therefore, rank at the level of sister species. Yet the average genetic identity between populations belonging to different types is $\bar{I}$ = .925, which is almost as high as the genetic identity between populations of the same type.

Within *Thomomys bottae,* which is distributed across the southwestern United States and northern Mexico, there is also extensive chromosomal variation, but of a form that creates no meiotic problems, and interbreeding is possible throughout the species. There is extensive allozyme variation among the populations of *Thomomys bottae:* $\bar{I}$ = .866, and at a number of loci different alleles are fixed in different populations. Geological information and the patterns of allozyme variation suggest that there was a long period in the history of this species during which the western and eastern portions were geographically isolated, and genetic differences accumulated without any reproductive isolating mechanisms arising. The present pattern of variation is presumably the result of reestablished contact throughout the species' range. In Arizona, however, there did occur a speciational event: The derived species *Thomomys umbrinus* separated from *Thomomys bottae.* Genetic similarities between reproductively isolated Arizonian populations of *Thomomys bottae* and *Thomomys umbrinus* are higher than the genetic similarities between populations of *Thomomys bottae* in Arizona and eastern New Mexico!

This interpretation of the situation in *Thomomys bottae* implies that allozyme differentiation may accumulate simply as a result of major geographical isolation. An example of such an event is found in the tailed frog, *Ascaphus truei,* which was once continuously distributed throughout the Pacific Northwest but now has a fragmented distribution restricted to moist mountain ranges. Comparisons between populations from different regions in the species' range are summarized in Table 25-5. Genetic similarity decreases with increasing length of time that the populations have been geographically isolated from each other.

These examples suggest that allozyme differentiation between two groups is primarily related to isolation of gene pools and only indirectly to speciation. The stories of *Thomomys bottae* and *Ascaphus truei* show that allozyme differentiation may occur as a result of long-term isolation whether or not speciation occurs. On the other hand, speciation necessarily results in genetic isolation; however, the stories of *Thomomys talpoides* and *Stephanomeria malheurensis* show that if the separation is recent, allozyme differentiation may be slight. John Avise assessed genetic similarity for allozyme loci in 69 species of North American minnows and 19 species of sunfish. These two families of fishes are of roughly equal age but appear to

Table 25-5   Genetic Identities for Allozyme Loci Within and Between Geographically Isolated Regions in the Range of the Frog *Ascaphus Truei.*

Geographically adjacent populations		$\bar{I}$ = .99
Rockies	*versus* Blue Mountains	$\bar{I}$ = .95
Olympics	*versus* Coastal–Cascade Mountains	$\bar{I}$ = .86
Coastal–Cascades *versus* Blue Mountains–Rockies		$\bar{I}$ = .71

Data from C. H. Daugherty and F. W. Allenforf, 1977, *Genetics* 86: s14.

have undergone grossly different rates of speciation. The minnows are extremely diverse (over 200 species) whereas the sunfish are represented by only 30 species. Avise found that average genetic similarities among the species in each of these two families is unrelated to the frequency with which speciation has occurred and, instead, appears to be directly related to the average length of time that the various lineages within each family have been separated.

EVOLUTIONARY DIVERGENCE

In this final section we shall further consider genetic differences between lineages that have become separated by speciation. Although the evolution of allozyme differences among species shows a pattern of independent vertical evolution in separate lineages, such need not always be the case. Horizontal and vertical evolution may differ in rate and pattern. Horizontal evolution can be expected to show an *episodic* pattern in which most changes occur rather rapidly during speciation. Vertical evolution, in contrast, need not be correlated with speciation and may be expected to show a more or less *uniform* pattern of change of the sort we have seen for allozymes. Consequently, when dealing with evolutionary changes in a particular characteristic, we should expect different patterns depending upon whether changes in the characteristic tend to be associated primarily with vertical evolution or with horizontal evolution. We can illustrate this point with three general examples.

Amino acid sequences in homologous proteins supply measures of long-term genetic divergence.

Any good biology textbook shows evolutionary relatedness by comparisons of homologous structures of living and fossil organisms. Determination of the accumulated genetic divergence between lineages that have been genetically isolated for long periods of time is, in general, difficult. But for a few genes, genetic divergence between distantly related organisms can be measured by comparing the number of amino acid differences in common proteins such as cytochrome *c* and hemoglobins (see Table 19-1).

For each amino acid site at which two species differ, there must have been at least one allele substitution within one or the other of the two lineages during genetic isolation. By comparing the entire amino acid sequence for a given protein, we can detect the cumulative effects of multiple allele substitutions in the same gene (although we might miss multiple substitutions at the same amino acid site). As a result, we can determine not only whether two species differ with respect to a particular gene but *how much* they differ in terms of the minimum number of genetic changes required to account for the difference.

If the standard sequences are known for some particular protein for a number of different organisms, the numbers and kinds of amino acid differences between pairs of organisms can be used to arrange the organisms into a phylogenetic tree. Figure 25-3 shows a tree constructed on the basis of amino acid sequences for cytochrome *c*. Cytochrome *c* is a principal component in the mitochondrial respiratory machinery of organisms as diverse as yeast, wheat, and humans. This figure not only shows that such diverse organisms have the same protein but that the differences in this protein directly reflect the evolutionary relationship of each pair of species. The biochemical structure of this protein provides unmistakable and remarkable evidence of universal biological relatedness.

Changes in amino acid sequences occur at rates roughly constant for a given protein but different for different proteins.

If we go one step further and compare the genetic relatedness in Figure 25-3 with known estimates of the times at which the various evolutionary lineages diverged from one another, we can estimate the rates at which

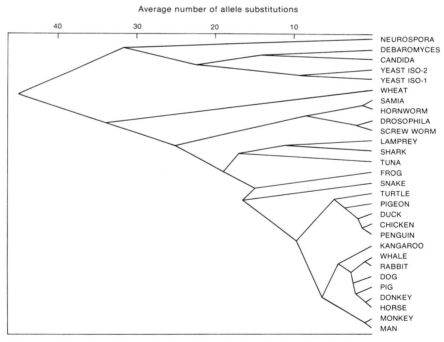

Figure 25-3   Phylogenetic tree based upon differences in the amino acid sequences of cytochrome *c* in a variety of organisms. The axis is measured in terms of the average number of allele substitutions in a lineage of the given length. From W. M. Fitch and E. Markowitz, *Biochemical Genetics*, v. 4, p. 581, 1970.

allele substitutions have occurred in the various branches of the tree. When this is done for cytochrome *c*, the rate of allele substitutions per unit time turns out to be remarkably similar in different, independent lineages. Put broadly, the number of allele substitutions in a given lineage is primarily a matter of the time span. There are, however, discrepancies. For example, the rate of evolution of cytochrome *c* has been higher than average among snakes, with the result that an anomalous bump appears in Figure 25-3 at the point at which the snake lineage joins with that of other terrestrial vertebrates.

Rates of changes in amino acid sequences have been determined for a number of proteins in addition to cytochrome *c*. Table 25-6 lists some of the best known examples. The rates of molecular evolution vary substantially from one gene to another, which should come as no surprise inasmuch as different proteins do different things and have different biochemical requirements. On the other hand, the rate of molecular evolution of a given protein is approximately constant from one lineage to another.

Gene duplication has played a major role in the evolution of new genes.

In comparisons of homologous proteins having the same biochemical function in a variety of species, the amino acid differences reflect primarily the evolutionary fine tuning without a change in protein function. In Figure 25-3, we observe the results of vertical evolution at the molecular level. The horizontal event of particular interest at the molecular level is the origin of new genes with products having new and different functions. New genes have often arisen by gene duplication, which results in a pair of identical genes that can, through subsequent vertical evolution, diverge in function.

Hemoglobins provide a well-known case of successive gene duplication, followed by gradual divergence in function. Humans normally have one gene for myoglobin and, as we saw in Chapter 17, separate genes for a variety of hemoglobin chains such as α, β, γ, and δ. The evolutionary histories of these genes (Figure 25-4) can be reconstructed both from amino acid differences in the various globin sequences and from information about the presence or absence of the different genes in other vertebrates along the line of human

Table 25-6   Rates of Allele Substitutions Leading to Amino Acid Changes in the Sequences of Various Proteins During the Course of Mammalian Evolution.

Protein	Substitutions per codon per year
Insulin A and B	3.3×10^{-10}
Cytochrome *c*	4.2 "
Hemoglobin α-chain	9.9 "
Hemoglobin β-chain	10.3 "
Ribonuclease	25.3 "
Immunoglobulin light chain	33.2 "
Fibrinopeptide A	42.9 "

From J. L. King and T. H. Jukes, 1969, *Science* 164: 789.

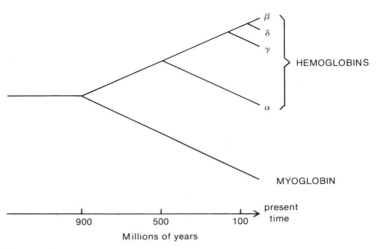

Figure 25-4   Evolution of the genes for human globins. Branch points indicate gene duplications, which, followed by changes in sequence, led to adaptive differentiation of the various globin genes. The numbers of amino acid differences between the various hemoglobin genes are: 78 for α and β; 83 for α and γ; 36 for β and γ; 6 for β and δ. After M. O. Dayhoff et al. For reference, see Table 19-1.

evolution. The earliest duplication occurred well before the emergence of vertebrates and yielded separate genes for myoglobin and for the hemoglobins. Subsequent duplication produced the different hemoglobin genes now present in the human genome.

Hemoglobin functions in red blood cells, whereas myoglobin functions in the muscles. Divergence of the α and β chains apparently led to the tetrameric structure of vertebrate hemoglobin, as neither myoglobin nor α chains alone readily form tetramers. The presence of γ chains in place of β chains in fetal hemoglobin is an apparent adaptation for oxygen transfer across the placenta. In addition to these functional differences, which are results of changes in duplicated structural genes, we saw in Chapter 17 (p. 471) that concomitant changes have occurred in the regulation of these various genes during the evolution of this co-adapted complex of genes from a single ancestral gene.

Numerous other cases of apparent gene duplication and divergence in function are known. Sequence data are available for enough proteins so that M. O. Dayhoff and her colleagues were able to arrange large numbers of genes into families based on sequence homologies. Human immunoglobulin genes are similar to the globins. In lactalbumin and lysozyme divergence following duplication has produced genes whose products differ greatly in chemical activity and specificity. In short, gene duplication and divergence have been frequent events in the course of molecular evolution.

Chromosomal evolution is generally associated with speciation.

Changes in numbers of chromosomes and chromosomal arms have been extensively studied. We know from such examples as *Drosophila pseudoobscura* and *Thomomys bottae* that chromosomal polymorphism can be widespread if no meiotic problems are associated with the chromosomal rearrangement. In most cases, however, chromosomal evolution involves fixation of rearrangements that do produce meiotic difficulties and loss of fertility in individuals heterozygous for the standard and the rearranged karyotypes. Because of such heterozygote inferiority, we can expect, for a number of reasons, that permanent establishment of chromosomal changes will tend to be associated with speciation. In the first place, heterozygote inferiority makes intraspecific polymorphism for normal and rearranged karyotypes an evolutionary unstable state of affairs. Consequently, permanent establishment of such chromosomal changes is unlikely unless the changes occur concomitantly with speciation. That is, speciation can be considered, in general, to be a necessary requirement for chromosomal evolution. Second, we know from Chapter 24 that changes involving heterozygote inferiority have little likelihood of becoming established except in a very small population. Founder effects, which are often associated with speciation, provide some of the best opportunities for chromosomal rearrangements. Third, if a chromosomal rearrangement should become established in a small population, we have seen (p. 646) that the associated segregational hybrid sterility may, in some cases, be the first step towards eventual speciation.

This is not, however, to say that chromosomal rearrangement is either a necessary or sufficient condition of speciation. For one thing, a population with a newly established rearrangement may well be swamped (and the rearrangement lost) before speciation can occur. For another thing, segregational hybrid sterility is only one of many kinds of reproductive isolating mechanisms, and a rather poor one at that. Known cases of sister species such as *Drosophila polimana* and *Drosophila aglaia* in which both species have identical polytene sequences demonstrate that speciation does not require any concomitant change in karyotype. Nevertheless, chromosomal evolution tends to be associated more with horizontal evolution than with vertical evolution. Guy Bush and his co-workers, for example, have shown that among mammalian genera changes in the numbers of chromosomes and chromosomal arms have occurred most frequently within those genera in which speciation has occurred at comparatively high rates.

The evolution of morphological differences is associated more with changes in adaptation than with length of divergence.

Morphological characteristics play a major role in making an organism what it is, and we can expect changes in morphology to be a principal component of horizontal evolution. In general, morphological differences between closely related groups tend to reflect similarities and differences in adaptations and in the environmental conditions experienced by the groups more than the length of time their gene pools have been isolated. For example, when California salamanders of the genus *Batrachoseps* are grouped according to morphological characteristics of body shape and size, populations fall into two major clusters that reflect whether the populations live in relatively moist coastal regions or in drier inland regions. Two populations belonging to different species may be grouped together, while two populations belonging to the same subspecies fall in different clusters. On the other hand, when the same populations are grouped according to allozyme similarities, the resulting clusters form a pattern that can be directly correlated with presumed history of speciation and changes in biogeographic distributions that have occurred over the past 10 million years in conjunction with major geological changes in the California coastline.

Morphological changes, being intimately connected to the particular conditions under which the organisms live, contrast with the uniformity of molecular evolution illustrated by allozymes and amino acid substitutions. This point is strikingly demonstrated in a broad survey by Allan C. Wilson and his co-workers of the comparative rates of evolution in frogs and placental mammals. Table 25-7 summarizes their findings. Both groups comprise a large number of species. However, frogs as a group are twice as old as the placental mammals and yet are represented by only two thirds as many species.

The most impressive difference between the two groups lies in morphological evolution. The spectrum of phenotypic diversity among frogs is sufficiently narrow that all 3,000 or so species are classified in a single taxonomic order, whereas the placental mammals are distributed among 16 to 20 orders. The frogs do not begin to approach the anatomical, behavioral, and ecological diversity displayed by mice, bats, cows, whales, and apes. Relative to their rates of speciation, biological diversification among mammals has been enormous and among frogs has been severely restricted.

On the other hand, molecular evolution, as measured by differences in the structure of albumin, has proceeded at the same rate in both groups, with the result that the albumin sequences of two frog species belonging to the same genus may differ from each other as much as the sequence in bats differs from the one in whales. Like morphological evolution, chromosomal evolution and loss of the ability to form viable interspecific hybrids have occurred at different rates in the two groups and are discordant with the accumulation of differences in albumin structure. Whereas mammalian species that have been isolated long enough to differ by more than 10 per cent of their albumin sequences invariably cannot form hybrids and have different numbers of chromosomes, many frog species that have been isolated more than twice as long as such mammals have the same number of chromosomes and can form viable hybrids.

In this comparison of frog and mammalian evolution, as in earlier examples, patterns of evolutionary change vary not only with the groups of organisms but also with the kinds of genetic changes. The second sort of variations can be ascribed, in part, to different patterns of vertical and horizontal evolution. Molecular evolution, from the evidence of differences

Table 25-7   Comparative Rates of Evolution in Frogs and Placental
Mammals.

Type of Change	Rate of Evolution	
	Frogs	Placental Mammals
Speciation	Moderate	Fast
Morphological Evolution	Slow	Very Fast
Albumin Evolution	Standard	Standard
Loss of Hybridization Potential	Slow	Fast
Chromosomal Evolution	Slow	Fast

As a group frogs are 150 million years old and comprise one order of approximately 3,000 species. Placental mammals are 75 million years old and comprise 16 to 20 orders of 4,600 species.

From A. C. Wilson, 1975, *Stadler Symposium* 7: 122.

in amino acid sequences and electrophoretic mobility (allozymes), appears to reflect almost entirely vertical evolution. The similar rates of molecular evolution in frogs and mammals are consistent with a view of vertical evolution as an ever-present, continuous circumstance in all gene pools. In contrast, changes in karyotypes appear almost wholly dependent upon and associated with horizontal evolution. Morphological and behavioral changes, which are frequently the basis of biological differentiation during speciation, likewise tend to show an irregular and episodic pattern more typical of horizontal evolution than of vertical evolution.

**Population genetics describes the "rules" of evolutionary change.
Application of the rules is determined by the problems that organisms face
and the genetic variation available to deal with those problems.**

In practice, horizontal and vertical evolution are not independent phenomena, and the generalizations that we have just made can be at best partial explanations. We are still left, for example, with the question of why the differences between frogs and mammals in morphological evolution greatly exceed their differences in speciation.

We must return to our discussion of biological fitnesses at the beginning of Chapter 21. In most instances, morphological change, in general, and horizontal evolution, in particular, are directly associated with changes in the adaptations of a species: that is, change in the collection of traits that are especially suited to accomplish particular tasks and functions in the life of the organism. In Chapter 21 we saw that differences in fitnesses—and, therefore, the effects of natural selection—pertain only to extant genetic variation and derive from differences in the consequences to the organism that those genetic variants have in a particular environment. Acquisition of a new adaptation, therefore, depends, first, upon the organism's being confronted with circumstances in which the task or function in question is important to the life of the organism and, second, upon there being present in the organism's gene pool genetic variation that affects the efficiency with which the task can be accomplished.

Therefore, diversification of a group into new species or genera—its horizontal evolution—is governed both by its ecological experiments and by the genetic variation available to provide solutions to new problems encountered in the course of those experiments. The ecological experiments and the genetic variation are the *causes of evolution.* Given the experiment and the genetic variation, the principles described by population genetics determine the *results of evolution,* which are the changes in diversity of organisms and in the composition of their gene pools.

SUMMARY

Biological evolution deals not only with changes in the genetic makeup of extant species (vertical evolution) but also with changes in the number and kinds of species that exist at any particular time (horizontal evolution). *Speciation* is the basic event of horizontal evolution. It is a two-part process that involves 1) genetic differentiation of a single kind of organism into two biologically distinct organisms and 2) establishment of genetically determined reproductive isolation between the two groups. Each part of the process requires genetic changes that occur within one, but not the other, of the groups. Both reproductive isolation and biological differentiation can be achieved in many ways, and speciation has many scenarios, depending upon the available genetic variation, environmental conditions, and the entire rest of the biology of the organisms involved. At one extreme, cumulative biological differentiation may lead, as a by-product of genetic divergence, to incompatible genomes and hybrid inviability between geographically isolated populations. At the other extreme, a fortuitous allele substitution within a small, temporarily isolated population may lead rather abruptly to reproductive isolation of the population, which if it avoids being swamped, may biologically differentiate into a new species.

Reproductive isolation commits the isolated groups to independent vertical evolution and progressive genetic divergence. Some genetic changes, notably molecular changes revealed by allozyme studies and comparisons of amino acid sequences in structural genes, tend to be associated almost entirely with vertical evolution and show patterns of evolution that directly reflect the time that the gene pools have been isolated from each other. Other kinds of genetic changes that are more directly involved with speciation show irregular and episodic patterns. The evolution of morphological differences, for example, is associated more with changes in adaptation and biological differentiation than with the time that groups have been isolated.

Epilogue

It is interesting to contemplate an entangled bank, clothed with many plants of many kinds, with birds singing on the bushes, with various insects flitting about, and with worms crawling through the damp earth, and to reflect that these elaborately constructed forms, so different from each other, and dependent on each other in so complex a manner, have all been produced by laws acting around us.

Charles Darwin

Unity-in-diversity is the great attribute of the natural world. The diversity in kind, condition, and mode of life exhibited by the plants and animals inhabiting the Earth is staggering. Amidst the diversity, numerous common characteristics testify to the relatedness of all living things. At the center of both the diversity and the unity stand the laws of genetics and genetic change.

Reproduction—like begetting like—gives continuity and a measure of durability to life, and the genetic material composed of nucleic acids is the essential medium by which instructions for life are transmitted from parent to offspring. Without nucleic acids, biological reproduction as we know it would be impossible. The use, organization, and operation of this ubiquitous genetic material is one of the profound unities of nature.

The enormous diversity of life exists because very different, but always coherent, instructions for life have come to be written in the genetic material of different organisms. The history of biological evolution is the history of alterations arising in genetic instructions, spreading among a group of individuals, and being partitioned by speciation into separate, but related, versions of the instructions for life. Over the past three billion years, these biological processes, obeying the laws to which Darwin referred, created the immense biological library that now contains the instructions for all the present victors of evolution.

Further Reading and References

General

GENETIC TERMINOLOGY

King, Robert. *A Dictionary of Genetics.* Oxford University Press, 1974. Particularly good for the beginning student of genetics.

Rieger, R., A. Michaelis, and Melvin M. Green. *A Glossary of Genetics and Cytogenetics.* 4th Edition, Springer-Verlag, 1978. Provides historical perspective.

OTHER TEXT BOOKS

Goodenough, U. *Genetics.* 2nd Edition. Holt, Rinehart and Winston, 1978, 840 pp. Emphasis on molecular genetics.

Stent, Gunther S. and Richard Calendar. *Molecular Genetics. An Introductory Narrative.* 2nd Edition. W. H. Freeman and Co., San Francisco, 1978, 773 pp. A somewhat advanced historical approach.

Strickberger, M. *Genetics.* 2nd Edition. Macmillan, New York, 1976, 914 pp. A broad general text.

Suzuki, D., and A. Griffiths. *An Introduction to Genetic Analysis.* Freeman, San Francisco, 1976, 468 pp. Emphasis on genetic analysis to deduce biological mechanisms.

Watson, J. *The Molecular Biology of the Gene.* 3rd Edition. W. A. Benjamin, Inc., 1976, 739 pp. Very lucid exposition. The state of knowledge in 1976.

Chapter 1

COLLECTIONS OF ORIGINAL PAPERS

Corwin, H., and J. Jenkins. *Conceptual Foundations of Genetics,* Houghton Mifflin, Boston, 1976.

Taylor, J. Herbert. *Selected Papers on Molecular Genetics.* Academic Press, 1965.

Two collections of primary papers on the discovery of DNA as a carrier of hereditary information.

REVIEWS

Oparin, A. *The Origin of Life.* Dover, New York, 1953. The classic work.

Orgel, L. *The Origin of Life: Molecules and Natural Selection.* Wiley, New York, 1973. A more recent origins book.

Penzias, A. "The origin of the elements." *Science.* v. 205, p. 549, 1979. An article by a recent Nobel Laureate.

Watson, James. *The Double Helix.* Atheneum, 1968. Watson's personal view of the discovery of the double helical structure of DNA. Fascinating.

Weinberg, Stephen. *The First Three Minutes.* Andre Deutsch, 1977. A physicist's exciting description of the origin of the universe that even a geneticist can understand.

RECENT RESEARCH PAPERS

Wang, A., G. Quigley, F. Kolpak, J. Crawford, J. van Boom, G. van der Marel, and A. Rich. "Molecular structure of a left-handed double helical DNA fragment at atomic resolution." *Nature,* v. 282, p. 680, 1979. Lest we get complacent, there are other DNA's that, in contrast to the right-handed Watson and Crick type, are left-handed.

Chapter 2

COLLECTIONS OF ORIGINAL PAPERS

Freifelder, D. *The DNA Molecule: Structural Properties.* Freeman, San Francisco, 1978. A collection of papers on the isolation and properties of DNA molecules.

REVIEWS

Comings, D. "Mechanisms of chromosome banding and implications for chromosome structure." *Annual Review of Genetics,* v. 12, p. 25, 1978. Possible structural significance of bands. An attempt to relate some bands produced by staining to chromomeres.

Felsenfeld, G. "Chromatin." *Nature,* v. 271, p. 115, 1978. A brief review on the structure of chromatin.

Kornberg, R. D. "Structure of chromatin." *Annual Review of Biochemistry,* v. 46, p. 931, 1977. An overview of chromatin structure with particular emphasis on nucleosomes.

Lilly, D., and J. Purdon. "Structure and function of chromatin." *Annual Review of Genetics,* v. 13, p. 197, 1979.

RECENT RESEARCH PAPERS

Bak, A., J. Zeuthen, and F. H. C. Crick, "Higher-order structure of human mitotic chromosomes." *Proceedings of the National Academy of Sciences, U.S.,* v. 74, p. 1595, 1977. A proposal for a multiple solenoid model for a chromosome.

"Chromatin." *Cold Spring Harbor Symposium on Quantitative Biology,* v. 42, 1978. An extensive collection of then-current reports on chromatin structure and function. Includes paper by Sedat.

Chapter 3

COLLECTIONS OF ORIGINAL PAPERS

Freifelder, D. *The DNA Molecule: Structural Properties.* Freeman, San Francisco, 1978. Contains papers on DNA denaturation and stability.

REVIEWS

Britten, R. J., D. Graham, and B. Neufeld. "Analysis of repeating DNA sequences by reassociation." in *Methods in Enzymology,* v. 39, p. 363, L. Grossman and K. Moldave, Eds. Academic Press, 1974.

Straus, N. "Repeated DNA in eukaryotes." in *Handbook of Genetics,* v. 5, *Molecular Genetics,* Robert King, Ed. Plenum Press, 1976, p. 3.

These two reviews deal with the methodologies of DNA renaturation kinetics with emphasis on repeated sequences.

Wu, R. "DNA sequence analysis." *Annual Review of Biochemistry,* v. 47, p. 607, 1978. Different procedures for sequencing DNA.

Chapter 4

COLLECTIONS OF ORIGINAL PAPERS

Carpenter, B. H. *Molecular and Cell Biology.* Dickenson, Belmont, California, 1967.

Raake, I. D. *Molecular Biology of DNA and RNA.* Mosby, St. Louis, Mo., 1971.

Taylor, J. H. *Selected Papers in Molecular Genetics.* Academic Press, Inc., New York, 1965.

Zubay, G., and J. Marmur, *Papers in Biochemical Genetics.* 2nd Edition. Holt, Rinehart and Winston, New York, 1973.

These four compendia contain original papers dealing with the mechanisms of DNA replication and enzymology of DNA synthesis.

REVIEWS

Alberts, B., and R. Sternglanz. "Recent excitement in the DNA replication problem." *Nature,* London, v. 269, p. 655, 1977. Insight into the methods of DNA replication.

Gefter, M. "DNA Replication." *Annual Review of Biochemistry,* v. 44, p. 45, 1975.

Hand, R. "Eukaryotic DNA: Organization of the genome for replication." *Cell,* v. 15, p. 317, 1978.

Kolter, R., and D. Helinski. "Regulation of initiation of DNA replication." *Annual Review of Genetics,* v. 13, p. 355, 1979.

Kornberg, A. *DNA Synthesis.* Freeman, San Francisco, 1976, 399 pp. The first twenty years of DNA replication by the pioneer in the field.

Rogers, J. "How viruses copy their DNA." *New Scientist,* v. 74, p. 526, 1977.

Sheinin, R., J. Humbert, and R. Pearlman. "Some aspects of eukaryotic DNA replication." *Annual Review of Biochemistry,* v. 47, p. 277, 1978.

RECENT RESEARCH PAPERS

"DNA: Replication and recombination." *Cold Spring Harbor Symposium on Quantitative Biology,* v. 43, 1978. A large collection of recent papers on DNA replication.

Wolff, S., and P. Perry. "Differential Giemsa staining of sister chromatids and the study of sister chromatid exchanges without autoradiography." *Chromosoma,* v. 48, p. 341, 1974. A detailed description of the harlequin staining technique.

Chapter 5

COLLECTIONS OF ORIGINAL PAPERS

Boyer, S. H. *Papers on Human Genetics.* Prentice-Hall, Englewood Cliffs, N.J., 1963.

Levine, L. *Papers on Genetics.* Mosby, St. Louis, Mo., 1971.

These two books contain original papers dealing with various aspects of chromosomal transmission including non-disjunction.

Voeller, B. *The Chromosome Theory of Inheritance*. Appleton-Century-Crofts, New York, 1968. A collection of turn-of-the-century papers on the role of the nucleus and chromosomes in inheritance.

REVIEWS

Baker, B., A. Carpenter, M. Esposito, R. Esposito, and L. Sandler. "The genetic control of meiosis." *Annual Review of Genetics,* v. 10, p. 53, 1976. Investigations of the genetics of meiosis, particularly in fungi and *Drosophila.*

Kurit, D., and H. Hoehn. "Prenatal diagnosis of human genome variation." *Annual Review of Genetics.* v. 13, p. 235, 1979. Current applications of amniocentesis.

Moens, P. "Ultrastructural studies of chiasma distribution." *Annual Review of Genetics,* v. 12, p. 433, 1978. Electron microscopy reveals structures involved in crossing over.

Chapter 6

COLLECTIONS OF ORIGINAL PAPERS

Moore, J. A. *Readings in Heredity and Development.* Oxford, N.Y., 1972.

Peters, J. A. *Classic Papers in Genetics.* Prentice-Hall, Englewood Cliffs, N.J., 1959.

Stern, C., and E. Sherwood. *The Origin of Genetics, A Mendel Source Book.* Freeman, San Francisco, 1966.

Voeller, B. *The Chromosome Theory of Inheritance,* Appleton-Century-Crofts, New York, 1968.

These four books all contain original papers dealing with the origins of genetics and with Mendelism. The book by Stern and Sherwood is particularly recommended for its excellent translation of Mendel's paper and because it contains the "rediscovery" papers and Fisher's article, "Has Mendel's Work Been Rediscovered?"

Chapter 7

COLLECTIONS OF ORIGINAL PAPERS

Hershey, A. D. *The Bacteriophage Lambda.* Cold Spring Harbor Laboratories, Cold Spring Harbor, New York, 1971. A collection of primary and review papers on the genetics of lambda. Everything you might want to know—at this time.

Stent, G. S. *Papers on Bacterial Viruses,* Little, Brown, Boston, 1965. Early papers on the genetics of bacteriophage.

REVIEWS

Radding, C. M. "Genetic recombination: strand transfer and mismatch repair." *Annual Review of Biochemistry,* v. 47, p. 847, 1978. Aspects of the molecular mechanisms of recombination.

RECENT RESEARCH PAPERS

"DNA: replication and recombination." Cold Spring Harbor Symposium on Quantitative Biology, v. 43, 1978. Several papers on molecular mechanisms of recombination.

Chapter 8

COLLECTIONS OF ORIGINAL PAPERS

Corwin, H. O., and J. B. Jenkins. *Conceptual Foundations in Genetics.* Houghton Mifflin, Boston, 1976.

Levine, L. *Papers on Genetics.* Mosby, St. Louis, Mo., 1971.

Voeller, B. *The Chromosome Theory of Inheritance.* Appleton-Century-Crofts, New York, 1968.

These three collections include papers on gene transmission in both lower and higher eukaryotes and mechanisms of genetic mapping.

Chapter 9

COLLECTIONS OF ORIGINAL PAPERS

Ledoux, L. *Genetic Manipulations with Plant Materials.* Plenum Press, New York, 1975.

Levine, L. *Papers on Genetics.* Mosby, St. Louis, Mo., 1971. Contains primary papers on non-nuclear inheritance.

REVIEWS

Birky, C. W., Jr. "Transmission genetics of mitochrondria and chloroplasts." *Annual Review of Genetics,* v. 12, p. 471, 1978.

Kleinhofs, A., and Behki, R. "Prospects for plant genome modification by nonconventional methods." *Annual Review of Genetics,* v. 11, p. 79, 1977.

McKusick, V., and Ruddle, F. H. "The status of the gene map of the human chromosomes." *Science,* v. 196, p. 390, 1977. A review of the current status of gene maps and gene mapping in humans. Excellent.

RECENT RESEARCH PAPERS

Esposito, M. "Evidence that spontaneous mitotic recombination occurs at the two-strand stage." *Proceedings of the National Academy of Sciences, U.S.,* v. 75, p. 4436, 1978. An advanced primary paper indicating that mitotic recombination occurs before DNA replication.

Chapter 10

COLLECTIONS OF ORIGINAL PAPERS

Adelberg, E. A. *Papers on Bacterial Genetics.* Little, Brown, Boston, 1966. Contains many of the original papers investigating the means of gene transfer in bacteria.

Stent, G. S. *Papers on Bacterial Viruses.* Little, Brown, Boston, 1965. Primary references on the role of phages in the transmission of bacterial genes.

REVIEWS

Clark, A. J., and G. Warren. "Conjugal transmission of plasmids." *Annual Review of Genetics,* v. 13, p. 99, 1979.

Fox, M. "Some features of genetic recombination in prokaryotes." *Annual Review of Genetics,* v. 12, p. 47, 1978.

Kleckner, N. "Transposable elements in prokaryotes." *Cell,* v. 11, p. 11, 1977. A brief review on IS elements, transposons, R elements and episomes (e.g., λ).

Low, K., and R. Porter. "Modes of gene transfer and recombination in bacteria." *Annual Review of Genetics,* v. 12, p. 249, 1978.

Sinsheimer, R. L. "Recombinant DNA." *Annual Review of Biochemistry,* v. 46, p. 415, 1977. A readable summation of recombinant DNA techniques.

RECENT RESEARCH PAPERS

Bukhari, A., J. Shapiro, and S. Adhya. "Insertion elements, plasmids and episomes." Cold Spring Harbor Laboratories, 1977. A collection of papers dealing with IS elements, transposons, and plasmids.

Hicks, J., J. Strathern, and A. Klar. "Transposable mating type genes in *Saccharomyces cerevisiae.*" *Nature,* v. 282, p. 478, 1979.

Chapter 11

COLLECTIONS OF ORIGINAL PAPERS

Boyer, S. H. *Papers on Human Genetics.* Prentice-Hall, Englewood Cliffs, New Jersey, 1963.

Taylor, J. H. *Selected Papers on Molecular Genetics.* Academic Press, New York, 1965.

These two collections contain early papers on biochemical genetics and genetics of human hemoglobins.

REVIEWS

Muruyama, M. "Molecular mechanism of human red cell sickling." in *Molecular Aspects of Sickle Cell Hemoglobin: Clinical Applications.* R. Nalbandian, Ed. Charles C Thomas, Springfield, Ill., 1971.

Perutz, M., J. Rosa, and A. Schechter. "Therapeutic agents for sickle cell disease." *Nature,* v. 275, p. 369, 1978. A brief report on a meeting in which current research on the development of chemicals to control the symptoms of sickle cell anemia was discussed.

Stamatoyannopoulos, G. "Molecular basis of hemoglobin disease." *Annual Review of Genetics,* v. 6, p. 47, 1972. A summary of inherited hemoglobin diseases with emphasis on the effects of amino acid substitutions.

Chapter 12

COLLECTIONS OF ORIGINAL PAPERS

Carpenter, B. H. *Molecular and Cell Biology.* Dickenson, Belmont, California, 1967.

Raake, I. *Molecular Biology of DNA and RNA.* Mosby, St. Louis, Mo., 1971.

Zubay, G., and J. Marmur. *Papers in Biochemical Genetics.* 2nd Edition. Holt, Rinehart and Winston, New York, 1973.

These compendia contain primary papers on transcription. The last is particularly good.

REVIEWS

Adhya, S., and Gottesman, M. "Control of Transcription Termination." *Annual Review of Biochemistry,* v. 47, p. 967, 1978.

Burgess, R. "RNA Polymerase." *Annual Review of Biochemistry,* v. 40, p. 711, 1971.

Darnell, J. "Implications of RNA·RNA splicing in evolution of eukaryotic cells." *Science,* v. 202, p. 1257, 1978. The evolutionary independence of prokaryotes and eukaryotes is proposed.

Losick, R. "In vitro transcription." *Annual Review of Biochemistry,* v. 41, p. 409, 1972.

Rosenberg, M., and D. Court. "Regulatory sequences involved in the promotion and termination of RNA transcription." *Annual Review of Genetics,* v. 13, p. 319, 1979.

RECENT RESEARCH PAPERS

Catterall, J., B. O'Malley, M. Robertson, R. Staden, Y. Tanaka, and G. Brownlee. "Nucleotide sequence homology at 12 intron-exon junctions in the chick ovalbumin gene." *Nature,* v. 275, p. 510, 1978. The junctions of coding and intervening sequences have short common sequences.

Cochet, M., F. Gannon, R. Heu, F. Perrin, and P. Chambon. "Organization and sequence studies of the 17-piece chicken conalbumin gene." *Nature,* v. 282, p. 567. 1979. The ultimate number of intervening sequences?

Gilbert, W. "Why genes in pieces?" *Nature,* v. 271, p. 501, 1978. A brief speculation on the function of intervening sequences.

Knapp, G., J. Beckmann, P. Johnson, S. Fuhrman, and J. Abelson. "Transcription and processing of intervening sequences in yeast tRNA genes." *Cell,* v. 14, p. 221, 1978. The discovery of an enzyme that removes intervening sequences.

Tilghman, S., P. Curtis, D. Tiemeier, P. Leder, and C. Weissman. "The intervening sequence of a mouse β-globin gene is transcribed with the 15S β-globin mRNA precursor." *Proceedings of the National Academy of Sciences, U.S.,* v. 75, p. 1309, 1978. An early demonstration that intervening sequences are transcribed and then removed.

van den Berg, J., A. van Ooyen, N. Mantei, A. Schambock, G. Grosveld, R. Flavel, and C. Weissman. "Comparison of cloned rabbit and mouse β-globin genes showing strong evolutionary divergence of two homologous pairs of introns." *Nature,* v. 276, p. 37, 1978. Nucleotide sequences in intervening sequences are not evolutionarily conserved.

Chapter 13

COLLECTIONS OF ORIGINAL PAPERS

Carpenter, B. *Molecular and Cell Biology.* Dickenson, Belmont, California, 1967.

"The Genetic Code." *Cold Spring Harbor Symposium on Quantitative Biology,* v. 31, 1966. A volume of primary papers on the code.

Zubay, G., and J. Marmur. *Papers in Biochemical Genetics.* 2nd Edition. Holt, Rinehart, and Winston, New York, 1973. The last of these compendia contains a particularly good collection of papers on protein synthesis.

REVIEWS

Kozack, M. "How do eucaryotic ribosomes select initiation regions in messenger RNA?" *Cell,* v. 15, p. 1109, 1978. Initiation in eukaryotes with a comparison with prokaryotes.

Woese, C. R. *The Genetic Code: The Molecular Basis for Genetic Expression.* Harper and Row, New York, 1967. An extensive examination of the conceptual and experimental bases for cracking the code.

RECENT RESEARCH PAPERS

Lagerkvist, V. "'Two out of three': An alternative method for codon reading." *Proceedings of the National Academy of Sciences, U.S.,* v. 75, p. 1759, 1978. An explanation of an alternative to "wobble."

Chapter 14

COLLECTIONS OF ORIGINAL PAPERS

Zubay, G. *Papers in Biochemical Genetics.* 1st Edition. Holt, Rinehart and Winston, New York, 1968. Contains papers on mutagenesis, repair, and suppression.

REVIEWS

Arlett, C. F., and A. Lehmann. "Human disorders showing increased sensitivity to the induction of genetic damage." *Annual Review of Genetics.* v. 12, p. 95, 1978. Xeroderma pigmentosum and others.

Auerbach, C., and B. Kilbey. "Mutation in eukaryotes." *Annual Review of Genetics,* v. 5, p. 163, 1971.

Drake, J. *The Molecular Basis of Mutation.* Holden-Day, San Francisco, 1970, 273 pp.

Drake, J., and R. Baltz. "The biochemistry of mutagenesis." *Annual Review of Biochemistry,* v. 45, p. 11, 1976.

Nagao, M., T. Sugimura, and T. Matsushima. *"Environmental mutagens and carcinogens." Annual Review of Genetics,* v. 12, p. 117, 1978.

Weatherall, D., and J. Clegg. "Molecular genetics of human hemoglobins." *Annual Review of Genetics,* v. 10, p. 157, 1976.

Chapter 15

COLLECTIONS OF ORIGINAL PAPERS

Zubay, G. *Papers in Biochemical Genetics.* 1st Edition. Holt, Rinehart, and Winston, New York, 1968.

REVIEWS

Benzer, S. "The Fine Structure of the Gene." *Scientific American,* v. 206, p. 70, 1962.

Fincham, J. *Genetic Complementation.* Benjamin, New York, 1966, 143 pp.

Kirschner, K., and H. Bisswanger. "Multifunctional proteins." *Annual Review of Biochemistry,* v. 45, p. 143, 1976. The prevalence and variation in multifunctional proteins.

RECENT RESEARCH PAPERS

"DNA:Replication and Recombination." *Cold Spring Harbor Symposium on Quantitative Biology,* v. 43, 1978. Contains papers on gene conversion.

Chapter 16

COLLECTIONS OF ORIGINAL PAPERS

Adelberg, E. *Papers on Bacterial Genetics.* Little, Brown, Boston, 1966.

Beckwith, J., and D. Zipser. *The Lactose Operon.* Cold Spring Harbor Laboratory, Cold Spring Harbor, N.Y., 1970. Reviews and a collection of original papers on the *lac* operon.

Hershey, A. D. *The Bacteriophage Lambda.* Cold Spring Harbor Laboratory, Cold Spring Harbor, New York, 1971. A collection of papers on lambda.

Zubay, G., and J. Marmur. *Papers in Biochemical Genetics.* 2nd Edition. Holt, Rinehart and Winston, New York, 1973. This collection has a particularly good set of papers on prokaryotic gene regulation.

REVIEWS

Echols, H. "Bacteriophage and bacteria: friend and foe." in *The Bacteria, A Treatise on Structure and Function.* v. 7, *Mechanisms of Adaptation,* J. Sokatch and L. Ornston, Eds., Academic Press, 1979, p. 487.

Ptashne, M., K. Backman, M. Humayun, A. Jeffrey, R. Maurer, B. Meyer, and R. Sauer. "Autoregulation and function of a repressor in bacteriophage lambda." *Science,* v. 194, p. 156, 1976. A view of regulation in lambda.

RECENT RESEARCH PAPERS

Barnes, W. "DNA sequence from the histidine operon control region: seven histidine codons in a row." *Proceedings of the National Academy of Sciences, U.S.,* v. 75, p. 4281, 1978. A common property of attenuators: They contain codons for the amino acid synthesized by the enzymes encoded in the operon.

Johnson, A., B. Meyer, and M. Ptashne. "Mechanism of action of the *cro* protein of bacteriophage λ." *Proceedings of the National Academy of Sciences, U.S.,* v. 75, p. 1783, 1978. The binding of *cro* protein to the right operator.

Kotewicz, M., S. Chung, Y. Takeda, and H. Echols. "Characterization of the integration protein of bacteriophage λ as a site-specific DNA-binding protein." *Proceedings of the National Academy of Sciences, U.S.,* v. 74, p. 1511, 1977. The protein product of the *int* gene binds specifically to the λ attachment site.

Sanger, F., G. Air, B. Barrell, N. Brown, A. Coulson, J. Fiddes, C. Hutchison III, P. Slocombe, and M. Smith. "Nucleotide Sequence of bacteriophage ØX174 DNA." *Nature,* v. 265, p. 687, 1977. The sequence of ØX174 and what it reveals.

Stauffer, G., Zurawski, and C. Yanofsky. "Single base-pair alterations in the *Escherichia coli* operon leader region relieve transcription termination at the *trp* attenuator." *Proceedings of the National Academy of Sciences, U.S.,* v. 75, p. 4833, 1978. The formation of a base-paired loop in the leader of the *trp* mRNA may be critical for attenuation.

Chapter 17

COLLECTIONS OF ORIGINAL PAPERS

Loomis, W. *Papers on Regulation of Gene Activity During Development.* Harper and Row, N.Y., 1970. Eukaryotic gene regulation.

REVIEWS

Chovnick, A., W. Gelbart, and M. McCarron. "Organization of the Rosy locus in *Drosophila melanogaster.*" *Cell,* v. 11, p. 1, 1977.

O'Malley, B., H. Towle, and R. Schwartz. "Regulation of gene expression in eucaryotes." *Annual Review of Genetics,* v. 11, p. 239, 1977.

Weatherall, D., and J. Clegg. *Recent developments in the molecular genetics of human hemoglobins. Cell,* v. 18, p. 467, 1979.

RECENT RESEARCH PAPERS

Ashburner, M., C. Chihara, P. Meltzer, and G. Richards. "Temporal control of puffing activity in polytene chromosomes." *Cold Spring Harbor Symposium on Quantitative Biology,* v. 38, p. 655, 1974. The control of transcription by ecdysone.

Kleene, K., and T. Humphreys. "Similarity of hnRNA sequences in blastula and pluteus stage sea urchin embryos." *Cell,* v. 12, p. 143, 1977. The apparent presence of posttranscriptional regulation during sea urchin development.

Wold, B., W. Klein, B. Hough-Evans, R. Britten, and E. Davidson. "Sea urchin embryo mRNA sequences expressed in the nuclear RNA of adult tissues." *Cell,* v. 14, p. 941, 1978. More on posttranscriptional regulation during sea urchin development.

Chapter 18

REVIEWS

Baker, W. "A genetic framework for *Drosophila* development." *Annual Review of Genetics,* v. 12, p. 451, 1978. A geneticist's analysis of compartment formation.

Dunnick, W. "Immunoglobulin genes: More facts to guide speculation." *Nature,* v. 276, p. 322, 1978. A brief review of immunoglobulin synthesis.

Garcia-Bellido, A., P. Lawrence, and G. Morata. "Compartments in Animal Development." *Scientific American,* p. 102, 1979. A readable description of compartment formation.

Gehring, W. "Developmental genetics of *Drosophila,*" *Annual Review of Genetics,* v. 10, p. 209, 1979. A particularly clear review.

Hall, J., and R. Greenspan. "Genetic Analysis of *Drosophila* Neurobiology." *Annual Review of Genetics,* v. 13, p. 127, 1979.

Klein, J. "The major histocompatability complex of the mouse." *Science,* v. 203, p. 516, 1979. The genetics of cell-surface antigens.

Seidman, J., A. Leder, M. Nau, B. Norman, and P. Leder. "Antibody Diversity, *Science,* v. 202, p. 11, 1978. Possible explanations for antibody diversity.

RECENT RESEARCH PAPERS

Brack, C., M. Hirama, R. Lenhard-Schuller, and S. Tonegawa. "A complete immunoglobulin gene is created by somatic recombination." *Cell,* v. 15, 1, 1978. The making of an immunoglobulin light chain.

Chan, H., M. Israel, C. Garon, W. Rowe, and M. Martin. "Molecular cloning of polyoma virus DNA in *Escherichia coli*: Lambda phage vector system." *Science,* v. 203, p. 887, 1979. The dangers of polyoma DNA in λWES.

Deppe, U., E. Schierenberg, T. Cole, C. Krieg, D. Schmitt, B. Yoder, and G. von Ehrenstein. "Cell lineage in the embryo of the nematode *Caernorhabditis elegans.*" *Proceedings of the National Academy of Sciences, U.S.,* v. 75, p. 376, 1978.

Crea, R., A. Kraszewski, T. Hirose, and K. Itakura. "Chemical synthesis of genes for human insulin." *Proceedings of the National Academy of Sciences, U.S.,* v. 75, p. 5765, 1978. Synthesizing the nucleotide sequences for insulin.

Goeddel, D., H. Heynecker, T. Hozumi, R. Arentzen, K. Itakura, D. Yansura, M. Ross, G. Miozzari, R. Crea, and P. Seeburg. "Direct expression in *Escherichia coli* of a DNA sequence coding human growth hormone." *Nature,* v. 281, p. 544, 1979. Synthesis of human growth hormone in bacteria.

Goeddel, D., D. Kleid, F. Bolivar, H. Heyneker, D. Yansura, D. Crea, T. Hirose, A. Kraszewski, K. Itakura, and A. Riggs. "Expression in *Escherichia coli* of chemically synthesized genes for human insulin." *Proceedings of the*

National Academy of Sciences, U.S., v. 76, p. 106, 1979. The synthesis of human insulin in bacteria.

Israel, M., H. Chan, W. Rowe, and M. Martin. "Molecular cloning of polyoma virus DNA in *Escherichia coli*: Plasmid vector system." *Science,* v. 203, p. 883, 1979. The dangers of polyoma DNA in pBR322.

Kauffman, S. A., R. Shymko, and K. Trabert. "Control of sequential compartment formation in *Drosophila.*" *Science,* v. 199, p. 259, 1978. The proposal of the zip code model for compartment formation.

Khorana, H. "Total synthesis of a gene." *Science,* v. 203, p. 614, 1979. Another method for the chemical synthesis of a gene.

Mantei, N., W. Boll, and C. Weissman. "Rabbit β-globin production in mouse L-cells transformed with cloned rabbit β-globin chromosomal DNA." *Nature,* v. 281, p. 40, 1979. Transformation of mammalian cells with cloned DNA—the future is now.

Chapter 19

BOOKS

Ford, E. B. *Ecological Genetics.* 4th Edition. John Wiley & Sons, New York, 1975.

Lewontin, R. C. *The Genetic Basis of Evolutionary Change.* Columbia University Press, New York and London, 1974.

REVIEWS

Da Cunha, A. B. "Chromosomal polymorphism in the Diptera." *Advances in Genetics,* v. 7, pp. 93–138, 1955.

Jones, J. S., B. H. Leith, and P. Rawlings. "Polymorphism in *Cepaea.*" *Annual Review of Ecology and Systematics,* v. 8, pp. 109–143, 1977.

Koehn, R. K. and W. F. Eanes. "Molecular structure and protein variation within and among populations." *Evolutionary Biology,* v. 11, pp. 39–103, 1978.

Powell, J. R. "Protein variation in natural populations of animals." *Evolutionary Biology,* v. 8, pp. 79–120, 1975.

Chapters 19–25

BOOKS

Bodmer, W. F. and L. L. Cavalli-Sforza. *Genetics, Evolution, and Man.* W. H. Freeman and Co., San Francisco, 1976.

Dobzhansky, T. *Genetics of the Evolutionary Process.* Columbia University Press, New York and London, 1970.

Fisher, R. A. *Genetical Theory of Natural Selection.* 2nd Edition. Dover, New York, 1958.

Haldane, J. B. S. *The Causes of Evolution.* Harper and Row, New York, 1932.

Shorrocks, B. *The Genesis of Diversity.* University Park Press, Baltimore, 1979.

Spiess, E. B. *Genes in Populations.* John Wiley & Sons, New York, 1977.

Wallace, B. *Topics in Population Genetics.* W. W. Norton & Co., Inc., New York, 1968.

REVIEWS

Hedrick, P., S. Jain, and L. Holden. "Multilocus Systems in Evolution." *Evolutionary Biology,* v. 11, pp. 104–184, 1978.

Wright, S. "Evolution in Mendelian populations." *Genetics,* v. 16, pp. 97–159, 1931. A classic work.

Chapter 25

BOOKS

Mayr, E. *Populations, Species, and Evolution.* Belknap Press. Cambridge, Massachusetts, 1970.

Stabbins, G. L. *Processes of Organic Evolution.* Prentice-Hall, Inc., Englewood Cliffs, New Jersey, 1966.

White, M. J. D. *Modes of Speciation.* W. H. Freeman and Company, San Francisco, 1978.

COLLECTIONS OF ORIGINAL PAPERS

Ayala, F. J. *Molecular Evolution.* Sinauer Associates, Inc. Sunderland, Massachusetts, 1976.

Evolution. A Scientific American Book. W. H. Freeman and Company, San Francisco, 1978.

REVIEWS

Ayala, F. J. "Genetic differentiation during the speciation process." *Evolutionary Biology,* v. 8, pp. 1–78, 1975.

Bush, G. L. "Modes of animal speciation." *Annual Review of Ecology and Systematics.* v. 6, pp. 339–364, 1975.

Throckmorton, L. H. "*Drosophila* systematics and biochemical evolution." *Annual Review of Ecology and Systematics,* v. 8, pp. 235–254, 1977.

Glossary

Adaptation. A trait especially suited for accomplishing some paricular function of an organism.

Additive Fitness (h = 0.5). A selection regime in which the relative fitness of the heterozygote is precisely midway between the fitnesses of the two homozygotes.

Allele Substitution. The establishment of a previously rare or absent allele as the predominant allele in a population.

Alleles. Alternative forms of a gene.

Allozyme. A protein that can be distinguished from other allelic forms of the protein by gel electrophoresis.

Amniocentesis. Technique of obtaining fetal cells for prenatal diagnosis of a genetic condition.

Aneuploidy. A chromosome number that is not an exact multiple of the haploid number, for example 2n + 1.

Antibody. Y-shaped proteins, immunoglobulins, that bind specifically to antigens.

Anticodon. The triplet of adjacent bases in tRNA that pairs with three bases in mRNA during translation.

Antigen. A substance that induces the formation of an antibody that specifically binds the substance.

Autosome. A chromosome other than a sex chromosome.

Bottleneck. A severe reduction during one or more generations in the number of individuals in a population.

CAP. Catabolite activator protein; a protein that binds cyclic AMP and regulates binding of RNA polymerase to promoter sites in inducible operons.

Carcinogen. A chemical or physical agent that causes cancer.

Centromere. The region of a chromosome, containing the kinetochore, to which spindle fibers attach during mitosis and meiosis.

Chiasma (pl. chiasmata). Cross-shaped structure seen in prophase I of meiosis that results from crossing over between homologous chromosomes.

Chromatids. The daughter strands of a replicated chromosome that are joined by a single centromere.

674

Chromatin. The complex of nucleic acid and protein that comprises much of a cell nucleus and chromosomes.

Chromomere. Darkly staining, condensed regions, beads, in euchromatic arms of chromosomes. Chromomeres, lying in register, produce the banded appearance of polyetene chromosomes.

Chromosome. A thread-like entity, containing, or composed entirely of, nucleic acid, that carries genetic information.

Cistron. A genetic unit of function, defined by complementation tests, often synonymous with gene.

Cline. The progressive, effectively uniform change in the frequency of a genetic variation across a geographic region.

Clone. A group of cells descended from one common ancestor.

Coding Sequence. The region of a transcript or gene that encodes a polypeptide sequence. In prokaryotes, coding sequence is synonymous with gene, in eukaryotes, with exon, the expressed region of a gene.

Codon. Three adjacent nucleotides in mRNA that code for an amino acid or a peptide termination signal.

Complementary Sequence. A nucleotide sequence that base pairs with another nucleotide sequence.

Complementation Test. The introduction of two mutations into a cell in a diploid state to test their functional relationship. If the mutations complement, that is, produce a normal phenotype, they are often non-allelic; if the mutations are non-complementing, they are often allelic.

Concatemer. A linear structure composed of repeating units, for example, a chain; a chromosome linearly repeated two or more times.

Conjugation. Process of mating involving temporary fusion between single cell organisms such as bacteria.

Crossing Over. Reciprocal exchange of parts of chromosome arms between homologous chromosomes.

Cytokinesis. Cytoplasmic or cell division.

Deficiency. The absence of a large number (hundreds, thousands, millions) of nucleotide pairs from a chromosome.

Degenerate Genetic Code. Having more than one codon for an amino acid.

Deletion Mapping. The ordering of a set of deficiencies or the placement of a mutation site using recombination between chromosomes carrying deficiencies.

Diploid. Having two of each chromosome.

DNA Ligase. The enzyme that joins the 3' end of a polydeoxynucleotide to the 5' end of a polydeoxynucleotide.

Effective Population Size (N^e). The number of individuals in a population corrected for biological factors such as the sex-ratio, that allows different populations to be directly compared with respect to genetic drift.

Endoreplication. Replication without separation of chromatids.

Episome. A plasmid that can replicate autonomously, separate from the bacterial chromosome, or integrate and replicate as part of the bacterial chromosome.

Euchromatin. The lowly condensed, lightly staining region(s) of a chromosome (see heterochromatin).

Fitness (relative fitness; biological fitness). The relative contribution to the next generation that is made, on average, by an individual with a given genotype.

Founder Effect. A form of genetic drift in which a new population is established by a small number of colonizers whose alleles are a small sample from the parental population's gene pool.

Frame-Shift Mutation. A mutation, a base-pair deletion or addition, that shifts the normal reading frame and produces a nonfunctional polypeptide.

Frequency Dependent Selection. A selection regime in which the relative fitnesses of genotypes are determined, in part, by their frequencies.

Gamete. A cell used in fertilization; a germ cell.

Gene Arrangement. A particular distribution of genes along a chromosome, typically identified by the distribution of bands along a polytene chromosome.

Gene Conversion. The change, as a result of heteroduplex formation and repair during recombination, of the nucleotide sequence of one allele to that of the other allele.

Gene Flow. Exchange of genetic variation between the gene pools of different populations.

Gene Pool. The total collection of all copies of genes existing within a population at a given time.

Genetic Drift. Random changes in allele frequencies that occur as a result of the alleles in any generation being a finite sample from the parental generation's gene pool.

Genetic Identity (Nei's I). A quantitative measure, between zero and one, of the similarity of two populations that is calculated from the allele frequencies of a given locus in each of the two populations.

Genome. Complete single set of genes or chromosomes.

Genotype. The genetic constitution of an organism (see phenotype).

Haploid. An organism having one of each chromosome.

Hardy-Weinberg Proportions. The particular set of genotype frequencies in which the frequency of each homozygote equals the square of its constituent allele frequency and the frequency of the heterozygote equals twice the product of its constituent allele frequencies.

Hemizygous. Gene present in one copy in a diploid.

Heritability. A quantitative estimate of the degree to which phenotypic differences for a trait are inherited among individuals in a *population*.

Heterochromatin. The highly condensed, darkly staining region of a chromosome, for example, centric heterochromatin. In some cases, such as a Barr body, entire chromosomes are heterochromatic.

Heteroduplex. A double helix formed between two strands that are not perfectly complementary.

Heterozygous. Carrying two different alleles at a particular locus.

Homozygous. Carrying two identical alleles at a particular locus.

Identical by Descent. The condition or property of two copies of a gene being historically related through direct lines of gene replication from a single ancestral copy of the gene.

Inbreeding. Sexual reproduction between related individuals or, in self-fertilization, in a single individual.

Inbred. Having parents who are related to each other.

Intercalating Agent. A planar compound that slides between stacked bases in DNA. Intercalating agents cause frame-shift mutations.

Intervening Sequence. A region interrupting the coding sequence of a eukaryotic gene that is transcribed but whose complementary RNA is removed during the processing of the pre-mRNA. Also called an intron.

Karyotype. The chromosome complement of an organism.

Kinetochore. The region in the centromere to which spindle fibers attach.

Leader Sequence. Region of mRNA preceding the coding sequence.

Linkage Disequilibrium. Non-random association between alleles of two genes; measured by the difference between the frequency in gametes and the product of the frequencies of the alleles.

Locus. Site in a chromosome where a gene is located.

Map Unit (Centimorgan). One per cent actual recombination.

Messenger RNA (mRNA). The RNA transcript of DNA that carries information for polypeptide synthesis.

Missense Mutation. A base-pair substitution that causes an amino acid substitution.

Monomorphic. A gene having a single form or trait in a population.

Mutagen. A chemical or physical agent that causes mutation.

Mutation. An inherited change in the nucleotide sequence of a chromosome.

N. Haploid genome.

Neutral Alleles. Two or more alleles of a gene whose relative fitnesses are indistinguishable.

Non-disjunction. Failure of chromatids to separate at anaphase.

Nucleosome. The "bead" composed of four pairs of histone molecules (H2A, H2B, H3 and H4) around which DNA winds.

Oncogenic Virus. A DNA or RNA virus that causes cancer.

Operator. The region of a chromosome that binds a repressor protein for the regulation of adjacent structural genes in an operon.

Operon. Chromosomal region containing structural genes within a single transcriptional unit and adjacent regulatory regions (operator, promoter).

Overdominance. The characteristics of a selection regime in which the fitness of the heterozygote is greater than the fitness of either homozygote.

Palindrome. DNA where identical, or nearly identical sequences, run in opposite directions on each strand.

Partial Dominance. The characteristic of a selection regime in which the fitness of the heterozygote is anywhere between the fitnesses of the two homozygotes.

Plaque. Region of clearing in a bacterial lawn caused by lytic phage growth.

Polygenes. A collection of genes for which each allelic difference for each gene has a small effect on the phenotypic variability of a given trait.

Polymorphic. A gene or trait for which two or more alleles occur commonly in a population.

Population. A group of intermating individuals living in the same area.

Probability of Ultimate Fixation [U(p)]. The probability that a given allele with frequency p will eventually become fixed within a population.

Phenotype.   The physical characteristics of an organism (see genotype).

Plasmid.   A cytoplasmic, autonomously replicating element in bacteria.

Point Mutation.   A mutation altering one or a few base pairs.

Polymerases.   Enzymes that synthesize polymers, for example, RNA polymerase.

Polysome.   A group of ribosomes linearly attached together by mRNA and engaged in polypeptide synthesis.

Polytene Chromosome.   A chromosome created by endoreplication and composed of hundreds of chromatids aligned with their chromomeres in register.

Postmeiotic Segregation.   Segregation during the first mitotic division following meiosis as a result of a heteroduplex created by recombination.

Primer.   An RNA or DNA molecule that serves to initiate DNA synthesis.

Promoter.   A region of DNA where RNA polymerase binds to initiate transcription.

Repair.   The correction of alterations in DNA structure before they are inherited as mutations.

Replication.   The process by which chromosomal DNA is duplicated.

Replication Fork.   Y-shaped part of replicating chromosome where DNA synthesis occurs; synonym, growing fork.

Replicon.   A single complete unit of DNA replication.

Repressor.   Protein that regulates transcription through reversible binding to an operator.

Reproductive Isolating Mechanisms.   Genetically determined characteristics that prevent individuals of a population from either mating with or producing fertile offspring by mating with individuals of another population.

Restriction Endonucleases.   Endonucleases that make cuts in single strands of double helical DNAs at specific points in palindromes.

Reverse Transcriptase.   A polymerase that can make DNA using an RNA template.

Reversion.   The inherited return of a mutant organism to a normal phenotype as a result of mutation. Reversion can result from correction of the original mutation (true reversion) or from suppression.

Ribosome.   A particle composed of RNA and protein that when associated with mRNA is the site of protein synthesis.

Ribosomal RNA (rRNA).   The RNA molecules that are part of the structure of ribosomes: 23 and 16S in prokaryotes; 28 and 18S in eukaryotes, as well as 5S rRNA.

Satellite DNA.   DNA that bands at a different position in a density gradient than the bulk of the DNA; usually composed of highly repeated sequences.

Selection Coefficients (s, h).   Quantitative measures that characterize a selection regime in terms of *differences* between relative fitnesses.

Selection Regime.   The particular set of relative fitnesses for alternative genotypes within a population living in a particular environment.

Sex Chromosome.   A chromosome such as an X or Y that plays a major role in determining the sex of an organism.

Sexual Selection.   A selection regime in which differences in the relative fitnesses are caused by differences among the genotypes with respect to their relative ability to attract mates.

Sigma (σ) Factor.   A polypeptide that causes bacterial RNA polymerase to bind to a promoter.

Speciation.   The process of forming a new species.

Suppressor.   A mutation that eliminates the effect of another mutation.

Syngamy.   The union of nuclei of two gametes following fertilization.

Terminally Redundant.   Having the same nucleotide sequence at both ends of a chromosome.

Tetrad.   A pair of synapsed, replicated homologous chromosomes.

Transduction.   Genetic recombination in bacteria mediated by phages.

Transformation.   Genetic Transformation: A process in which free DNA from one cell integrates by recombination into the chromosome of another cell and effects a genotypic change. Oncogenic Transformation: A process by which a virus converts a normal cell into a cancerous one with excessive growth.

Transition Mutation.   Base substitution in which the purine-pyrimidine orientation is maintained (purine → purine; pyrimidine → pyrimidine).

Transcription.   The synthesis of RNA on a template of DNA using base-pairing.

Transcriptional Unit.   A region of DNA that is transcribed as a single entity.

Transfer RNA (tRNA).   An RNA molecule with about 80 nucleotides that covalently binds a specific amino acid and, through base pairing with mRNA, delivers the amino acid for inclusion in a growing polypeptide.

Translation.   Polypeptide synthesis, that is, the conversion of a nucleotide sequence in RNA into an amino acid sequence.

Transversion Mutation.   A base substitution that reverses the purine-pyrimidine orientation (purine → pyrimidine; pyrimidine → purine).

Variance (σ^2).   A statistical measure of the extent to which individual measurements within a collection of measurements vary about the mean of the collection.

Index